Computer Simulation and the Statistical Mechanics of Adsorption

Computer Simulation and the Statistical Mechanics of Adsorption

D. NICHOLSON and N. G. PARSONAGE
Department of Chemistry, Imperial College, London, UK

1982

ACADEMIC PRESS

A Subsidiary of Harcourt Brace Jovanovich, Publishers
London New York
Paris San Diego San Francisco São Paulo
Sydney Tokyo Toronto

ACADEMIC PRESS INC. (LONDON) LTD.
24/28 Oval Road
London NW1

United States Edition published by
ACADEMIC PRESS INC.
111 Fifth Avenue
New York, New York 10003

British Library Cataloguing in Publication Data

Nicholson, D.
Computer simulation and the statistical mechanics of adsorption.
1. Adsorption—Mathematical models
2. Solids
I. Title II. Parsonage, N. G.
541.3′453′0724 QD506

ISBN 0–12–518060–8

Printed in Great Britain by
Page Bros (Norwich) Ltd

Preface

The field of interfacial science, and in particular that associated with physisorption, is one which for many years was dominated by chemists, whose main interest was often in the characterization of highly divided powdered materials with ill-defined surface properties. However, the increasing availability of well characterized and homogeneous adsorbents and the highly accurate results gained with traditional techniques in the last 10–15 years have attracted workers from other disciplines, particularly physics and mathematics. Along with newer experimental techniques these groups have introduced theoretical approaches which have contributed greatly to our understanding of the phenomena involved. The advent of computer simulation in this field, and the stimulation this has provided, has served to widen the gap between the older theories and these newer approaches at an ever-increasing rate; and many workers, not new to the field, must feel a reluctance to make the considerable effort to appreciate the new methods. It is a main aim of this book to interpret to fellow chemists some of the new theoretical methods and to indicate the results which have been obtained by their use. Although our target is principally the chemist we would hope that other scientists and mathematicians will find the book useful. The level has been pitched such that a graduate embarking on research in this area could use the book as a means of entry to the field.

Since the subject does not have very obvious limits, the choice of topics to be included has been, inevitably, influenced by the interests of the authors. An example is the inclusion of the final chapter on hydrophobic behaviour. An important contributory reason for including that section is that it is a particularly good example of an area in which modern computer methods have already answered many questions which have puzzled chemists for decades. As we have already indicated, we have concentrated on phenomena arising from physical, rather than chemical, interactions between fluids and substrates. It follows that, although workers in the field of chemisorption may find something to interest them in some parts of the book, their interest has not been our main object.

We have, in the main, adhered to SI units. However, in some places the use of atomic units leads to simpler (and more transparent) equations. In those cases we have used atomic units and indicated this by an asterisk against the equation number. Readers interested in following the thread of the argument should find this presentation more appealing. Those interested in applying the equations to real systems may refer to Appendix 1, where the factors required for conversion to SI units are given.

We are indebted to K. Binder, W. E. Carlos and J. W. Perram for helpful correspondence, to Milton Cole, Nicos Christou and John Whitehouse for reading parts of the manuscript and making useful comments. We also thank those who sent us manuscripts in advance of publication and allowed us to use material from them. Finally we would like to thank those nearest to us who have endured "lost weekends" and working "holidays" with patience and tolerance.

September 1982
London

D. Nicholson
N. G. Parsonage

Contents

List of Symbols

α	bulk elasticity parameter (Eqn (5.17))
α, α'	critical indices (Eqns (2.81–82))
$\alpha_{\alpha\beta}$	polarizability; component of the tensor
β	critical index; $(kT)^{-1}$; shear elasticity parameter (Eqn (5.17))
γ	surface tension (Eqn (2.26))
γ, γ'	critical indices (Eqns (2.82–83))
δ	critical index (Eqn (2.84)); "actual mismatch" (Fig. 5.19)
$\delta_{\alpha\beta}$	Kronecker delta
$\delta(x)$	Dirac delta function
ε	energy parameter
$\varepsilon_{\alpha\beta}$	permittivity tensor
ε_0	permittivity of free space
ε_r	relative permittivity
ε_w	energy parameter (Eqn (3.81a))
$\varepsilon^{[a,b..]}$	$=\exp(-\beta u_{ij}^{[a,b\ldots]})$ (Eqn (3.145c, d))
$\zeta^{[a]}$	defined by (Eqn (3.145b))
η	reduced density (Eqn (7.1f)); order parameter (Ch. 8); critical index
$\eta^{(i)}$	domain of *i* (Ch. 4)
θ	coverage
$\theta^{(e)}$	excess coverage
$\theta(n)$	coverage in *n*th layer
κ	reciprocal of correlation length (Eqn (2.35)); isothermal compressibility; reciprocal of double-layer thickness
κ^{Σ}	compressibility of the adsorbed phase
λ	absolute activity; thermal conductivity
λ_H, λ_t	renormalization group scaling factors
$\lambda^{(a,b)}$	parameter in the Weiner–Hopf equations (Eqn (7.25))
μ	chemical potential
μ_α	component of the dipole vector
ν	critical index (Eqn (2.85))
ξ	correlation length (Ch. 8); coupling parameter (Ch. 4); field conjugate to an order parameter (Eqn (2.84))
ξ_j	density function in a mixture (Eqn (7.21c))
ρ_a	uniform density in a fluid of "a" particles
$\rho^{(a,b,\ldots)}$	*n*-particle density for species a, b, . . .
$\rho_\tau^{(a,b,\ldots)}$	angle-averaged *n*-particle density for species a, b . . .
$\rho_{\mathbf{k},n}$	Fourier coefficients (Eqn (3.73))
$\sigma^{(a,b)}$	distance parameter or hard-sphere diameter in a potential function

σ_a	as $\sigma^{(a,b)}$ but for an a,a pair
σ_w	see Eqn (6.1)
$\boldsymbol{\tau}$	reciprocal lattice vector of adsorbate (Ch. 5)
$\boldsymbol{\tau}$	2-vector giving x, y positions (Ch. 3, §3(e))
ϕ	surface pressure (Eqn (2.26)); continuous symmetry parameter (Ch. 5)
ϕ_c	surface pressure at the surface of tension for cylindrical ($c \equiv$ CYL) or spherical ($c \equiv$ SPH) meniscus
$\phi^{(a,b)}$	potential function (Eqn (7.22))
ϕ_G, $\tilde{\phi}_G$	Fourier coefficient of potential (Ch. 5) and derivative see below Eqn (5.4)
$\boldsymbol{\phi}$	eigenvector of p
χ	magnetic susceptibility
ψ	sticky-sphere potential parameter (Eqn (7.26))
$\Delta\psi$, $\Delta\psi'$	potential barriers (Ch. 6)
ω	oscillator frequency (rad s^{-1})
$\tilde{\omega}$	dominant oscillator frequency (Eqn (3.42))
$\boldsymbol{\omega}_{ia}$	normalized orientational coordinates for the ith particle of species a
$\omega^{\circ}(\rho_1)$	grand free energy per unit volume (Eqn (3.201))
Γ_a	surface excess number of molecules per unit area of species a
Δ	equivalent to chemical potential (Eqn (5.39)); r.m.s. displacement of atoms from 0 K position (Ch. 6); non-ordering field (Ch. 2)
$\Theta_{\alpha\beta}$	component of the quadrupole tensor
Λ_a	$h/\{(2\pi m_a kT)^{1/2} Z^{1/2}_{\text{rot, a}}\}$
Ξ	grand partition function
Π	disjoining pressure (Eqn (8.5)); osmotic pressure (Eqn (8.10)); isothermal-isobaric partition function
Σ	(as superscript) surface excess
Φ, Φ_{ij}	force-constant matrix for adatom–adatom interactions
Ω	grand thermodynamic potential or grand free energy
a	Helmholtz free energy density; nn distance (in solid); periodicity of the substrate potential; lattice parameter
a_c	area of a unit cell
$a^{(a,a)}$	coefficients in Eqn (7.1)
$\boldsymbol{a}_1$, $\boldsymbol{a}_2$	vectors specifying a two-dimensional cell in a solid adsorbent
$\bar{b}$	mismatch (Eqn (5.13))
$\boldsymbol{b}_1$, $\boldsymbol{b}_2$	reciprocal lattice vectors (Ch. 5)
$c^{(2)}(\boldsymbol{r}_1, \boldsymbol{r}_2)$, c_{12}	direct correlation function
c_L, c_T	longitudinal and transverse sound velocities
d	hard-sphere diameter in perturbation theory (Ch. 3, §5(g)); dimensionality
d, d_s	lattice parameters of adsorbate and substrate (Ch. 5)
e	electronic charge
f	force/unit area (Ch. 8)
f_s	oscillator strength (Ch. 3, 3(b))

$f(X, Y)$	$= \langle XY \rangle - \langle X \rangle \langle Y \rangle$
$f(t)$	memory function (Ch. 4)
g	Gibbs free energy density
$g^{(a,b)}$	pair correlation function (radial distribution function) for particles (a, b) (Eqn (3.20))
$g^{(2)}$	pair correlation function for identical particles
$g_{\parallel}^{(2)}$	as $g^{(2)}$ with inter-particle vector approximately parallel to the wall
$g_{\perp}^{(2)}$	as $g^{(2)}$ with inter-particle vector approximately perpendicular to the wall
g_K	Kirkwood g-factor
g_I	degeneracy factor (Eqn (7.64))
$h^{(2)}(\boldsymbol{r}_1, \boldsymbol{r}_2)$, h_{12}	total correlation function
k	Boltzmann constant
$\boldsymbol{k}$	wave vector $= 2\pi/\lambda$ where λ = wavelength
l	side of simulation box
l_x, l_y, l_z	length of rectangular box in x, y, z directions
m	mass of particle; magnetic moment of single spin
n	number of components in order parameter (or degrees of freedom)
n_f	density of free dislocations
p_N, p_T	pressure normal and transverse to an interface
p_0	saturated vapour pressure
$\boldsymbol{p}_i$	momentum of ith particle
$\boldsymbol{p}$	transition probability matrix
$p_{\alpha\beta}$	pressure tensor
q	charge; number of states for each particle in Potts model
q_{st}	isosteric heat of adsorption
q_w	isosteric heat of adsorption wall contribution to
q_I, q_r	vibrational partition functions (Eqn (7.64))
$\boldsymbol{q}$	$= \{\boldsymbol{q}_a, \ldots \boldsymbol{q}_\sigma\}$, see Ch. 3, §2(a)
$\boldsymbol{q}_a$	$= \{\boldsymbol{q}_{1a} \ldots \boldsymbol{q}_{N_a a}\}$
$\boldsymbol{q}_{ia}$	position coordinates for ith particle of species a
$q^{(a,b)}$	Weiner–Hopf factorization function (Eqn (7.19))
r_p	cylinder radius (Eqn (2.27))
$\boldsymbol{r}_{ia}$	position of ith molcule of species a
$\boldsymbol{r}_{ij}$	$= \boldsymbol{r}_j - \boldsymbol{r}_i$
$\boldsymbol{\delta r}$	displacement
s_a	number of degrees of freedom for particle of species a
s_i	Ising variable
t	film thickness (Ch. 2, §1); time; transform parameter (Eqn (7.2))
t_i	occupation variable (Ch. 5)
t_c	mean time between intermolecular collisions; film thickness in a cylinder (Ch. 2)
$u^{(1)}$	singlet potential between a molecule and a wall
$u^{(2)}$	pair potential between like species
$u_{ij\ldots}^{(a,b,\ldots)}$	n-body interaction potential between particles
$u^{[a,\ldots]}$	n-body adsorbate–adsorbent interaction which contains contributions of all orders (Eqns (3.26, 3.27))

u_I	Ith layer energy in a lattice model
$\boldsymbol{u}$	probability vector (Ch. 4)
$\mathbf{v}$	velocity
w	as superscript, species in the adsorbent (= wall)
$\mathscr{W}$	as superscript, water (Ch. 8)
$y^{(2)}$	$= \exp(\beta u^{(2)})\, g^{(2)}$
$y_{\mathscr{H}}, y_t$	see below Eqn (2.103)
z	activity ($= \lambda\Lambda^{-3}$); coordination number
$z_{\min}$	distance from wall of minimum in $u^{(1)}$
A	Helmholtz free energy
$A_{\alpha\beta}^{(a,b)}$	2-body interaction integral (Eqn (3.40))
$A^{(a,b,c)}$	3-body interaction integral (Eqn (3.86))
B	as superscript, bulk phase
B_s, B_{ss}	binding energies to water of monomer and dimer solute molecules
C_1, C_2	intensive variables (Eqn (2.20a))
C_p, C_v	heat capacity at constant pressure and constant volume
$C_{N^{(e)}}$, C_{N_s}	heat capacity at constant Gibbs excess
$C(t)$	autocorrelation function
$C_6^{(a,b)}, C_8^{(a,b)}, \ldots$	dispersion potential coefficients of inverse 6th, 8th, ... power interactions between species a and b
D	self-diffusion coefficient
$D_\parallel$, $D_\perp$	self-diffusion coefficient for motion parallel and perpendicular to the wall (Ch. 6)
D_τ	two-dimensional diffusion coefficient from (Eqn (6.12)) after time τ
E	Derjaguin–Landau–Verwey–Overbeek potential
$E_0, E_1, \ldots$	contributions to the periodic adsorption potential (Eqn (3.82))
E_c, E_I, E_D, E_R	electrostatic, inductive, dispersion and repulsive contributions to the interaction energy
$E_3^{(a,a)}$	pair-potential with third-body contribution from the adsorbent
E_L	lattice energy in absence of distortion (Ch. 5)
F	as superscript, film
$\boldsymbol{F}$, F_α	force acting on a molecule
$F_{\alpha\alpha}^{(a)}$	dominant oscillator strength in the α-direction for a particle of species a
G	Gibbs free energy; as superscript, gas phase
$\boldsymbol{G}$	reciprocal lattice vector of graphite (0001) face
H	enthalpy
$H(x)$	Heaviside step function
HW	hard-wall
HS	hard-sphere
H_N	classical Hamiltonian for a set of N particles
I	moment of inertia
I_a	ionization potential for species a
$I(k)$	X-ray interference function (Eqn (6.13))
J	Ising interaction parameter
Jnn, Jnnn	Ising interaction parameter for nn and nnn interactions

$J_1, J_2, \ldots$	Ising interaction parameter for 1st, 2nd, ... neighbour pairs
K	$= J/kT$
K_m	mth order Bessel function (Eqn (3.75))
k_N	kinetic energy for an N-molecule system
LJ	Lennard-Jones potential; system having LJ interactions between the particles
M	magnetization; general thermodynamic variable (Eqn (2.51))
$M_{\alpha\beta\gamma\ldots\sigma}$	general 2^σ-th multipole moment
MC	Monte Carlo
MD	molecular dynamics
MI	minimum image
$\boldsymbol{N}$	torque (Ch. 4)
N_a	number of molecules of species a
N_{a}^{Σ}	surface excess number of molecules of species a
$N_{\alpha\beta\ldots}$	general normalized multipole moment
$\mathcal{N}$	total number of molecules
$\boldsymbol{P}$	polarization
Q_N	canonical partition function for set $\boldsymbol{N}$
$Q_{\alpha\beta}$	component of the quadrupole tensor (Eqn (3.56))
$Q^{(\mathrm{a,b})}$	Weiner–Hopf factorization function (Eqn (7.186))
R	radius of a spherical drop or cavity
R_{c}	cut-off radius
$\boldsymbol{R}$	$= l_x\boldsymbol{i} + l_y\boldsymbol{j} + l_z\boldsymbol{k}$
S	entropy
$S^{(\mathrm{a,b})}$	hard-sphere limit (Eqn (7.19b))
SC	spherical cut-off
$T_{\mathrm{c}}, T_{\mathrm{c}}^*$	critical and reduced critical temperatures
T_{SK}	Stanley–Kaplan temperature
$T_{12}^{\alpha\beta\ldots}$	multipole interaction tensor for particles at r_{12}; coordinate components $\alpha, \beta, \ldots$
U	internal energy
U^{LR}	long-range energy contribution
V	volume $= \mathcal{A}l_z$
V_{m}	molar volume
$\overline{V}$	partial molar volume
W	work (Ch. 2, §2); potential of mean force (Ch. 8)
Z	collision rate
Z_{N}	configuration integral
$\mathcal{A}$	interfacial area
$\mathcal{B}$	number of adsorption sites
$\mathcal{C}(\boldsymbol{a}; \boldsymbol{R}, \boldsymbol{R}')$	orientational correlation function for nn pairs (Eqn (5.15))
$\mathcal{C}_k(\boldsymbol{R})$	correlation function $= \langle \rho_k(\boldsymbol{O})\ \rho_k(\boldsymbol{R}) \rangle$
$\mathcal{C}_1(t)$	orientational autocorrelation function (Eqn (6.9))
$\mathcal{F}^{(\mathrm{a,b})}$	2-particle Ursell distribution function (Eqn (3.23))
$\mathcal{H}$	magnetic field
$\mathcal{L}$	Laplace transform (Eqn (7.2))
$\mathcal{P}$	probability of given generic state in GCE
$\mathcal{P}_N$	probability of given state in canonical ensemble of N molecules

$\tilde{}$	over a symbol denotes a property of a uniform phase (except where stated otherwise)
$\hat{}$	over a symbol denotes a Fourier transformation or an operator

Chapter 1

Introduction

1. Adsorption theory and experiment

Applications of adsorption are very far-reaching, extending from such areas of industrial importance as adhesives, catalysts, colloids, surface coatings and thin films, wetting phenomena and various aspects of electrochemistry through to biochemistry, where interfaces play a crucial role, for example in membranes and enzyme binding. In many of these fields a more detailed understanding at a microscopic or molecular level is becoming increasingly necessary. Such an understanding can only derive from an extensive and self-consistent theoretical framework within which the interpretation of new experimental observations can proceed.

In recent years two major new developments have contributed to these general areas. One is the rapid expansion and refinement of experimental techniques and the other is the introduction of computer simulation into the study of interfaces. It is the latter which provides the central motif for this book. Ever since the classical studies of the liquid state by Wood using Monte Carlo methods and by Alder and Wainwright using molecular dynamics, it has been recognized that computer simulation could be of immense importance in exploring natural phenomena. Its primary value in this respect is that it bestows on the investigator the Janus property of being able to look two ways at once. On the one hand, since computer simulation, at least ideally, solves the statistical mechanics exactly, the only unknown which stands between it and experimental data is the nature of the intermolecular force laws in the system of interest. On the other hand, since these intermolecular force laws are exactly known for a simulation, the approximations necessary in a statistical mechanical theory can be refined by reference to the data resulting from a computer simulation.

The experimental side, of course, is of overriding importance, since it not only provides the stimulus for the development of better theories, but must obviously be the final arbiter in deciding the utility and validity of a theory (the two are not necessarily judged with equal rigour).

The utility of the data from many measurements on adsorption systems has been lessened by uncertainties introduced by experimental artefacts. This has been particularly true in chemisorption studies, where clean metal surfaces could not be obtained before the arrival of ultra-high vacuum techniques. Physisorption studies are also afflicted by uncertainties associated not only with surface impurities, but also with heterogeneity and micropores. It is therefore not too surprising that valuable advances in theory have originated with the introduction of one or two materials of high purity and well defined structure. Notable amongst these are the so-called exfoliated graphites such as grafoil and papyex, whose predominant adsorbent surface is the (0001) graphite plane. In a sense these materials, and associated graphites such as graphon which likewise exhibits a very low degree of energetic heterogeneity, have provided paradigms for the more precise physical adsorption studies. The rich array of phase transitions which adsorbates can undergo on these surfaces, has provided a constant source of fascination and challenge to experimentalists and theoreticians alike. Of course the wide range of experimental techniques now available has by no means been confined to studies of these graphite adsorbents alone and other materials, including micas and zeolites, are of great importance in fundamental studies.

2. Survey of experimental methods

For the purpose of making a brief survey, those experimental methods which are of most relevance to the subject of this book can be conveniently classified under a few headings.

(a) Thermodynamic methods

The more traditional measurements which have provided basic adsorption data, such as isosteric heats, heat capacities and isotherms of the amount adsorbed versus concentration in an external bulk phase, have benefited from technical advances of a general kind such as improvements in cryogenic methods and electronic devices. For example, in the pioneering work of Thomy and Duval (1969) with krypton, xenon and methane adsorbed on graphite, very accurate temperature control was employed with otherwise essentially standard volumetric adsorption apparatus, but the enhanced accuracy thereby attained, and the introduction of grafoil as an adsorbent,

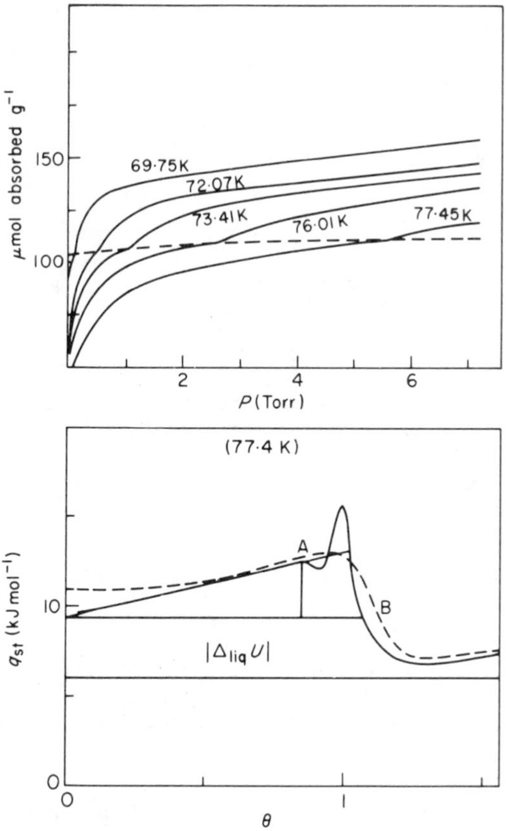

Fig. 1.1. Adsorption isotherms and isosteric heat as a function of fractional coverage in the Ar/Sterling system graphitized at 2700°C. Solid curves are the data of Grillet *et al.* (1979); dashed curves are from Beebe and Young (1954) for Ar on spheron. (Adapted from Grillet *et al.*, 1979.)

revealed for the first time the existence of sub-monolayer steps in the isotherms. The facility for studying a range of temperatures enabled a detailed phase diagram to be constructed. Following their initial work, a spate of papers has appeared, particularly from French groups; for example the microcalorimetric measurements stemming from the development of the Tian–Calvet differential calorimeter (Rouquerol *et al.*, 1977) in conjunction with automatic recording has exposed in a dramatic way hitherto unsuspected structure in the isosteric heat curves (Fig. 1.1, Grillet *et al.*, 1979). Similarly, in heat capacity measurements the incorporation of highly accurate pressure transducers and computer control (Bretz *et al.*, 1973;

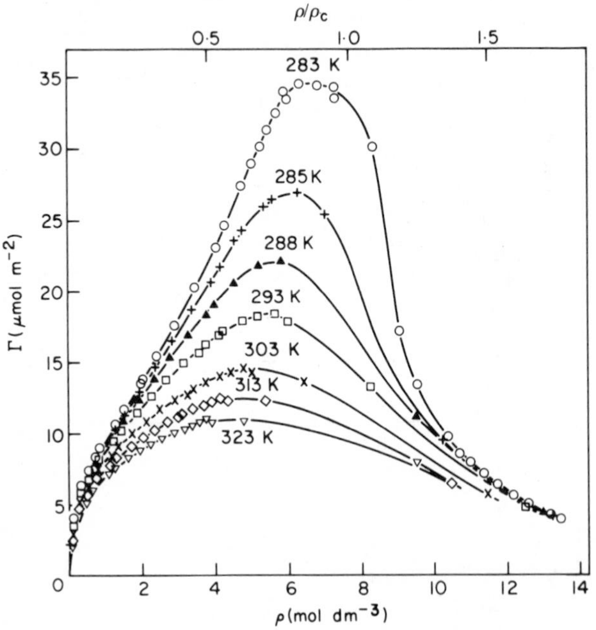

Fig. 1.2. Surface excess concentration Γ as a function of bulk density ρ of ethylene on graphitized carbon black. Densities are along the 283·15 K isotherm (T_c = 282·35 K, ρ_c = 7·635 mol dm^{-3}). (From Specovius and Findenegg, 1980.)

Butler *et al.*, 1974) has shown up a number of λ-anomalies and less exaggerated maxima which are characteristic signatures of particular systems (Butler *et al.*, 1979). High-precision work is important also in determining properties affected by the structuring and adsorption of liquids at interfaces (Clint, 1973).

Other methods have benefited from a great increase in precision and the concomitant increase in sensitivity, and from the wider range of temperatures and pressures which can be measured accurately. Among these may be mentioned the vacuum microbalance (Khan, 1972) and gas chromatography. High-pressure measurements of adsorption are also becoming increasingly common following the work of Menon (1968) and new behaviour, such as maxima in isotherms, has been observed (e.g. Specovius and Findenegg, 1980) under these conditions (Fig. 1.2).

(b) Diffraction and scattering methods

Perhaps the most significant contributions in this category are associated with LEED and neutron scattering. The former certainly takes precedence, since it goes back to the classical experiments of Davisson and Germer, but again it was only with the advent of modern electronics, which within the last decade have enabled a direct display of diffraction patterns to be made, that electron diffraction has become a widely used tool for the examination of surface structures.

The capabilities of the method in the field of physisorption have been vastly increased by the use of a channel electron multiplier which enables the incident current to be lowered to a level at which there is no risk of desorption by the beam (Chinn and Fain, 1977). In this way information can be obtained not only about the structure of adsorbed layers but also about their orientation and misfit properties with respect to the substrate (Shaw *et al.*, 1978). The latter have also been determined by photometrically analysing the diffraction spots obtained in transmission high-energy electron diffraction (THEED) with a computer-controlled densitometer (Price and Venables, 1976).

It is convenient to combine LEED with Auger spectra measurements in the same vessel (Suzanne *et al.*, 1973), and in this way it becomes possible to measure adsorption isotherms at extremely low pressures (10^{-4}–10^{-9} Torr), thereby extending the range previously available to traditional methods (10^{-5}–100 Torr).

Neutron diffraction has also been widely applied (Hall and Wright, 1978; McTague *et al.*, 1978) to the determination of adsorbate structure, not only of rare gases, but also of larger molecules, such as methane and ammonia where advantage can be taken of constituent atoms with high scattering cross-sections (White *et al.*, 1978). Because of the variety of conditions pertaining to the scattering of neutrons—elastic, inelastic, coherent and incoherent—the technique is very versatile and in addition to structural properties it can yield information on dynamical aspects. For example, diffusion coefficients can be found from incoherent inelastic scattering measurements by fitting a Lorentzian function to plots of scattering intensity versus wave vector.

Neutron scattering offers the great advantage that background scattering can often be reduced to a negligible proportion of the total by a suitable choice of materials. More recently the armoury of tools available for the study of surface films has been augmented by the addition of X-ray scattering, for which the same is true in some cases. For example, in the Xe-graphite system, the graphite background is very much less than the signal from the adsorbate and can readily be subtracted out. In conjunction with the first measurements of this kind (Brady *et al.*, 1977), a method employing Fourier transformation was also developed which enabled a "two-dimensional" distribution function to be extracted. In view of the direct importance which these functions have for theoretical interpretation, it is unfortunate that this approach has not yet gained more widespread application.

(c) Spectroscopy

Conventional infrared and NMR spectroscopic methods have been extensively used for a long time in adsorption studies quite apart from their more recent use to study adsorption in the well defined adsorbate–graphite systems (Boddenberg and Moreno, 1977). They have also been employed to provide evidence in support of the belief that liquids are "structured" in the vicinity of solid surfaces and large molecules (Roberts and Zundel, 1980). It is perhaps inevitable that in many of these cases there is considerable controversy about the actual existence of structure due to adsorption (Everett and Podoll, 1979; Clifford, 1975), and quite extravagant claims have been made on behalf of interfacial aqueous systems (Drost-Hansen, 1969). It may be that, in these more controversial areas, computer simulations and theoretical work can help to distinguish what is reasonable from what must be regarded with scepticism.

(d) Direct measurement of interaction

It is an appealing idea, because of its simplicity, that the force between two solids might be measured directly. However, at separations of the order of a few molecular diameters, great experimental skill is required to do this (Israelachvili and Adams, 1978) and it is again only during the last decade or so, and with modern instrumentation, that measurements of the necessary precision have become feasible. Here it is invariably mica which is the favoured solid material, in part because, in common with graphite, it can be cleaved to expose uniform and atomically smooth surfaces. The mica is glued to cylindrical rods orientated at right angles to each other. The direct force between the rods can be measured by spring deflection to better than 10^{-7} N and their separation, by multiple beam interferometry, to an accuracy of about 0·1 nm. More recent work by Horn and Israelachvili (1980), in which the rods are immersed in a liquid phase, has established unequivocally that the force between them oscillates as their separation is reduced, and this can be taken as very direct evidence for the adsorption or "structuring" of the liquid normal to the solid surface.

This brief survey does no more than highlight a few of the regions of interaction between theory and experiment which are of particular current interest. Many more examples and methods could be added; however our primary concern here is with the development and application of the theory.

3. Summary of later chapters

We begin in the traditional way by discussing some aspects of the thermodynamics of interfaces. The discussion serves to introduce some basic ideas and notation applied to equilibrium, non-uniform, systems in a general sense. At the same time considerable attention has been paid to several phenomena—such as capillary condensation, the film description of the adsorbate and the perturbation of the adsorbent by the adsorbate phase—which expose the limitations of macroscopic thermodynamic methods in the face of essentially microscopic phenomena. In such areas there is a need to develop microscopic theories and, as pointed out already, computer simulation can play an important role in their development. Nevertheless, the formalism of thermodynamics provides an invaluable framework on which to construct any discussion of these theories as well as the necessary link between them and experimental results. A number of topics relating to these links are discussed, including some aspects of phase transitions and critical phenomena which are becoming increasingly important in interfacial studies.

In Chapter 3 we consider non-uniform systems from the standpoint of classical equilibrium statistical mechanics, beginning with a consideration of the Hamiltonian and the analysis of the kinetic and potential energy terms. The treatment of the latter depends on the model adopted for the interface under study; the subdivision of this term in the case when one phase is assumed to be a completely inert and rigid solid is set in a general context. This discussion leads naturally to the calculation of the interaction potentials which, as already mentioned, have a particular importance in simulation studies. Following this we turn to the determination of the thermodynamic quantities from an assumed knowledge of distribution functions or from the fluctuations which can be accumulated in computer simulations. In the remainder of the chapter some theoretical approaches to the calculation of distribution functions themselves are considered. These usually take the form of integral equations. Unfortunately, all of these equations require some kind of approximation in their application and it is therefore important to set them in a general context and attempt to highlight the inter-relation between them, as well as the limitations imposed by the approximations. A very effective mathematical tool for this purpose is the method of functional differentiation, which we have applied extensively here. However, the BBGKY hierarchy, although it can be developed in this way, is first derived in the more usual fashion by direct differentiation of the distribution functions. This treatment, and the discussion of the non-uniform Ornstein–Zernike equation which follows it, serves to introduce the problem in the wider context of multicomponent mixtures. The chapter ends with a discussion of van der Waals and perturbation theories as applied to non-uniform systems in general.

Chapter 4 takes up the discussion of computer simulation techniques starting with a consideration of the information available from their application. There follow fairly detailed accounts of the two methods used: Monte Carlo and molecular dynamics. In both cases some attention is paid to methods of improving the efficiency of the process, whether these be by simple book-keeping procedures or by more subtle sampling methods such as the force bias method of Berne and co-workers. The anisotropic and non-uniform systems which are characteristic of adsorption make great demands on computer resources and so it is necessary to exploit to the full any improvements in technique. One such technique which has been found to have great advantages for gas–solid adsorption, the grand canonical ensemble Monte Carlo method, is examined in some detail. A good deal of attention is also devoted to special devices for obtaining values of the free energy, including umbrella sampling methods. The correction for long-range effects outside the simulation box, which is inevitably restricted in size, is treated in some detail, both for dispersion interaction and for the still incompletely solved problem of longer-ranged electrostatic forces. Finally some of the extra difficulties encountered in dealing with assemblies of di- or poly-atomic molecules are discussed.

Chapter 5, which is concerned with monolayers of structureless particles, starts by considering the relevance of such idealized models to real systems. A great deal of fundamental work has been devoted to two-dimensional systems in zero field. This is discussed particularly with respect to the possibility of the occurrence of a "melting" transition. Attention is then turned to the effects which can be produced by the intervention of a periodic field from the adsorbent, which introduces the possibility of transitions between epitaxial and non-epitaxial adsorbed phases. This topic is a lively one, but it has not progressed to the stage where the effects to be expected from the presence of large numbers of vacancies in the layer can be adequately understood. Nevertheless, some interesting work has been carried out on lattice models of adsorption in which vacancies are permitted and the chapter closes with some examples of the phase diagrams obtained for models of this type.

Chapter 6 is devoted to simulations of three-dimensional model systems which represent fairly closely some of the real situations which are of importance in interfacial science. Although most of the work described deals with spherical molecules, there is no reason why the successful methods should not be applied to more complicated systems, and some work of that character has been done and is included. Gas–solid adsorption is first examined in detail. In spite of the considerable difficulties deriving from the severe non-uniformity of density which is encountered here, this is an area in which there has been much progress with regard to the understanding of bulk (static and dynamic) as well as structural properties. Liquid–solid interfaces, a field in which there has been great activity recently, are then examined. Attention is focused on the structural properties, as has been done in the original studies. A short discussion deals with the surprisingly small amount of simulation work which has been done on gases in pores. The chapter ends with a discussion of microclusters, that is, clusters of less than about 100 atoms. Most of the interest is in the differences between the properties of the microcluster and those of the bulk phase, the importance of this for the understanding of catalysis by such clusters requiring no emphasis.

In Chapter 7, we consider the results which can be obtained for non-uniform systems which are essentially three-dimensional, from the application of theoretical methods. Such systems include liquid interfaces, either with solid surfaces or with other fluid phases and multilayer adsorbates. The theoretical methods are subdivided into three sections, going from the exact analytical results which can be found for certain integral equations with special potentials, through numerical solutions to model systems. This gradation covers not only the newer approaches which enable singlet distributions as well as isotherms to be determined, but also, in the third section, encompasses a discussion of some of the older theories such as the BET model and the slab theory of Frankel Halsey and Hill which have a long record of service to surface science. Computer simulation has been of assistance in helping to understand the well known limitations of these theories and their discussion in this light has proved to be revealing. The theoretical treatment in this section is set within the framework of a rather general lattice theory which not only helps to explore interrelationships in a unified way but also shows how extension to more realistic model theories can be made.

Chapter 8 consists, in the main, of a detailed discussion of the contribution which simulation studies have made to the understanding of hydrophobic effects. In the Preface we have already made our apologia for including what many may regard as a topic which is not strongly connected with the main theme of the book, and we shall not debate this point further here. After a short examination of some general results and the work which has been done on very simple fluids, attention is turned to water. The various formulations of the interaction potential for pairs of water molecules are discussed. The effect of the solute on the structure of water (a topic which has excited interest among chemists for at least thirty years) and the forces between solute atoms which can be attributed to these structural effects are examined in considerable detail. Although most of the results so far have been for spherical solute molecules there has been important progress in the development of the treatments for application to more complex molecules and indeed even to biological molecules, and this is the subject on which we end our consideration of the hydrophobic effect.

References

Beebe, R A. and Young, D. M. (1954). *J. Phys. Chem.* **58**, 93.

Boddenberg, B. and Moreno, J. A. (1977). *J. de Phys.* **38**, C-4, 52.

Brady, G., Fein, D. B. and Steele, W. A. (1977). *Phys. Rev. B* **15**, 1120.

Bretz, M., Dash, J. G., Hickernell, D. C., McLean, E. O. and Vilches, O. E. (1973). *Phys. Rev. A* **8**, 1589.

Butler, D. M., Huff, G. B., Toth, R. W. and Stewart, G. A. (1974). *Phys. Rev. Lett.* **32**, 724.

Butler, D. M., Litzinger, J. A., Stewart, G. A. and Griffiths, R. B. (1979). *Phys. Rev. Lett.* **42**, 1289.

Chinn, M. D. and Fain Jr., S. C. (1977). *J. Vac. Technol.* **14**, 314.

Clifford, J. (1975). *In* "Water: A Comprehensive Treatise" (F. Franks, ed.), Vol. 5, p 75. Plenum Press, New York.

Clint, J. H. (1973). *J. Chem. Soc. Faraday Trans. I* **69**, 1320.

Drost-Hansen, W. (1969). *Chem. Phys. Lett.* **2**, 647.

Everett, D. H. and Podoll, R. T. (1979). *Chem. Soc. Specialist Periodical Reports*: "Colloid Science", Vol. 3.

Grillet, Y., Rouquerol, J. and Rouquerol, F. (1979). *J. Colloid Interf. Sci.* **70**, 239.

Hall, P. G. and Wright, C. J. (1978). *Chem. Soc. Specialist Periodical Reports*: "Chemical Physics of Solid Surfaces", Vol. 7.

Horn, R. G. and Israelachvili, J. N. (1980). *Chem. Phys. Lett.* **71**, 192.

Israelachvili, J. N. and Adams, G. E. (1978). *J. Chem. Soc. Faraday Trans. I* **74**, 975.

Khan, G. M. (1972). *Rev. Sci. Instrum.* **43**, 117.

Landman, U. and Kleinman, C. G. (1977). *Chem. Soc. Specialist Periodical Reports*: "Surface and Defect Properties of Solids", Vol. 6.

McTague, J. P., Nielsen, M. and Passell, L. (1978). *Crit. Rev. Solid State and Mater. Sci.* **8**, 439.

Menon, P. G. (1968). *Chem. Rev.* **68**, 277.

Price, G. L. and Venables, J. A. (1976). *Surf. Sci.* **59**, 50.

Roberts, N. K. and Zundel, G. (1980). *J. Phys. Chem.* **84**, 3655.

Rouquerol, J., Partyka, J. and Rouquerol, F. (1977). *J. Chem. Soc. Faraday Trans. I* **73**, 306.

Shaw, C. G., Fain Jr., S. C. and Chinn, M. D. (1978). *Phys. Rev. Lett.* **41**, 955.

Specovius, J. and Findenegg, G. H. (1980). *Ber. Bunsenges. Phys. Chem.* **84**, 690.

Suzanne, J., Coulomb, J. P. and Bienfait, M. (1973). *Surf. Sci.* **40**, 414.

Thomy, A. and Duval, X. (1969). *J. Chim. Phys.* **66**, 1966.

White, J. W., Thomas, R. K., Trewern, T., Marlow, I. and Bomchill, G. (1978). *Surf. Sci.* **76**, 13.

Chapter 2

Thermodynamics of Interfaces

1. Gibbs excess and thickness of the interface

The local density and hence non-uniformity of a single species a is expressed by the singlet number density function $\rho^{(a)}(\boldsymbol{r})$ at the position $\boldsymbol{r}$. The quantity $\rho^{(a)}(\boldsymbol{r})\,\mathrm{d}\boldsymbol{r}$ can be defined as the average number of molecules of species a to be found in the volume element $\mathrm{d}\boldsymbol{r}$ over a long period of time. In the vicinity of an interface the singlet density will vary with $\boldsymbol{r}$. At a large distance (on the molecular scale) from any interface the phases become uniform and the density of component a then has the same value ρ_a everywhere.

The adsorption, or surface excess number of molecules of component a in the interfacial region, is given by

$$N_a^{\Sigma} = \int_V (\rho^{(a)}(\boldsymbol{r}) - \rho_a)\,\mathrm{d}\boldsymbol{r}. \tag{2.1}$$

In a more general case the singlet density of the adsorbate may not be uniform anywhere in the system and may not even be anywhere equal to ρ_a. Such a situation might arise, for example, when a fluid adsorbate (liquid or gas) is contained between solid surfaces in a microporous region. Nevertheless, the adsorption can still be expressed by an equation of the above form in which ρ_a now stands for a reference density (e.g. the external bulk phase in equilibrium with the adsorbate) and the domain of the integration could comprise the whole of the adsorbate fluid.

Since the choice of the domain of integration is arbitrary, the surface excess of at least one component in an interfacial system may always be chosen to be zero. We consider as an example, the more general situation depicted in Fig. 2.1. with two components a and b, and for simplicity, the singlet density varying in the z-direction only. The interfacial area in a

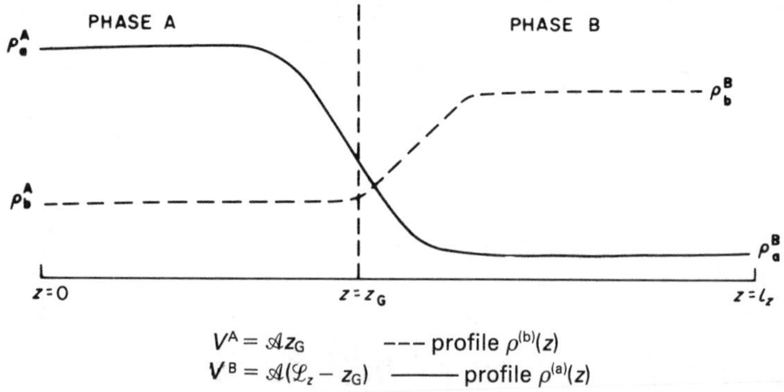

Fig. 2.1

plane normal to the z-axis, is $\mathcal{A}$. The origin is in phase A at a sufficiently large distance from the interface for the densities of both phases to have attained their uniform values ρ_a^A and ρ_b^B, where the superscript identifies the phase. The surface excesses of component a in phases A and B are

$$N_a^{A,\Sigma} = \mathcal{A} \int_0^{z_G} (\rho^{(a)}(z) - \rho_a^A) \, dz. \tag{2.2a}$$

$$N_a^{B,\Sigma} = \mathcal{A} \int_{z_G}^{\infty} (\rho^{(a)}(z) - \rho_a^B) \, dz. \tag{2.2b}$$

From Fig. 2.1 it is clear that $N_a^{A,\Sigma}$ and $N_a^{B,\Sigma}$ have opposite signs and therefore z_G can be chosen such that the overall surface excess of component a, $N_a^\Sigma = N_a^{A,\Sigma} + N_a^{B,\Sigma}$, is zero. The surface excess of component b, defined in an analogous manner, may then be either positive or negative. The hypothetical dividing surface at z_G is known as the Gibbs interface.

To a reasonable first approximation, the solid in a solid–fluid system may be regarded as being uniform up to its surface. However, this is never strictly true. In the first place the symmetry of the potential field around the atoms of the solid will be different at the surface from that in the bulk material, giving rise to relaxation effects. Secondly the adsorbate fluid itself is likely to modify the structure of the solid, just as the converse is possible. The assumption that the solid is rigid must therefore be regarded with caution, especially for example, in biological systems. With this in mind the position of the surface of the solid from a thermodynamic point of view should be chosen to be a Gibbs surface (in the above sense) which makes the surface excess of the solid equal to zero. The surface excess of adsorbate molecules in the system may then be defined with respect to this solid surface using Eqn (2.1). For a single-component adsorbate, denoted by a, it is (see Fig. 2.2(*a*))

$$N_a^\Sigma = \mathscr{A} \int_{z=0}^{l_z} (\rho^{(a)}(z) - \rho_a^G)\, dz. \tag{2.3}$$

In the case of gas adsorption, an adsorbate thickness may be introduced by placing an interface between dense adsorbate and the gas phase such that the adsorbate excess is zero with respect to this interface, as shown in Fig. 2.2(*b*) for a planar adsorbent. If the dense adsorbate is denoted by F (film) and the less dense region by G (gas) there is now an obvious analogy with the situation depicted in Fig. 2.1. The condition for zero adsorbate excess at $z = t$ is given by

$$N_a^{F,\Sigma} + N_a^{G,\Sigma} = 0 \tag{2.4a}$$

$$N_a^{F,\Sigma} = \mathscr{A} \int_{z=0}^{t} (\rho^{(a)}(z) - \rho_a^F)\, dz \tag{2.4b}$$

$$N_a^{G,\Sigma} = \mathscr{A} \int_{t}^{l_z} (\rho^{(a)}(z) - \rho_a^G)\, dz. \tag{2.4c}$$

If the last two equations are added and combined with (2.4a) and (2.3), t may be found from

$$t = N_a^\Sigma / \mathscr{A}(\rho_a^F - \rho_a^G) = \Gamma_a/(\rho_a^F - \rho_a^G). \tag{2.5}$$

The density ρ_a^F may be regarded as a reference density and still remains to be specified. A natural choice for ρ_a^F might be the density of a uniform liquid of molecules of a at the same temperature as the adsorbate and under saturated vapour pressure. However, as indicated in the figure, there is no reason to suppose that the adsorbate will possess this density over any extensive region and will not in general converge to this value. Indeed since $\rho^{(a)}$ usually oscillates, a bulk reference density chosen in this way is somewhat arbitrary.

A similar analysis may be made of adsorbate thickness in a cylinder which can serve as an idealized model for a pore in a porous material. If the cylinder is of radius R and length l and we again assume that $\rho^{(a)}$ only varies in a direction normal to the surface, then Eqn (2.1) becomes

$$N_a^\Sigma = 2\pi l \int_{0}^{R} (\rho^{(a)}(r) - \rho_a^G) r\, dr \tag{2.6a}$$

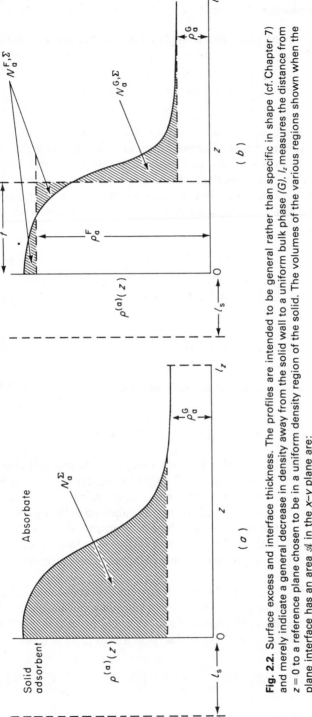

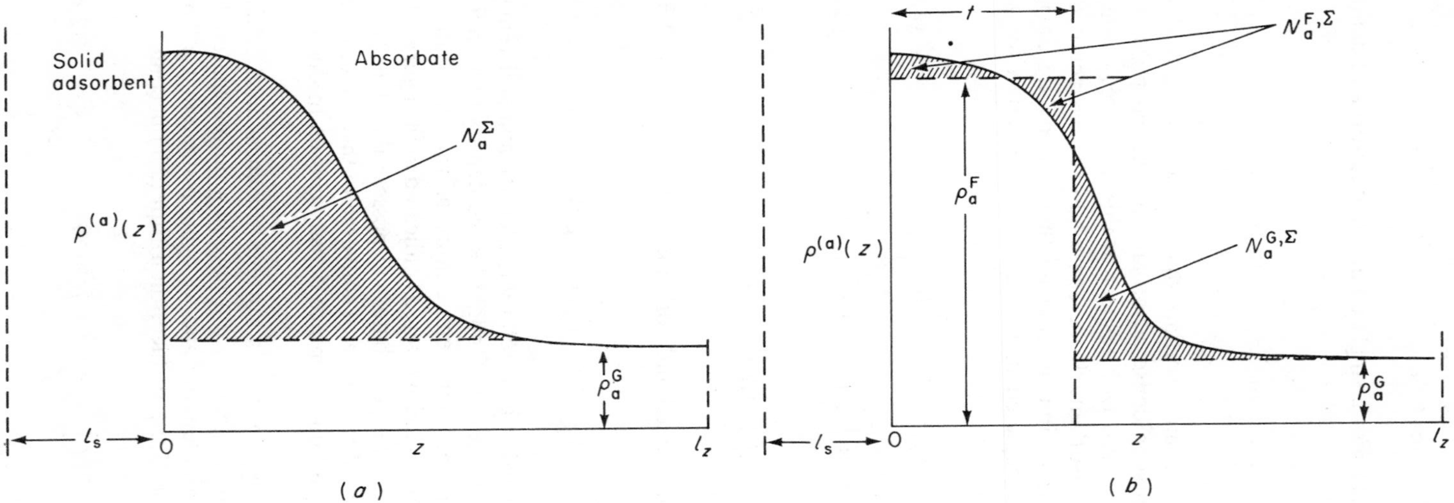

Fig. 2.2. Surface excess and interface thickness. The profiles are intended to be general rather than specific in shape (cf. Chapter 7) and merely indicate a general decrease in density away from the solid wall to a uniform bulk phase *(G)*. l_z measures the distance from $z = 0$ to a reference plane chosen to be in a uniform density region of the solid. The volumes of the various regions shown when the plane interface has an area $\mathscr{A}$ in the *x–y* plane are:

(*a*)	adsorbent	$V^S = \mathscr{A} l_s$
	adsorbate	$V^A = \mathscr{A} l_z$
(*b*)	adsorbent	$V^S = \mathscr{A} l_s$
	adsorbate "film"	$V^F = \mathscr{A} t$
	"gas" phase	$V^G = \mathscr{A}(l_z - t)$

and Eqns (2.4b) and (2.4c) are replaced by

$$N_{\mathrm{a}}^{\mathrm{B},\Sigma} = 2\pi l \int_{r-t_c}^{R} (\rho^{(\mathrm{a})}(r) - \rho_{\mathrm{a}}^{\mathrm{F}}) r \, \mathrm{d}r \tag{2.6b}$$

$$N_{\mathrm{a}}^{\mathrm{G},\Sigma} = 2\pi l \int_{0}^{r-t_c} (\rho^{(\mathrm{a})}(r) - \rho_{\mathrm{a}}^{\mathrm{G}}) r \, \mathrm{d}r. \tag{2.6c}$$

The thickness t_c of the adsorbate layer in this geometry can be found from

$$2(t_c/R) - (t_c/R)^2 = N_{\mathrm{a}}^{\Sigma}/V(\rho_{\mathrm{a}}^{\mathrm{F}} - \rho_{\mathrm{a}}^{\mathrm{G}}) \tag{2.7}$$

in which V is the volume of the cylinder. It is interesting to note, however, that in this case *neither* of the reference state densities may exist over any extended region of the adsorption volume because potential fields from adsorbent and dense adsorbate may modify the density from its gas phase value in the rarefied parts of the adsorbate near the cylinder axis. Near to the wall, considerations similar to those already mentioned with plane surfaces will operate.

Inside a confined space such as a cylinder or the slit between two planar adsorbent walls, the thickness of the adsorbate increases to a point where the intensity of the potential field in the remaining volume causes immediate filling with dense material on further increase of the chemical potential. This phenomenon of capillary condensation is considered from the thermodynamic standpoint in the following section.

The point of view in which an adsorbate on a solid surface in equilibrium with the gas phase is regarded as if it were a film of uniform fluid is sometimes useful, but it should be borne in mind that, as the foregoing shows, such a picture is not rigorous and that deductions based upon it must be viewed with some reservation. A discussion of films from the molecular point of view is made in Chapter 7.

2. The pressure tensor

Much of the practical interest in interfacial phenomena is concerned with curved interfaces such as liquid drops on solid surfaces or fluids confined within pore spaces which, as in the preceding section, may often be idealized as being circular in cross-section. It is therefore useful to begin our discussion of the pressure tensor in a sufficiently general way as to encompass such examples and then to examine specific cases. Although the term hydrostatic pressure is normally well understood, some confusion may arise in discussing this in the context of interfacial systems. This is because, in the case of a fluid at a solid surface, the pressure within the fluid can be considered as being solely due to the fluid molecules, or alternatively the external field terms, due to the solid adsorbent, can be implicitly included. The latter view, although unconventional, offers advantages when only fluid–solid interfaces are being discussed and is commonly adopted in such cases (Steele, 1973). Here we shall retain the conventional usage for greater generality.

In a uniform fluid the hydrostatic pressure can be represented by a single scalar quantity which is independent of position or direction. In non-uniform regions of the fluid, such as occur at interfaces, this is no longer the case and the work term dW which appears in the basic equation of thermodynamics needs to be reconsidered. The theory of elasticity provides a natural starting point for the discussion and has been developed, for example, by Buff (1951). Useful reviews treating this topic include those by Goodrich (1971) and by Navascués (1979). The latter also covers molecular aspects.

Consider a body which undergoes deformation such that a point at position $\boldsymbol{r}$ is displaced by an infinitesimal amount to $\boldsymbol{r} + \delta\boldsymbol{r}$, where $\delta\boldsymbol{r}$ is known as the displacement vector and in a cartesian coordinate system, for example, would have the components δx, δy, δz. The local strain in the body is completely described by a 9-component strain tensor, $e_{\alpha\beta}$ in which α, β can be any pair of the three space coordinates. In particular when $\alpha = \beta$ the diagonal components of the strain tensor are related to the displacement vector by

$$e_{\alpha\alpha} = \frac{\partial \delta r_\alpha}{\partial r_\alpha} \tag{2.8}$$

when $\alpha \neq \beta$, $e_{\alpha\beta}$ is a more complicated expression which is symmetrical in α, β so that in fact only 6 components of the strain tensor are independent.

Corresponding to the strain tensor there is a stress tensor, again with 6 independent components, $p_{\alpha\beta}$, known as the pressure tensor in hydrostatics. The virtual work done on the fluid during a small strain is

$$\delta W = -\int_V p_{\alpha\beta} e_{\alpha\beta} \, d\boldsymbol{r}. \tag{2.9}$$

The usual convention in tensor notation of summing over repeated subscriptions has been adopted.

In the particular case of an isotropic fluid we have

$$p_{\alpha\beta} = 0 \; (\alpha \neq \beta) \qquad p_{\alpha\alpha} = p \tag{2.10}$$

and Eqn (2.9) becomes

$$\delta W = -p \int e_{\alpha\alpha} \, d\boldsymbol{r}. \tag{2.11a}$$

With the aid of (2.8) the usual expression for the work of expansion can be easily obtained from (2.11a) in the infinitesimal limit

$$\mathrm{d}W = -p\,\mathrm{d}V. \tag{2.11b}$$

Since for anisotropic systems the pressure tensor is still symmetrical it is always possible to find a transformation which reduces it to a diagonal form whose three components, $p_{\alpha\alpha}$ lie along the mutually orthogonal axes of the coordinate system appropriate to the problem. If the system is invariant to a rotation about the direction normal to the interface, then these three components may be further reduced to two; one of them, p_N normal to the interface, and two equal components, p_T along the mutually orthogonal directions parallel to the interface.

The condition for mechanical equilibrium in a deformable body may be expressed in the general form

$$\sum_{\beta} \frac{\partial p_{\alpha\beta}}{\partial r_{\alpha}} - \rho^{(a)}(\boldsymbol{r})F_{\alpha} = 0 \tag{2.12}$$

in which F_α is the force acting on a molecule in the $(+\alpha)$ direction due to the external fields.

We now consider some specific examples of (2.12) which are of interest.

(a) An adsorbate fluid in contact with a planar solid adsorbent

Suppose the solid surface is in the x–y plane. In general the external field due to the solid may vary in x, y and z directions, and (2.12) applies with $\boldsymbol{r} = (x, y, z)$. If the x, y variation is neglected

$$\frac{\partial p_{xx}}{\partial x} = \frac{\partial p_{yy}}{\partial y} = 0 \tag{2.13a}$$

$$\frac{\partial p_{zz}(z)}{\partial z} - \rho^{(a)}(z)F_z = 0. \tag{2.13b}$$

If the external field term is absent or negligible, then Eqn (2.13b) implies that p_{zz} is independent of z and therefore everywhere equal to the pressure in the adjacent uniform phase. A similar result applies when the adsorbent is an infinitely repulsive hard wall since $-F_z$, which is the gradient of the potential, is zero in this case. When the adsorbent is regarded as an internal part of the system, the pressure terms would arise as a result of all possible interactions in the total system. In this case there would of course be no external field and the z-component of the pressure, $p'_{zz}(z)$, say, would be constant throughout the system and equal to the pressure in the adjacent uniform phases. This approach is the natural one to adopt for adsorption at fluid interfaces.

(b) A fluid enclosed in a cylindrical space with solid walls

This example is important in connection with adsorption in porous materials. The relevant coordinates are (r, φ, z) with r normal to the interface and z along the axis of the cylinder. Application of Eqn (2.12) leads to the results,

$$\frac{\partial p_{zz}}{\partial z} - \rho^{(a)}(\mathbf{r})F_z = 0 \tag{2.14a}$$

$$\frac{1}{r}\frac{\partial p_{\varphi\varphi}}{\partial \varphi} - \rho^{(a)}(\mathbf{r})F_\varphi = 0 \tag{2.14b}$$

$$\frac{\partial p_{rr}}{\partial r} - \frac{(p_{rr} - p_{\varphi\varphi})}{r} - \rho^{(a)}(\mathbf{r})F_r = 0. \tag{2.14c}$$

If the periodicity in the external field can be neglected, $(\partial p_{zz}/\partial z) = (1/r)(\partial p_{\varphi\varphi}/\partial \varphi) = 0$, and the transverse components of the pressure tensor are equal to each other over the surface of concentric annuli of radius r in the fluid. We may therefore write

$$p_{\mathrm{T}}(r) = \tfrac{1}{2}(p_{zz} + p_{\varphi\varphi}). \tag{2.14d}$$

The normal component $p_{\mathrm{N}}(r)$ is equivalent to p_{rr} and with (2.14d), Eqn (2.14c) becomes

$$\frac{\partial p_{\mathrm{N}}(r)}{\partial r} - \rho^{(a)}(r)F_r = 0 \tag{2.14e}$$

which is equivalent to (2.13b).

(c) A spherical interface

Possible examples of this geometry include: a spherical meniscus or bubble, a droplet on a solid substrate, or a fluid adsorbed in a spherical cavity.

The coordinates are now (r, θ, φ) with r normal to the interface. Equation (2.12), after appropriate transformation, gives

$$\frac{1}{r\sin\theta}\frac{\partial p_{\varphi\varphi}}{\partial \varphi} - \rho^{(a)}(\mathbf{r})F_\varphi = 0 \tag{2.15a}$$

$$\frac{1}{r}\frac{\partial p_{\theta\theta}}{\partial \theta} + \frac{(p_{\theta\theta} - p_{\varphi\varphi})}{r}\cot\theta - \rho^{(a)}(\mathbf{r})F_\theta = 0 \tag{2.15b}$$

and

$$\frac{\partial p_{rr}}{\partial r} + \frac{(p_{rr} - p_{\varphi\varphi})}{r} + \frac{(p_{rr} - p_{\theta\theta})}{r} - \rho^{(a)}(\mathbf{r})F_r = 0 \tag{2.15c}$$

If the system is invariant to rotation about a radius, which would be the case when the external field is either absent or subject to appropriate symmetries, the tangential part of the pressure tensor, $p_T(r) = \frac{1}{2}(p_{\theta\theta} + p_{\varphi\varphi})$ and (2.15c) becomes

$$\frac{\partial p_N}{\partial r} + \frac{2(p_N - p_T)}{r} - \rho^{(a)}(r)F_r = 0. \tag{2.15d}$$

The virtual work of expansion δW in an anisotropic system is given by

$$\delta W = -\int p_{\alpha\alpha} e_{\alpha\alpha}\, d\mathbf{r}. \tag{2.16}$$

This is the work done on the system against the intermolecular forces in the fluid but does not include any body forces such as the external forces due to a solid adsorbent.

With the aid of (2.8), Eqn (2.16) can be expanded and rearranged to give explicit expressions for δW. For example, for the interface between a fluid and a planar solid we may write,

$$\delta W = -l_y \delta x \int_0^{l_z} (p_{xx} - p_{zz})\, dz - l_x \delta_y \int_0^{l_z} (p_{yy} - p_{zz})\, dz$$
$$- \int_V p_{zz}(z)\, d\delta V \tag{2.17}$$

where

$$d\delta V = l_y \delta x\, dz + l_x \delta y\, dz + l_x l_y (\partial \delta V/\partial z)\, dz$$

and if the solid gives rise to a potential field varying in the z-direction only, $p_{zz}(z)$ may be obtained from Eqn (2.13b) by integration as

$$p_{zz}(z) - p = \int_{l_z}^{z} \rho^{(a)}(z)\, F_z\, dz \tag{2.18}$$

when $p_{zz}(z)$ from (2.18) is substituted into (2.17) the external field term on the right-hand side of (2.18) appears multiplied by $l_z \delta \mathcal{A} (= \delta V)$ from the first two integrals in (2.17) and therefore cancels with the external field term from the third integral, leaving the result

$$\delta W = -l_y \delta x \int_0^{l_z} (p_{xx} - p)\, dz - l_x \delta y \int_0^{l_z} (p_{yy} - p)\, dz - p\delta V. \tag{2.19}$$

The virtual work for curvilinear systems can also be obtained from Eqn (2.16) (Melrose, 1970), but these systems do not yet appear to have received detailed analysis for the case when the fluid is in contact with a solid. This is also true for planar systems with fields varying periodically in the x, y directions.

3. Surface pressure and surface tension

The principle of the Gibbs dividing surface and the associated concept of surface excess quantities introduced in §1 can easily be extended to other thermodynamic variables. For example the mechanical work done by two homogeneous phases A, B, if they were uniform up to an interface between them, would be $-p^{\mathrm{A}}\delta V^{\mathrm{A}} - p^{\mathrm{B}}\delta V^{\mathrm{B}}$. The work done by the actual nonuniform system, on the other hand, includes additional terms which result from changes in interfacial shape or area. To account for these possibilities, δW may be written

$$\delta W = -[p^{\mathrm{A}}\delta V^{\mathrm{A}} + p^{\mathrm{B}}\delta V^{\mathrm{B}} + \phi\delta\mathcal{A} + C_1\delta R_1 + C_2\delta R_2] \qquad (2.20a)$$

in which R_1, R_2 are principal radii. Equation (2.20a) provides the thermodynamic definitions of the surface pressure ϕ and the two intensive quantities C_1, C_2. For an interface with constant curvature δR_1, δR_2 are both zero and Eqn (2.20a) becomes

$$\delta W = -[p^{\mathrm{A}}\delta V^{\mathrm{A}} + p^{\mathrm{B}}\delta V^{\mathrm{B}} + \phi\delta\mathcal{A}]. \qquad (2.20b)$$

Equations (2.20) can be used to examine the relationship between bulk phase pressure and surface pressure for the different geometries discussed in §2 by introducing the condition that δW is an invariant with respect to the position of the interface.

(a) A planar interface

The general case is illustrated in Fig. 2.1, with the interface lying between phases A and B. It is easily seen that, because $V(=V^{\mathrm{A}} + V^{\mathrm{B}})$ is fixed, $(\partial\delta V^{\mathrm{A}}/\partial z) = -(\partial\delta V^{\mathrm{B}}/\partial z)$ and since the surface pressure ϕ and the area $\mathcal{A}$ are also independent of the position of the interface, the condition $(\partial\delta W/\partial z) = 0$ implies

$$p^{\mathrm{A}} = p^{\mathrm{B}} \qquad (2.21)$$

As already pointed out in §1, an adsorbate at a plane, unperturbed solid surface may be regarded, either as a single phase of a volume V^{A} with density decreasing to that of the uniform gas phase, or as a two-phase system with an interface between a dense phase of volume V^{F} adjacent to the wall, and a gas phase of volume V^{G}. In the latter description, of course, (2.21) applies with $p^{\mathrm{F}} = p^{\mathrm{G}}$. In the former case the condition $(\partial\delta W/\partial z) = 0$ is equivalent to the statement $(\partial p/\partial z) = 0$ which is compatible with the hydrostatic result in (2.13b) and (2.18)

The hydrostatic equation (2.19) can easily be extended to the more general case considered here in which two different homogeneous fluid phases F, G meet at an interface, by introducing

$$p^{\mathrm{FG}} = [1 - H(z - z_{\mathrm{G}})]p^{\mathrm{F}} + H(z - z_{\mathrm{G}})p^{\mathrm{G}} \tag{2.22}$$

where the Heaviside step function is

$$\begin{aligned} H(x) &= 0 \qquad x < 0 \\ H(x) &= 1 \qquad x \geqslant 0 \end{aligned} \tag{2.23}$$

Eqns (2.17), (2.19) now become

$$\begin{aligned} \delta W = -[l_y \delta x \int (p_{xx} - p^{\mathrm{FG}})\, \mathrm{d}z + l_x \delta y \int (p_{yy} - p^{\mathrm{FG}})\, \mathrm{d}z \\ + p^{\mathrm{F}} \delta V^{\mathrm{F}} + p^{\mathrm{G}} \delta V^{\mathrm{G}}] \end{aligned} \tag{2.24}$$

Comparing (2.24) with (2.20b) gives the relationship between the mechanical and thermodynamic expressions for the surface pressure,

$$\phi \delta \mathcal{A} = l_y \delta x \int (p_{xx} - p^{\mathrm{FG}})\, \mathrm{d}z + l_x \delta y \int (p_{yy} - p^{\mathrm{FG}})\, \mathrm{d}z \tag{2.25}$$

If the system is rotationally invariant, then $p_{xx} = p_{yy} = p_{\mathrm{T}}$ and, since $\delta \mathcal{A} = l_y \delta x + l_x \delta y$, we have the more familiar result,

$$\phi = (-\gamma) = \int (p_{\mathrm{T}} - p^{\mathrm{FG}})\, \mathrm{d}z \tag{2.26}$$

When liquid or liquid–vapour interfaces are being considered, it is more usual to replace $-\phi$ by the surface tension γ, as indicated in (2.26).

Equation (2.26) provides an essential link between macroscopic phenomenological theories of the interface, and molecular theories (Goodrich, 1971). This result requires careful consideration since it involves the assumption that $\delta \mathcal{A}$ may be factorized from the hydrostatic work equation as the term $\int\int (\partial \delta x/\partial x + \partial \delta y/\partial y)\, \mathrm{d}x\, \mathrm{d}y$, which is justified only if the expression in brackets changes slowly over the region of the interface. The existence of an external field does not further invalidate (2.26) provided that variations in the x, y directions can be neglected.

(b) Curved interfaces

We consider the two examples of cylindrical and spherical surfaces discussed above. In the former the film description of the adsorbate fluid leads to a picture of this as an annular dense fluid inside a cylinder of unperturbed solid adsorbent as illustrated in Fig. 2.3(a). The condition for invariance of δW may now be written $\partial \delta W/\partial (r_{\mathrm{p}} - t) = 0$.

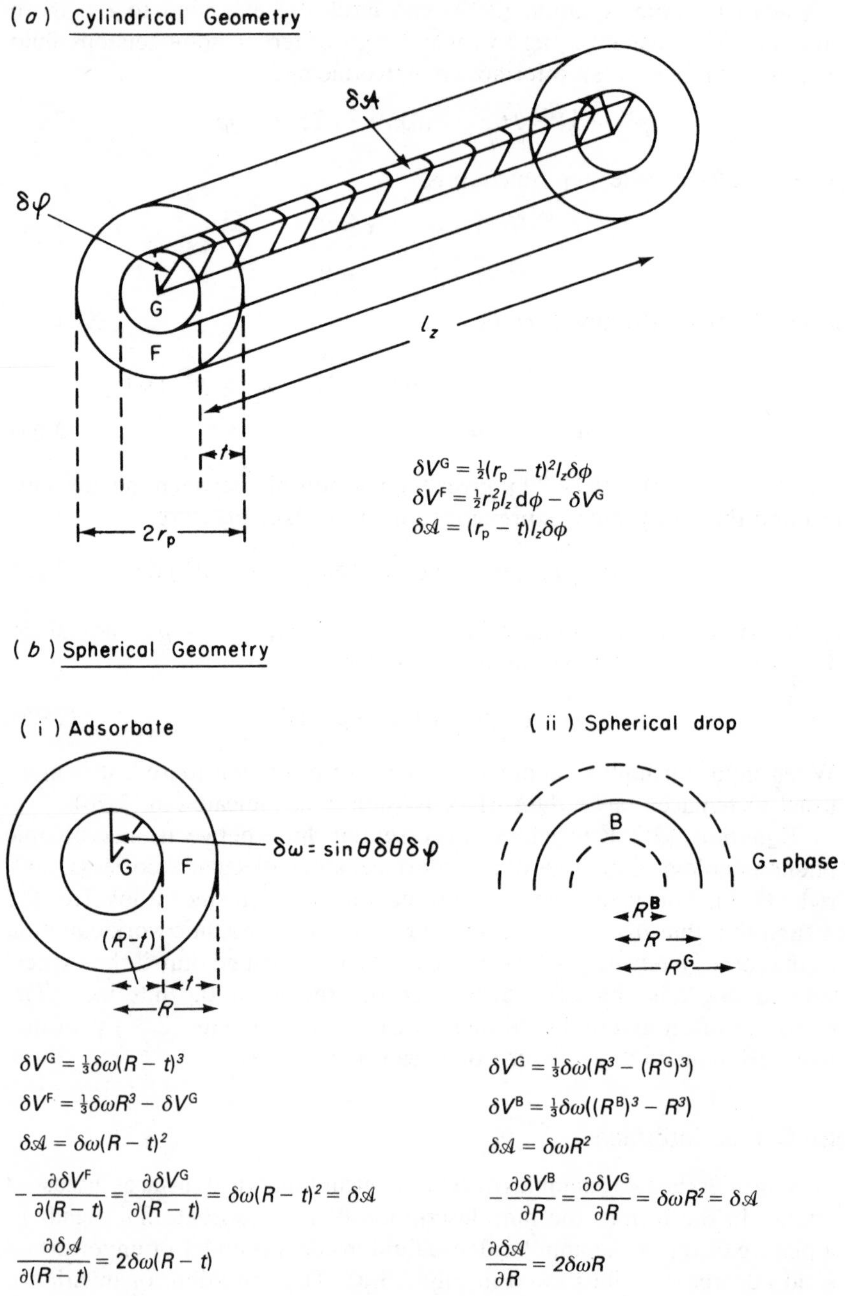

(a) Cylindrical Geometry
$\delta\mathcal{A}$
$\delta\varphi$
G
F
l_z
t
$2r_p$
$\delta V^G = \frac{1}{2}(r_p - t)^2 l_z \delta\phi$
$\delta V^F = \frac{1}{2}r_p^2 l_z \, d\phi - \delta V^G$
$\delta\mathcal{A} = (r_p - t) l_z \delta\phi$
(b) Spherical Geometry
(i) Adsorbate
G
F
$\delta\omega = \sin\theta\,\delta\theta\,\delta\varphi$
$(R-t)$
t
R
(ii) Spherical drop
B
G-phase
R^B
R
R^G
$\delta V^G = \frac{1}{3}\delta\omega(R - t)^3$
$\delta V^F = \frac{1}{3}\delta\omega R^3 - \delta V^G$
$\delta\mathcal{A} = \delta\omega(R - t)^2$
$-\frac{\partial \delta V^F}{\partial (R-t)} = \frac{\partial \delta V^G}{\partial (R-t)} = \delta\omega(R-t)^2 = \delta\mathcal{A}$
$\frac{\partial \delta\mathcal{A}}{\partial (R-t)} = 2\delta\omega(R-t)$
$\delta V^G = \frac{1}{3}\delta\omega(R^3 - (R^G)^3)$
$\delta V^B = \frac{1}{3}\delta\omega((R^B)^3 - R^3)$
$\delta\mathcal{A} = \delta\omega R^2$
$-\frac{\partial \delta V^B}{\partial R} = \frac{\partial \delta V^G}{\partial R} = \delta\omega R^2 = \delta\mathcal{A}$
$\frac{\partial \delta\mathcal{A}}{\partial R} = 2\delta\omega R$

Fig. 2.3

Applying this to (2.20b) we find, with the aid of Fig. 2.3(a),

$$p^{\mathrm{F}} - p^{\mathrm{G}} = \frac{\phi}{(r_{\mathrm{p}} - t)} + \frac{\partial \phi}{\partial (r_{\mathrm{p}} - t)} \tag{2.27}$$

For the spherical geometry, illustrated in Fig. 2.3(b), again adopting the film model for the adsorbate, the condition $\partial \delta W/\partial R$ leads to the result

$$p^{\mathrm{F}} - p^{\mathrm{G}} = \frac{2\phi}{R} + \frac{\partial \phi}{\partial R} \tag{2.28}$$

Although it is usual to identify the less dense phase, denoted here by G, with the external gas phase, it must be borne in mind, as mentioned earlier, that the two do not necessarily have the same properties.

For curved interfaces, as the examples in Eqns (2.27) and (2.28) illustrate, ϕ can vary with the position of the interface. It is conventional to set $[\partial \phi/\partial (r_{\mathrm{p}} - t)]$ or $(\partial \phi/\partial R)$ to zero, and use this condition to locate that particular surface in the fluid known as "the surface of tension" (Melrose, 1970). When this convention is adopted, Eqn (2.28) simplifies to the Laplace equation and (2.27) takes a similar form. It is convenient to refer collectively to these and similar expressions, relating surface pressure (or tension) across curved interfaces, to the difference in bulk phase pressures, as Laplace equations.

Procedures similar to those used for the planar geometry, may be used to relate these equations to the pressure tensor components. However, a number of difficulties are involved, and although these have been recognized for some time (Defay *et al.*, 1966; Tolman, 1949; Kirkwood and Buff, 1949), they are often overlooked in making practical applications of the Laplace equation. It is particularly worthwhile to discuss these equations in more detail here, since simulation techniques offer the opportunity to resolve these problems at a molecular level.

As a first example, consider Eqn (2.15b) applied to a spherical drop. We introduce the assumption, already discussed, that $p_{\theta\theta} = p_{\varphi\varphi}$ and define $p_{\mathrm{N}} = p_{rr}$; $p_{\mathrm{T}} = p_{\theta\theta} = p_{\varphi\varphi}$. If it is further assumed that the inside of the drop is a uniform bulk phase B up to the surface and that the external phase G is also uniform, integration of (2.15b) from a region in the B phase to a region in the G phase leads to (Goodrich, 1971)

$$p^{\mathrm{B}} - p^{\mathrm{G}} = -2 \int \frac{(p_{\mathrm{N}} - p_{\mathrm{T}})}{r}\, \mathrm{d}r. \tag{2.29}$$

When a liquid drop rather than an adsorbate is considered it is more usual to discuss the surface tension γ, rather than the surface pressure. Comparison of (2.29) and (2.28) ($\mathrm{F} \equiv \mathrm{B}$) at the surface of tension leads to an expression for the surface tension in a form reminiscent of Eqn (2.26),

$$\gamma_{\mathrm{SPH}} = R \int \frac{(p_{\mathrm{N}} - p_{\mathrm{T}})}{r} \,\mathrm{d}r \tag{2.30}$$

where the subscript SPH emphasizes that this is for a spherical interface. If γ_{SPH} is to be equated with γ in Eqn (2.26), it is necessary to assume that r can be taken from under the integral sign and identified with the radius of the drop, and also that the remaining integral, $\int (p_{\mathrm{N}} - p_{\mathrm{T}})\,\mathrm{d}r$, is identical with that in Eqn (2.26). If these approximations are accepted then Eqn (2.29) can be written

$$p^{\mathrm{B}} - p^{\mathrm{G}} = -2\gamma/R \tag{2.31}$$

which now has the form of the Laplace equation. The assumptions used to reach this result are not perhaps as demanding as might appear. Its validity rests on the requirements that the non-uniformities associated with an interface are confined to a region which is small compared to the size of the drop (i.e. a sharp transition from the liquid to the vapour phase), and that the surface tension at the spherical surface γ_{SPH} can be approximated by that at a plane surface. An alternative view might be taken that, since most experimental determinations of γ make use of (2.31), it is this quantity which is actually obtained from experimental measurements. Comparisons between experimental data and calculations carried out for plane interfaces should be valid provided that the above requirements have been met.

As an example of a system with cylindrical interface, we consider an annular adsorbate on the walls of a cylinder. Here we write $p_{\mathrm{N}} \equiv p_{rr}$ and $p_{\mathrm{T}} \equiv p_{\varphi\varphi}$ and integration of (2.14c) over the radius of the cylinder results in

$$p^{\mathrm{F}} - p^{\mathrm{G}} = \int_0^{r_{\mathrm{p}}} \frac{(p_{\mathrm{T}} - p_{\mathrm{N}})}{r} \,\mathrm{d}r + \int_0^{r_{\mathrm{p}}} \rho^{(\mathrm{a})}(r) F_r \,\mathrm{d}r. \tag{2.32}$$

A comparison of (2.27) at the surface of tension with (2.32) gives an expression for the surface pressure, which will be distinguished here by the subscript CYL.

$$\phi_{\mathrm{CYL}} = (r_{\mathrm{p}} - t) \int_0^{r_{\mathrm{p}}} \frac{(p_{\mathrm{T}} - p_{\mathrm{N}})}{r} \,\mathrm{d}r + (r_{\mathrm{p}} - t) \int_0^{r_{\mathrm{p}}} \rho^{(\mathrm{a})} F_r \,\mathrm{d}r \tag{2.33}$$

where r_p is the radius of the cylinder and the thickness t of the adsorbate layer was discussed in §2(a).

The identificiation of ϕ_{CYL} with ϕ given in Eqn (2.26) involves assumptions similar to those just discussed for spherical geometry, but here the question of the external field term must also be considered. If the above arguments relating to spherical drops are followed, then only the first part of Eqn (2.33) would be replaced by ϕ. However, it is often assumed, without justification, that the whole of ϕ_{CYL} can be equated to ϕ. Although both replacements involve approximations, those implied in the latter case would appear to be more serious.

Equations (2.27) or (2.28) with the convention $\partial\phi/\partial(r_p - t)$ or $\partial\phi/\partial R = 0$ locating the surface of tension, may be written

$$p^F - p^G = k\phi_c/(r_p - t) \tag{2.34}$$

with $k = 1, 2$ and $c \equiv$ CYL, SPH for cylindrical or spherical menisci respectively. Equation (2.34) forms the basis for the calculation of the pressure at which condensation occurs in model capillaries and hence the use of adsorption to determine pore size distribution in porous materials. It is clear from the above discussion that the surface pressure or surface tension term appearing in this equation may be affected both by geometry and by the external field due to the adsorbent and cannot therefore be identified with the corresponding quantity for a plane liquid–vapour interface. Geometrical effects become more important for values of r_p of the order of 10 nm, which is often the size range of most interest. Corrections have been discussed by several authors (Defay *et al.*, 1966; Tolman, 1949; Buff, 1955, 1956, 1960; Melrose, 1970). However, a purely phenomenological approach, based on continuum models, is not likely to be wholly satisfactory and a molecular theory is more appropriate. Similarly, the effect of the external field is perhaps best considered from a molecular point of view, although once again an approximate macroscopic theory is feasible (Derjaguin, 1957, Broekhof and de Boer, 1967; Nicholson, 1968). We return to this point in Chapter 7.

The Kelvin or Cohan equations may be derived from (2.34) using the standard thermodynamic relationship

$$\left(\frac{\partial\mu}{\partial p}\right) = \overline{V}. \tag{2.35}$$

Where $\overline{V}$ is a partial molecular volume. If the phase G is identified with the external gas phase (assumed ideal) at pressure p, integration of (2.35) leads to

$$\mu^G(p) - \mu^G(p_0) = RT \ln (p/p_0) \tag{2.36}$$

where p_0 is the pressure of a saturated vapour at temperature T.

It is unlikely that the identification of p^G with p can be strictly valid, since the state of the fluid phases in a confined capillary space will be influenced by the surrounding dense phases, but this assumption is probably not too serious. Applying (2.35) to the dense phase (F) and assuming that, in this case $\overline{V}^{\mathrm{F}}$ is not a function of p^{F}, leads to

$$\mu^{\mathrm{F}}(p^{\mathrm{F}}) - \mu^{\mathrm{F}}(p_0) = \overline{V}^{\mathrm{F}}(p^{\mathrm{F}} - p_0). \tag{2.37}$$

The condition for equilibrium between the G and F phases, with both at the same temperature T, is

$$\mu^{\mathrm{G}}(p) - \mu^{\mathrm{G}}(p_0) = \mu^{\mathrm{F}}(p^{\mathrm{F}}) - \mu^{\mathrm{F}}(p_0) \tag{2.38}$$

which, with the right-hand sides of (2.36) and (2.37), and after introducing (2.34), gives the result

$$RT \ln (p/p_0) = \frac{K\overline{V}^{F}\phi_{\mathrm{c}}}{r_{\mathrm{p}} - t} - \overline{V}^{\mathrm{F}}(p^{\mathrm{G}} - p_0) \tag{2.39}$$

If the second term on the right-hand side of this equation is neglected, the remaining expression is the Kelvin ($K = 2$) or Cohan ($K = 1$) equation. The significance of the assumptions leading from (2.34) to (2.39) has been discussed by Melrose (1966, 1968) who concluded that compressibility of the fluid phase can be important in some cases. The replacement of ϕ_{c} in Eqn (2.39) by ϕ (or $-\gamma$) for an isolated planar liquid–vapour interface has already been discussed. Equations (2.28), (2.30) and (2.33) show the way in which a surface pressure, defined at the Gibbs dividing surface, which is a purely thermodynamic quantity, is related to the pressure tensor obtained solely from mechanical considerations. The final link which needs to be made, between the components of the pressure tensor and molecular quantities, will be discussed in Chapter 3.

4. Generalized thermodynamic equations

The fundamental equation of thermodynamics for a system containing the species a, b, . . . , α, . . . , σ, may be written for the whole of a region which encloses the interface and extends into adjacent uniform domains at its boundaries,

$$\mathrm{d}U = T\,\mathrm{d}S + \mathrm{d}W + \sum_{\alpha} \mu_\alpha \mathrm{d}N_\alpha \tag{2.40}$$

where the summation extends over the species present, and $\mathrm{d}W$ includes the mechanical work done on the system from Eqn (2.20) as well as contributions to electrical work which may arise from the presence of ions or polar species, and contributions to magnetic or other work if these are relevant to the system under consideration.

Several thermodynamic potentials may be defined by appropriate Legendre transformations from Eqn (2.40) when different specific expressions have been substituted for dW. A particularly useful one in dealing with interfacial systems is defined by the equation

$$\Omega = U - TS - \sum_{\alpha} \mu_\alpha N_\alpha. \tag{2.41}$$

As will be seen later, this is directly related to the grand ensemble and we shall therefore refer to it as the grand thermodynamic potential.

In a system of neutral, non-polar molecules, mechanical work alone needs to be considered and Ω is a function of T, l_z, $\mathscr{A}$ and μ_α only. Thus when the interface is planar, the expression for dW is given by Eqn (2.19) and integration can be carried out by observing that the number of molecules in the system is proportional to the area of the interface $\mathscr{A}$, and that the system must be extended by increasing $\mathscr{A}$ keeping l_z constant. Since $\int(l_y \, dx + l_x \, dy) = \mathscr{A}$, this procedure leads to

$$\Omega = -\mathscr{A} \int \tfrac{1}{2}(p_{xx} + p_{yy}) \, dz. \tag{2.42}$$

Under rotational invariance, $p_T = p_{xx} = p_{yy}$, and Eqn (2.42) may be combined with Eqn (2.26) to give

$$\Omega = -\phi\mathscr{A} - p^F V^F - p^G V^G. \tag{2.43}$$

This process of integration can also be applied to curved interfaces; here dW contains the additional terms involving principal radii given in Eqn (2.20a) which account for the work done in changing the curvature. When the Gibbs dividing surface is chosen to be at the surface of tension, these terms vanish and ϕ has the special value associated with that surface. If the system is now permitted to extend in such a way that the curvature of the interface remains constant, then Eqn (2.43) once again results.

If electrical or other properties need to be considered, additional terms also appear in the expression for Ω, which can be written in an extended form as

$$\Omega = -p^A V^A - p^B V^B - \phi\mathscr{A} - \sum_j P_j X_j \tag{2.44}$$

When ionic or polar species for example are present in the system, P_j would be the electrical potential $\mathscr{E}$ or the dielectric polarization P, and the conjugate capacity factors X_j would be the quantity of charge Q or the polarizing field E (Böttcher, 1973) respectively. The interaction between surface tension and electrical charge affecting the shape of a meniscus is one example of a situation which can be analysed from this starting point. It is to be noted, however, that a full analysis for non-uniform systems would also have to take into account the anisotropy and non-uniformity of these quantities, and can lead to rather complicated thermodynamic equations.

It is often convenient to treat the properties of the interfacial region in isolation from those of the adjacent uniform phases. We choose examples where only mechanical work can occur although obvious extensions to other cases can be made. As a first example consider the situation where the interface is between two fluid phases, usually a liquid–gas interface; the thermodynamic potential for a uniform X-phase in isolation is

$$\tilde{\Omega}^{\mathrm{X}} = -p^{\mathrm{X}}V^{\mathrm{X}} \tag{2.45}$$

where we have introduced the notation of a tilde (~) over a symbol to denote a uniform phase, and V^{X} is the volume it occupies.

If the liquid (F) and gas (G) phases in (2.43) are supposed uniform up to the dividing surface, then Eqn (2.45) can be written for $\tilde{\Omega}^{\mathrm{F}}$ and $\tilde{\Omega}^{\mathrm{G}}$ respectively and after subtracting from (2.43) the remaining expression for the thermodynamic potential of the interface alone is

$$\Omega^{\Sigma} = -\phi\mathscr{A} = \gamma\mathscr{A}. \tag{2.46}$$

A second example of importance occurs when a fluid adsorbate is in contact with a solid phase. Here, as already mentioned in §1, the solid itself may be perturbed due to the presence of the adsorbate. The solid adsorbent phase is specified by a superscript S and, for a free crystal surface (indicated by a zero subscript) we may write

$$\Omega_0^{\mathrm{S}} = -pV_0^{\mathrm{S}} - \phi_0^{\mathrm{S}}\mathscr{A}_0 \tag{2.47}$$

in which the surface pressure term is to be interpreted (cf. Eqn (2.25)) as

$$\phi_0^{\mathrm{S}}\,\mathrm{d}\mathscr{A}_0 = l_y\,\mathrm{d}x\int(p_{xx}^{\mathrm{S}} - p)\,\mathrm{d}z + l_x\,\mathrm{d}y\int(p_{yy}^{\mathrm{S}} - p)\,\mathrm{d}z \tag{2.48}$$

and depends on the interactions amongst the crystal atoms at their rest sites. At the surface of the solid relaxation occurs; that is to say the atoms vibrate about mean positions which are different from those which would be predicted from the structure of the bulk crystal. Lattice dynamic calculations have been extensively applied to this problem (see e.g. Benson and Yun, 1967), but simulation methods offer an alternative approach (Barber *et al.*, 1979). Relaxation of the crystal surface is equivalent to the view, advanced in §1, that $\rho^{(\mathrm{a})}(z)$ for the solid is not a step function but is continuously variable across the boundary.

In the presence of adsorbate, the relaxation of the solid phase may be modified. If we imagine this modified crystal surface to be still in contact with a vacuum, Eqn (2.47) would become

$$\Omega^{\mathrm{S}} = -pV^{\mathrm{S}} - \phi^{\mathrm{S}}\mathscr{A}. \tag{2.49}$$

The boundary defining V^S can be established with the aid of a Gibbs dividing surface criterion as described in §1. Referring to Fig. 2.2, it is seen that V^S would be equal to $\mathscr{A}l_s$, and if this boundary is placed at $z = 0$, the remaining volume of the system, $V^A = l_z\mathscr{A}$, is the volume of the adsorbate. Thus for the adsorbate phase A only we have, in place of Eqn (2.43)

$$\Omega^A = -pV^A - (\phi - \phi^S)\mathscr{A} \tag{2.50}$$

where ϕ is the surface pressure of the whole system, including any contributions from the adsorbent phase. Equation (2.45) applies to the adsorptive gas phase at pressure p in equilibrium with the adsorbate, therefore subtracting $\tilde{\Omega}^G = -pV^A$, from (2.50) for a uniform gas phase occupying the whole of the adsorbate volume once again leads to Eqn (2.46), provided that ϕ in this equation is now reinterpreted as $(\phi - \phi^S)$.

These procedures may be stated in a generalized form. We first introduce the surface excess quantities M^Σ; $(M \equiv U, S, N,$ etc.) which are

$$M^\Sigma = M - \sum_X \tilde{M}^X \tag{2.51}$$

where M without superscript refers to an extensive property of the whole system, $\tilde{M}^X$ refers to the same property for uniform bulk phases, and the number of terms in the sum is one more than the number of interfaces. In the case, $N \equiv M$ the consistency of (2.51) and (2.3) may easily be verified. It is also an easy matter to show from (2.51) that, as derived above, $\Omega^\Sigma = -\phi\mathscr{A}$ for a planar gas–solid system.

Alternatively a system of this type may be considered as a fluid adsorbate film in contact with a solid adsorbent at one interface and gaseous adsorbate at the other. Here the total volume can be divided, as in Fig. 2.2, into the zones $V^S(=\mathscr{A}l_s)$, $V^F(=\mathscr{A}t)$ and $V^G(=V^A - V^F)$, and the surface excess quantities would refer to the dense fluid adsorbate–gas interface. Writing $\tilde{\Omega}^F(= -pV^F)$ for the uniform bulk fluid in the volume V^F and $\tilde{\Omega}^G$ for bulk gas (or vapour) in V^G, we would have

$$\Omega^{F,\Sigma} = \Omega - \tilde{\Omega}^S - \tilde{\Omega}^F - \tilde{\Omega}^G \tag{2.52a}$$

where Ω is now given by

$$\Omega = -\phi\mathscr{A} - pV \tag{2.52b}$$

in which ϕ contains transverse pressure components for the whole system and p is the pressure in the contiguous bulk gas phase.

An equation for Ω^Σ resembling (2.41) can be written for the interfacial phase, in which N_α^Σ is the excess number of molecules of species α,

$$\Omega^\Sigma = U^\Sigma - TS^\Sigma - \sum_\alpha \mu_\alpha n_\alpha^\Sigma \tag{2.53}$$

and an analogous expression could be written for $\Omega^{F,\Sigma}$, bearing in mind, from §1 and from (2.51) that N_α^Σ is the same in both cases. Similarly, the fundamental equation of thermodynamics may be written in a form which applies to the interfacial phase only:

$$dU^\Sigma = T\,dS^\Sigma + dW^\Sigma + \sum_\alpha \mu_\alpha\,dN_\alpha^\Sigma \tag{2.54}$$

in which

$$dW^\Sigma = -\Sigma\,P_j^\Sigma\,dX_j^\Sigma \tag{2.55}$$

where the p_j^Σ are potentials and X_j^Σ the conjugate capacity factors; thus for mechanical work ($j = 1$, say) $P_1^\Sigma = \phi$ and $X_1^\Sigma = \mathscr{A}$.

Differentiation of (2.53) and substitution of (2.54) leads to an expression for ϕ:

$$\left(\frac{\partial\Omega^\Sigma}{\partial\mathscr{A}}\right)_{T,\mu,X_j^\Sigma(j\neq1)} = -\phi. \tag{2.56a}$$

A similar, slightly more general, expression can also be obtained in the same way, by using (2.41) and (2.40):

$$\left(\frac{\partial\Omega}{\partial\mathscr{A}}\right)_{T,\mu,V,\ X_j(j\neq1)} = -\phi. \tag{2.56b}$$

Euler integration carried out on (2.54) with (2.55) gives

$$U^\Sigma = TS^\Sigma - \Sigma P_j^\Sigma X_j^\Sigma + \Sigma\mu_\alpha N_\alpha^\Sigma \tag{2.57}$$

and a Gibbs–Duhem equation may be obtained by differentiation of (2.56) and substitution of (2.54) and (2.55) as

$$S^\Sigma\,dT + \sum_\alpha N_\alpha^\Sigma\,d\mu_\alpha - \sum X_j^\Sigma\,dP_j^\Sigma = 0. \tag{2.58}$$

As one example of the use of (2.58), consider an isothermal system in which only mechanical work can occur; in this case

$$\mathscr{A}\,d\phi = \sum_\alpha N_\alpha^\Sigma\,d\mu_\alpha. \tag{2.59}$$

In a two-component system, with the Gibbs dividing surface placed so that the surface excess of one component is zero, (e.g. an adsorbate with the Gibbs dividing surface at the surface of the solid), we have, for the adsorbate phase (a),

$$d\phi = (N_a^\Sigma/\mathscr{A})\,d\mu_a. \tag{2.60}$$

This is one form of the Gibbs adsorption isotherm. For adsorption from a gas phase at activity λ, Eqn (2.60) can be integrated to give

$$\phi - \phi_0^{\mathrm{S}} = \int_0^{\lambda} \frac{N_{\mathrm{a}}^{\Sigma}}{\mathscr{A}} kT\, \mathrm{d}(\ln \lambda). \tag{2.61a}$$

Equations (2.60) and its integrated form (2.61a) provide alternative expressions for ϕ, the first being useful as a link to theoretical and numerical results, while the second is more important as a link to experimental data. If the gas phase is ideal or nearly ideal, Eqn (2.61a) can be written

$$\phi - \phi_0^{\mathrm{S}} = \int_0^{p} \frac{N_{\mathrm{a}}^{\Sigma} V_{\mathrm{m}}^{\mathrm{G}}}{\mathscr{A}}\, \mathrm{d}p \tag{2.61b}$$

where $V_{\mathrm{m}}^{\mathrm{G}}$ is the molar volume of the gas phase.

5. Classification of films

The picture of an adsorbate as a film of uniform liquid on a solid surface has been discussed at several points in this chapter. The thermodynamic equations subsequently developed can be used to analyse the way in which such films might grow as the pressure in the contiguous gas phase is increased (Dash, 1977a, b; Peierls, 1978).

The grand free energy of the film is given in Eqn (2.52), where the terms for the uniform phases are given by $\tilde{\Omega}^{\mathrm{S}} = -pV^{\mathrm{S}}$, $\tilde{\Omega}^{\mathrm{G}} = -pV^{\mathrm{G}}$ and $\tilde{\Omega}^{\mathrm{F}} = -pV^{\mathrm{F}}$, the pressure being equal in all phases since this is a planar system. When these are substituted into (2.52), this becomes

$$\Omega^{\mathrm{F},\Sigma} = -\phi\mathscr{A}. \tag{2.62}$$

The criterion for wetting of a surface by a liquid is derived from the classical Young's equation which expresses the free force balance at a three-phase contact between a liquid drop (F), a solid adsorbent surface (S) and a gas phase (G), in terms of the contact angle θ (which is measured through the liquid). Young's equation is expressed as

$$\cos\theta = (\gamma^{\mathrm{SG}} - \gamma^{\mathrm{SF}})/\gamma^{\mathrm{FG}}. \tag{2.63}$$

It should be mentioned that this is certainly an over simplification of the real situation existing at a three-phase contact, but an experimental demonstration of this assertion is very difficult (Fox and Zisman, 1950). This problem is therefore an example of a case where computer simulation can be helpful in analysing the limitations of an established theory and has demonstrated (Saville, 1976) that Eqn (2.63) is indeed inaccurate at the microscopic level (White, 1976; Berry, 1974). However, the level of approximation implied in (2.63) is probably quite in keeping with the uniform density film model under discussion. It follows from Eqn (2.63) that when the contact angle becomes zero, wetting of the surface by an evenly spread film of liquid occurs. The condition for complete wetting is therefore

$$\gamma^{SG} > \gamma^{FG} + \gamma^{SF}. \tag{2.64}$$

In Eqn (2.62), ϕ is the surface pressure for the whole system, including the solid–film interface. It was shown in (2.50) that under certain assumptions, this can be separated into a part ϕ^S due to the solid vacuum interface, and a part, which we designate ϕ^F, due to the adsorbate film, thus

$$\phi^F = \phi - \phi^S \tag{2.65}$$

and since $\phi = -\gamma^{SG}$ and $\phi^S = -\gamma^S$, this equation can be rewritten

$$\gamma^{SG} = \gamma^S - \phi^F. \tag{2.66}$$

The condition for complete wetting of the adsorbent by the adsorbate film is obtained by replacing γ^{SG} in (2.64) by this expression, giving

$$\phi^F(p) < \gamma^S - \gamma^{FG} - \gamma^{SF} \tag{2.67}$$

where it has been emphasized the ϕ^F is a function of the gas phase pressure and can be related to it through the Gibbs isotherm, Eqn (2.61b). $\phi^F(p)$ will increase towards a maximum value at the saturated vapour pressure p_0. Three classes of film can be distinguished, depending on how this limit is approached.

Class I. If the inequality (2.67) holds at $p = p_0$, then the film wets the surface completely and it grows to infinite thickness.

Class II. The inequality holds up to a certain film thickness but is violated above this coverage. Here the contact angle becomes positive and drops may form on top of the adsorbed film. If γ^{SG} is large enough for Eqn (2.63) to have a real solution for θ, no further adsorption onto the surface can occur (Peierls, 1978). Class II isotherms, which approach $p/p_0 = 1$ at a finite angle, instead of asymptotically have been observed experimentally in several systems. Some examples, for NO adsorbed on lamellar solids, are illustrated in Fig. 2.4 (Matecki *et al.*, 1974).

Class III. Here the inequality of Eqn (2.67) is violated for all values of p and no adsorption can occur until p_0 is reached. If γ^{SG} is sufficiently large, drops may form on the substrate, otherwise there will be no adsorption at all; Kr adsorbed on Na metal at 75 K (Pierotti and Halsey, 1959) is an example of this type of behaviour.

Dash (1977a) has considered the growth of droplets in class II films in some detail and has shown, from thermodynamic considerations, that they increase in size mainly by lateral growth, eventually coalescing to form larger irregular patches.

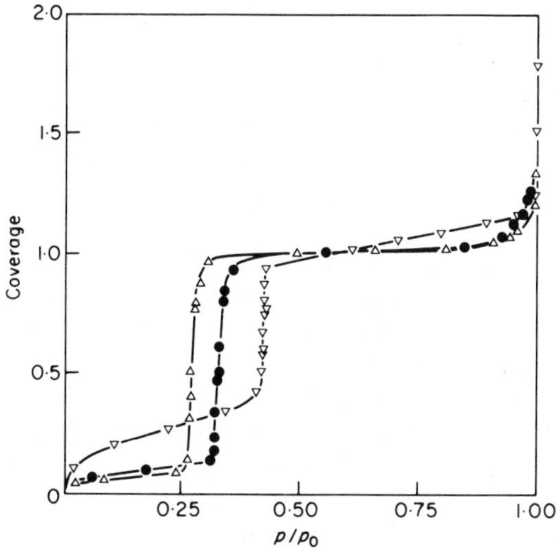

Fig. 2.4. Adsorption of NO onto $CdBr_2$ at 83 K (△) $CdCl_2$ at 77 K (●) and CdI_2 at 79 K (▽). The coverage is expressed in fractional monolayers; the inequality of Eqn (2.67) holds up to approximately monolayer coverage for all three substrates, but is then violated; little or no adsorption is then observed until saturation is reached. (From Matecki *et al.*, 1974).

The isotherm class to which a particular system belongs will depend on the relative strength of adsorbate–adsorbate and adsorbate–absorbent interactions and on temperature, and a cross over from one class to another is possible as the temperature changes.

The relationship with the older BET classification of isotherms is also apparent from this consideration. Thus when the adsorbate–adsorbent interaction is relatively large and the temperature low, a type II isotherm is likely to be observed. Under these conditions class II behaviour would also be anticipated. When the converse conditions prevail, type III isotherms would be expected (i.e. isotherms initially convex to the pressure axis). However it would appear that the BET type is not necessarily always associated with a particular class of film.

6. Heats of adsorption

A change in the integral heat of adsorption $-\mathrm{d}Q$ associated with an interfacial system may be defined as

$$-\mathrm{d}Q = T\,\mathrm{d}S \tag{2.68}$$

where S is the total entropy for a closed system comprising bulk and interfacial phases in equilibrium.

A differential heat of adsorption is defined by

$$q = \left(\frac{\partial Q}{\partial N^{\Sigma}}\right) \tag{2.69}$$

where Eqn (2.51) and the condition $\mathrm{d}N = 0$ can be used to express $\mathrm{d}N^{\Sigma}$ in terms of the virtual changes in numbers of molecules present in the different phases of the system. Various heats can be defined according to the specific constraints on the system.

Taking physical adsorption at a gas–solid interface as an example, the most important quantity from the experimental point of view is the isosteric heat q_{st} which corresponds to the differential heat change when an infinitesimal number of molecules $\mathrm{d}N$ are transferred at constant pressure from the bulk gas phase to the adsorbed phase. Assuming that the adsorbent remains unperturbed, we have $\mathrm{d}N^{\Sigma} = -\mathrm{d}N^{\mathrm{G}}$ and, since $S = S^{\mathrm{G}} + S^{\Sigma} + S^{\mathrm{S}}$,

$$q_{\mathrm{st}} = \left(\frac{\partial Q}{\partial N^{\Sigma}}\right)_{p,T,\mathcal{A}} = T\left(\frac{\partial S}{\partial N^{\Sigma}}\right) = T\left[\left(\frac{\partial S^{\mathrm{G}}}{\partial N^{\mathrm{G}}}\right)_{p,T} - \left(\frac{\partial S^{\Sigma}}{\partial N^{\Sigma}}\right)_{p,T,\mathcal{A}}\right] \tag{2.70}$$

which may also be put in the form

$$q_{\mathrm{st}} = \left(\frac{\partial U^{\mathrm{G}}}{\partial N^{\mathrm{G}}}\right)_{p,T} - \left(\frac{\partial U^{\Sigma}}{\partial N^{\Sigma}}\right)_{p,T,\mathcal{A}} + p\overline{V}^{\mathrm{G}} - p\overline{V}^{\mathrm{A}} \tag{2.71}$$

in which the partial molar volume of the adsorbate phase $\overline{V}^{\mathrm{A}}$ is usually considered to be negligible compared to that of the adsorptive phase $\overline{V}^{\mathrm{G}}$.

A link between q_{st} and experimental properties is made through the Clapeyron equation which enables q_{st} to be found from adsorption isotherm data. To establish the Clapeyron equation we write the total derivatives of the chemical potential as a function of p, T, $\mathcal{A}$ for the adsorbed phase

$$\mathrm{d}\mu = \left(\frac{\partial \mu}{\partial T}\right)_{p,\mathcal{A}} \mathrm{d}T + \left(\frac{\partial \mu}{\partial p}\right)_{T,\mathcal{A}} \mathrm{d}p + \left(\frac{\partial \mu}{\partial \mathcal{A}}\right)_{p,T} \mathrm{d}\mathcal{A} \tag{2.72a}$$

$$= \left(\frac{\partial S^{\Sigma}}{\partial N^{\Sigma}}\right)_{p,T,\mathcal{A}} \mathrm{d}T + \overline{V}^{\mathrm{A}}\,\mathrm{d}p + \left(\frac{\partial \phi}{\partial N^{\Sigma}}\right)_{p,T,\mathcal{A}} \mathrm{d}\mathcal{A} \tag{2.72b}$$

where the Maxwell relations have been introduced through the thermodynamic potential $G^\Sigma(=U^\Sigma - TS^\Sigma + pV^A)$ whose total derivative is

$$\begin{aligned} dG^\Sigma &= dU^\Sigma - T\,dS^\Sigma - S^\Sigma\,dT + p\,dV^A + V^A\,dp \\ &= -S^\Sigma\,dT + V^A\,dp - \phi\,d\mathscr{A} + \mu\,dN^\Sigma \end{aligned} \tag{2.73}$$

For a bulk gas phase in equilibrium with the adsorbate

$$d\mu = -\left(\frac{\partial S^G}{\partial N^G}\right)_{p,T} dT + \overline{V}^G\,dp. \tag{2.74}$$

At a fixed amount adsorbed, and for a constant area of the adsorbent surface, Eqns (2.72) and (2.74) give

$$\left(\frac{dp}{dT}\right)_{N,\mathscr{A}} = \frac{(\partial S^G/\partial N^G) - (\partial S^\Sigma/\partial N^\Sigma)}{\overline{V}^G - \overline{V}^A}. \tag{2.75}$$

Inserting Eqn (2.70) we have

$$\left(\frac{dp}{dT}\right)_{N,\mathscr{A}} = \frac{q_{st}}{T(\overline{V}^G - \overline{V}^A)}. \tag{2.76}$$

As mentioned above, the adsorbed phase will be very much denser than the gas phase, and so $\overline{V}^A$ can be reasonably neglected in comparison with $\overline{V}^G$ and this term can be evaluated from the equation of state of the gas phase. It is usual to assume a perfect gas, when (2.76) leads to the standard result

$$q_{st} = RT^2\left(\frac{d\ln p}{dT}\right)_{N,\mathscr{A}}. \tag{2.77}$$

A set of isotherms at different temperatures can be transformed into plots of $\ln p$ against temperature at different coverages and q_{st} determined from these using (2.77).

When experimental data are measured directly by adiabatic calorimetry, for example, S in Eqn (2.68) must include the heat change in the calorimeter.

7. Thermodynamics of phase transitions

(a) The Ehrenfest approach

The classification of transitions proposed by Ehrenfest (1933) has left its mark on this topic both by virtue of its position as the first of a series of such attempts at systematization and also by the terminology which was introduced with it. Ehrenfest suggested that a transition should be classified as being nth order if there was a discontinuity in the nth derivatives of the free energy. First-order and second-order transitions would, therefore, have discontinuities in $(\partial G/\partial T)_p = -S$ and $(\partial^2 G/\partial T^2)_p = (-C_p/T)$, respectively. Corresponding discontinuities in the pressure derivatives could also occur. This treatment of first-order transitions is satisfactory, but that for the second-order transitions has serious flaws. Apart from the fact that the theory seemed to predict that the same phase (of the two involved) would be the more stable both above and below the transition (Justi and von Laue, 1934), it also gave C_p–T graphs in the neighbourhood of the transition which corresponded to only a very small number of real systems. The prediction of a finite drop in the heat capacity at the transition is a result which is found for $TmVO_4$, $TbVO_4$, and the superconducting–normal transition in, for example, tin, but for almost no other systems (Parsonage and Staveley, 1979). For this reason the term second-order is usually used sparingly, the more vague "higher-order", "gradual" or "continuous" being more common. Related, but more elaborate, treatments have since been put forward by Pippard (1957) and Tisza (1961) which overcome most of these problems. But of more importance still has been the work of Griffiths and others on higher critical points, which will be discussed later.

(b) Critical indices

Many of the comparisons between theoretical predictions and experimental results have been concerned with ways in which various quantities go to zero or diverge to infinity as a critical point is approached. For an order parameter (η) which goes to zero as the critical temperature is approached from below it is convenient to represent this behaviour as

$$\eta \sim (T_c - T)^{\beta} \tag{2.78}$$

from which it is clear that the critical index β can be found from:

$$\beta = \lim_{T \to T_c} \frac{\ln \eta}{\ln(T_c - T)}. \tag{2.79}$$

Other commonly used indices are defined by the following asymptotic equations:

$$C_\xi \sim (T - T_c)^{-\alpha} \qquad \text{for } T > T_c \tag{2.80}$$

$$C_\xi \sim (T_c - T)^{-\alpha'} \qquad \text{for } T < T_c \tag{2.81}$$

$$\chi_T \sim (T - T_c)^{-\gamma} \qquad \text{for } T > T_c \tag{2.82}$$

$$\chi_T \sim (T_c - T)^{-\gamma'} \qquad \text{for } T < T_c \tag{2.83}$$

$$|\xi| \sim |\eta|^{\delta} \qquad \text{for } T = T_c,\ \xi \to 0 \tag{2.84}$$

$$\kappa^{-1} \sim (T - T_c)^{-\nu} \qquad \text{for } T > T_c \tag{2.85}$$

where C_ξ, χ_T and κ^{-1} are the heat capacity at constant, zero value of the field (ξ) conjugate to η, the isothermal "susceptibility" $(\partial\eta/\partial\xi)_T$, and the correlation length, respectively. In order to obtain the true limiting values of the critical indices it is usually necessary to make measurements very close to the critical point indeed, e.g. at $|(T_c - T)/T_c| < 10^{-3}$. Even for idealized theoretical models the wrong values will be obtained if the fitting is made to results for states which are more remote from the critical point than this.

It is found that the values of the critical indices are dependent on the dimensionality (d) of the system and the number of components (or degrees of freedom) (n) of the ordering parameter, but often on nothing else. Thus

Table 2.1. Values of the critical indices for various common models

	α	α'	β	γ	γ'	δ	ν
Ising ($d = 2$)	ln	ln	1/8	1·75	1·75	15	1
Ising ($d = 3$)	~0·125	0·06–0·13	0·31	1·25	1·28	5	0·63
Heisenberg ($d = 3$)	0·1	—	0·36	1·40	—	5	0·71
3-state Potts ($d = 2$)	0·33	0·33	1·11	1·45	1·45	11·2	0·83[a]
4-state Potts ($d = 2$)	0·67	0·67	0·083	1·17	1·17	15	0·67[a]
Landau (mean-field)	discontinuous		1/2	1	1	3	$\frac{1}{2}$

[a] Pearson, R. B. (1980) *Phys. Rev. B* **22**, 2579.

all systems which can be well represented by the two-dimensional Ising model would have $d = 2$, $n = 1$, since only the "spin" component in a single direction is involved; a one-dimensional Heisenberg model would have $d = 1$, $n = 3$, since three "spin" components are relevant. Some of the accepted values of the critical constants for some simple models are shown in Table 2.1. Systems which have critical constants which fit in with this kind of simple classification, the critical constants being independent of the values of the interaction coefficients, are said to exhibit universality. There are exceptions to the general (universality) rule described above. Such systems are considered to be in a state of "cross-over" between one set of critical index values and another, the values of the interaction constants determining the extent to which the "cross-over" has proceeded.

The great majority of the experimental work on critical indices has concerned either magnetic or gas–liquid critical phenomena, for both of which it is possible, if sufficient care is taken, to obtain meaningful results. In adsorbed phases the problems are much more severe, since the transition will be rounded by surface heterogeneity, which affects especially the parameter region within which the indices would normally be studied. Computer simulations are also unreliable in the critical region and so critical indices derived from such studies should be treated with scepticism. Nevertheless, an increasing number of values of critical indices for adsorption are becoming available and some of these will be mentioned in their particular context.

(c) The Landau theory of transitions

Landau (1937), like Ehrenfest, based his treatment on a thermodynamic, rather than statistical mechanical, view of transitions. In particular, he assumed that the free energy remained analytic at the critical point and in its neighbourhood, and that it was therefore possible to make a Taylor expansion for the free energy about the critical point:

$$A(T, \eta) = a_0(T) + a_2(T)\eta^2 + a_4(T)\eta^4 + a_6(T)\eta^6 + \dots \qquad (2.86)$$

where η is the order parameter which becomes zero at the critical point, the term in η^1 must be zero since $\eta = 0$ minimizes A for all $T > T_c$, and the other terms in odd powers of η may be omitted if A is symmetric about $\eta = 0$. At the critical point $(\partial^2 A/\partial\eta^2)_{T=T_c} = 0$, so $a_2(T_c) = 0$ and an expression for the field (ξ) conjugate to η may be obtained:

$$\xi = (\partial A/\partial\eta)_T = 4a_4(T_c)\eta^3 + 6a_6(T_c)\eta^5 + . \qquad (2.87)$$

Thus, the critical index δ is predicted to be 3.

A more important situation is that with $\xi = 0$ at T slightly below T_c. Expanding $a_2(T)$ about T_c, we then have from the equation for ξ obtained by differentiating (2.86):

$$0 = 2\{a_{21}(T_c - T) + a_{22}(T_c - T)^2 + \dots\}\eta + 4a_4(T)\eta^3 + \dots \qquad (2.88)$$

so that as $T \to T_c$

$$4a_4(T)\eta^2 \sim 2a_{21}(T_c - T) \qquad (2.89)$$

which corresponds to a value of the critical index $\beta = \frac{1}{2}$.

In a similar way, it is easy to show that the heat capacity at constant $\eta(C_\eta)$ should have a finite discontinuity of the kind suggested by Ehrenfest for a second-order transition. Again, the analogue of the susceptibility $(\partial\eta/\partial\xi)_T$ should diverge at $T = T_c$ with the index $\gamma = 1$.

Most of the attention which has been given to Landau's theory has been devoted to the results which have been discussed above. However, there is another prediction of the theory which will be of importance in this book: the existence of another type of situation, a tricritical point. This arises if both a_2 and a_4 become zero at the same temperature, which can, in general, occur if the coefficients a_i are functions of a second field (Δ) (not ξ) as well of T. Suitable choice of Δ and T can then lead to a tricritical point. In adsorption systems there are usually several different parameters describing the interaction of the adsorbate with the adsorbent and each of these may be considered as a field. There are, therefore, many opportunities for tricritical points to occur in adsorbed phases.

If the symmetry of the system allows the presence of an η^3 term in (2.86) a first-order transition will be predicted. Another situation which leads to a prediction of a first-order transition is that in which the formula for the Helmholtz free energy is correctly given by (2.86), but a_4 is negative.

It is now known that the fundamental assumption of Landau theory, that the free energy is analytic at the critical point, is incorrect. For this reason the predictions of the theory for the critical point and, in particular, the critical indices, are likely to be wrong also. Nevertheless, this type of approach is valuable for regions of the phase diagram not too close to the critical point. It also appears that the theory is reasonably good at the tricritical point.

Corrections for fluctuations, neglected in Landau theory, can be applied to the predictions, and these act in such a way as to tend to drive the transition more first-order. Transitions predicted by Landau theory to be first-order, remain so: some transitions predicted to be second-order become first-order.

(d) Multicritical points

Griffiths (1970) showed that tricritical behaviour could be well represented by diagrams of the kind shown in Fig. 2.5. For $\Delta = 0$, $\xi = 0$ the two-phase region gives way to a single phase at an ordinary critical point. For $\xi = 0$, T_c decreases as Δ increases until at Δ_t, T_t the transition changes its nature, becoming first-order. This change-over is the way in which tricritical points are most frequently detected.

For the remainder of this section we shall discuss the more complicated situations which may occur mainly in terms of the magnetic analogue, as this is invariably the simplest way in which they can be envisaged. Real adsorption systems are often more complicated still, with many contributing interactions and possible ways of ordering. In what follows it may be helpful to think of antiferromagnetic and paramagnetic phases as corresponding respectively to adsorbed phases in which the admolecules avoid close contact with each other or are randomly distributed.

The best known example of a tricritical point is that which occurs in

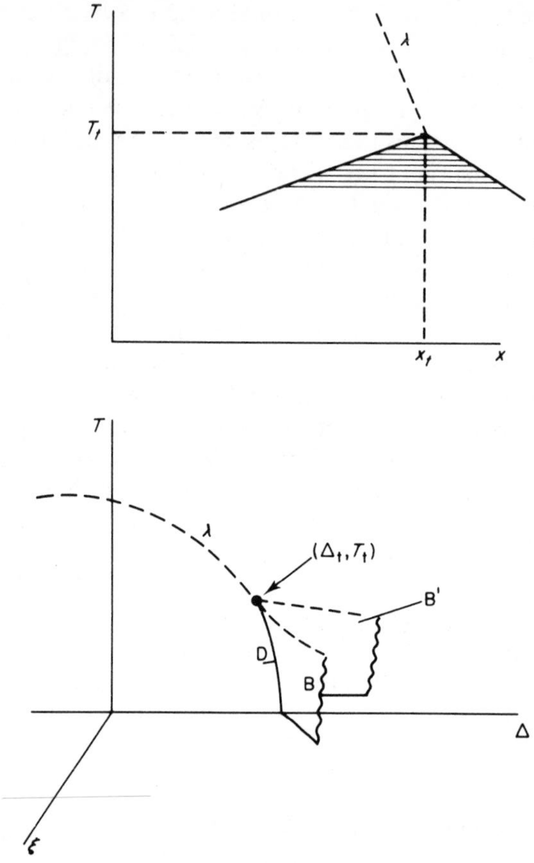

Fig. 2.5. Phase diagrams showing a tricritical point ($\xi_t = 0$, T_t, x_t, Δ_t). ξ and Δ are the ordering and non-ordering fields, respectively. _ _λ_ _, continuous transition; full line, first-order transition. Shaded area; tie-lines join co-existent phases. (After Griffiths, 1970).

metamagnets, a type of layer antiferromagnet in which there is strong ferromagnetic coupling within layers but anti-parallel alignment between one layer and the next, in the absence of an applied magnetic field. The ordering field (ξ) in this example would be the so-called staggered magnetic field, which acts so as to orient spins on alternate sub-lattices in opposite directions; the actual applied uniform magnetic field plays the role of the non-ordering field (Δ). The adsorption analogue of the applied magnetic field is the chemical potential of the adsorbate; the analogue of the staggered magnetic field is not any simply adjustable parameter, but rather the difference in site adsorption energies of the sub-lattices. A high value of

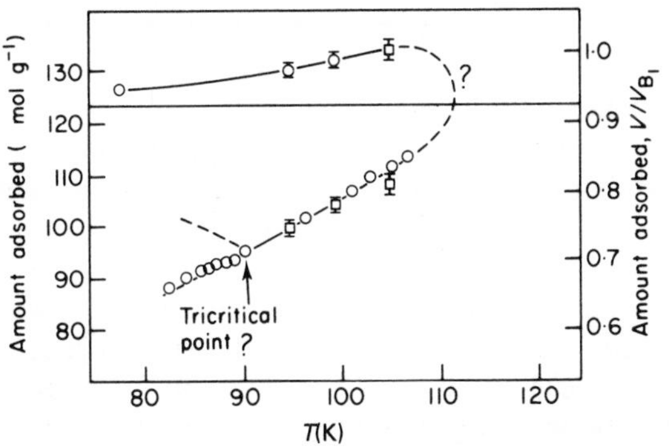

Fig. 2.6. Adsorption of krypton on graphite, showing a tricritical point. (After Putnam *et al.*, 1977).

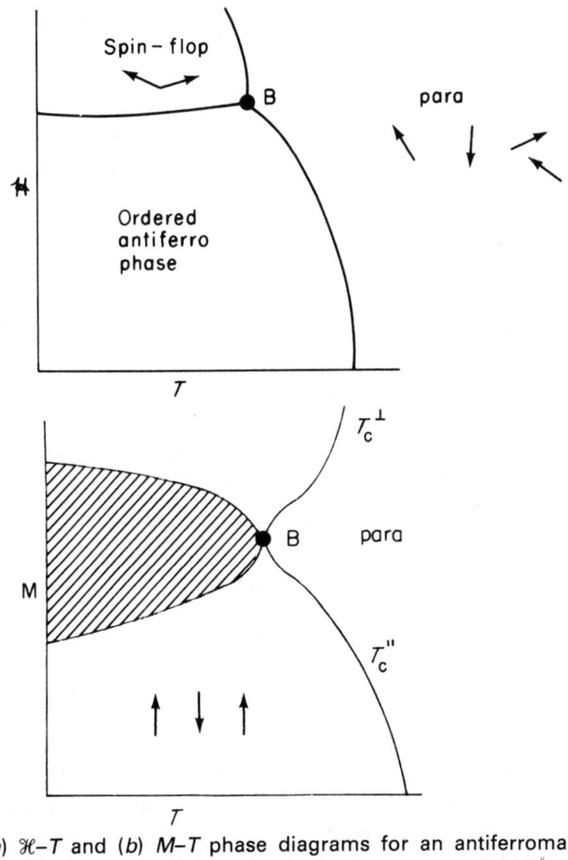

Fig. 2.7. (*a*) $\mathcal{H}$–T and (*b*) M–T phase diagrams for an antiferromagnet in a magnetic field, showing a bicritical point (B). The lines marked $T_c^{\parallel}$ and $T_c^{\perp}$ correspond to continuous transitions (schematic). (After Fisher and Nelson, 1974.)

this latter difference will favour occupation of one sub-lattice at the expense of the other, and so lead to ordering. A high value of the chemical potential will tend to overcome this selective occupation of one sub-lattice by encouraging the occupation of all sites. An important example of a tricritical point in surface chemistry is that which occurs for the adsorption of krypton on graphite (Fig. 2.6, see also Fig. 5.32) (Putnam and Fort, 1975; Putnam *et al.*, 1977).

Other types of multicritical point can also occur. More weakly anisotropic antiferromagnets when subjected to a uniform magnetic field along the uniaxial direction can show "spin-flopping", in which the magnetic spins take up orientations approximately perpendicular to the applied field. If the anisotropy were too strong, then the field would be insufficient to cause the spins to take up the unfavoured directions. The transition between the antiferromagnetic and the spin-flop phases, which is first-order, meets two other critical lines at a so-called bicritical point (B in Fig. 2.7). The critical line for disordering of the antiferromagnetic phase is of the usual Heisenberg type. By contrast, the disordering of the spin-flop phase concerns the x- and y-components of the spins and so is similar to the transition in an XY-model. At the bicritical point, the two ordered phases (antiferromagnetic and spin flop) become identical with the disordered (paramagnetic) phase. Fisher and Nelson (1974) have studied this kind of behaviour by the renormalization group method (§ 7(g)) and have concluded that a somewhat more complicated type of diagram may even be possible. They suggest a so far unobserved possibility, that the transition between the antiferromagnetic and spin-flop phases may proceed via an intermediate phase, the bicritical point being then replaced by a tetracritical point.

A further type of transition, which may have a greater chance of occurrence in real systems, is the Lifshitz point (Hornreich *et al.*, 1975). Just as the existence of tricritical points can be deduced from the Landau equation (2.86) so the Lifshitz point is a consequence of the Landau–Ginzburg equation, which is a development of Eqn (2.86) in which extra terms have been included arising from the spatial derivatives of the order parameter:

$$A(\eta) = a_0 + a_2(T)\eta^2 + a_4(T)\eta^4 + a_6(T)\eta^6 + \ldots + c_1(\nabla\eta)^2 + c_2(\nabla^2\eta)^2 + \ldots \quad (2.90)$$

where ∇ and ∇^2 are the first and second spatial derivatives, respectively. In ordinary critical phenomena the only terms of importance are those with coefficients a_2, a_4 and c_1, the first being zero at the critical point and the last acting to suppress variations of the order parameter from place to place. A Lifshitz point occurs if c_1 becomes zero, thereby making $c_2(\nabla^2\eta)^2$ the leading term in the derivatives (cf. the deduction of the tricritical point from Eqn. (2.87)). In these circumstances the structures which are favoured are those known as "helicoidal", which includes the helix and the conal spiral, in which $\nabla\eta$ is non-zero but $\nabla^2\eta$ is zero. A hypothetical phase diagram which includes a Lifshitz point for a magnetic system is shown in Fig. 2.8. The wave vector of the helicoidal structure increases from zero

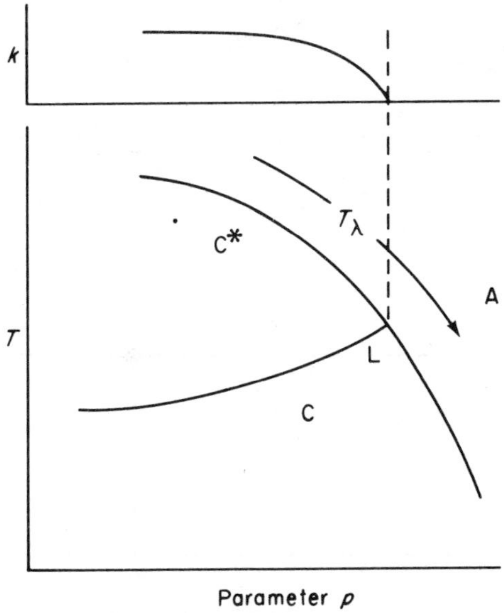

Fig. 2.8. Suggested phase diagram showing a Lifshitz point (L). The parameter *p* could be the pressure or alloy composition, according to the nature of the system; *k* is the wave vector of the helicoidal structure (C*). C and A are, respectively uniformly ordered and disordered phases. (After Hornreich *et al.*, 1975.)

at the Lifshitz point (L). Many such helicoidal structures are known to occur in magnetic materials, although no definite identification of a Lifshitz point has been made. Luban *et al.* (1977) have, however, suggested that such points may be present in the phase diagrams of the Gd–Y and Gd–Sc alloys. Michelson (1977) has suggested that a better chance of finding a Lifshitz point would be in the *T–H* diagram of a chiral liquid crystal. Possible phase diagrams for such a system are shown in Fig. 2.9, these corresponding to different ranges of values for the ratios of the interaction constants. The phases A, C and C* are, respectively, disordered, uniformly ordered (as a ferromagnet) and modulated ordered ("helicoidal").

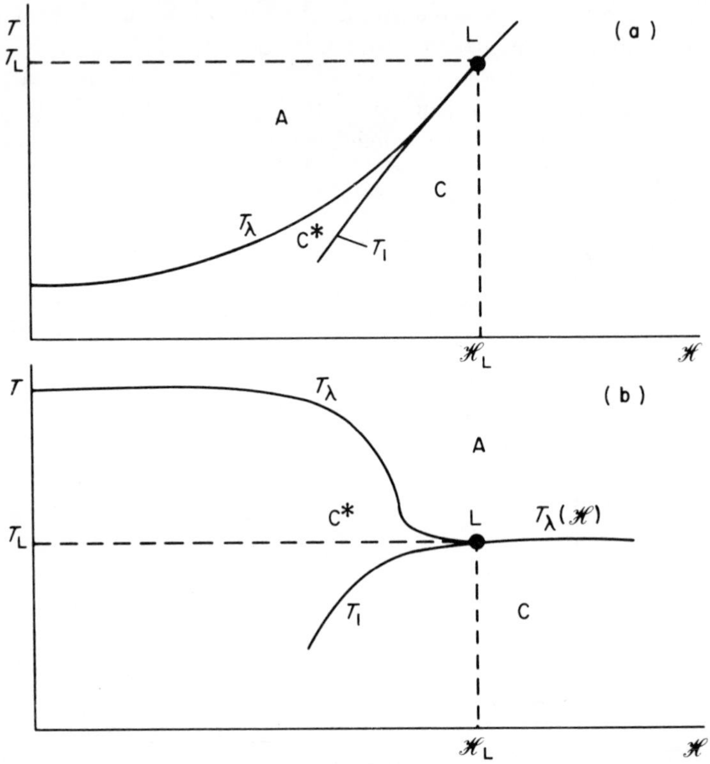

Fig. 2.9. Suggested phase diagram for a liquid crystal, showing a Lifshitz point (L). T_1 and T_λ are lines of first-order and continuous transitions, respectively. A, disordered; C, ferromagnetically ordered; C*, helicoidally ordered. (After Michelson, 1977.)

(c) Landau–Lifshitz symmetry theory (Landau and Lifshitz, 1969)

This concerns itself with the symmetry of the two phases at a transition and the implications this has for the order of the transition, rather than with any quantitative aspects of the transition. One rule requires that for a second-order transition the symmetry group of one phase must be a sub-group of that of the other phase. No such restriction applies to a first-order transition: the symmetries may be completely unrelated or, at the other extreme, they may be the same. It is an interesting example of the application of this rule that if both phases have the same symmetry the transition cannot be second-order. In assessing the symmetry of the phases it is necessary to take account of non-crystallographic symmetry, as for example that of the magnetic spins when dealing with a magnetic transition. It is interesting that at a second-order transition it is possible for the high-temperature form to have the lower symmetry, although this is, of course, very unusual. A well known example of this exceptional behaviour is the lower Curie transition in Rochelle salt, $NaOOC(CHOH)_2COOK.4H_2O$, where the structure passes from orthorhombic to monoclinic on raising the temperature through the transition point. For a gradual transition, the distortion on going from the more to the less symmetric phase must be representable in terms of the members of a single irreducible representation, the coefficients of each member being an order parameter component. Thus it is possible to have a number of components up to the dimension of the irreducible representation to which the parameter corresponds. If the components refer to different irreducible representations the transition would be first-order.

Another rule forbids the existence of any third-order combinations of the order parameters which are invariant under the symmetry group. If such invariants did exist, then they would be expected to form part of a third-power term in the Landau equation (2.86) and hence to cause the transition to be first-order. Domany *et al.* (1978) noted that it does not seem to hold in two-dimensional systems, examples of its failure being for the 3- and 4-state Potts models (§ 7(f)). For both of these cases third-order invariants can be constructed, yet the order–disorder transition is known to be gradual (second-order). For this reason they rejected this rule in their considerations on the implications of symmetry restrictions for transitions on two-dimensional lattices.

A further rule, which Domany *et al.* accepted, was one primarily due to Lifshitz that may be summarized as forbidding the existence of any invariants which contain the first power of spatial derivatives of the order parameter. If such invariants existed and were present in the free energy equation, then the free energy could be reduced by giving the system either a positive or negative gradient of the order parameter at each point. This rule would veto the possibility of the many helicoidal structures which are known: rare earth metals (Ho, Er, Tm), spinels ($MnCo_2O_4$, $CoCr_2O_4$), and many others. Goshen *et al.* (1974) have shown how the treatment of such helicoidal structures by a separate procedure can be justified. They wrote the free energy as

$$G(T,p,C_{li}) = G_0(T,p) + \sum_l A_l(T,p) \sum_i C_{li}^2 \qquad (2.91)$$

where l enumerates the irreducible representations and i the members of a representation, C_{li} being the associated order parameter component. For $T > T_c$ all the factors $A_l(T,p)$ must be positive, and at $T = T_c$ just one of these factors (that with index l_0) must be on the point of changing sign.

Goshen *et al.* distinguished two types of model: those for which l_0 varies as $T_c(p)$ varies, and those for which l_0 does not change, being fixed by symmetry. The first kind includes the helicoidal transitions which have been mentioned above; the second includes phase changes in epitaxially adsorbed layers, with which Domany *et al.* were concerned. For the second type, Goshen *et al.* showed that the free energy contained a term of the form $\sum_{ijl} \gamma_{ijl} a_i b_j k_l$ with $\gamma_{ijl} = -\gamma_{jil}$ or $\sum_{i,l>j} \gamma_{ijl}(a_i b_j - a_j b_i) k_l$ where $\boldsymbol{k}$ is the wave vector describing the spatial variation of the order parameter. If the antisymmetric part of T^2 (where T is the operator describing the transition) has a representation in common with a vector representation then it would be possible to achieve a lower free energy for the system by allowing a spatial distortion corresponding to the particular wave vector. Thus for the second kind of symmetry a contradiction arises which can only be resolved by forbidding the antisymmetric part of T^2 from including the vector representation. For the first kind of model this restriction does not apply.

Having rejected only the second of these symmetry rules, Domany *et al.* determined the effect on the nature of the transitions which such restrictions would have by writing the Landau–Ginzburg Hamiltonian in terms of each order parameter component in turn and comparing it with the known Hamiltonians for various well studied models. If, for example, the Ham-

Table 2.2. Substrates (first column) of various symmetries. Universality classes that can be realized and the expected critical behaviour are listed in the first row. The entries identify the corresponding ordered superlattice structures.

	Ising $\alpha = 0$ (log) $\beta = 0{\cdot}125$	X–Y with cubic anisotropy nonuniversal	Three-state Potts $\alpha \sim 0{\cdot}42 \pm 0{\cdot}05$ $\beta \sim 0{\cdot}1$	Four-state Potts $\alpha = 0{\cdot}6667$ $\beta = 0{\cdot}0833$
P2mm Ex fcc (110)	(2 × 1) (1 × 2) c(2 × 2)			
C2mm Ex bcc(110)	c(2 × 2)	(2 × 1) (2 × 2)		
P4mm Ex fcc (100) bcc(100)	c(2 × 2)	(2 × 1) (2 × 2)		
p6mm Ex bcc(111) fcc(111) Graphite			$(\sqrt{3} \times \sqrt{3})$ *R*30°	(2 × 2)

(After Domany, E., Schick, M., Walker, J. S. and Griffiths, R. B. (1978). *Phys. Rev. B* **18**, 2209).

iltonian is of the same form as that for an Ising model, then the critical properties would be those for that model (of the same universality class) and the transition would be characterized as being Ising-like. Table 2.2 summarizes their results. The second category employed is one which does not display universality at all, i.e. the critical indices vary with the values of the interchain constants. The most famous example of such behaviour is the 8-vertex model which was solved by Baxter (1972).

(f) Potts transitions

With the study of the standard forms of the Ising and Heisenberg models almost complete, some of the effort previously devoted to them has been turned towards a model first studied by Potts in 1952. Each site (spin) may be in any of q states with equal *a priori* probability, and the interaction energy between two nearest neighbour sites takes the value K if the sites are in the same state, whatever that may be, but zero otherwise. In the presence of a field ($\mathcal{H}$) acting to favour state 1, the Hamiltonian for N particles could be written:

$$H_N(s) = \mathcal{H} \sum_i \delta_{s_i 1} + K \sum_{i>j} \delta_{s_i s_j}. \tag{2.92}$$

If $q = 2$ this is equivalent to the Ising lattice-gas model. Both q and the dimensionality (d) of the system are important in deciding the nature of the order–disorder transition. For $d = 2$, Baxter (1973) predicted that the transition would be continuous (second-order) for $q \leqslant q_c$ and first-order for larger values of q, with $q_c = 4$. Nienhuis *et al.* (1979), using a renormalization group technique (§ 7(g)), concluded that $q_c \approx 4{\cdot}7$. The results for $d = 2$ models have been employed in the interpretation of results for adsorbed layers.

Table 2.3 summarizes the prevalent beliefs on the order of the Potts transitions for various values of d and q.

Table 2.3. Transition types for q-state Potts models in d dimensions

d \ q	2	3	4	5
2	cont.[b]	cont.[b]	cont.[b]	first[a,b]
3	cont.[f]	first[c,d,f]	first[e,f]	first[e,f]
4	second (Landau)[e]	first[e]	first[e]	first[e]

[a] Baxter (1973). [b] Nienhuis *et al.* (1979). [c] Golner (1973). [d] Zia and Wallace (1975). [e] Aharony, A. and Pytte, E. (1981). [f] Kogut, J. B. and Sinclair, D. K. (1981). Baxter, R. J. in Zia and Wallace (1975): 1st order for large q for all d.

(g) Renormalization group method

In view of the increasingly complex range of transitions which have been discovered or proposed it is fortunate that a unifying treatment has been developed. This is the renormalization group method, which was earlier used in field theoretical discussions of nuclear problems and has been adapted by Wilson (1971) and Wilson and Fisher (1972) for the study of phase phenomena. Apart from the insight into the problem which it provides, it can also yield some valuable quantitative information. For example, values of the critical indices and complete mappings of the phase diagram have been obtained. It should be stressed that, because of approximations which it is necessary to make at earlier stages of the calculation, these values are approximate.

The starting point for this treatment is the realization that near critical points it is unnecessary to make a detailed study of the correlations at atomic distances, since the phenomena of interest are governed by very long-range correlations. The procedure consists, therefore, of removing the short-range correlations from consideration by repeated application of a so-called renormalization group.

Suppose the Hamiltonian of the spins (in units of $kT = 1/\beta$) is $H_N(s)$, then the partition function is

$$\sum_s \exp(-H_N(s)). \tag{2.93}$$

Now, for a system such as a simple Ising or Heisenberg array, $\mathcal{H}$ will be made up of interactions between nearest-neighbour pairs of spins. To obtain a more coarse-grained view the spins may be formed into blocks and a single spin assigned to each block (Kadanoff, 1966). The decision as to the block spin may be made on the basis of a simple majority of the spins in the block, or a more complicated method may be used. In principle, it is necessary to retain the correct value of the partition function:

$$\sum_{\boldsymbol{\sigma}} \exp(-H_{N'}(\boldsymbol{\sigma})) = \sum_s \exp(-H_N(s)) \tag{2.94}$$

where $\boldsymbol{\sigma}$ is the set of block spins and $H_{N'}$ the corresponding Hamiltonian which involves only interactions between the block spins. The interaction constants between the block spins will not, of course, have the same value as those for the original spins. In other words, the coarse-graining involves changes of the type

$$J \Rightarrow J' \tag{2.95}$$

for the interaction constants. The elucidation of the correct form of Eqn. (2.95) involves both mathematical and physical approximations which call for a great deal of insight into the problem. Often, instead of Eqn (2.94), the condition set is that the resultant partition function should differ from the initial one by a factor $\exp(G)$, which does not change as the subsequent iteration proceeds. This extra flexibility is used to make $H_{N'}$ traceless. We then have

$$A' = A - G \tag{2.96}$$

where A and A' are the initial and final free energies (in units of kT), respectively. Even when the original Hamiltonian included only nearest-neighbour interactions the renormalized Hamiltonian ($H_{N'}$) would in general involve non-nearest-neighbour interactions also. It is necessary, therefore, to truncate the Hamiltonian in order to prevent it from becoming more and more complicated during the iterative application of the transformation, and this introduces approximations. More will be said about this problem below, but, for the moment, it will be assumed that the relationship for the transformation (2.95) and correspondingly

$$H_N \Rightarrow H_{N'} \tag{2.97}$$

has been found. It may appear that nothing has been achieved, since the original problem has been replaced by another for which the solution is of similar difficulty. However, this is not so. The correlations in the new problem are of the same range when expressed in absolute terms, but they are of shorter range when expressed in terms of the current lattice parameter (the new lattice parameter is determined by the linear dimensions of the block). Thus, a system which is nearly critical originally becomes less critical on renormalization. Repeated application of the renormalization procedure will cause the system to drift ("flow") further and further from criticality. But this would not be the case if the original system was actually critical. For then the original correlation length would be infinite and would remain so after renormalization. This is then spoken of as a critical fixed point:

$$H_{N',\mathrm{c}} = H_{N,\mathrm{c}} = H_{N^*}. \tag{2.98}$$

The course of these transformations can be represented on a diagram in which the coordinates are the interaction parameters (in units of kT: $K_i = J_i/(kT) = \beta J_i$). For the three-parameter system represented in Fig. 2.10 there is a critical surface ($H_N = H_{N^*}$) with two degrees of freedom separating the two phases, and this boundary surface may be located by finding the fixed points, which are unchanged by repeated application of the renormalization procedure. If the starting point is nearly but not exactly critical, the renormalization trajectory will approach the critical fixed point but then veer away towards another fixed point not corresponding to critical

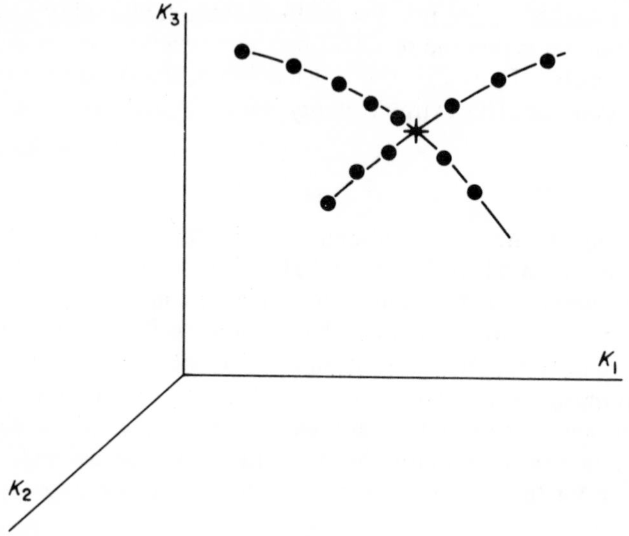

Fig. 2.10. Renormalization flow diagram.

behaviour. In practice, when studies of this kind are made by numerical means, the initial state will never be exactly critical and so the position of the critical fixed point is obtained from a judgment as to the point to which the trajectory was initially attracted.

Information on the critical indices is obtained by a consideration of the effect of the change of scale on the correlation length. This method may be illustrated by its application to the case of a simple magnetic system. If the ratio of the lattice parameters after and before renormalization is b, then

$$\xi' = \xi/b \tag{2.99}$$

and if the reduction in the "distance" of T from T_c is written as

$$(T - T_c)' = \lambda_t(T - T_c) \tag{2.100}$$

then

$$\xi \sim (T - T_c)^{-\ln b/\ln \lambda_t}. \tag{2.101}$$

No term in H is present in Eqn (2.100) because $(T - T_c)$ is an eigenvector for this system. Hence the critical parameter ν may be obtained as:

$$\nu = \ln b/\ln \lambda_t. \tag{2.102}$$

By rather similar arguments it is possible to show that

$$\beta = (d - y_{\mathcal{H}})/y_{\mathrm{t}} \tag{2.103}$$

where d is the dimensionality, $y_{\mathcal{H}} = \ln \lambda_{\mathcal{H}}/\ln b$, $y_{\mathrm{t}} = \ln \lambda_{\mathrm{t}}/\ln b$, and $\lambda_{\mathcal{H}}$ is the eigenvalue corresponding to the eigenvector $\mathcal{H}$. Likewise, the heat capacity exponent α is given by

$$\alpha = 2 - d/y_{\mathrm{t}} \tag{2.104}$$

and the exponent δ by

$$\delta = y_{\mathcal{H}}/(d - y_{\mathcal{H}}). \tag{2.105}$$

With regard to first-order transitions, the position of the renormalization group technique is somewhat more uncertain. Several groups of workers have used the method to deduce complete phase diagrams, many of them of a very complicated nature and including both first- and second-order transitions. The absence of a stable critical fixed point has been frequently used to suggest that any transition between the phases concerned must be first-order (Bruce, 1980; Bruno and Sak, 1980). The stability referred to here need be with respect to only one parameter: the fixed point may be unstable with respect to other parameters. Bruno and Sak in their study of the Ising model on a two-dimensional compressible lattice rejected as being unphysical the only stable fixed point that they were able to find: they therefore concluded that the disordering occurred by a first-order transition. Pytte's (1980) results for a q-state Potts model on a six-dimensional lattice seem to raise some doubts about the validity of this method of recognizing first-order transitions, at least as far as that dimensionality is concerned: he found that in the range $q = 2$ to $10/3$ the transition is first-order even though there is an accessible fixed point.

The existence of first-order (or discontinuity) fixed points, to which trajectories starting from first-order phase points would flow, was proposed by Nienhuis and Nauenberg (1976). These have $\xi = 0$ (which remains unaffected by the scaling procedure), but are to be distinguished from "phase sink" fixed points which also have $\xi = 0$. The first-order fixed points are recognized in numerical studies from the way in which the flow first moves towards them, "hovers", and finally moves away. Confirmation of the nature of the first-order fixed point is provided by the eigenvalue—which should be equal to b^{-1}, the scale change.

A considerable amount of quantitative information has been obtained by summation along the trajectory. Berker and Wortis (1976) have shown how values of generalized densities (ρ_i) (conjugate to a field K_i) can be determined at any state point provided that ρ_i is known for some other point to which the original point is connected by a trajectory. For example,

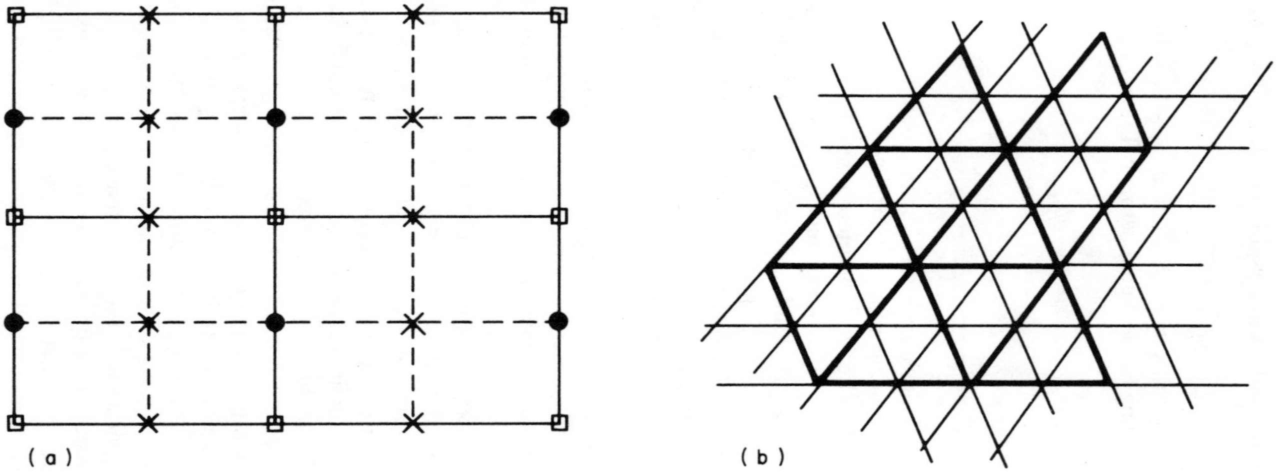

Fig. 2.11. Migdal decimation diagram. (*a*) Quadratic lattice. (*b*) Triangular lattice.

if points on either side of a phase boundary are joined by trajectories to their respective "phase sinks" then ρ_i at each of the original points can be calculated by means of the relation:

$$\rho_i = \frac{1}{b^d} \sum_j \rho_j' \frac{\partial K_j'}{\partial K_i'}. \tag{2.106}$$

A discontinuity in ρ_i could then provide a further important characterizing feature for a first-order transition. The free energy per site can, likewise, be written as a sum of terms along the trajectory:

$$A/N = \sum_{j=0}^{\infty} b^{-dj} g(K^{(j)}) \tag{2.107}$$

where $g(K^{(j)}) = G/N^{j\prime}$, and G is the term in Eqn (2.96) (Van Leeuwen, 1977).

Since much of the interest in the theory of adsorption systems has centred on the question of the "structure" in the fluid phase, it is disappointing that calculations of correlation functions are rarely carried out. The relationship between the pair distribution function after and before a renormalization step has been given by Van Leeuwen (1977) as

$$\rho^{(2)\prime}(r_{i'j'}) = b^2 \sum_{i\in i'} \sum_{j\in j'} \rho^{(2)}(r_{ij}). \tag{2.108}$$

However, the equation which has to be used is one similar in form to the Ornstein–Zernike equation (cf. Ch. 3, § 5(b)):

$$\rho^{(2)}(r_{ij}) = c(r_{ij}) + \sum_{i'} \sum_{j'} T_{i'i}\, \rho^{(2)}(r_{i'j'}) T_{j'i} \tag{2.109}$$

where

$$c(r_{ij}) = \delta^2\rho^{(2)}/\delta H_i \delta H_j \quad \text{and} \quad T_{i'i} = \delta H_{i'}/\delta H_i.$$

We turn now to a further consideration of the actual processes involved in "setting up" the renormalization group. The block spin procedure of Kadanoff which has been briefly described above is an example of position space renormalization. Another example is the decimation procedure, mainly due to Migdal (1976). Starting with the simple quadratic lattice (Fig. 2.11) with spins on all sites (□, × and ●) it is desired to "remove" the spin variables represented by × and ●. A bond-moving procedure is first carried out on the vertical bonds through the × spins. Each such line of vertical bonds is "split" into two equal parts and one part is added to each of the adjacent remaining lines of vertical bonds (those through the □ and ● spins. For the arrangement shown in Fig. 2.11(a), each of the remaining vertical bonds would now have double "strength" ($2J$). The spins

× are now merely "decorating" the lattice and may be "removed" by summation over all possible configurations of them. The same procedure is now applied to the horizontal bonds, leaving a quadratic array of spins with bonds of strength $2J$ between each pair. The justification for this procedure is more difficult than the use of the technique itself. The example given above is a particularly simple one. The method is not restricted to the elimination of alternate rows and columns. All vertical bonds could be moved leaving only every nth vertical line, thereby causing every residual bond to be of strength nJ. The same procedure could then be applied to the horizontal bonds. For the triangular lattice (which is of great interest in adsorption as it is the structure of commensurately adsorbed layers on the (0001) face of graphite) alternate bonds in each of the three lattice directions are moved as shown in Fig. 2.11(b), the spins which then decorate the bonds of the enlarged triangular lattice being removed by summation. An indication of the complexities which are introduced even by such an apparently simple transformation is the renormalization relationship found by Berker *et al.* (1978) for an initial Potts Hamiltonian:

$$-\beta H_N = -\beta \sum_{ij} V_{ij} t_i t_j - \Delta \Sigma t_i \tag{2.110}$$

where

$$-\beta V_{ij} = J(3\delta_{s_i s_j} - 1) + K.$$

The constants of the transformed system were found to be:

$$J' = \tfrac{1}{3}\ln(R_3/R_4) \qquad K' = \tfrac{1}{3}\ln(R_1^3 R_3 R_4^2)/R_2^6 \qquad \Delta' = 6\ln(R_1/R_2)$$

$$R_1 = 1 + 3z^2 \quad R_2 = z + (2x^{-1} + x^2)yz^3 \quad R_3 = z^2 + (2x^{-2} + x^4)y^2z^4$$

$$R_4 = z^2 + (x^{-2} + 2x)y^2z^4$$

$$x = \exp(bJ) \qquad y = \exp(bK) \qquad z = \exp(-b\Delta/6).$$

As before, the method could be modified so as to leave less than one-half of each type of bond.

A transformation method has also been employed by Kadanoff (1975, 1976), in which a variational parameter is chosen so as to minimize the free energy. This means, of course, that it is necessary to evaluate the free energy (to enable it to be minimized) even though it is not required for any other reason.

All the methods so far described are of the types known as position space renormalization procedures. A second kind of renormalization group process is that used in field theoretical work, where a continuous spin density, rather than an array of discrete spins, is considered. The basic Hamiltonian used is that of Landau–Ginzburg, Eqn (2.90). The method, which has been reviewed by Brézin *et al.* (1976), has not been used to any great extent for adsorption problems so far.

References

Aharony, A. and Pytte, E. (1981). *Phys. Rev. B* **23**, 362.

Barber, M., Heyes, D. M. and Clark, J. H. R. (1979) *J. Chem. Soc. Faraday Trans II* **75**, 1469.

Baxter, R. J. (1972). *Ann. Phys.* **70**, 193.

Baxter, R. J. (1973). *J. Phys. C: Solid State Physics* **6**, L445.

Benson, G. C. and Yun, K. S. (1967). *In* "The Solid–Gas Interface" (E. A. Flood, ed.), Vol. 1, p. 203. Arnold, London.

Berker, A. N. and Wortis, M. (1976). *Phys. Rev. B* **14**, 4946.

Berker, A. N., Ostlund, S. and Putnam, F. A. (1978). *Phys. Rev.* **B17**, 3650.

Berry, M. V. (1974). *J. Phys. A: Mathematical and General* **7**, 231.

Böttcher, C. J. F. (1973). "Theory of Electric Polarization", Vol. 1. Elsevier, Amsterdam.

Brézin, E., Le Guillou, J. C. and Zinn-Justin, J. (1976). *In* "Phase Transitions and Critical Phenomena" (C. Domb and M. S. Green, eds) Vol. 6, p. 125. Academic Press, London and New York.

Broekhoff, J. C. P. and de Boer, J. H. (1967). *J. Catalysis* **9**, 8.

Bruce, A. D. (1980). *Adv. Phys.* **29**, 111.

Bruno, J. and Sak, J. (1980). *Phys. Lett. A* **77**, 46.

Buff, F. P. (1951). *J. Chem. Phys.* **18**, 1591.

Buff, F. P. (1955). *J. Chem. Phys.* **23**, 419.

Buff, F. P. (1956). *J. Chem. Phys.* **25**, 146.

Buff, F. P. (1960). Theory of Capillarity, *in* "Handbuch der Physik", Vol. 10, p. 281. Springer-Verlag, Berlin.

Dash, J. G. (1977a). *Phys. Rev. B* **15**, 3136.

Dash, J. G. (1977b). *J. de Phys.* **38**, C4–201.

Defay, R., Prigogine, I., Bellemans, A. and Everett, D. H. (1966). "Surface Tension and Adsorption". Wiley, New York.

Derjaguin, B. V. (1957). *In* "Proceedings of the 2nd International Congress on Surface Activity" (J. Schulman, ed.), Vol. 2, p. 153. Butterworth, London.

Domany, E., Schick, M., Walker, J. S. and Griffiths, R. B. (1978). *Phys. Rev. B* **18**, 2209.

Ehrenfest, P. (1933). *Commun. Kamerlingh Onnes Laboratorium, Leiden*, Suppl. 75b.

Fisher, M. E. and Nelson, D. R. (1974). *Phys. Rev. Lett.* **32**, 1350.

Fox, H. W. and Zisman, W. A. (1950). *J. Colloid Sci.* **5**, 514.

Golner, G. R. (1973). *Phys. Rev.* B**8**, 3419.

Goodrich, F. C. (1971). *In* "Surface and Colloid Science" (E. Matijevic, ed.), Vol. 1, p. 1. Wiley, New York.

Goshen, S., Mukamel, D. and Shtrikman, S. (1974). *Int. J. Mag.* **6**, 221.

Griffiths, R. B. (1970). *Phys. Rev. Lett.* **24**, 715.

Hornreich, R. M., Luban, M. and Shtrikman, S. (1975). *Phys. Rev. Lett.* **35**, 1678.

Justi, E. and von Laue, M. (1934). *Z. Tech. Phys.* **15**, 521.

Kadanoff, L. P. (1966). *Physics* **2**, 263.
Kadanoff, L. P. (1975). *Phys. Rev. Lett.* **34**, 1005.
Kadanoff, L. P. (1976). *J. Stat. Phys.* **14**, 171.
Kirkwood, J. G. and Buff, F. P. (1949). *J. Chem. Phys.* **17**, 338.
Kogut, J. B. and Sinclair, D. K. (1981). *Phys. Lett. A* **81**, 149.
Landau, L. D. (1937). *Phys. Z. Sowjetunion* **11**, 26. Also in "Collected Papers of L. D. Landau" (1965). (D. ter Haar, ed.), p. 193. Pergamon Press, Oxford.
Landau, L. D. and Lifshitz, E. M. (1969). "Statistical Physics", 3rd edn, p. 424. Pergamon Press, Oxford.
Luban, M., Hornreich, R. M. and Shtrikman, S. (1977). *Physica* **86–88B**, 605.
Matecki, M., Thomy, A. and Duval, X. (1974). *J. Chim. Phys.* **71**, 1484.
Melrose, J. C. (1966). *A.I.Ch.E.J.* **12**, 986.
Melrose, J. C. (1968). *Ind. Eng. Chem.* **60**, 53.
Melrose, J. C. (1970). *Pure Appl. Chem.* **22**, 273.
Michelson, A. (1977). *In* "Statphys 13", IUPAP Conference on Statistical Physics, Haifa, 1977 (D. Cabib, C. G. Kuper and I. Riess, eds), Vol. 2, p. 459. Adam Hilger, Bristol.
Migdal, A. A. (1976). *Sov. Phys. JETP* **42**, 743.
Navascués, G. (1979). *Rep. Prog. Phys.* **42**, 1131.
Nicholson, D. (1968). *Trans. Faraday Soc.* **64**, 3416.
Nienhuis, B. and Nauenberg, M. (1976). *Phys. Rev. B* **13**, 2021.
Nienhuis, B., Berker, A. N., Riedel, E. K. and Schick, M. (1979). *Phys. Rev. Lett.* **43**, 737.
Parsonage, N. G. and Staveley, L. A. K. (1978). "Disorder in Crystals", p. 14. Clarendon Press, Oxford.
Peierls, R. (1978). *Phys. Rev. B* **18**, 2013.
Pierotti, R. A. and Halsey Jr., G. D. (1959). *J. Phys. Chem.* **61**, 1158.
Pippard, A. B. (1957). "The Elements of Classical Thermodynamics", Ch. 9. Cambridge University Press.
Potts, R. B. (1952). *Proc. Camb. Phil. Soc.* **48**, 106.
Putnam, F. A. and Fort, Jr. T. (1975). *J. Phys. Chem.* **79**, 459.
Putnam, F. A., Fort, T. J., and Griffiths, R. B. (1977). *J. Phys. Chem.* **81**, 2171.
Pytte, E. (1980). *Phys..Rev.* B**22**, 4450.
Saville, G. (1977). *J. Chem. Soc. Faraday Trans. II* **73**, 1221.
Steele, W. A. (1973). "The Interaction of Gases with Solid Surfaces". Pergamon Press, Oxford.
Thomy, A. and Duval, X. (1970). *J. Chim. Phys.* **67**, 1101.
Tisza, L. (1961). *Annals Phys.* **13**, 1.
Tolman, R. C. (1949). *J. Chem. Phys.* **17**, 333.
Van Leeuwen, J. M. J. (1977). *In* "Statphys. 13", IUPAP Conference on Statistical Physics, Haifa, *1977*. (D. Cabib, C. G. Kuper and I. Riess, eds.), Vol. 1, p. 173. Adam Hilger, Bristol.
White, L. R. (1976). *J. Chem. Soc. Faraday Trans. I* **73**, 390.
Wilson, K. G. (1971). *Phys. Rev. B* **4**, 3174, 3184.
Wilson, K. G. and Fisher, M. E. (1972). *Phys. Rev. Lett.* **28**, 248.
Zia, R. K. P. and Wallace, D. J. (1975). *J. Phys. A: Mathematical and General* **8**, 1495.

Chapter 3

Statistical Mechanics and Intermolecular Forces

1. Introduction

In this chapter we discuss the basic theoretical material needed in simulation studies of adsorption phenomena and for the development of approximate theoretical models. Very roughly this material falls into two categories: in simulation studies attention is directed towards the selection of an appropriate intermolecular potential and towards the need to extract thermodynamic quantities from the raw data obtained from the simulation. Theoretical models also require the development of equations from which the properties of the system can be calculated fairly rapidly after the introduction of appropriate, and preferably well controlled, approximations. Here we have incorporated both of these requirements into a single framework so that inter-relationships can be better discerned.

In the opening section the fundamental statistical mechanical equations are given. We introduce a notation which will support an exposition sufficiently general to be applicable to any adsorption system. In fact it turns out that fluid–solid systems, of particular interest in their own right, are in some ways more general. This is because the interaction energy in this case needs to include an external field term in addition to the higher-order interaction potentials.

A discussion of the consequent subdivision of the partition function and the limitations inherent in this subdivision then leads naturally towards a discussion of the inter-molecular potentials themselves. Much has been written on this subject and the discussion here stresses the adsorbate–solid adsorbent aspects of the problem but from the general point of view of pairwise and higher-order interactions between individual atoms and molecules. The rather different approach, initiated by Lifshitz, which combines inter-molecular interactions and statistical mechanical aspects of the problem, is also given some consideration at this stage.

In § 3 we develop expressions for the calculation of thermodynamic properties and this links, through a discussion of the pressure tensor to the Bogoliubov, Born, Green, Kirkwood, Yvon(BBGKY) integral equation. We have discussed this equation in some depth, using it as a paradigm to illustrate, for example the manipulation of coordinate transformations. This equation is of course only one among many integral equations which have proved to be the most useful entities in theoretical studies. A second equation, the Ornstein–Zernike (OZ) equation is of equal if not greater importance and its fundamental significance has been carefully considered from the point of view of non-uniform systems.

A mathematical tool of great power and elegance which can be applied to the discussion of these and other theoretical equations is that of functional differentiation. Its full exploitation in statistical mechanics, and the revelation of its particular aptness for adsorption problems is to be credited to Percus. In the remainder of this chapter, in order to bring out relationships between various theories and between the approximations which have been used to establish working expressions, we have used this technique as a common source for the derivation and development of a number of integral equations and as a basis for the explanation of perturbation theories.

2. Partition functions and distribution functions

(a) Basic definitions

Any adsorption system will contain at least two components and it is therefore natural to choose a multicomponent formulation for its molecular description. Furthermore, experimental adsorption data are often in the form of isotherms which relate the number of molecules in the system to the chemical potential (or a closely related quantity such as gas phase pressure in gas–solid adsorption); it is therefore convenient to use the grand ensemble which readily leads to such relationships.

We consider a system containing a total of $\mathcal{N}$ molecules of species a, b, . . . , in the composition set $\boldsymbol{N} = N_a, N_b, \ldots$. The momentum of the molecule labelled i and of type a is $\boldsymbol{p}_{ia}$ and its position and orientation are specified by the co-ordinate $\boldsymbol{q}_{ia}$. For spherically symmetrical (structureless) molecules $\boldsymbol{q}_{ia} \equiv \boldsymbol{r}_{ia}$ and simply locates the centre of mass of the ith molecule; the corresponding volume element is $d\boldsymbol{r}_{ia}$. For molecules of lower symmetry, $\boldsymbol{q}_{ia}$ also contains a specification $\boldsymbol{\omega}_{ia}$ of the orientation of the molecule and the corresponding volume element is $\mathrm{d}\boldsymbol{q}_{i\mathrm{a}} = \mathrm{d}\boldsymbol{r}_{i\mathrm{a}}\,\mathrm{d}\boldsymbol{\omega}_{i\mathrm{a}}$ where the rotational part $\mathrm{d}\boldsymbol{\omega}_{i\mathrm{a}}$ is normalized so that $\int \mathrm{d}\boldsymbol{\omega}_{i\mathrm{a}} = 1$.

The probability density, in the grand ensemble, of finding a molecule labelled 1 which is also of type a at the point in phase space $\boldsymbol{p}_{1\mathrm{a}}, \boldsymbol{q}_{1\mathrm{a}}$; a molecule labelled 2 of type b at $\boldsymbol{p}_{2\mathrm{b}}, \boldsymbol{q}_{2\mathrm{b}}; \ldots; \mathcal{N}, \sigma$ at $\boldsymbol{p}_{\mathcal{N}\sigma}, \boldsymbol{q}_{\mathcal{N}\sigma}$ is

$$\mathcal{P} = \frac{\exp[\beta(\boldsymbol{\mu}\cdot\boldsymbol{N} - H_N)]}{\boldsymbol{N}!\,\Xi\, h^{\boldsymbol{s}\cdot\boldsymbol{N}}} \tag{3.1}$$

in which $\beta = 1/kT$, and the Hamiltonian H_N is the sum of the kinetic energy term (T_N) and potential energy term (U_N). The index $\boldsymbol{s}\cdot\boldsymbol{N} = \sum_\alpha s_\alpha N_\alpha$, where s_α is the number of degrees of freedom per molecule for species α (e.g. 3 for atoms, 6 for an asymmetric rotator), and similarly $\boldsymbol{\mu}\cdot\boldsymbol{N} = \sum_\alpha \mu_\alpha N_\alpha$.

The normalization factor in (3.1) contains the grand partition function Ξ, which can be written as a sum over the canonical partition function Q_N for $\mathcal{N}$ molecules,

$$\Xi = \sum_N Q_N \boldsymbol{\lambda}^N \tag{3.2}$$

where

$$\boldsymbol{\lambda}^N = \prod_\alpha \lambda_\alpha^{N_\alpha}$$

and the absolute activity of species a is

$$\lambda_\mathrm{a} = \exp(\beta\mu_\mathrm{a}). \tag{3.3}$$

The canonical partition function is

$$Q_N = \frac{\int \ldots \int \exp(-\beta H_N)\,\mathrm{d}\boldsymbol{q}\,\mathrm{d}\boldsymbol{p}}{\boldsymbol{N}!\, h^{\boldsymbol{s}\cdot\boldsymbol{N}}} \tag{3.4}$$

in which we introduce the following notation for the co-ordinates:

(i) The set of co-ordinates for all the N_a molecules of type a labelled $1, \ldots, N_\mathrm{a}$, is

$$\boldsymbol{q}_\mathrm{a} = \boldsymbol{q}_{1\mathrm{a}}; \boldsymbol{q}_{2\mathrm{a}}; \ldots; \boldsymbol{q}_{N_\mathrm{a},\mathrm{a}}$$

where the additional subscript a may be dropped when this type of molecule is understood.

(ii) The co-ordinate set for all species a, b, . . . , σ, is

$$\boldsymbol{q} = \boldsymbol{q}_\mathrm{a}; \boldsymbol{q}_\mathrm{b}; \ldots; \boldsymbol{q}_\sigma$$

and the corresponding elements of volume are represented by $\mathrm{d}\boldsymbol{q}_\mathrm{a}$ and $\mathrm{d}\boldsymbol{q}$ respectively, or in terms of separate positional and rotational components, $\mathrm{d}\boldsymbol{q} = \mathrm{d}\boldsymbol{r}\,\mathrm{d}\boldsymbol{\omega}$, etc. The notation is readily extended to the momentum co-ordinates. In the denominator of (3.4), $\boldsymbol{N}! = \prod_\alpha N_\alpha!$.

In some cases it may be desirable to denote the phase by an additional superscript, as in Chapter 2, but since this is cumbersome we shall avoid it where possible.

For systems in equilibrium, the integrals over the translational and angular momentum co-ordinates in (3.4) may be taken directly to give

$$Q_N = Z_N/(N!\boldsymbol{\Lambda}^{\nu N}) \tag{3.5}$$

where the set $\boldsymbol{\Lambda}$ contains

$$\Lambda_a = h/(2\pi m_a kT)^{1/2} Z_{\mathrm{rot,a}}^{1/\nu} \tag{3.6}$$

with the molecular mass, m_a and the partition function for the ath type of free rotator, $Z_{\mathrm{rot,a}}$ in ν-dimensional space.

The configurational integral Z_N, in Eqn (3.5) now depends on spatial co-ordinates alone and contains only the potential energy from the Hamiltonian; it may be written in a compact form using the total co-ordinate set $\boldsymbol{q}$ as,

$$Z_N = \int \dots \int \exp(-\beta U_N)\, \mathrm{d}\boldsymbol{q}. \tag{3.7}$$

Momentum-integrated forms of the probability density functions can be likewise obtained from Eqn (3.1) and expressed in equivalent ways:

$$P_N = \frac{\exp(-\beta U_N)}{Z_N} = \frac{\exp(-\beta U_N)}{Q_N N!\boldsymbol{\Lambda}^{\nu N}} = \frac{\sum \boldsymbol{\lambda}^N \exp(-\beta U_N)}{\Xi N!\boldsymbol{\Lambda}^{\nu N}}. \tag{3.8a}$$

In the grand ensemble there is, in addition to P_N, a momentum integrated probability density $P_{;N}$ that the system contains $\mathcal{N}$ molecules in the set $\boldsymbol{N}$ with specified co-ordinates:

$$P_{;N} = \frac{\boldsymbol{\lambda}^N \exp(-\beta U_N)}{\Xi N!\boldsymbol{\Lambda}^{\nu N}}. \tag{3.8b}$$

The canonical partition function leads to thermodynamic quantities through the usual relationship for the Helmholtz free energy A:

$$\beta A = -\ln Q_N. \tag{3.9}$$

Similarly the grand free energy Ω, defined in Eqn (2.41) is related to the grand partition function in Eqns (3.1) and (3.2) by

$$\beta\Omega = -\ln \Xi. \tag{3.10}$$

The mean number of molecules of type a in the grand ensemble can be found by differentiating Eqn (3.2) with respect to λ_a and making use of the last member of (3.8a) to yield

$$\beta\langle N_a\rangle = \left(\frac{\partial \ln \Xi}{\partial \mu_a}\right)_{V,T,\boldsymbol{\mu}'} \tag{3.11}$$

where the primed set $\boldsymbol{\mu}'$ excludes the species named in the subscript. further differentiation of (3.11) with respect to μ_a gives the fluctuation in N_a

$$\beta[\langle N_a^2\rangle - \langle N_a\rangle^2] = \left(\frac{\partial\langle N_a\rangle}{\partial \mu_a}\right)_{V,T,\boldsymbol{\mu}'} = \beta\left(\frac{\partial\langle N_a\rangle}{\partial \ln \lambda_a}\right)_{V,T,\boldsymbol{\lambda}'} \tag{3.12}$$

Fluctuation expressions of this type will occur fairly frequently and it is convenient to introduce a compact notation $f(X, Y)$, where

$$f(X, Y) = \langle XY\rangle - \langle X\rangle\langle Y\rangle. \tag{3.13}$$

A second route to thermodynamic quantities is available through the n-particle distribution functions which also provide valuable information about adsorbate structures. The generic n-particle distribution function in the grand ensemble will be denoted by the symbol $\rho^{(a,b,\ldots,\alpha)}(\boldsymbol{q}_{1a}, \ldots, \boldsymbol{q}_{n\alpha})$ and is defined as the probability of finding a particle of type a at $\boldsymbol{q}_{1a}$ (that is at position $\boldsymbol{r}_{1a}$ in orientation $\boldsymbol{\omega}_{1a}$), a second particle of type b at $\boldsymbol{q}_{2b}$, etc., where it is understood that some of the particles may be of the same type.

The generic distribution function is found from P_N in Eqn (3.8) by integration over the non-relevant co-ordinates and summation over the non-relevant particles. However, since there are N_a ways of selecting the first particle of type a, $(N_a - 1)$ of selecting the second of these particles, and so on, it is necessary to introduce correction factors such as $N_a!/(N_a - n_a)!$ where n_a is the number of particles of the same type a; the resulting expression is

$$\rho^{(a,b,\ldots,\alpha)}(\boldsymbol{q}_{1a}, \ldots, \boldsymbol{q}_{n\alpha}) = \sum_{N \geqslant n} \frac{\boldsymbol{\lambda}^N \int \ldots \int \exp(-\beta U_N)\, \mathrm{d}\boldsymbol{q}^{n+1}}{\Xi \boldsymbol{\Lambda}^{\nu N}(N - \boldsymbol{n})!} \tag{3.14}$$

in which $\mathrm{d}\boldsymbol{q}^{n+1}$ indicates an integration over the co-ordinate set for those molecules not specified in the distribution function. It is sometimes useful to have angle-averaged functions which depend only on the translational part of the co-ordinate. These may be obtained by integrating over $\boldsymbol{\omega}$, the set of orientational co-ordinates, and are denoted by the subscript T. For example:

$$\rho_T^{(a,b,\ldots,\alpha)}(\boldsymbol{r}_{1a}, \ldots, \boldsymbol{r}_{n\alpha}) = \int \ldots \int \rho^{(a,b,\ldots,\alpha)}(\boldsymbol{q}_{1a}, \ldots, \boldsymbol{q}_{n\alpha})\, \mathrm{d}\boldsymbol{\omega}^n. \tag{3.15}$$

The somewhat cumbersome notation for the n-particle distribution functions may be simplified in many cases of interest. For example, if two species (a, b) are present, with particle 1 of type a and particle 2 of type b, the pair distribution $\rho^{(a,b)}$ $(\boldsymbol{q}_{1a}, \boldsymbol{q}_{2b})$ may be written $\rho_{12}^{(a,b)}$, etc. When a single species only is considered, the superscript serves merely to indicate the multiplicity of the distribution and the pair distribution function could be written in either of the equivalent forms $\rho_{12}^{(a,a)}$ or $\rho_{12}^{(2)}$ when the identity of the species a is understood.

From the definition (3.14) it follows that, for all σ species present,

$$\int \dots \int \rho_{12,\dots,n}^{(a,b,\dots,\sigma)} \, d\boldsymbol{q}_{1a}, \dots, d\boldsymbol{q}_{n\sigma} = \langle N!/(N-n)! \rangle. \tag{3.16}$$

The singlet and pair distribution functions are of particular interest; the former, for species a, may be written

$$\rho^{(a)}(\boldsymbol{q}_{1a}) = \sum_{\substack{N_a \geq 1 \\ \{N_b, N_c \dots\} \geq 0}} \frac{N_a \boldsymbol{\lambda}^N \int \dots \int \exp(-\beta U_N) \, d\boldsymbol{q}_{2a} \dots d\boldsymbol{q}_{N_a,a} \, d\boldsymbol{q}_b \dots d\boldsymbol{q}_\sigma}{\Xi \boldsymbol{\Lambda}^{\nu N} N!}. \tag{3.17}$$

The correlation between one particle of type a from $\{N_a\}$ and one particle of type b in $\{N_b\}$ numbered $N_a + 1$ is expressed by the pair distribution function,

$$\rho^{(a,b)}(\boldsymbol{q}_{1a}, \boldsymbol{q}_{N_a+1,b})$$
$$= \sum_{\substack{N_a \geq 1, N_b \geq 1 \\ \{N_c, \dots\} \geq 0}} \frac{N_a N_b \lambda^N \int \dots \int \exp(-\beta U_N) \, d\boldsymbol{q}_{2a} \dots d\boldsymbol{q}_{N_a,a}, \, d\boldsymbol{q}_{N_a+2,b} \dots d\boldsymbol{q}_\sigma}{\Xi \boldsymbol{\Lambda}^{\nu N} N!}. \tag{3.18}$$

When only a single species comprising N molecules is to be considered, this equation simplifies to

$$\rho^{(2)}(\boldsymbol{q}_1, \boldsymbol{q}_2) = \sum_{N \geq 2} \frac{N(N-1)\lambda^N \int \dots \int \exp(-\beta U_N) \, d\boldsymbol{q}_3 \dots d\boldsymbol{q}_N}{\Xi \Lambda^{\nu N} N!} \tag{3.19}$$

and Eqn (3.17) can also be simplified in this case in a similar way.

The generic distribution functions provide a starting point for the definition of various correlation functions. The most important of these, the pair correlation function $g^{(a,b)}(\boldsymbol{q}_{1a}, \boldsymbol{q}_{2b})$ is defined by the equation,

$$\rho^{(a,b)}(\boldsymbol{q}_{1a}, \boldsymbol{q}_{2b}) = \rho^{(a)}(\boldsymbol{q}_{1a})\rho^{(b)}(\boldsymbol{q}_{2b})g^{(a,b)}(\boldsymbol{q}_{1a}, \boldsymbol{q}_{2b}). \tag{3.20}$$

For a single species this equation, in simplified notation, becomes

$$\rho_{12}^{(2)} = \rho_1^{(1)} \rho_2^{(1)} g_{12}^{(2)}. \tag{3.21}$$

In a uniform system, where $\rho^{(1)}$ is the same everywhere and equal to the mean density ρ, Eqn (3.21) further simplifies to

$$\tilde{\rho}^{(2)}_{12} = \rho^2 \tilde{g}^{(2)}_{12} \tag{3.22}$$

where the tilde will be used when it is necessary to emphasize the fact that we are considering a uniform system so that $\tilde{\rho}^{(2)}_{12}$ and $\tilde{g}^{(2)}_{12}$ depend only on the separation and relative orientation of a molecular pair and not on their location and orientation with respect to an external frame of reference, which is the case for non-uniform fluids.

From the above definitions of the pair correlation functions it is clear that $g^{(a,b)} \to 1$ when the pair separation $\to \infty$. It is therefore useful to introduce the two-particle Ursell distribution:

$$\mathscr{F}^{(a,b)}_{12} = \rho^{(a,b)}_{12} - \rho^{(a)}_1 \rho^{(b)}_2. \tag{3.23}$$

The total correlation function $h^{(a,b)}_{12}$ is related to $g^{(a,b)}_{12}$ and the Ursell function through

$$\mathscr{F}^{(a,b)}_{12} = \rho^{(a)}_1 \rho^{(b)}_2 h^{(a,b)}_{12} = \rho^{(a)}_1 \rho^{(b)}_2 [g^{(a,b)}_{12} - 1]. \tag{3.24}$$

(b) Decomposition of the partition function

The potential energy U_N, appearing in the equations of § 2(a) can be expressed in terms of summations over sets of n-body potential energy functions $u^{(a,b\ldots)}$ as

$$U_N = \sum_{i>j} u^{(a,b)}_{ij} + \sum_{i>j>k} u^{(a,b,c)}_{ijk} + \ldots \tag{3.25}$$

Here the subscripts refer to the co-ordinates of the molecules in the system so that $u^{(a,b)}_{12} = u^{(a,b)}(\boldsymbol{q}_{1a}, \boldsymbol{q}_{2b})$, for example, is a pairwise interaction potential between a molecule at $\boldsymbol{r}_{1a}$ in orientation $\boldsymbol{\omega}_{1a}$ and one at $\boldsymbol{r}_{2b}$ in orientation $\boldsymbol{\omega}_{2b}$ and will of course depend on the types of molecules which are interacting. The higher-order terms in Eqn (3.25) are of less importance, but nevertheless it has been convincingly demonstrated that 3-body potentials are not negligible.

The potential energy may now be subdivided at the domain boundaries defined in Chapter 2 in terms of Gibbs dividing surfaces. The procedure is quite general, but in order to avoid undue abstraction we shall refer specifically to fluid–solid systems with a single phase boundary in what follows.

As in Chapter 2 the total volume V of the system is subdivided into an "adsorbent" phase of volume V^S and the remaining "adsorbate" phase of volume V^A. The quotes serve as a reminder that this is a Gibbsian subdivision and that some atoms from the solid adsorbent may actually be located in the volume V^{A} and vice versa. The summation for the interaction between a molecule of species a within the adsorbate volume and the subset within the adsorbent volume comprising species b, c, . . . at locations denoted by upper case subscripts is

$$u_i^{[\mathrm{a}]} = \sum_{J \geqslant 1} u_{iJ}^{(\mathrm{a,b})} + \sum u_{iJK}^{(a,b,c)} + \dots \tag{3.26}$$

In the same way the pairwise interaction energy takes a modified form; for example for interactions between adsorbate molecules of the same species (a):

$$u_{ij}^{[\mathrm{a,a}]} = u_{ij}^{(\mathrm{a,a})} + \sum_{K \geqslant 1} u_{ijK}^{(\mathrm{a,a,w})} + \dots \tag{3.27}$$

where species w is in the adsorbent phase. Analogous equations could be written for the higher-order potentials. In most cases, theories dealing with fluid–solid systems assume that only "adsorbent" species $\mathrm{b}_1, \mathrm{b}_2 \dots$ are present in V^{S} and only adsorbate species $\mathrm{a}_1, \mathrm{a}_2, \dots$ are present in V^A, in which case potential functions in (3.26) will be of the type $u_{iJ}^{(\mathrm{a_1,w_1})}$, $u_{iJ}^{(\mathrm{a_2,w_1})}$, $u_{iJK}^{(\mathrm{a_1,w_1,w_2})}$, etc.

The potential $u^{[\mathrm{a}]}$ may now be regarded as the external field due to the adsorbent experienced by a molecule of species a. The contribution to this from pairwise terms has received considerable attention and is discussed in more detail in the following section. Relatively little is known about the possible importance of higher-order contributions to $u^{[\mathrm{a}]}$. On the other hand, the higher-order terms in Eqn (3.27) have been considered and are discussed in detail in § 3(f).

With the definitions of Eqns (3.26) and (3.27), the potential energy may be written in a modified form for an "adsorbate" phase with a total of $N_{\mathrm{a}}^{\mathrm{A}}$ molecules of type a, $N_{\mathrm{b}}^{\mathrm{A}}$ of type b, etc., as

$$U_{N^{\mathrm{A}}} = \sum_{i>1} u_i^{[1]} + \sum_{i>j} u_{ij}^{[2]} + \sum_{i>j>k} u_{ijk}^{[3]} + \dots \tag{3.28}$$

where summations over relevant species are implied in the superscripts [1], [2], ...

The remaining part of the potential energy comprises only interactions among the N^{S} atoms within the "adsorbent" volume and will be denoted by $U_{N^{\mathrm{S}}}$; thus Eqn (3.25) may be written

$$U_N = U_{N^{\mathrm{A}}} + U_{N^{\mathrm{S}}}. \tag{3.29}$$

Note that this subdivision of the potential introduces an asymmetry into the equations since $U_{N^{\mathrm{A}}}$ contains the adsorbate–adsorbent and adsorbate interaction terms, whereas $U_{N^{\mathrm{S}}}$ contains only the adsorbent interactions.

Corresponding to the subdivision of the potentials, it is possible to factorize the configurational integral, provided that the co-ordinates of the adsorbent atoms, which are needed to calculate $u^{[1]}$, etc., are fixed. This approximation is an underlying assumption in (nearly) every treatment of fluid–solid adsorption and is often applied in other adsorption systems. With this proviso, Eqn (3.7) can be written

$$Z_N = Z_{N^A} Z_{N^S} \tag{3.30}$$

where

$$Z_{N^A} = \int_{q \subset V^A} \cdots \int \exp(-\beta U_{N^A})\, d\boldsymbol{q} \tag{3.31}$$

in which the co-ordinates now refer to the location and orientation of molecules in the adsorbate volume V^A only. The corresponding canonical and grand partition functions for the adsorbate fluid are Q_{N^A} and Ξ^A respectively where

$$Q_{N^A} = Z_{N^A} / \boldsymbol{\Lambda}^{\nu N^A} N^A! \tag{3.32}$$

$$\Xi^A = \sum_{N^A \geq 0} Q_{N^A} \boldsymbol{\lambda}^{N^A}. \tag{3.33}$$

Expressions similar to (3.31), (3.32) and (3.33) may be written for a solid adsorbent except that (3.32) must include a factor N^S in the numerator with the integral giving Z_{N^S} being restricted to a single cell, since the atoms in the solid are localized. It should be noted that in general both localized and non-localized states are also possible for the adsorbate. This question has been considered by Hill (1960).

The factorization of the partition function also of course implies, from (3.7), that the free energy may be separated into components appropriate to the phases present in the system using the equations discussed in § 4 of Chapter 2. To consider this in more detail we take the example of a fluid–solid system which is subdivided, as before, into an "adsorbent" volume V^S and and "adsorbate" volume V^A. The grand partition function may be written

$$\exp(-\beta\Omega) = \Xi(V, \boldsymbol{\mu}, T) = \Xi^A(V^A, \boldsymbol{\mu}, T)\Xi^S(V^S, \boldsymbol{\mu}, T) \tag{3.34}$$

where Ξ^A is given by (3.33) and Ξ^S by an analogous expression. From Eqn (2.51) the surface excess grand free energy function is

$$\Omega^\Sigma = \Omega - \tilde{\Omega}^S - \tilde{\Omega}^G = -\phi\mathscr{A} \tag{3.35}$$

where $\tilde{\Omega}^S$ has been discussed in Eqn (2.47) *et seq.* If we construct a hypothetical system comprising perturbed adsorbent in V^S and uniform adsorptive (G) in volume V^{A} with no interaction between phases, its grand partition function would be

$$\tilde{\Xi}(V, \boldsymbol{\mu}, T) = \tilde{\Xi}^{\mathrm{S}}(V^{\mathrm{S}}, \boldsymbol{\mu}^{\mathrm{S}}, T)\tilde{\Xi}^{\mathrm{G}}(V^{\mathrm{A}}, \boldsymbol{\mu}^{\mathrm{A}}, T). \quad (3.36)$$

From (3.34), (3.35) and (3.36) the surface excess quantities may be expressed in terms of the appropriate grand partition functions, thus

$$\beta\phi\mathcal{A} = -\beta\Omega^{\Sigma} = \ln\left\{\frac{\Xi^{\mathrm{A}}(V^{\mathrm{A}}, \boldsymbol{\mu}, T)\Xi^{\mathrm{S}}(V^{\mathrm{S}}, \boldsymbol{\mu}, T)}{\tilde{\Xi}^{\mathrm{G}}(V^{\mathrm{A}}, \boldsymbol{\mu}^{\mathrm{A}}, T)\tilde{\Xi}^{\mathrm{S}}(V^{\mathrm{S}}, \boldsymbol{\mu}^{\mathrm{S}}, T)}\right\}. \quad (3.37a)$$

When the assumption is added that the solid remains unperturbed by the presence of the adsorbate, the adsorbent phase terms in Eqn (3.37a) cancel out and subsequent equations may be written for the adsorbate alone. Equation (3.37a) then becomes

$$\begin{aligned}\beta\phi\mathcal{A} = -\beta\Omega^{\Sigma} &= -U^{\Sigma} + TS^{\Sigma} - \mu N^{\Sigma}\\ &= \ln \Xi^{\mathrm{A}}(V^{\mathrm{A}}, \boldsymbol{\mu}, T) - \ln \tilde{\Xi}^{\mathrm{G}}(V^{\mathrm{A}}, \boldsymbol{\mu}^{\mathrm{A}}, T) \quad (3.37b)\end{aligned}$$

where the uniform phase partition function is that for the same system in the absence of adsorbent forces. It would be possible of course to consider Eqn (3.37a) from the point of view of more general systems in which the adsorbate–adsorbent interaction is more severe. Simulation studies are particularly well suited to investigations of this type because the prior assumption of rigidity in the solid phase is not essential and it is equally feasible to simulate such systems as solid–liquid interfaces where disruption or dissolution are possible; the interface between a liquid and an adsorbed monolayer in which the structure of both phases is likely to be modified; or the interface between immiscible fluids in which a clear separation of species at the boundary region would be obviously unrealistic. As pointed out above, the factorization of (3.34) requires that no species can cross the domain boundaries, so that (3.37a) can be used in the form given provided this limitation is met. When dissolution is permitted, the factorized numerator of (3.37a) must be replaced by Ξ.

3. Potential energy functions

(a) Introduction

In this section we consider the potential energy functions which contribute to the summation in Eqn (3.28). The topic is a broad one which can be set first of all in the context of the general calculation of intermolecular forces starting from a knowledge of the fundamental particles present in a given system and employing wave mechanics. Obviously it will be inappropriate to deal with the subject in great depth and we refer the reader to other reviews (e.g. Amos and Crispin, 1976; Buckingham and Utting, 1970; Margenau and Kestner, 1969) which provide suitable background material. At the same time it will be necessary to consider some details of the problem at the electronic level in order to expose some of the difficulties which attach specifically to interfacial systems. Our emphasis will be towards the calculation of $u^{[1]}$ and of the effect of a dense substrate on $u^{[2]}$ (essentially via the three-body term in Eqn (3.27)). We consider a specific example, that of the interaction potential between water molecules, in Chapter 7 in connection with hydrophobic systems.

The calculation of gas–solid potentials has been discussed by several authors, the reviews by Steele (1974), and Pierotti and Thomas (1971) being particularly useful.

The physical interaction between a collection of molecules is conveniently regarded as being composed of two summable parts: a short-range repulsive part E_R, which arises at separations close enough for orbital overlap to become energetically significant; and a long-range attractive part. Since the long-range interaction energy is small compared to the total energy of the system, it can be conveniently treated via perturbation theory. The perturbation energy is written as a multipole expansion of the coulombic interactions between all the charged particles of the system and it can be shown that the interaction potential may be expressed as the sum of two terms: (i) an electrostatic term E_c, arising from the first-order perturbation, which accounts for the interaction between fixed charge distributions on the molecules and which can be written as a classical multipole expansion in which the appropriate dipole vectors, quadrupole tensors, etc., appear (Böttcher, 1973); (ii) a polarization term E_p, arising from the second-order perturbation, which subdivides into an induction energy E_I, due to the distortion of electron configurations by permanent electrical fields, and the so-called dispersion energy E_D, resulting from the interaction of fluctuating charge distributions. The dispersion energy is of course the only contribution to the long-range interactions when permanent charge separations are absent.

The development of this approach enables the two-body, three-body (and higher-order) terms of Eqn (3.25) to be expressed separately and explicitly. More recent developments of the perturbation method have made allowance for Pauli exclusion (since orbital overlap will occur at all separations). The resulting exchange energy terms appear as additive corrections to the coulombic and polarization energy terms already mentioned; they can become dominant at separations of the order of one or two Ångstroms and their significance has been reviewed, for example, by Amos and Crispin (1976).

(b) The pairwise dispersion energy

The dispersion energy contribution to the interaction energy between an isolated pair of molecules of types a and b in specified orientations with their centres positioned at $\boldsymbol{r}_1, \boldsymbol{r}_2$ can be written as a series in inverse powers of the separation r_{12}. The leading term in r_{12}^{-6} is due to the interaction between fluctuating dipoles and higher-order contributions appear as a result of interactions between other fluctuating multipoles. Without further approximation these terms may be expressed as functions of the frequency-dependent polarizabilities (or hyperpolarizabilities) and the appropriate multipole interaction tensors. for example, the leading dipole–dipole term is

$$E_{D,\mathrm{d-d}}^{(\mathrm{a,b})} = -T_{12}^{\alpha\beta}T_{12}^{\gamma\delta}\frac{1}{2\pi}\int_0^\infty \alpha_{\alpha\beta}^{(\mathrm{a})}(i\omega)\alpha_{\gamma\delta}^{(\mathrm{b})}(i\omega)\,\mathrm{d}\omega \qquad (3.38^*)\dagger$$

in which $T_{12}^{\alpha\beta}(\boldsymbol{r}_{12})$ is the dipole interaction tensor between the molecules at $\boldsymbol{r}_1, \boldsymbol{r}_2$. This, and other multipole interaction tensors may be defined through the equations

$$T^{\alpha} = \frac{\partial}{\partial R_\alpha}\left(\frac{1}{\boldsymbol{R}}\right) = -\frac{R_\alpha}{R^3} \qquad (3.39\mathrm{a}^*)$$

$$T^{\alpha\beta} = \frac{\partial}{\partial R_\alpha}\frac{\partial}{\partial R_\beta}\left(\frac{1}{\boldsymbol{R}}\right) = (3R_\alpha R_\beta - R^2\delta_{\alpha\beta})/R^5 \qquad (3.39\mathrm{b}^*)$$

$$T^{\alpha\beta\gamma} = \frac{\partial}{\partial R_\alpha}\frac{\partial}{\partial R_\beta}\frac{\partial}{\partial R_\gamma}\left(\frac{1}{\boldsymbol{R}}\right)$$

$$= -\,(15R_\alpha R_\beta R_\gamma - 3R^2(R_\alpha\delta_{\beta\gamma} + R_\beta\delta_{\gamma\alpha} + R_\gamma\delta_{\alpha\beta}))/R^7 \qquad (3.39\mathrm{c}^*)$$

etc., where $\boldsymbol{R}$ here stands for $\boldsymbol{r}_{12} = \boldsymbol{r}_1 - \boldsymbol{r}_2$ and $\alpha, \beta, \ldots$ denote the x, y, z components. From (3.39b) it can be seen that (3.38) contains the factor r_{12}^{-6}. The term $\alpha_{\alpha\beta}^{(\mathrm{a})}(i\omega)$ appearing under the integral sign in (3.38) is the frequency-dependent polarizability tensor for the molecule of type a, at the imaginary frequency $i\omega$; and summation over all permutations of the superscripts is implied by the tensor notation.

In many cases an axis of rotational symmetry exists in the system of interest so that the polarizability tensor may be diagonalized and the integral appearing in (3.38) written

$$A_{\alpha\beta}^{(\mathrm{a,b})} = \frac{1}{2\pi}\int_0^\infty \alpha_{\alpha\alpha}^{(\mathrm{a})}(i\omega)\alpha_{\beta\beta}^{(\mathrm{b})}(i\omega)\,\mathrm{d}\omega. \qquad (3.40^*)$$

† Quantities in starred equations are in atomic units. See Appendix I for further details.

In these circumstances the product of the dipole–dipole interaction tensors reduces to the corresponding square term $(T_{12}^{\alpha\beta})^2$ or trace $(\boldsymbol{T}\cdot\boldsymbol{T})$ which is 6 for isotropic species. Some consequences of polarization anisotropy, particularly in relation to adsorption problems, were first discussed by de Boer and Heller (1937); we shall return to this question in more detail below.

The mean polarizability, $\bar{\alpha} = \frac{1}{3}$ trace $(\boldsymbol{\alpha})$, in an isotropic molecule may be expressed as a sum over the states s, involving transition frequencies and oscillator strengths f_s which can be obtained from X-ray data (Leonard and Barker, 1975). Assuming that this procedure can be extended to anisotropic molecules, the α-direction component of the polarizability is

$$\alpha_{\alpha\alpha}(i\omega) = \sum_s \frac{f_{s\alpha}}{(\omega_{s\alpha}^2 + \omega^2)}. \tag{3.41}$$

A table comparing values of $A^{(a,b)}$ for rare gas pairs derived by different authors using (3.41) with (3.40) has been given by Leonard and Barker (1975).

A further simplification can be made if it is possible to replace the finite sum by a single composite term. Such a replacement would be justified, for example, if one oscillator strength predominates over the others. An approximation for the static polarizability may then be written

$$\alpha_{\alpha\alpha} = \frac{F_{\alpha\alpha}}{\bar{\omega}_\alpha^2 + \omega^2}. \tag{3.42}$$

With this approximation (3.40) becomes

$$A_{\alpha\beta}^{(a,b)} = \frac{\frac{1}{4}F_{\alpha\alpha}^{(a)}F_{\beta\beta}^{(b)}}{\bar{\omega}_\alpha^{(a)}\bar{\omega}_\beta^{(b)}(\bar{\omega}_\alpha^{(a)} + \bar{\omega}_\beta^{(b)})}. \tag{3.43*}$$

Two useful limiting expressions are obtained as further approximations from (3.43):

(i) $\omega \to 0$. In this limit Eqn (3.42) becomes $F_{\alpha\alpha} = \alpha_{\alpha\alpha}\bar{\omega}_\alpha^2$ and (3.43) becomes the London formula,

$$A_{\alpha\beta,\mathrm{LOND}}^{(a,b)} = \frac{\alpha_{\alpha\alpha}^{(a)}\alpha_{\beta\beta}^{(b)}\bar{\omega}_\alpha^{(a)}\bar{\omega}_\beta^{(b)}}{4(\bar{\omega}_\alpha^{(a)} + \bar{\omega}_\beta^{(b)})}. \tag{3.44*}$$

An alternative form which neglects the directional properties of $\bar{\omega}$, is obtained by replacing $\bar{\omega}$ by the ionization potential I. Buckingham (1965) arrives at this result by a different route which suggests that the neglect of directional properties in $\bar{\omega}$ may not be serious.

(ii) $\omega \to \infty$. It has been shown (Mavroyannis and Stephen, 1962) that for this limit the isotropic polarizability is $\bar{\alpha}(i\omega) = N/\omega^2$ where N is the number of electrons in the molecule (usually interpreted as "those which contribute to the dispersion interactions"). If this approximation is also extended to anisotropic systems, the Slater–Kirkwood formula is obtained:

$$A^{(a,b)}_{\alpha\beta,S-K} = \frac{\frac{1}{4}\alpha^{(a)}_{\alpha\alpha}\alpha^{(b)}_{\beta\beta}}{[(\alpha^{(a)}_{\alpha\alpha}/N_a)^{1/2} + (\alpha^{(b)}_{\beta\beta}/N_b)^{1/2}]}. \tag{3.45a*}$$

A third approximation, the Kirkwood–Müller formula, can be derived from this result by introducing the diamagnetic susceptibilities χ via the approximate formula

$$N^{1/2} = \frac{4\chi m_e^{3/2} c_0^2 \varepsilon_0}{\hbar e \alpha^{1/2}}. \tag{3.45b}$$

It is clear that the London and Slater–Kirkwood approximations represent opposite extremes to the correct value of A. Calculations, particularly for rare gases, using refractive index and oscillator strength data in Eqn (3.41) seem to support this view (Amos and Crispin. 1976).

The higher-order pairwise interaction terms may be obtained by including contributions additional to the dipole–dipole interaction in the multipole expansion. These give rise to a series of terms resembling (3.38) in which the higher-order interaction tensors and polarizabilities appear (e.g. Buckingham, 1965). If the molecules are able to rotate freely, only even powers of r_{12}^{-1} appear in the expansion, however Buckingham (1965, 1967) has shown that inverse 7th power and higher-order contributions can be significant if rotations are hindered and it is possible that this may be an important consideration in adsorption systems. Approximate expressions for the coefficients of r_{12}^{-8} and r_{12}^{-10} have been derived by Kiselev and Poshkus (1958) as an extension of the Kirkwood–Müller equation, and have been given for the London equation by Hornig and Hirschfelder (1952).

The theory of pairwise dispersion forces discussed up to this point includes only attractive interactions between molecules, and cannot therefore account for the existence of an equilibrium distance of separation as evidenced, for example, by the finite volume and low compressibility of solids. This region of the intermolecular potential is particularly difficult to treat theoretically and a variety of techniques have been developed which may roughly be considered as being range-dependent. Thus at the larger separations the perturbation theory which has been modified for the possibility of electron exchange, and which was referred to earlier, can provide some information about the energy minimum. At very close separations a wave mechanical, united-atom, method can be used to investigate the repulsive side of the potential well (Stevens *et al.*, 1974; Wagner *et al.*, 1974). Perhaps the most interesting methods are those employing a conserved density approximation in which the total electron density of a pair of atoms is assumed equal to the sum of the electron densities of the individual atoms. This approach has been developed with some success by Gordon and Kim (1972) who included correlation energy and found good agreement with empirical potentials in the region of the minimum. Rae (1973, 1975) has improved the method in the long-range part of the interaction; further modifications have been made by Cohen and Pack (1974). With these extensions the Gordon and Kim method is capable of describing the whole of the interaction. Its weakness lies in the need for accurate wave functions and for adjustment at larger separations. Applications to adsorbate–adsorbent interaction have not yet achieved the level of success reached by the best summation calculations (Bennett, 1974; Freeman, 1975).

Evidently a purely theoretical calculation of the whole of the potential energy curve for dispersion interactions is a formidable task and the best compromise solution probably lies in the development of expressions which exploit available theory and experimental data. Computer simulations of molecular assemblies play a central role in this requirement because, on the one hand, relatively simple potential functions are desirable where long repetitive calculations are undertaken and, on the other hand, results from simulations can often provide the impetus to improve existing potential functions. The asymmetric feature of adsorption systems may prove to be particularly demanding in this respect.

Wave mechanical treatments of the purely repulsive part of the interaction potential in isolation have led to the conclusion that an exponential dependence of E_R on separation is the most appropriate choice. Nevertheless inverse nth power functions are still widely employed and prove to be very satisfactory in applications to properties which are not too sensitive to the repulsive part of the potential well. In practice these include most equilibrium and many transport properties of molecular assemblies. However, this consideration does not extend to scattering experiments, and consequently these are especially useful in developing empirical expressions to represent the repulsive part of the interaction function. The combination of these expressions with carefully calculated coefficients for the attractive dispersion forces has led to the development of accurate semi-empirical, semi-theoretical potentials.

Amongst the most successful of these is that developed by Barker and co-workers (see, e.g., Barker *et al.*, 1971) for rare gas interactions. This has the form

$$E_{\mathrm{DR}}^{(\mathrm{a,a})} = E^{(\mathrm{a,a})}\left[\sum_{i=0}^{5} B_i^{(\mathrm{a,a})}(R-1)^i \exp[-\kappa^{(\mathrm{a,a})}(1-R)] - \sum_{j=0}^{2} \mathrm{C}_{2j+6}^{(\mathrm{a,a})}/(\mathrm{R}^{2j+6}+\delta)\right]. \quad (3.46)$$

In this expression R is $(r_{12}/r_{\min}^{(a,a)})$ where $r_{\min}^{(a,a)}$ is the separation at the potential minimum. The first summation accounts for the repulsive part of the potential and incorporates the accumulated evidence that this is better represented by an exponential type of function than by inverse power terms of higher order. The second summation in (3.46) contains essentially the fluctuating multipole interactions for which the coefficients were found using X-ray data (the constant $\delta = 0.01$ is necessary to eliminate a spurious minimum in the repulsive part of the potential). Values for the constants appearing in (3.46) have been given by Barker (1975). The equation might be regarded as a development of previous simplified forms such as the Buckingham exp–6, Lennard-Jones 12–6 or the extended forms used by Kiselev and co-workers (AKLP) (Avgul *et al.*, 1970).

For interactions between an unlike pair (a, b) at $(\boldsymbol{r}_1, \boldsymbol{r}_2)$ these may be written respectively as

$$\text{exp–6(B):}\ E_{\mathrm{DR}}^{(a,b)} = \frac{\varepsilon^{(a,b)}}{1 - 6/B^{(a,b)}}\left[\frac{6}{B^{(a,b)}}\exp[-\kappa^{(a,b)}(1-R)] - R^{-6}\right] \qquad (3.47)$$

$$\text{12–6(L}-\text{J):}\ E_{\mathrm{DR}}^{(a,b)} = 4\varepsilon^{(a,b)}\left[\left(\frac{\sigma^{(a,b)}}{r_{12}}\right)^{12} - \left(\frac{\sigma^{(a,b)}}{r_{12}}\right)^{6}\right] \qquad (3.48)$$

$$\text{AKLP:}\ E_{\mathrm{DR}}^{(a,b)} = B_0^{(a,b)}\exp[-\kappa^{(a,b)}(1-R)] - \sum_{j=0}^{2} C_{2j+6}^{(a,b)}/R^{2j+6} \qquad (3.49)$$

where $\sigma^{(a,b)}$ is the hard-sphere diameter and the potential minimum in (3.47) and (3.48) is at $-\varepsilon^{(a,b)}$. $B_0^{(a,b)}$, $C_{2j+6}^{(a,b)}$ and $\kappa^{(a,b)}$ are constants which do not necessarily have the same values as those used in (3.46). (Note that an induction energy term has been omitted from the AKLP equation.)

In contrast to these earlier equations, Eqn (3.46) has been shown to be close to being a "true" pairwise potential (Barker *et al.*, 1971). The former apparently incorporate three-body and higher-order contributions quite satisfactorily, at least as far as liquid state properties are concerned, but since some of this success may be due to the uniformity of the environment in bulk liquids it is necessary to regard effective potentials with caution in accurate work with non-uniform systems. Further advances in the development of accurate dispersion potentials have been made by Koide and co-workers (1980).

Pairwise dispersion energy equations for species other than rare gases are less reliable and it appears that atom–atom and/or bond–bond pairwise summations of the type discussed below often offer the most useful route towards a suitable intermolecular potential (Hirschfelder *et al.*, 1961).

The C–C interaction for non-bonded carbon atoms in graphite is of particular interest in physical adsorption work, and has been discussed in

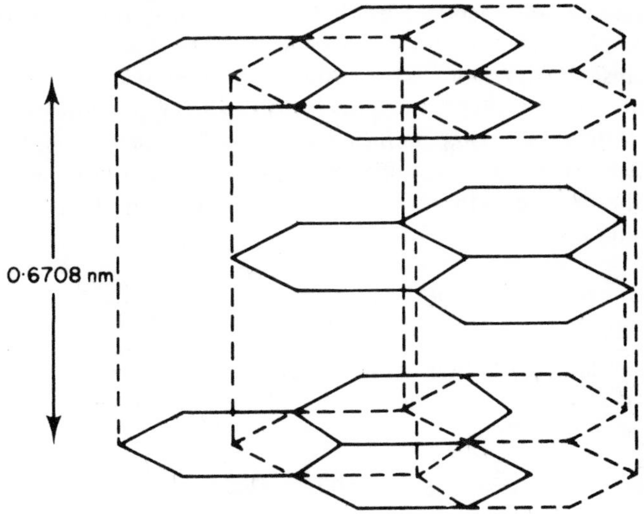

Fig. 3.1. XYXY stacking in graphite.

detail by Ricca and co-workers (Pisani *et al.*, 1973). The most stable arrangement for the stacked layers of carbon hexagons in graphite is found experimentally to be the XYXY sequence illustrated in Fig 3.1, but XYZ and XXX sequences are possible alternatives. In addition to this observation, experimental data are available for the layer separation (0·3354 nm), the cohesion energy ($-0{\cdot}238\ \mathrm{J\ m^{-2}}$) and the compressibility normal to the basal planes ($2{\cdot}75\ 10^{-11}\ \mathrm{m^2\ N^{-1}}$). Using pairwise summation over the carbon atoms, it was found that neither the exp–6 nor the 12–6 potentials could satisfactorily account for the data, since XYXY and XYZ stacking gave identical lattice energies with these simple potential functions. This state of affairs persisted when the observed anisotropy of polarizability of the carbon in graphite was taken into account. Ricca and co-workers declare a preference for the exp–6 potential; it is of interest that the extended forms of this equation, Eqns (3.46) and (3.48) have not been subjected to the test of accounting for the observed properties of graphite.

(c) Combining rules

To calculate $u^{[1]}$ it is necessary to know the dispersion energy between two unlike molecules. From theoretical expressions such as (3.46) and (3.48) the calculation of the interaction potential between unlike species is, in principle, no more difficult than for identical species. However, when semi-empirical potentials are employed, combining rules are needed to obtain the parameters.

If we consider Eqn (3.46) as the basic form of expression for the combined dispersion and repulsive contributions to the pair potential energy, the problem is to replace the coefficients $B_i^{(a,a)}$, $C_{2j+6}^{(a,a)}$ for a single species by appropriate expressions for unlike pairs. For example, for the coefficient $C_6^{(a,b)}$ of the inverse sixth power term we can use Eqn (3.43) to eliminate $F_{\alpha\alpha}^{(a)}$, $F_{\beta\beta}^{(b)}$ and obtain

$$C_6^{(a,b)} = [C_6^{(a,a)} C_6^{(b,b)}]^{1/2} \left[\frac{r_{\min}^{(a,a)} r_{\min}^{(b,b)}}{(r_{\min}^{(a,b)})^2}\right]^3 \left[\frac{2(\tilde{\omega}^{(a)}\tilde{\omega}^{(b)})^{1/2}}{\tilde{\omega}^{(a)} + \tilde{\omega}^{(b)}}\right] \qquad (3.50^*)$$

which is similar in form to the result derived by Crowell (1961). It is usually considered satisfactorily to take the combining rule for $r_{\min}^{(a,b)}$ (or an analogous form for $\sigma^{(a,b)}$) as

$$r_{\min}^{(a,b)} = \tfrac{1}{2}(r_{\min}^{(a,a)} + r_{\min}^{(b,b)}). \qquad (3.51)$$

With this choice for $r_{\min}$, the first square bracket in (3.50) can be expanded to give

$$\left[\frac{r_{\min}^{(a,a)} r_{\min}^{(b,b)}}{(r_{\min}^{(a,b)})^2}\right]^3 = 1 - \tfrac{3}{4}Y^2 + \tfrac{1}{2}Y^3 + \tfrac{3}{16}Y^4 + \ldots \qquad (3.52)$$

where $(r_{\min}^{(a,a)}/r_{\min}^{(b,b)} - 1) = Y$ and a, b are assigned to make $Y > 0$. Clearly this term is fairly close to unity in many cases of interest. Similarly, the second square bracket can be expanded in the form

$$\left[\frac{2(\tilde{\omega}^{(a)}\tilde{\omega}^{(b)})^{1/2}}{\tilde{\omega}^{(a)} + \tilde{\omega}^{(b)}}\right] = 1 - \tfrac{1}{8}W^2 + \tfrac{1}{8}W^3 + \tfrac{11}{128}W^4 + \ldots \qquad (3.53)$$

(where $|(\tilde{\omega}^{(a)}/\tilde{\omega}^{(b)} - 1)| = W$); and is once again seen to be close to unity provided that W is not too large.

When anisotropy of polarizability is to be taken into account a similar approach to that outlined above, but using the London equation (Eqn (3.44)) can be employed. In this case $\tilde{\omega}$ is eliminated between expressions for like and unlike pairs to give an equation which contains the components $\alpha_{\alpha\alpha}$, etc., of the polarizability tensor,

$$C_6^{(a,b)} = \frac{2C_6^{(a,a)} C_6^{(b,b)} \alpha_{\alpha\alpha}^{(a)} \alpha_{\beta\beta}^{(b)}}{Y^3 C_6^{(a,a)} [\alpha_{\alpha\alpha}^{(a)}]^2 + Y^{-3} C_6^{(b,b)} [\alpha_{\beta\beta}^{(b)}]^2} \left[\frac{r_{\min}^{(a,a)} r_{\min}^{(b,b)}}{(r_{\min}^{(a,b)})^2}\right]^3. \qquad (3.54^*)$$

This is equivalent to the expression employed by Pisani *et al.*, (1973) since their A_{GC} can be equated to $(r_{\min}^{(a,b)})^6 C_6^{(a,b)}$. Procedures similar to these may be applied to $C_8^{(a,b)}$ and $C_{10}^{(a,b)}$ coefficients.

For the coefficients B, in the repulsive part of the potential a geometric mean combining rule, $B^{(a,b)} = (B^{(a)}B^{(b)})^{1/2}$ is usually adopted where simple forms such as the exp–6 or 12–6 potential are employed. For purely repulsive interaction between rare gases this proposal does not appear to be very satisfactory, but an alternative form based on the conserved density model of Gordon and Kim (1972) shows no significant improvement (Murrell, 1976).

Maitland and Wakeham (1978) point out that the geometric mean combining rule, although valid in the asymptotic r_{12}^{-6} region for pairwise dispersion interactions, cannot be expected to hold over the whole range for an empirical potential such as the Lennard-Jones 12–6 potential, characterized by only two parameters $\varepsilon^{(a,b)}$ and $\sigma^{(a,b)}$. They find that a harmonic mean combining rule (derivable from the London formula, Eqn (3.44) if $\varepsilon^{(a,b)}$ is proportional to $A^{(a,b)}/(\alpha^{(a)}\alpha^{(b)})$, and given by

$$\varepsilon^{(a,b)} = 2\varepsilon^{(a,a)}\varepsilon^{(b,b)}/[\varepsilon^{(a,a)} + \varepsilon^{(b,b)}] \tag{3.55}$$

is more successful for a number of rare gas pairs. A similar conclusion may be drawn from comparisons with scattering data on He and Ne rare gas pairs, but it must be stressed that the agreement with experimental data is often no better than within 10% even when the harmonic rule is used.

(d) Electrostatic and inductive pair interactions

The contributions from dispersion forces and from repulsion comprise the most important part of the interaction energy in the sense that they occur in any system and can rarely be neglected, except perhaps where the interaction is between ionic species. The remaining components of the interaction E_c and E_I are of course only relevant where permanent charge distributions are present. Their treatment can be approached via the classical multipole expansion. For the interaction of a polar molecule with an external field having a potential ϕ, the electrostatic part of the interaction potential at r can be written from Taylor expansion, in the form

$$E_c = q\phi + \mu_\alpha \frac{\partial \phi}{\partial r_\alpha} + \frac{1}{2!} Q_{\alpha\beta} \frac{\partial^2 \phi}{\partial r_\alpha \partial r_\beta} + \frac{1}{3!} R_{\alpha\beta\gamma} \frac{\partial^3 \phi}{\partial r_\alpha \partial r_\beta \partial r_\gamma} \tag{3.56*}$$

where q is the ionic charge and the permanent dipole, quadrupole, octopole, etc., moments belong to a series of general $2m$-pole moments $M_{\alpha\beta\gamma\ldots}$ defined by the mth degree tensors:

$$M_{\alpha\beta\gamma\ldots\mu} = \sum e_i s_{i\alpha} s_{i\beta} s_{i\gamma} \ldots s_{i\mu} \tag{3.57*}$$

where $s_{i\alpha}$ is the α-component of the vector distance s_i, in the presence of the field, of the ith charge e_i from the origin. The origin is taken to be within the molecule, and such that s is small compared with $\boldsymbol{r}$. Often it is convenient to define the multipoles in normalized form with zero trace so that the quadrupole moment tensor, for example, becomes

$$\Theta_{\alpha\beta} = \frac{1}{2}\sum_i e_i(3s_{i\alpha}s_{i\beta} - s_i^2\delta_{\alpha\beta}) \qquad (3.58^*)$$

Unless the particles are ions, the dipole moment remains unchanged under normalization and all orders of multipole moments, from the dipole moment with $m = 1$ upwards, are given by the general definition (Stogryn and Stogryn, 1966)

$$N_{\alpha\beta\ldots\mu} = \frac{(-1)^m}{m!}\sum_i e_i s^{2m+1}\frac{\partial^m}{\partial s_\alpha \partial s_\beta \ldots \partial s_\mu}\left(\frac{1}{s}\right). \qquad (3.59^*)$$

By employing Laplace's equation, $\partial^2\phi/\partial r_\alpha^2 = 0$, and its higher-order analogues (Böttcher, 1973), it can be verified that Eqn (3.56) becomes

$$E_c = q\phi + \mu_\alpha\frac{\partial\phi}{\partial r_\alpha} + \frac{1}{3}\Theta_{\alpha\beta}\frac{\partial^2\phi}{\partial r_\alpha \partial r_\beta} + \frac{1}{15}\Omega_{\alpha\beta\gamma}\frac{\partial^3\phi}{\partial r_\alpha \partial r_\beta \partial r_\gamma}. \qquad (3.60^*)$$

In many cases of interest the number of components required to specify a multipole is considerably reduced due to the symmetry of the system. For example, in molecules with axial symmetry the off-diagonal elements of $\Theta_{\alpha\beta}$ are zero and $\Theta_{xx} = \Theta_{yy} = -\frac{1}{2}\Theta_{zz}$, so that only one quantity, usually $\Theta = \frac{1}{2}\Theta_{zz}$, is required to specify the quadrupole moment. In general, if a molecule possesses an n-fold axis of symmetry only one scalar quantity is required to determine any multipole moment with a tensor of rank less than n (Stogryn and Stogryn, 1966). Similar considerations apply to the polarizability tensors and the number of constants required in their specification is determined by the symmetry group of the system in question (Buckingham, 1967). The presence of an external field also causes a distortion of the charge distribution in a molecule, giving rise to an induced potential which can be written as

$$E_{\mathrm{I}} = -\tfrac{1}{2}\alpha_{\alpha\beta}F_\alpha F_\beta - \tfrac{1}{3}A_{\alpha,\beta\gamma}F_\alpha F_{\beta\gamma} - \ldots \qquad (3.61^*)$$

where $F_\alpha = -(\partial\phi/\partial r_\alpha)$, $\alpha_{\alpha\beta}$ is the dipole polarizability tensor and higher-order terms involve higher polarizabilities. The polarizabilities have a wave mechanical interpretation as expectation values of the total (i.e. field-dependent) multipole moments. Usually E_{I} is the smallest part of the long-range interaction energy so that the customary neglect of the higher-order and hyperpolarizability contributions would appear to be justified as a first approximation. However, it is to be noted that, when the surface of an ionic solid relaxes, the fields may become large and these terms may no longer be negligible (see § 3(e)).

The external field terms appearing in Eqns (3.56), (3.60) can be considered from two points of view in dealing with adsorption systems. Either the total field due to an adsorbent, which contains ionic or polar species, may be calculated, or else the pair interactions may be written down and subsequently summed over all pairs. For the latter approach, and also for consideration of adsorbate intermolecular interactions, it is necessary to have expressions for the relevant pair interactions. Here the external field is that due to another molecule. The electrostatic interaction energy is

$$\begin{aligned} E_{\mathrm{c}}^{(\mathrm{a,b})} = {} & \frac{q^{(\mathrm{a})}q^{(\mathrm{b})}}{r_{12}} - T_{12}^{\alpha}(q^{(\mathrm{a})}\mu_{\alpha}^{(\mathrm{b})} - q^{(\mathrm{b})}\mu_{\alpha}^{(\mathrm{a})}) \\ & - T_{12}^{\alpha\beta}(\mu_{\alpha}^{(\mathrm{a})}\mu_{\beta}^{(\mathrm{b})} - \tfrac{1}{3}q^{(\mathrm{b})}\Theta_{\alpha\beta}^{(\mathrm{a})} - \tfrac{1}{3}q^{(\mathrm{a})}\Theta_{\alpha\beta}^{(\mathrm{b})}) \\ & + T_{12}^{\alpha\beta\gamma}(\tfrac{1}{3}\mu_{\alpha}^{(\mathrm{a})}\Theta_{\beta\gamma}^{(\mathrm{b})} - \tfrac{1}{3}\mu_{\alpha}^{(\mathrm{b})}\Theta_{\beta\gamma}^{(\mathrm{a})} - \tfrac{1}{15}q^{(\mathrm{a})}\Omega_{\alpha\beta\gamma}^{(\mathrm{b})} + \tfrac{1}{15}q^{(\mathrm{b})}\Omega_{\alpha\beta\gamma}^{(\mathrm{a})}) \\ & - \ldots \end{aligned} \qquad (3.62^*)$$

in which the multipole–multipole interaction tensors were given in Eqn (3.39). The terms involving the ionic charge of course vanish from (3.62) when it is applied to the interaction between neutral molecules. If two dipolar, axially symmetric molecules interact, the leading term in (3.62) is that containing the dipole–dipole term and can be written in the familiar form involving the angles θ_1, θ_2 between the molecular axes and $\boldsymbol{r}_{12}$, and the dihedral angle φ between projections of the molecular axes onto a plane normal to $\boldsymbol{r}_{12}$. Writing $\cos\theta_1 = c_1$, $\sin\theta_1 = s_1$, etc., we have:

$$E_{\mathrm{c}}^{(\mathrm{a,b})} = \mu^{(\mathrm{a})}\mu^{(\mathrm{b})}[-2c_1c_2 + s_1s_2\cos\varphi]/r_{12}^3. \qquad (3.63^*)$$

The induction energy for a pair of molecules is

$$E_{\mathrm{I}}^{(\mathrm{a,b})} = -\tfrac{1}{2}\alpha_{\alpha\beta}^{(\mathrm{a})}F_{\alpha}^{(\mathrm{b})}F_{\beta}^{(\mathrm{b})} - \tfrac{1}{2}\alpha_{\alpha\beta}^{(\mathrm{b})}F_{\alpha}^{(\mathrm{a})}F_{\beta}^{(\mathrm{a})} \qquad (3.64^*)$$

where $F_{\alpha}^{(\mathrm{b})}$, etc., is the field at the origin of molecule a due to the permanent moments of molecule b and is given by

$$F_{\alpha}^{(\mathrm{b})} = T_{12}^{\alpha}q^{(\mathrm{b})} + T_{12}^{\alpha\beta}\mu_{\beta}^{(\mathrm{b})} + \tfrac{1}{3}T_{12}^{\alpha\beta\gamma}\Theta_{\beta\gamma}^{(\mathrm{b})}. \qquad (3.65^*)$$

The higher-order terms omitted from (3.64) involve hyperpolarizabilities. Again the interaction of two linear dipolar molecules furnishes a familiar example in which the interaction energy of the molecules of type a at $\boldsymbol{r}_1$ due to the molecule of type b at $\boldsymbol{r}_2$ is given by

$$\begin{aligned} E_{\mathrm{I}}^{(\mathrm{a})} = {} & -\tfrac{1}{2}\alpha^{(\mathrm{a})}(\mu^{(\mathrm{b})})^2(3c_1^2 + 1)r_{12}^{-6} - \tfrac{1}{6}(\alpha_{\parallel}^{(\mathrm{a})} - \alpha_{\perp}^{(\mathrm{a})}) \\ & \times (\mu^{(\mathrm{b})})^2[12(c_1c_2)^2 + 3(s_1s_2\cos\varphi)^2 - 3c_1^2 - 1 - 12s_1c_1s_2c_2\cos\varphi]r_{12}^{-6}. \end{aligned} \qquad (3.66^*)$$

The first term in this equation was used by Kiselev and co-workers (1958) to account for the interaction between dipolar species and the carbon atoms in graphite; their neglect of the second term amounts to the assumption that the graphite polarizability is isotropic. They further assumed that the angle dependence could be averaged over all orientations to give $\langle c_1^2 \rangle = \frac{1}{3}$. Since the induction contribution is often small compared to the total interaction energy, the error due to these assumptions may not be serious.

(e) Calculation of the summation terms in $u^{[1]}$

Having established suitable pairwise potentials for the interaction of two dissimilar molecules, the first summation in Eqn (3.26) can be evaluated. The feasibility of pairwise summation in many-body systems has been discussed by Amos and Crispin (1976) who propose wave mechanical criteria for making a subdivision of the system into regions of localised charge distribution. In many cases these can be reasonably taken as being centred on the atomic nuclei, although, in the case of hydrocarbons in particular, orbitals located in the vicinity of the bonds are probably a better choice. Pairwise summation in the specific context of adsorption systems has also been considered by Hobson (1974) who concluded that the approach should be applicable to rare gas–graphite systems in spite of the collective effects of the intraplanar electrons.

Often the adsorbent possesses regular periodic properties in the x–y plane, parallel to the interface. This would be the case, for example, for

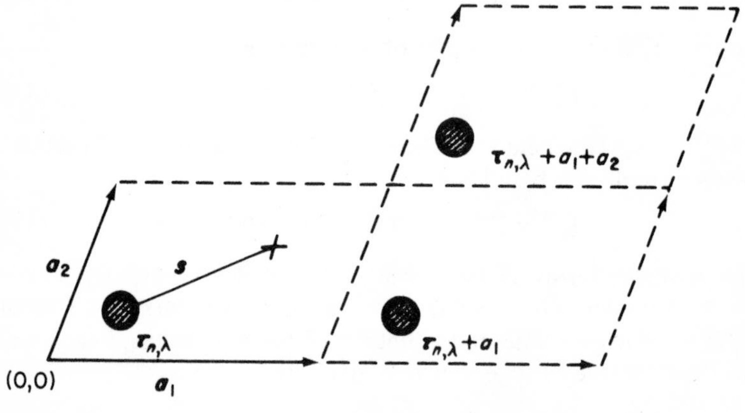

Fig. 3.2. A two-dimensional cell in plane λ showing an adsorbent atom at $\tau_{n,\lambda}$ which occurs in every cell; no other atoms are shown. The position of the adsorbate atom τ in the x–y plane is shown as a projection at + onto the cell containing the origin. The vector s is shown for the case $L_{1,n,\lambda} = L_{2,n,\lambda} = 0$.

a solid adsorbent, if it can be assumed to be unperturbed in this plane by the proximity of adsorbate atoms. The pairwise summation procedures can then be formalized by introducing a lattice vector $\boldsymbol{L}$ which automatically generates the positions of the adsorbent atoms.

We consider a plane of atoms or ions at a distance z_λ from the adsorbate, where $\lambda = 1, 2, \ldots$, and define 2-D unit cells by the vectors $\boldsymbol{a}_1, \boldsymbol{a}_2$. The corresponding reciprocal vectors $\boldsymbol{b}_j\,(j = 1, 2)$ are defined such that $\boldsymbol{a}_i \cdot \boldsymbol{b}_j = \delta_{ij}$. The adsorbate is placed at a position $(\boldsymbol{\tau}, z_\lambda)$ above the cell containing the origin and we consider its interaction with the nth type of atom or ion in the cell (Fig. 3.2) which is positioned at $\boldsymbol{\tau}_{n,\lambda}$.

The pairwise potential for a single adsorbate–adsorbent pair depends on the distance R between them. This can be written as $R = (s^2 + z_\lambda^2)^{1/2}$, where s is the magnitude of a vector $\boldsymbol{s} = (\boldsymbol{\tau}_{n,\lambda} + \boldsymbol{L} - \boldsymbol{\tau})$ and the lattice vector $\boldsymbol{L}$ is designated by $\boldsymbol{a}_i$ and a set of integers $\{L_{i,n,\lambda}\}$ chosen so as to place adsorbent atoms in appropriate cells. Thus

$$\boldsymbol{L} = L_{1,n,\lambda}\boldsymbol{a}_1 + L_{2,n,\lambda}\boldsymbol{a}_2. \tag{3.67}$$

It has been seen that components of the pairwise potential are either of the form

$$P_n^{(\mathrm{a,b})}(s, z_\lambda) = R^{-q} \tag{3.68a}$$

or

$$P_n^{(\mathrm{a,b})}(s, z_\lambda) = R^{q}\exp(-BR). \tag{3.68b}$$

The corresponding components for a whole plane of adsorbent are

$$P_n^{(\mathrm{a,b})}(\boldsymbol{r}_\lambda) = \sum_{L_{1,n,\lambda};\,L_{2,n,\lambda}} P_n(s, z_\lambda) \tag{3.69}$$

where $\boldsymbol{r}_\lambda$ here stands for $(\boldsymbol{\tau}, z_\lambda)$. The component of the potential at $\boldsymbol{r} = (\boldsymbol{\tau}, z)$ for all the type-n atoms or ions in the lattice may be found by summing P_n over $\lambda = 1, 2, \ldots$,

$$\mathcal{P}_n^{(\mathrm{a,b})}(\boldsymbol{r}) = \sum P_n^{(\mathrm{a,b})}(\boldsymbol{r}_\lambda). \tag{3.70}$$

It is to be noted that these procedures could be generalized to three- and higher-body interactions of the type (a, a, . . . , b), where a here designates adsorbent and b adsorbate, by appropriate subscripting of $R, \boldsymbol{s}, \boldsymbol{L}, n$ and λ and by nesting the summations of Eqn (3.69) for appropriate lattice vectors and sets of planes.

The computational effort required for pairwise summation can be considerably reduced by the use of Fourier transform methods (Steele, 1973; Price 1974). This is particularly desirable for long-range interactions such as those due to ionic charges in a crystal lattice, since pairwise summation is otherwise very slowly convergent, and it is similarly advantageous for the shorter-range dispersion interactions. For the very short-range repulsive interactions both inverse power and exponential forms can be evaluated by Fourier methods. However it should be borne in mind that pairwise summation is less easy to justify for repulsive forces and that only a very few terms need to be included in the sums in this case.

A Fourier representation of the pairwise interaction of the adsorbate with a whole plane of adsorbent atoms may be written

$$P_n^{(a,b)}(\boldsymbol{r}_\lambda) = \sum_{\boldsymbol{k}} \rho_{\boldsymbol{k},n}(z_\lambda) \exp[i\boldsymbol{k} \cdot (\boldsymbol{\tau}_{n,\lambda} - \boldsymbol{\tau})] \tag{3.71}$$

in which the wave vector $\boldsymbol{k}$ is related to the reciprocal lattice vectors by

$$\boldsymbol{k} = 2\pi(l\boldsymbol{b}_1 + m\boldsymbol{b}_2). \tag{3.72}$$

The Fourier coefficients are given by

$$\rho_{\boldsymbol{k},n}(z_\lambda) = \frac{1}{a_c} \sum_{L_{i,n,\lambda}} \exp(i\boldsymbol{k} \cdot \boldsymbol{L}) \int P_n^{(a,b)}(\boldsymbol{s}, z_\lambda) \exp(-i\boldsymbol{k} \cdot \boldsymbol{s})\, \mathrm{d}\boldsymbol{s} \tag{3.73}$$

where $a_c(=\boldsymbol{a}_1 \times \boldsymbol{a}_2)$ is the area of a unit cell and also the domain of the integral, and Eqn (3.69) has been introduced. From (3.67) and (3.72), $\exp(i\boldsymbol{k} \cdot \boldsymbol{L})$ is unity and the summation over the integral in (3.73) can be replaced by an integral over the domain of the whole plane of adsorbent particles.

Since the problem is symmetrical in the x–y plane, it is convenient to transform to cylindrical co-ordinates (s, φ, z_λ) with the z-axis passing through the adsorbent atom and the x and y axes chosen such that $\boldsymbol{k} \cdot \boldsymbol{s} = ks \cos \varphi$. Then for a plane of infinite extent, Eqn (3.73) becomes

$$\begin{aligned} \rho_{\boldsymbol{k},n}(z_\lambda) &= \frac{1}{a_c} \int_0^\infty P_n^{(a,b)}(s, z_\lambda) s\, \mathrm{d}s \int_0^{2\pi} \exp(-iks \cos \varphi)\, \mathrm{d}\varphi \\ &= \frac{2\pi}{a_c} \int_0^\infty J_0(ks) P_n^{(a,b)}(s, z_\lambda) s\, \mathrm{d}s \end{aligned} \tag{3.74}$$

where J_0 is a zeroth-order Bessel function.

When $P_n^{(a,b)}$ has the reciprocal qth power dependence of (3.68a), this integral can be expressed in terms of mth-order modified Bessel functions of the second kind, K_m, as

$$\rho_{\boldsymbol{k},n}(z_\lambda) = \frac{2\pi}{a_c[(q-2)/2]!} \left(\frac{k}{2z_\lambda}\right)^{(q-2)/2} K_{(q-2)/2}(kz_\lambda) \tag{3.75}$$

and the appropriate component of the pairwise interaction potential for a whole plane of atoms is the real part of (3.71),

$$P_n^{(a,b)}(\boldsymbol{r}_\lambda) = \sum_k \frac{2\pi}{a_c[(q-2)/2]!}\left(\frac{k}{2z_\lambda}\right)^{[(q-2)/2]} K_{[(q-2)/2]}(kz_\lambda)$$
$$\times \exp[i\boldsymbol{k}\cdot(\boldsymbol{\tau}_{n,\lambda} - \boldsymbol{\tau})]. \qquad (3.76)$$

With $q = 1$, Eqn (3.76) when multiplied on the right-hand side by a factor of valency × electronic charge (e), expresses the coulombic potential at z_λ due to the nth type of ion in the plane λ. Here the $\frac{1}{2}$-order modified Bessel function is

$$K_{-1/2}(kz_\lambda) = K_{1/2}(kz_\lambda) = \left(\frac{\pi}{2kz_\lambda}\right)^{1/2} e^{-kz_\lambda}.$$

The singularity at $k = 0$ can be removed by imposing the condition of charge neutrality which causes positive and negative contributions to cancel out so that only terms with $k > 1$ need to be included in the sum in (3.76). Steele (1974) has applied this approach to the NaCl lattice and shown that it leads to the following result, originally derived by Lennard-Jones and Dent (1928) using a different method, for the electrostatic field $\phi(\boldsymbol{r})$ outside the (100) plane,

$$\phi(\boldsymbol{r}) = \frac{4e}{a}\left[\sum \frac{(-1)^{(l+m)/2}}{\sqrt{l^2+m^2}} \cos 2\pi\left(\frac{lx}{a} + \frac{my}{a} - \frac{l+m}{4}\right) \times \frac{\exp[-2\pi\sqrt{l^2+m^2}z/a]}{1+\exp(-\pi\sqrt{l^2+m^2})}\right] \qquad (3.77a)$$

where $a = |\boldsymbol{a}_1| = |\boldsymbol{a}_2|$, and the summation is over odd values of l, m only. Convergence of the summation is very rapid, and to within an accuracy of 1% Eqn (3.77a) can be approximated by its leading term, giving

$$\phi(\boldsymbol{r}) = \frac{16e}{\sqrt{2}a}\frac{\exp(-2\sqrt{2}\pi z/a)}{1+\exp(-\pi\sqrt{2})}\cos(2\pi x/a)\cos(2\pi y/a). \qquad (3.77b)$$

The derivatives required to obtain the electric field terms from Eqn (3.77b) have been given for several directions by Gready *et al.* (1978).

In more recent work, allowance has been made for relaxation of the crystal lattice in accordance with the calculations of Benson and co-workers (1957, 1967). These calculations, carried out for a number of mono-monovalent solids (in the absence of adsorbate), took into account the displacement of ions normal to the surface in layer λ and also the dipoles induced in the ions by the local electric field. It was found that, with the exception of KCl, the displacements of ions in the (100) face in layer λ, $(z_\lambda^+, z_\lambda^-)$ of the positive and negative ions were in opposite directions (the positive ion being displaced outwards) and large enough to have a considerable effect on the surface energy. Modified forms of the Lennard-Jones and Dent equation which include allowance for this behaviour have been derived (House and Jaycock, 1975; BenEphraim and Folman, 1976) and used in the estimation of Ar,Kr,H_2 and N_2 potentials at the surface of NaCl and KCl crystals. It should be emphasized, however, that although the relaxation will itself be modified by the presence of the adsorbate molecules, no calculations which make allowance for this have yet been carried out. Certainly without allowance for this additional adsorbate effect, surface distortion can make an important contribution—sufficient, for example to double the magnitude of $u^{(1)}$ in the case of Ar adsorbed onto NaI (Rogowska, 1978).

Possibly a more serious difficulty than that of adsorbent distortions can arise when molecules with permanent multipoles interact with an electrostatic field. Gready and co-workers (1978) have considered contributions to E_c for CO, N_2 and H_2 on the non-relaxed (100) surface of NaCl. They found that higher-order contributions, up to hexadecapole in the expansion of Eqn (3.60), were by no means negligible at around equilibrium separations; indeed the sum of octopole and hexadecapole terms often exceeds the quadrupole term in magnitude. Furthermore, it is possible that field gradients close to the surface are large enough for higher polarizability terms in E_I to be significant. No quantitative assessment of this possibility has yet been made.

With values of $q > 1$, the first term ($k = 0$) in the summation of (3.76) is obtained from the limiting value of the Bessel function K_m as

$$\frac{2\pi}{a_c[(q-2)/2]!}\left(\frac{k}{2z_\lambda}\right)^{(q-2)/2} K_{(q-2)/2}(kz_\lambda) = \frac{2\pi}{a_c(q-2)z_\lambda^{q-2}}. \quad (3.78)$$

An important example occurs when dispersion forces only contribute to the interaction, and the adsorbent and adsorbate atoms (which are each of one type only, labelled w and a) respectively, interact through the 12–6 potential of Eqn (3.48). The pairwise contribution to $u^{[a]}$, made by atoms in plane λ, when only the zero-wavelength term from Eqn (3.78) is included, can then be written

$$u_\lambda^{(a)} = 4\varepsilon^{(w,a)}2\pi\rho_w\left[\frac{(\sigma^{(w,a)})^{12}}{\Delta^9}\frac{1}{10}\frac{1}{(z/\Delta+\lambda-1)^{10}} - \frac{(\sigma^{(w,a)})^6}{\Delta^3}\frac{1}{4}\frac{1}{(z/\Delta+\lambda-1)^4}\right] \quad (3.79)$$

in which Δ is the spacing between planes in the lattice, and the number density $\rho_w = n_c/a_c\Delta$, where n_c is the number of atoms in the unit surface cell. In a FCC lattice, with $\Delta = 2^{1/6}\sigma_w$, ρ_w is $1/\sigma_w^3$.

When interactions only between the adsorbate and the first layer are considered to be important (and periodicity is neglected), Eqn (3.79) is the 10–4 interaction potential first obtained by Crowell for the graphite lattice using a direct integration over a continuum of C-atoms in the plane. When employed with Eqn (3.70), this leads to the summed 10–4 potential of Crowell.

If the higher-order terms in (3.76) are included, the resulting potential has a "summed 10–4" leading term and additional terms which represent lateral periodicity. Steele (1974) has applied (3.76) to 12–6 interactions and compared the approximate expressions with potentials obtained by direct computer summation over all atoms in the adsorbent. It was found that the two methods gave nearly identical results when the adsorbate atom was larger than, or at least comparable in size with, the adsorbent atom; but direct summation is to be preferred where the adsorbate is rather the smaller of the two.

A useful approximation to the non-periodic part of the energy function can be obtained by applying the Euler–Maclaurin theorem to the summation in (3.70) over planes of atoms (Steele, 1978). For example, when the terms $P_n^{(a,b)}(\boldsymbol{r}_\lambda)$ in this equation are of the form $(z/\Delta + \lambda - 1)^{-q}$, as in Eqn (3.79), the Euler–Maclaurin theorem gives

$$\sum_{\lambda=2} (z/\Delta + \lambda - 1)^{-q} = 1/[(q-1)(z/\Delta + 1)^{q-1}] + \tfrac{1}{2}(z/\Delta + 1)^{q} + \sum_m (-1)^m \frac{B_m}{(2m)!}\left[\frac{\partial^{2m-1}}{\partial\lambda^{2m-1}}(z/\Delta + \lambda - 1)^{-q}\right]. \quad (3.80a)$$

The summation over terms containing the Bernouilli numbers B_m can usually be ignored and a sufficiently accurate approximation obtained in the form

$$\sum_{\lambda=1} \frac{1}{(z/\Delta + \lambda - 1)^q} = \left(\frac{\Delta}{z}\right)^q + \frac{1}{(q-1)(z/\Delta + k_q)^{q-1}} \quad (3.80b)$$

where k_q ranges from $k_3 = 0{\cdot}61$ to $k_{10} = 0{\cdot}72$.

With these approximations the "summed 10-4" potential can be written

$$u^{(a)}(z) = \frac{2\pi n_c[\sigma^{(w,a)}]^2\varepsilon^{(w,a)}}{a_c} \times \left\{\frac{2}{5}\left[\left(\frac{\sigma^{(w,a)}}{z}\right)^{10} + \frac{[\sigma^{(w,a)}]^{10}}{9\Delta(z+0{\cdot}72\Delta)^9}\right] - \left(\frac{\sigma^{(w,a)}}{z}\right)^4 - \frac{[\sigma^{(w,a)}]^4}{3\Delta(z+0{\cdot}61\Delta)^3}\right\}. \quad (3.81a)$$

Another approach is to approximate the whole 3 D array of adsorbent atoms as a continuum which leads to a 9–3 potential which can be expressed in terms of the pair parameters for a dimer as

$$u^{(a)} = \frac{2\pi[\sigma^{(w,a)}]^3\varepsilon^{(w,a)}\rho_w}{3}\left\{\frac{2}{15}\left(\frac{\sigma^{(w,a)}}{z}\right)^9 - \left(\frac{\sigma^{(w,a)}}{z}\right)^3\right\}. \tag{3.81b}$$

Other forms of this equation can be written in which the equivalent parameters for the gas–solid interaction are used. Thus if the zero potential separation between the gas and the solid is s_0 and the well depth is ε_w, it is easily shown that Eqn (3.81b) can be written as

$$u^{(a)} = \frac{\varepsilon_w 3\sqrt{3}}{2}\left\{\left(\frac{s_0}{z}\right)^9 - \left(\frac{s_0}{z}\right)^3\right\} \tag{3.81c}$$

where

$$\varepsilon_w = (2\pi\sqrt{10}/9)[\sigma^{(w,a)}]^3\varepsilon^{(w,a)}\rho_w. \tag{3.81d}$$

However, values of ε_w calculated from (3.81d) are rather low and it is customary to treat ε_w as an adjustable parameter when the 9–3 form is used (see, e.g., Putnam and Fort, 1977). A comparison between the potentials calculated from these equations is illustrated in Fig. 3.3.

Provided that ε_w is adjusted in this way, the attractive inverse cube law part of this potential is in fairly good agreement with direct summation at distances greater than about two hard-sphere diameters, but it can be seen from the figure that this is not the case at closer separations.

An inverse cube dependence of the van der Waals interaction between a molecule and a solid surface also results from the treatment of this problem by Lifshitz (1955). In this approach the dispersion interaction is attributed to quantum fluctuations of the electromagnetic field which has been shown to be equivalent to fluctuations of electronic charge density (Dzyaloshinskii *et al.*, 1961). The material of the system is characterized by its dielectric constants, which are taken from data for bulk isotropic matter. The Maxwell equations for the system are solved to yield an equation containing the dispersion of the normal modes of the electromagnetic field and these can be written in terms of dielectric constants. The free energy of interaction is then found using thermodynamic fluctuation theory and the force between media is obtained from this result. Lifshitz theory and its subsequent development has been widely discussed (Landman and Kleiman, 1977; Richmond, 1975; Mahanty and Ninham, 1976; Linder, 1967), particularly in the context of colloid science. It embraces both the statistical mechanical and intermolecular force aspects of the problem, which are usually considered separately in adsorption. The theory has a particular advantage in that it incorporates many-body effects

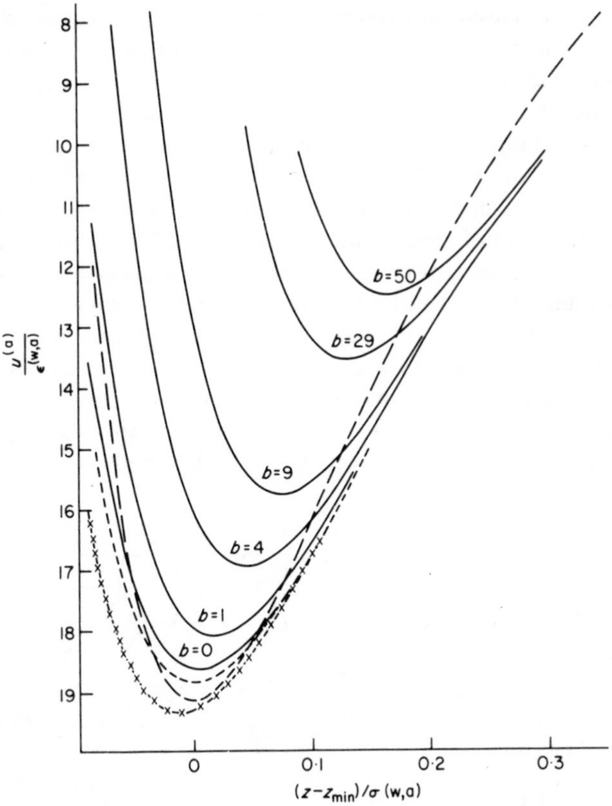

Fig. 3.3. Steele potential and 9–3 potential for Ar over graphite as a function of $z/\sigma^{(w,a)}$. (i) – – – –, summed 10–4 potential, (ii) ——–—, 9–3 potential Eqn (3.81c); (iii) –×–×–×–, potential over a hexagon site; (iv) ———, potential over an atom with different values of the barrier height as determined by the parameter b defined in Eqn (6.8).

Curves (i) and (ii) are plotted so that the minimum falls at zero and other potentials are shifted by the same amount as (ii).

and the collective nature of the wave functions in a solid. Because of the latter feature, the method is especially suitable for metal adsorbents for which the assumptions underlying the summation approach, that the wave functions are localized on individual atoms, will not be valid. A second advantage of Lifshitz theory is that it is readily extended to incorporate retardation effects. These occur when the distance of separation between molecules exceeds the characteristic wavelength of the radiation emitted due to dipolar transitions. When this situation prevails, the induced dipole–dipole potential falls off as r_{12}^{-7} rather than r_{12}^{-6}. Retardation effects are of particular interest in colloid science where the interaction of solid particles separated by relatively large distances is often important. It should be pointed out, however, that the advantages of Lifshitz theory are acquired at the expense of detailed information relating to the structure of adsorbed fluids in the vicinity of solid surfaces, which is potentially a major gain from the adsorption approach developed here. In its usual form the theory only takes into account dipolar interactions and this may be a considerable drawback in the treatment of orientational and induction interactions.

Some attempts to bridge the intervening territory have been made and are referred to in chapter 8, § 2(d). Of particular interest in the context of this section is the work of Zaremba and Kohn (1976), who have incorporated a Fourier method, similar to that described earlier, into a wave-mechanical perturbation theory. Their treatment of the interaction part of the Hamiltonian is consistent with the Lifshitz method. At zero wave-

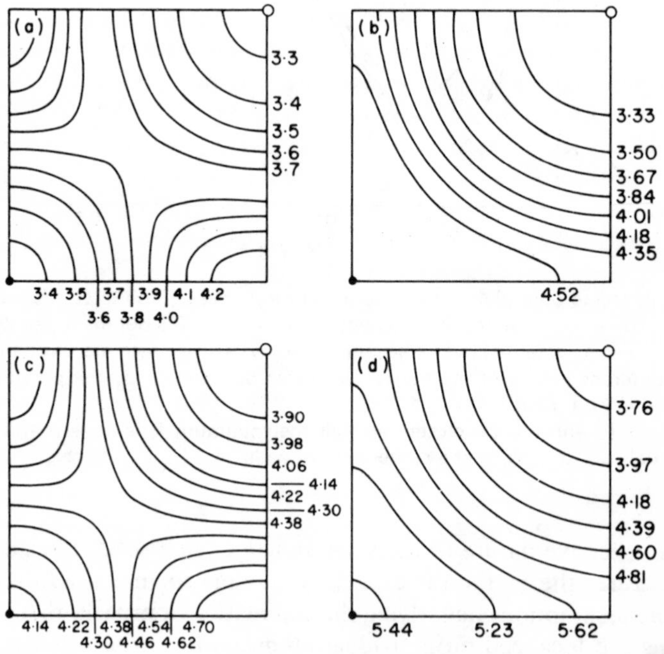

Fig. 3.4. Isopotential energy curves (values in kJ mol^{-1}) for the interaction of (*a*) Ar on the unrelaxed (100) face of NaCl, (*b*) Ar on the relaxed face (100) of NaCl, (*c*) Kr on the unrelaxed (100) face of NaCl, (*d*) Kr on the relaxed (100) face of NaCl, ●, Na^+; ○, Cl^-. (From House and Jaycock, 1975.)

number, therefore, their equations reduce to the inverse cubic potential of Lifshitz rather than the summed inverse 4th power potential appearing in Eqn (3.81b, c).

The existence of periodicity in the fluid–solid adsorption potential can obviously be important in determining localization or mobility (both translational and rotational) of the adsorbate and this controls the possible existence of epitaxial adsorbate structures whose effect might persist beyond the first adsorbed layer. Figure 3.4 illustrates some examples of isopotential contours at the potential minimum for Ar and Kr on relaxed and unrelaxed NaCl (House and Jaycock, 1975). Minimum barriers between sites can range from $\frac{1}{2}$ to 3 kJ mol^{-1} for rare gases on unrelaxed NaCl, but it is to be noted that these values can vary appreciably with the choice of potential functions and parameters.

Even on graphite, which is considered to be a relatively homogeneous adsorbent, the barriers between the sites at the centres of the carbon hexagons can be significant and, as noted above for NaCl, the barrier height in particular can be sensitive to the method of calculation. Some examples from the literature are collected in Table 3.1. Three positions over the C hexagons have been selected: the centre of the hexagon (=site) where the energy has an absolute minimum; the carbon atom (= C) and the mid-point between two carbons on a hexagon edge (=SP) which is the saddle point of the potential contour. The barrier height parameter is defined as $-\Delta\psi_2^{[a]} = u_{2,\mathrm{SITE}}^{[a]} - \frac{1}{2}(u_{2,\mathrm{C}}^{[a]} + u_{2,\mathrm{SP}}^{[a]})$, where the subscript 2 indicates that these are pairwise summation terms, and the superscript [a] that they are contributions to $u^{[a]}$. The following observations can be made in regard to the different methods of calculating these quantities:

A: Avgul and Kiselev (1970) used Eqn (3.49) as a starting point. In the exponential repulsive term they obtained B_0 as an arithmetic mean and defined k by $\Sigma \exp[-\kappa^{(\mathrm{C,a})}(1 - R)] = k \exp(-z/l)$. The constants $C_{2j+6}^{(\mathrm{C,a})}$ were calculated from the Kirkwood–Müller equation and appropriate extensions, and k was found by minimizing the summed rare gas–graphite potential at a separation z_0. Here z_0 is the equilibrium separation between C and noble gas (a) calculated from the arithmetic mean (Eqn (3.51)) with $r_{\mathrm{min}}^{(\mathrm{C,C})}$, the equilibrium separation between graphite planes (0.3354 nm). Price (1974) criticized this approach on the grounds of inconsistency and pointed out that k should be invariant over the graphite lattice.

B: Steele (1973) used the 12–6 potential (3.48) with $\varepsilon^{(\mathrm{C,a})}$ calculated from a geometric mean combining rule. Following Crowell (1958) a value of 0.340 nm was used for $\sigma^{(\mathrm{C,C})}$ determined from the properties of bulk graphite, and $\sigma^{(\mathrm{C,a})}$ calculated from the arithmetic mean. These and the other parameters used in these calculations were from different sources (Pauly and Toennies, 1965) to those chosen by Price who otherwise used

Table 3.1. Energies (K) for rare gas adsorbates over different positions on an ideal graphite surface

	He					Ne					Ar					Kr					Xe				
	A	B	C	D	E	A	B	C	D	E	A	B	C	D	E	A	B	C	D	E	A	B	C	D	E
$-E^{[1]}_{2SITE}/k$		252	221	131		558	514	440	272	351	1328	1140	971	773	935	1746	1373	1160	1008	1223		1838	1558	1534	1773
$-E^{[1]}_{2SP}/k$		235	207	124		423	485	406	254	322	1041	1107	930	742	888	1399	1335	1129	978	1177		1805	1519	1490	1709
$-E^{[1]}_{2C}/k$		232	205	124		387	481	402	252	319	981	1103	926	738	884	1303	1332	1124	975	1173		1802	1514	1484	1702
$\Delta\psi^{[1]}_2$		19	15	7		153	31	36	19	31	317	35	43	33	49	395	40	34	32	48		35	41	47	68
$u^{[a]}/k(\theta = 0)$ (experimental)			244					382					1113					1467					1928		

the same approach. The similarity between these two calculations confirms a conclusion reached earlier (Sams, 1964) that choice of these parameters is not a very critical factor.

The periodic potential resulting from the application of Eqn (3.76) to the 12–6 equation can be expressed in the form

$$u^{(a)} = E_0 + \sum_n E_n f_n \qquad (3.82)$$

where E_0 is the zero-wavelength ($k = 0$) term in (3.76) and is represented, either by a "summed 10–4" expression, or by a suitable approximation to it, such as that given by Eqn (3.81a). The second member of (3.82) comes from the $k = n$ terms in (3.76) and is the product of an expression f_n giving the lateral periodicity and a function E_n expressing the attenuation of its amplitude as z increases. For the graphite basal plane, only the first term ($E_1 f_1$) needs to be considered from this sum. Detailed expressions for up to $n = 5$ for graphite and for the (100) and (111) faces of an FCC lattice have been given by Steele (1973).

In the work of Pisani *et al.* (1973) on the rare gas–graphite interaction, a number of methods were investigated and their results, summarized here in columns C, D, E of the table, were given for the three considered to be the best by these authors. The different procedures used in these calculations were as follows:

C: These results were obtained using Eqn (3.47) with the parameters and a geometric mean combining rule for $\varepsilon^{(C,a)}$ as proposed by Srivastava (1958). Pisani *et al.* also took into account the relaxation of the outer layers of graphite, but found this to be negligibly small.

D: In their second calculation Pisani *et al.* used (3.47) with geometric mean combining rules for the attractive and repulsive parameters separately and an arithmetic mean rule for $\kappa^{(C,a)}$.

E: In their third method, account was taken of the anisotropy in the C-atom polarizability. Theoretical calculations, supported by experimental measurements have shown that $\alpha_{\parallel}$ in the graphite basal plane is very much greater than $\alpha_{\perp}$ normal to it. Recent work (Carlos and Cole, 1979) on He scattering from graphite has corroborated this finding. The treatment of the adsorbate–adsorbent potential follows the work of de Boer and Heller (1937) and of Meyer and Deitz (1967) in which the London formula (Eqn (3.44)) was used in Eqn (3.38). Evaluation of the square of the dipole tensor from Eqn (3.39b) with calculated anisotropic polarizabilities leads to an attractive component of the dispersion potential

$$E_D^{(C,a)} = \frac{C_6^{(C,a)}}{r_{12}^6}\left[2 + \frac{3s_{12}^2}{r_{12}^2}\right] \qquad (3.83)$$

where $C_6^{(C,a)}$ is given by Eqn (3.54) and s_{12} is the projection of r_{12} onto the graphite basal plane.

As a basis for comparison, some values of $u^{[a]}$ (Rybolt and Pierotti, 1979) estimated from accurate zero coverage data (Sams *et al.*, 1961, 1962), are also given in Table 3.1. The table shows that the choice of potential function and combining rules are quite significant factors in calculating the interaction energy. The rather high site energy and barrier heights resulting from the calculations of Avgul *et al.* can probably be rejected on the grounds of Price's criticism. However, close agreement between theoretical and experimental values should not be considered as final vindication of a particular theory unless that theory is also consistent with other well validated knowledge relating to intermolecular potentials. In this respect it may be noted that excellent effective potentials have been obtained using 3–*n* and 3–exp functions (Rybolt and Pierotti, 1979). The success of the inverse-3 attractive law as an effective potential makes an interesting contrast with Steele's conclusion that pairwise summation is better represented by a summed 10–4 potential, especially as the inverse–3 law also results from the Lifshitz treatment which incorporates higher orders of interaction than pairwise.

The effect of polarization anisotropy, as judged for example by comparison of columns D and E, seems to be of sufficient importance to need to be taken into account. Meyer and Deitz (1967) found that adsorption energies at the gaps in graphite edge planes were considerably higher than basal plane site energies, and showed that this effect can be largely attributed to anisotropy effects. The differences in barrier heights resulting from anisotropy would be manifested experimentally for example, as a shift in the position of localized to mobile and other transitions in the adsorbed phases (Nicholson *et al.*, 1981).

Finally the possible importance of electrostatic fields at graphite surfaces has been ignored in all the calculations cited. This seems to be substantially justified by experiment (Kiselev, 1965).

(f) Three-body and higher-order interactions

In the preceding section, although the methods discussed there are of general application, we placed particular emphasis on the rare gas–graphite system. This was because many high-quality data are available for adsorption on well characterized graphites and because considerable effort has gone into investigating different possible methods for calculating the pairwise potential in these systems. In discussing their wide-ranging *a priori* calculations, Pisani *et al.* (1973) conclude that it is not possible to obtain adequate potential laws for the physical adsorption of rare gases on graphite by rigorous combination of self-interaction potentials, independently defined, and consider that the arbitrary nature of combining rules, or the inadequacy of a pairwise model, could be possible causes. This prompts the question of the importance of three-body and higher-order interactions in adsorption. The work of Barker *et al.* (1971) on uniform liquid argon has shown that contributions from three-body terms are of the order of 10%, while those from higher-order terms are probably negligible. Similar conclusions have been reached from a consideration of many-body interactions in rare gas solids (Bell and Zucker, 1975) and it appears that a fortuitous cancelling of quadrupole triple-body terms with higher-order many-body terms may occur in this case. It would not, however, be safe to conclude that this benificence extends to other species or to non-uniform systems.

The most significant contribution to triple-body dispersion interactions comes from mutually induced dipoles and can be obtained in a form similar to that of Eqn (3.38). For isotropic atoms of types a, b, c positioned at $\boldsymbol{r}_1$, $\boldsymbol{r}_2$, $\boldsymbol{r}_3$ the dispersion potential is

$$E_{\mathrm{D,d-d-d}}^{(\mathrm{a,b,c})} = (T_{12}^{\alpha\beta})^3 A^{(\mathrm{a,b,c})} \tag{3.84}$$

where $T_{12}^{\alpha\beta}$ can be obtained from Eqn (3.39b) which for the isotropic case reduces to

$$(T_{12}^{\alpha\beta})^3 = \frac{3(1 + 3\cos\theta_1 \cos\theta_2 \cos\theta_3)}{r_{12}^3 r_{23}^3 r_{31}^3} \tag{3.85a}$$

where θ_1, θ_2, θ_3 are the angles of the triangle formed by the triplet of molecules. The polarizability term may be expressed in a form similar to Eqn (3.40) as

$$A^{(\mathrm{a,b,c})} = \frac{1}{\pi}\int_0^\infty \alpha^{(\mathrm{a})}(i\omega)\alpha^{(\mathrm{b})}(i\omega)\alpha^{(\mathrm{c})}(i\omega)\, \mathrm{d}\omega \tag{3.85b*}$$

Estimates of $A^{(\mathrm{a,b,c})}$ (tabulated as $3A^{(\mathrm{a,b,c})}$) for rare gases have been given by various authors and are summarized by Leonard and Barker (1975). Approximate forms of Eqn (3.85b) can be obtained in the same way as for two-body interactions so that, for example, the isotropic three-body analogue of the London equation (Eqn (3.44)) is (Kihara and Midzuno, 1956),

$$A^{(\mathrm{a,b,c})} = \frac{1}{2}\frac{\bar{\omega}^{(\mathrm{a})}\bar{\omega}^{(\mathrm{b})}\bar{\omega}^{(\mathrm{c})}[\bar{\omega}^{(\mathrm{a})} + \bar{\omega}^{(\mathrm{b})} + \bar{\omega}^{(\mathrm{c})}]\alpha^{(\mathrm{a})}\alpha^{(\mathrm{b})}\alpha^{(\mathrm{c})}}{[\bar{\omega}^{(\mathrm{a})} + \bar{\omega}^{(\mathrm{b})}][\bar{\omega}^{(\mathrm{a})} + \bar{\omega}^{(\mathrm{c})}][\bar{\omega}^{(\mathrm{b})} + \bar{\omega}^{(\mathrm{c})}]} \tag{3.86}$$

Higher-order triple-body interactions involving terms such as d − d − q, d − q − q, etc., have been discussed by Bell (1970).

Three-body effects can enter into adsorption problems in two ways: either as a contribution to $u^{[1]}$ if a = b = adsorbent (w), or as a contribution to $u^{[2]}$ if a = adsorbent species, and b, c = adsorbate. Some preliminary calculations for the former case have been carried out by Schmit (1976), who found that contributions of the order of 5% of the total potential could be due to three-body interactions.

The question of the mediation of adsorbate–adsorbate interaction by the substrate has a longer history, going back to the work of Sinanoğlu and Pitzer (1960). A lattice summation approach, based on the above equations, was investigated by McRury and Linder (1972). For an adsorbate pair (a, a) interacting with adsorbent atoms of type w at position $\boldsymbol{r}_J$ the contribution to $u^{[a,a]}$ from the triple-dipole terms of Eqn (3.84), can be written

$$E_3^{(a,a)} = \sum E_{D,d-d-d}^{(a,a,w)} = A^{(a,a,w)}(T^{\alpha\beta})^3 \tag{3.87}$$

$$= A^{(a,a,w)} \cdot \frac{1}{r_{12}^3} \sum_J \frac{3(1 + 3\cos\theta_1 \cos\theta_2 \cos\theta_J)}{r_{1J}^3 r_{2J}^3} \tag{3.88}$$

We take the isotropic case of Eqn (3.85a), and consider first those configurations where both adsorbate atoms lie at the same distance z from the plane of the adsorbent surface. If, moreover, the summation over individual atoms is replaced by an integration over a continuum solid with mean number density ρ_w, there results (Takaishi, 1975),

$$E_{3,ML}^{(a,a)} = \frac{A^{(a,a,w)}}{4r_{12}^3} \cdot \frac{4{\cdot}08\pi\rho_w}{z^3} f(\zeta) \tag{3.89}$$

in which $\zeta = r_{12}/2z$, and

$$f(\zeta) = \frac{2 - \zeta^2}{3 + \zeta^2} + \frac{0{\cdot}320\zeta^2}{(1 + \zeta^2)^3} - \frac{0{\cdot}422}{(1 + \zeta^2)^2} + \frac{\zeta^2}{5} - \frac{\zeta^4}{7} + \ldots \tag{3.90}$$

From Eqn (3.87) the first term in (3.89) can be expressed in the form,

$$\frac{A^{(a,a,w)}}{4r_{12}^3} = \left[\frac{\alpha^{(a)}(2\tilde{\omega}^{(a)} + \tilde{\omega}^{(w)})}{4r_{12}^3(\tilde{\omega}^{(a)} + \tilde{\omega}^{(w)})}\right]\left[\frac{\tilde{\omega}^{(a)}\tilde{\omega}^{(w)}\alpha^{(a)}\alpha^{(w)}}{4(\tilde{\omega}^{(a)} + \tilde{\omega}^{(w)})}\right] \tag{3.91}$$

in which the term in the second bracket will be recognized, from Eqn (3.44), as $A_{LOND}^{(a,w)}$ for isotropic species. It will be convenient to define a coefficient D by equating the first bracket to D/r_{12}^3.

Sinanoğlu and Pitzer (SP) calculated the mediation of the interaction between two adsorbate molecules by considering the solid as a third body in the interaction, thereby avoiding the direction evaluation of the tensor term. Their result for the above in-plane configuration may be written

$$E_{3,SP}^{(a,a)} = \frac{A_{LOND}^{(a,w)} D}{r_{12}^3} \cdot \frac{\pi\rho_w}{z^3} \tag{3.92}$$

where it has been assumed that $\bar{\omega}$ is proportional to the excitation energy Δ used in the SP theory.

A third approach to the problem was developed by McClachlan (1964) using the field theory methods pioneered by Lifshitz, in which the solid is treated as a continuum characterized by an isotropic permittivity and the adsorbate molecules interact with electrical images in the dielectric medium. The Clausius–Mosotti equation is used to introduce polarizability, and an harmonic oscillator model to relate this to $\bar{\omega}$. The final result may be put in the form

$$E_{3,\mathrm{M}}^{(\mathrm{a,a})} = \frac{A_{\mathrm{LOND}}^{(\mathrm{a,w})} D}{r_{12}^3} \cdot \frac{\pi\rho}{z^3} \cdot \frac{4\zeta^2 + 1}{(\zeta^2 + 1)^{5/2}} \tag{3.93}$$

It is quite clear from these three treatments that $E_3^{(\mathrm{a,a})}$ is repulsive for molecules lying in the same plane parallel to the surface, but a more general treatment (Sinanoğlu and Pitzer, 1960) of configuration dependence shows that $E_3^{(\mathrm{a,a})}$ depends on $(1 - 3\cos^2\psi)$ where ψ is the acute angle between a line joining the two molecules and a normal to the surface. According to this work, therefore, $E_3^{(\mathrm{a,a})}$ becomes attractive for $\psi < 54{\cdot}7°$.

A comparison of Eqns (3.89), (3.91) and (3.93) for noble gas pairs separated by a distance of 0·414 nm on a solid Xe adsorbent is given in Fig. 3.5 (Takaishi, 1975) where $E_3^{(\mathrm{a,a})}(z)$ is plotted as a percentage of the

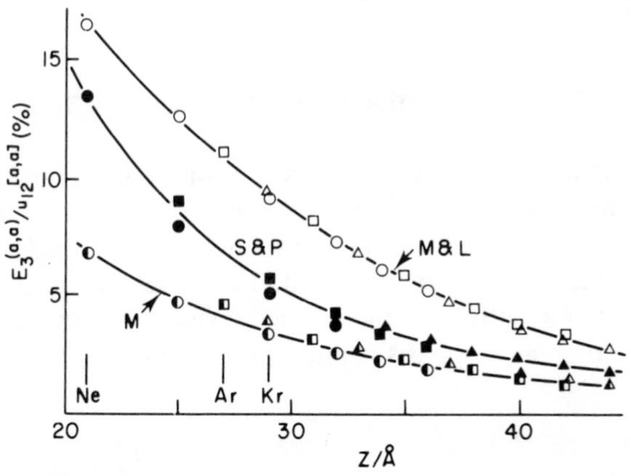

Fig. 3.5. The ratio of the adsorbent three-body term to the two-body potential as a function of the distance *z* from the (100) surface of xenon for a pair of rare gas atoms separated by a distance of 0·441 nm (the lattice constant for Xe). The atoms are positioned over the most favourable sites which are centred between four Xe atoms. The three curves are taken from the theories of McRury and Linder (1972), Sinanoğlu and Pitzer (1960) and McClachlan (1964). Ne: ○, ◐, ●. Ar: □, ◧, ■. Kr: △, ◭, ▲. (From Takaishi, 1975.)

total interaction $u_{12}^{[a,a]}$. The equilibrium separations from the surface for various admolecules are indicated on the curve.

It is to be seen that three-body effects, like polarization anisotropy, are not negligible in comparison to the accuracy of presently available experimental data. From a general point of view the examples discussed here show that, although much progress has been made, the accurate calculation of interaction potentials between unlike species is not easily achieved and that even in the most extensively studied systems further advances need to be made.

Calculations of potentials inside pore spaces are no different in principle from those already discussed here, but obviously results that differ in detail from those obtained for plane surfaces are to be expected. This can be easily seen if one considers the space between two slabs of continuum solid. Each slab gives rise to an inverse cube attractive potential as the leading term in $u^{[1]}$ and the overlap of the potentials from the two slabs intensifies the interaction. If the slabs are moved closer together, this intensification increases and the two potential minima associated with opposite walls eventually merge. Similar effects exist in pore spaces of different geometry, such as in zeolites which contain cavities capable of holding perhaps a few dozen molecules at most. Although, unlike many porous materials, the geometry of the adsorbent surface in zeolites is well defined, the location and possible relaxation of cations in the lattice is usually a serious complicating factor.

4. Thermodynamic quantities from molecular equations

We have seen in the preceding section how the potential energy contribution U_N to the Hamiltonian H_N, can be built up. We now proceed to the equations from which thermodynamic properties, providing a direct link with experimental data, can be obtained. The development is once again made in terms sufficiently general to encompass a variety of adsorption systems, but particular emphasis is placed on fluid–solid systems.

(a) Internal energy and differential heat of adsorption

The internal energy U is the mean value of the total energy. In the grand ensemble with the probability density $\mathscr{P}$ given by Eqn (3.1), this is

$$\langle H_N \rangle = \sum \int \ldots \int H_N \mathscr{P} \, \mathrm{d}\boldsymbol{q} \, \mathrm{d}\boldsymbol{p} \qquad (3.94)$$

Integration of the kinetic energy over the conjugate momenta may be carried out immediately to give a term $s_\alpha kT/2$ for species α where s_α, as before, is the number of degrees of freedom. Thus

$$U = \frac{1}{\Xi} \sum_{N \geq 0} \frac{\boldsymbol{\lambda}^N}{N! \boldsymbol{\Lambda}^{\nu N}} \int \ldots \int \{\tfrac{1}{2} kT \sum_\alpha s_\alpha + U_N\} \exp(-\beta U_N) \, d\boldsymbol{q} \quad (3.95)$$

The internal energy thus divides into a kinetic component U_K, the only one when the system comprises ideal gas alone, and a configurational component U_c

$$U = U_K + U_c \quad (3.96a)$$

where

$$U_K = \tfrac{1}{2} kT \boldsymbol{s} \,.\, \langle \boldsymbol{N} \rangle \quad (3.96b)$$

and

$$U_c = \frac{1}{\Xi} \sum_{N \geq 0} \frac{\boldsymbol{\lambda}^N}{N! \boldsymbol{\Lambda}^{\nu N}} \int \ldots \int U_N \exp(-\beta U_N) \, d\boldsymbol{q}. \quad (3.96c)$$

To develop this equation further, we take the example of a single-component adsorbate with the potential energy given by Eqn (3.26). Since the integrations over the n co-ordinates of an n-body potential will give identical results no matter which co-ordinates are named, Eqn (3.96c) may be expressed in terms of the distribution functions as,

$$U_c = \int u_1^{[1]} \rho^{(1)}(\boldsymbol{q}_1) \, d\boldsymbol{q}_1 + \frac{1}{2} \iint u_{12}^{[2]} \rho^{(2)}(\boldsymbol{q}_1, \boldsymbol{q}_2) \, d\boldsymbol{q}_1 \, d\boldsymbol{q}_2. \quad (3.97)$$

For a single-component fluid–vapour interface, the first term in (3.97) is of course zero. No useful further simplification is possible for non-uniform fluids, but for uniform fluids the co-ordinates can be referred to a frame of reference which is fixed with respect to a given molecule and, after substitution from (3.21), there results the familiar expression

$$U_c = 2\pi \rho \langle N \rangle \iint u^{(2)} \tilde{g}^{(2)}(r, \boldsymbol{\omega}) r^2 \, dr \, d\boldsymbol{\omega}. \quad (3.98)$$

As mentioned in Chapter 2, various differential heats of adsorption may be defined and obtained experimentally. These may be related to molecular quantities with the aid of the above equations for U. We first focus attention on the configurational part of the energy and introduce a differential heat for component a, defined (cf. Eqn (2.69)) by the equation

$$q_{c,a} = \left(\frac{\partial U_c^\Sigma}{\partial N_a^\Sigma} \right)_{N^{\Sigma'}, T, \mathcal{A}, \ell_z} \quad (3.99)$$

where primes added to a subscript denote all species not named in the derivative. Equation (3.99) may also be written

$$q_{c,a} = \sum_{\alpha} \left(\frac{\partial U_c^{\Sigma}}{\partial \lambda_\alpha}\right)_{\lambda',T,\mathscr{A}} \left(\frac{\partial \lambda_\alpha}{\partial N_a^{\Sigma}}\right)_{N^{\Sigma'},T,\mathscr{A}}. \tag{3.100}$$

The second term in this expression may be related to number fluctuations in a grand ensemble. We first introduce the identity

$$\left(\frac{\partial N_b^{\Sigma}}{\partial N_a^{\Sigma}}\right)_{N^{\Sigma'},T,\mathscr{A}} = \delta_{ab} = \sum_{\alpha} \left(\frac{\partial N_b^{\Sigma}}{\partial \lambda_\alpha}\right)_{\lambda',T,\mathscr{A}} \left(\frac{\partial \lambda_\alpha}{\partial N_a^{\Sigma}}\right)_{N^{\Sigma'},T,\mathscr{A}} \tag{3.101}$$

and use (2.51) with an obvious generalization of (3.12), equating N_α, etc., with the mean numbers $\langle N_\alpha \rangle$, etc.,

$$\left(\frac{\partial N_b^{\Sigma}}{\partial \ln \lambda_\alpha}\right)_{\lambda',T,\mathscr{A}} = \frac{\partial}{\partial \ln \lambda_\alpha}(N_b - \sum_X \tilde{N}_b^X)$$
$$= f(N_\alpha, N_b) - \sum_X (\partial \tilde{N}_b^X / \partial \ln \lambda_\alpha) \tag{3.102}$$

where $\tilde{N}_b^X$ is the number of molecules of type b which would be present if the phase of type X were uniform up to the interface. The second term therefore relates to the properties of uniform phases. For example, in the case of gas adsorption onto a rigid solid, only the term $(\partial \tilde{N}_b^G / \partial \ln \lambda_\alpha)$ remains and for an ideal gas phase this is simply $\tilde{N}_\alpha^G$. Thus Eqn (3.101) provides a set of linear equations, with matrix elements given by (3.102), which can be solved for $(\partial \lambda_\alpha / \partial N_a^{\Sigma})_{N^{\Sigma'},T,\mathscr{A}}$. The first term in (3.100) can also be expressed as a fluctuation using (3.96c) and (2.51) with $M \equiv U_c$ to obtain

$$\left(\frac{\partial U_c^{\Sigma}}{\partial \ln \lambda_\alpha}\right)_{\lambda',T,\mathscr{A}} = f(N_\alpha, U_N) - \sum_X \left(\frac{\partial \tilde{U}_c^X}{\partial \ln \lambda_\alpha}\right)_{\lambda',T,\mathscr{A}}. \tag{3.103}$$

The kinetic part of the differential heat for component a, q_{Ka}, is found in a straightforward way from (3.96) and (2.51) as

$$q_{K,a} = \left(\frac{\partial U_K^{\Sigma}}{\partial N_a^{\Sigma}}\right)_T = \tfrac{1}{2}kTs_a. \tag{3.104}$$

The use of (3.100), (3.102), (3.103) and (3.104) to obtain the differential heat can be quite complicated in general, since it involves a matrix inversion. However, in the case of single-component adsorption from an ideal gas phase, an expression for the differential heat is readily obtained. This can be combined with the expressions in Chapter 2, §6, to give results which may be compared with experiment. For example, using (2.71) the isosteric heat in the above case would be

$$q_{\rm st} = -\frac{f(N, U_N)}{f(N, N) - \tilde{N}^G} + kT \tag{3.105}$$

where, in the ideal gas phase,

$$(\partial \tilde{U}^{\rm G}/\partial \tilde{N}^{\rm G}) = skT/2, \qquad \partial \tilde{U}^{\rm G}_{\rm c}/\partial \mu = 0$$

and $p\tilde{V}^{\rm G} = kT$ and where also the term $p\tilde{V}^{\rm A}$ has been assumed to be negligible in comparison to kT.

(b) Compressibility

The compressibility κ, in a non-uniform system may be defined in relation to the components of the pressure tensor, using the surface pressure introduced in Chapter 2:

$$\frac{1}{\kappa} = -\frac{\partial}{\partial V}(p^{\rm A}V^{\rm A} + p^{\rm B}V^{\rm B} + \phi\mathscr{A})_{N,T|\mathscr{A},l_z}. \tag{3.106}$$

When a curved interface is considered, a condition of constant curvature must be imposed and likewise any other intensive variables must be held constant. Using the Gibbs–Duhem equation (2.58) with (3.106) then leads to the relationship

$$\frac{1}{\kappa} = -\sum N_\alpha \left(\frac{\partial \mu_\alpha}{\partial V}\right)_{N,T|\mathscr{A},l_z}. \tag{3.107}$$

Since $\mathrm{d}N = 0$ we have for a specified component a,

$$\left(\frac{\partial N_{\rm a}}{\partial V}\right)_{N,T} = \frac{N_{\rm a}}{V} = -\sum_\alpha \left(\frac{\partial N_{\rm a}}{\partial \mu_\alpha}\right)_{\mu',V,T} \left(\frac{\partial \mu_\alpha}{\partial V}\right)_{N,V,T} \tag{3.108a}$$

and introducing the fluctuation $f(N_{\rm a}, N_\alpha)$ with the aid of Eqn (3.12) and again identifying N_α with $\langle N_\alpha \rangle$,

$$(\partial N_{\rm a}/\partial \mu_\alpha)_{T,V,\mu,\mathscr{A},l_z} = \beta f(N_{\rm a}, N_\alpha) = \beta[\langle N_{\rm a} N_\alpha \rangle - \langle N_{\rm a}\rangle\langle N_\alpha\rangle]. \tag{3.108b}$$

Equations (3.108a, b) constitute a set of linear equations for $(\partial \mu_\alpha/\partial V)_{\mu',T}$ which may be substituted into (3.107) to give κ in terms of the fluctuations in N. Thus, for example, for a single-component fluid, Eqns (3.107), (3.108a), (3.108b) lead to

$$\langle N \rangle \rho\kappa = \beta f(N, N) \tag{3.108c}$$

where $\langle N \rangle / V = \rho$. This is a general thermodynamic result for κ valid for both uniform and non-uniform systems.

When the system can be approximated as a fluid in contact with an incompressible solid, it is often more useful to consider a compressibility for the adsorbate phase alone. Here the Gibbs–Duhem equation for the surface excess quantities, Eqn (2.58), may be used and the compressibility defined as

$$\frac{1}{\kappa^{\Sigma}} = -\left(\frac{\partial \phi \mathscr{A}}{\partial \mathscr{A}}\right)_{N,T,|\mathscr{A}} = -\sum N_{\alpha}^{\Sigma}\left(\frac{\partial \mu_{\alpha}}{\partial \mathscr{A}}\right)_{N^{\Sigma},T,|\mathscr{A}}. \tag{3.109}$$

Equations (3.108a, b) are now replaced by (3.110a, b) respectively where

$$\left(\frac{\partial N_{\mathrm{a}}^{\Sigma}}{\partial \mathscr{A}}\right)_{\mu,T} = \frac{N_{\mathrm{a}}^{\Sigma}}{\mathscr{A}} = -\sum \left(\frac{\partial N_{\mathrm{a}}^{\Sigma}}{\partial \mu_{\alpha}}\right)_{\mu',T}\left(\frac{\partial \mu_{\alpha}}{\partial \mathscr{A}}\right)_{N,T} \tag{3.110a}$$

$$\left(\frac{\partial N_{\mathrm{a}}^{\Sigma}}{\partial \mu_{\alpha}}\right)_{\mu',T} = \beta\left[f(N_{\mathrm{a}}, N_{\alpha}) - \left(\frac{\partial \tilde{N}^{x}}{\partial \mu_{\alpha}}\right)_{V,T}\right]. \tag{3.110b}$$

The last equation follows from (3.37a) after cancellation of the Ξ^{s} terms and introduction of the fluctuation $f(N, N)$, defined in Eqns (3.12) and (3.13).

In a system of this type it is straighforward to derive the analogue of the compressibility equation for a single-component non-uniform fluid. With the aid of Eqn (3.16), the fluctuation $f(N, N)$ may be written in terms of the distribution functions,

$$f(N, N) = \iint \mathscr{F}_{12}^{(2)}\, \mathrm{d}\boldsymbol{q}_1\, \mathbf{d}\boldsymbol{q}_2 + \langle N \rangle \tag{3.111}$$

where the 2-particle Ursell function is related to the molecular distribution functions by Eqn (3.23). It should be noted that $\rho^{(n)}$ here is defined for the adsorbate fluid only so that Ξ^{A} from (3.33), and similar terms, appropriate for this phase only, now appear in the defining equation (3.14). Equation (3.111) with (3.110) and (3.109) now leads to the required result

$$\kappa^{\Sigma} = \frac{\beta \mathscr{A}}{\langle N^{\Sigma}\rangle^{2}}\left[\langle N \rangle + \iint \mathscr{F}_{12}^{(2)}\, \mathrm{d}\boldsymbol{q}_1\, \mathrm{d}\boldsymbol{q}_2 - \left(\frac{\partial \tilde{N}^{X}}{\partial \mu}\right)_{V,T}\right]. \tag{3.112a}$$

In the same way, equation (3.108c) leads to an analogous equation for a single-component non-uniform system.

$$\kappa = \frac{\beta V}{\langle N \rangle^{2}}\left[\langle N \rangle + \iint \mathscr{F}_{12}^{(2)}\, \mathrm{d}\boldsymbol{q}_1\, \mathrm{d}\boldsymbol{q}_2\right]. \tag{3.112b}$$

In a uniform system this reduces to the familiar result

$$kT\kappa\rho = 1 + \rho \int \bar{h}^{(2)}(r_{12})\, \mathrm{d}\boldsymbol{r}_{12}. \tag{3.112c}$$

(c) Heat capacity

The heat capacity of an excess phase at constant area may be defined as

$$C_{\mathscr{A},N^\Sigma} = \left(\frac{\partial U^\Sigma}{\partial T}\right)_{N^\Sigma,\mathscr{A}} = -\left(\frac{\partial U^\Sigma}{\partial \beta}\right)_{N^\Sigma,\mathscr{A}}\left(\frac{1}{kT^2}\right) \tag{3.113}$$

where the configurational part of the derivative with respect to β is

$$\left(\frac{\partial U_c^\Sigma}{\partial \beta}\right)_{N^\Sigma,\mathscr{A}} = \left(\frac{\partial U_c^\Sigma}{\partial \beta}\right)_{\boldsymbol{\mu},\mathscr{A}} + \sum_\alpha \left(\frac{\partial U_c^\Sigma}{\partial \mu_\alpha}\right)_{\boldsymbol{\mu}',T,\mathscr{A}}\left(\frac{\partial \mu_\alpha}{\partial \beta}\right)_{N^\Sigma,\mathscr{A}}. \tag{3.114}$$

The derivatives $(\partial\mu_\alpha/\partial\beta)$ here may once again be found from a set of linear equations using the condition $(\partial N^\Sigma/\partial\beta)_{N^\Sigma,\mathscr{A}} = 0$ which leads to

$$\left(\frac{\partial N_a^\Sigma}{\partial \beta}\right)_{\boldsymbol{\mu},\mathscr{A}} + \sum_\alpha \left(\frac{\partial N_a^\Sigma}{\partial \mu_\alpha}\right)_{\boldsymbol{\mu}',T,\mathscr{A}}\left(\frac{\partial U_\alpha}{\partial \beta}\right)_{N^\Sigma,\mathscr{A}} = 0. \tag{3.115}$$

The coefficients in this equation are found from Eqns (3.108) and (3.103) via the auxiliary condition

$$\left(\frac{\partial N_\alpha^\Sigma}{\partial \beta}\right)_{\boldsymbol{\mu},\mathscr{A}} = -\left(\frac{\partial U_c^\Sigma}{\partial \ln \lambda_\alpha}\right)_{\boldsymbol{\mu}',T,\mathscr{A}}. \tag{3.116}$$

Equation (3.116) also provides the derivative $(\partial U_c^\Sigma/\partial\mu_\alpha)$ in (3.114), since $\ln\lambda_\alpha = \beta\mu_\alpha$. Finally, the term $(\partial U_c^\Sigma/\partial\beta)$ in (3.114) may be found as a fluctuation from (3.96) as

$$-\left(\frac{\partial U_c^\Sigma}{\partial \beta}\right) = f(U_N, U_N) - \sum_{\mathrm{X}}\left(\frac{\partial \tilde{U}_c^{\mathrm{X}}}{\partial \beta}\right)_V. \tag{3.117a}$$

Once again we take a single-component ideal gas as an example, in this case the heat capacity reduces to

$$kT^2 C_{\mathscr{A},N^\Sigma} = f(U_N, U_N) - [f(N, U_N)]^2/[f(N,N) - \tilde{N}^{\mathrm{G}}] + \tfrac{3}{2}k\langle N^\Sigma\rangle \tag{3.117b}$$

where the kinetic part $-(\partial U_{\mathrm{K}}^\Sigma/\partial\beta)$, has been calculated assuming a classical monatomic gas phase.

(d) The pressure tensor

According to Eqns (2.56b) and (3.34), the surface pressure can be obtained from

$$\phi = -\left(\frac{\partial \Omega}{\partial \mathscr{A}}\right)_{V,T,\boldsymbol{\mu}} = \frac{1}{\beta}\left(\frac{\partial \ln \Xi}{\partial \mathscr{A}}\right) \tag{3.118}$$

where the partition function is for the total system and, in the case of fluid–solid adsorption, for example, would include contributions from the non-uniform regions of the solid, as well as those arising from the fluid part alone. Similar considerations would apply in the case of a monolayer adsorbed on a liquid, or to the interface between two immiscible liquids. The molecular level treatment of such cases does not present any difficulties in principle, but the extensive literature (Buff, 1949, 1960; McClellan, 1952; Harasima, 1953; Lekner and Henderson, 1977; Navascués and Berry, 1977; Rowley *et al.*, 1976) appears to be confined at present to discussion of planar fluid–vapour and fluid–solid systems, although, as pointed out by Navascués and Berry (1977) the important differences between these two systems have often been overlooked.

The area dependence in (3.118) resides solely in the translational part $\boldsymbol{r}_i$ of the co-ordinates $\boldsymbol{q}_i$ and may be displayed by introducing dimensionless quantities chosen so that a virtual change in area can be carried out at constant volume. When the interface is planar, for example, with the interfacial non-uniformity contained within a volume $V = l_x l_y l_z$, we may write

$$\frac{x}{l_x} = \frac{x}{\sqrt{\mathscr{A}}} \qquad \frac{y}{l_y} = \frac{y}{\sqrt{\mathscr{A}}} \qquad \frac{z}{l_z} = \frac{z\mathscr{A}}{V}. \tag{3.119}$$

It is convenient to introduce the vector $r_i^* = (x_i/2, y_i/2, -z_i)$ and it is then easily shown that

$$\frac{\partial \boldsymbol{r}_i}{\partial \mathscr{A}} = \frac{r_i^*}{\mathscr{A}} \quad \text{and} \quad \frac{\partial r_{ij}}{\partial \mathscr{A}} = \frac{r_{ij}^2 - 3z_{ij}^2}{2\mathscr{A} r_{ij}}. \tag{3.120}$$

We take as an example the potential of Eqn (3.27), with a rigid solid wall composed of species w giving rise to the external field for a single-component adsorbate fluid, a. An extension to a multicomponent system, which included adsorbent particle correlations, could be made at the expense of some simplicity in the equations. Without this extension, ϕ can be identified with $\phi^{\mathrm{F}} = \phi - \phi^{\mathrm{S}}$, the spreading pressure associated with the adsorbate phase alone, which was introduced in Chapter 2. Equation (3.20) with (3.118) and the standard equations (3.14–3.19), defining the distribution functions, leads to the result

$$\phi^{\mathrm{F}}\mathscr{A} = -\int \boldsymbol{r}_1^* \cdot \frac{\partial u_1^{[\mathrm{a}]}}{\partial \boldsymbol{r}_1} \rho^{(\mathrm{a})}(\boldsymbol{q}_1)\, \mathrm{d}\boldsymbol{q}_1 - \frac{1}{4}\int\!\!\int \frac{(r_{12}^2 - 3z_{12}^2)}{r_{12}} \frac{\partial u_{12}^{[\mathrm{a,a}]}}{\partial r_{12}}$$

$$\times\, \rho^{(\mathrm{a,a})}(\boldsymbol{q}_1, \boldsymbol{q}_2)\, \mathrm{d}\boldsymbol{q}_1\, \mathrm{d}\boldsymbol{q}_2. \tag{3.121}$$

The separate contributions to ϕ^{F} from adsorbate–adsorbent interactions $\phi^{\mathrm{F(a,w)}}$ and from the interactions among the fluid particles alone $\phi^{\mathrm{F(a,a)}}$ may be written (Navascués and Berry 1977),

$$\phi^{\mathrm{F(a,w)}}\mathscr{A} = -\int \boldsymbol{r}_1^* \cdot (\partial u_1^{[\mathrm{a}]}/\partial \boldsymbol{r}_1)\, \rho^{(\mathrm{a})}(\boldsymbol{q}_1)\, \mathrm{d}\boldsymbol{q}_1 \tag{3.122}$$

and

$$\phi^{\mathrm{F(a,a)}}\mathscr{A} = -\frac{1}{4}\iint \frac{(r_{12}^2 - 3z_{12}^2)}{r_{12}} \frac{\partial u_{12}^{[\mathrm{a,a}]}}{\partial r_{12}} \times \rho^{(\mathrm{a,a})}(\boldsymbol{q}_1, \boldsymbol{q}_2)\, \mathrm{d}\boldsymbol{q}_1\, \mathrm{d}\boldsymbol{q}_2. \tag{3.123}$$

The external field term (3.122) would be absent from (3.121) in a system containing fluid species only. In particular, an expression for the liquid–vapour surface tension is obtained from (3.121) in the form of (3.123) when the first term in (3.121) is omited. However, it is important to note that the introduction of a solid adsorbent would modify the pair distribution function. This last consideration will of course also extend to curved surfaces, so that even when approximate allowance for the external field is made in the thermodynamic treatment of capillary condensation in pores (Deryaguin, 1957; Broekhoff and de Boer, 1967; Nicholson, 1968), it is still incorrect to use surface tensions obtained for free fluid–vapour interfaces. The additional complications introduced here by curvature effects were discussed in Chapter 2, and some further discussion of the molecular aspects of this problem is made in Chapter 7.

The separate components of the pressure tensor may be deduced from a simple force balance argument (Steele, 1974). Consider, for example, the forces acting across unit area of a plane at z: the contribution from momentum transfer due to molecules moving towards the solid is $\rho_{\mathrm{T}}^{(\mathrm{a})}kT$, where $\rho_{\mathrm{T}}^{(\mathrm{a})}$ is the angle-averaged function defined in (3.15); in addition there is a contribution from the interaction between fluid molecules in volume elements $\mathrm{d}\boldsymbol{r}_1$, $\mathrm{d}\boldsymbol{r}_2$ located on opposite sides of the plane. The force per unit area in the z-direction from these two sources may therefore be written

$$p_{zz}(\boldsymbol{r}) = \rho_{\mathrm{T}}^{(\mathrm{a})}(\boldsymbol{r})\, kT - \iint \mathrm{d}\omega_1\, \mathrm{d}\omega_2 \iint_{\mathscr{A}} \mathrm{d}x_2\, \mathrm{d}y_2 \int_0^z \mathrm{d}z_2 \int_z^{l_z} \mathrm{d}z_1 \times \frac{\partial u_{12}^{[\mathrm{a,a}]}}{\partial z_{12}} \rho_{\mathrm{T}}^{(\mathrm{a,a})}(x, y, z_1; \boldsymbol{r}_2). \tag{3.124}$$

Equation (3.124) may be combined with the hydrostatic result from Eqn

(2.18) to give the pressure in the uniform external phase (which is of course equivalent to p'_{zz})

$$p(=p'_{zz}) = \rho_{\mathrm{T}}^{(\mathrm{a})}(\boldsymbol{r})\,kT - \int \mathrm{d}\boldsymbol{\omega}_1 \int_z^{l_z} \left(\frac{\partial u_1^{[\mathrm{a}]}}{\partial z_1}\right) \rho_{\mathrm{T}}^{(\mathrm{a})}(\boldsymbol{r}_1)\,\mathrm{d}z_1$$

$$- \iint \mathrm{d}\boldsymbol{\omega}_1\,\mathrm{d}\boldsymbol{\omega}_2 \iint_{\mathscr{A}} \mathrm{d}x_2\,\mathrm{d}y_2 \int_0^z \mathrm{d}z_2 \int_z^{l_z} \mathrm{d}z_1 \frac{z_{12}}{r_{12}} \left(\frac{\partial u_{12}^{[\mathrm{a,a}]}}{\partial r_{12}}\right)$$

$$\times\, \rho_{\mathrm{T}}^{(\mathrm{a,a})}(x, y, z_1; \boldsymbol{r}_2) \qquad (3.125)$$

the relation $\partial r_{12}/\partial z_{12} = z_{12}/r_{12}$ has also been used in obtaining (3.125). The integrals with respect to z_1 and z_2 in Eqn (3.125) may be readily transformed using an integration by parts, to give an equation which is similar to that first obtained by Irving and Kirkwood (1950):

$$\int_0^{l_z} p\,\mathrm{d}z_1 = \int_0^{l_z} \rho_{\mathrm{T}}^{(\mathrm{a})}(\boldsymbol{r}_1)\,kT\,\mathrm{d}z_1 - \int d\boldsymbol{\omega}_1 \int_0^{l_z} z_1 \left(\frac{\partial u_1^{[\mathrm{a}]}}{\partial z_1}\right) \rho_{\mathrm{T}}^{(\mathrm{a})}(\boldsymbol{r}_1)\,\mathrm{d}z_1$$

$$- \iint \mathrm{d}\boldsymbol{\omega}_1\,\mathrm{d}\boldsymbol{\omega}_2 \iint_{\mathscr{A}} \mathrm{d}x_2\,\mathrm{d}y_2 \int_0^{l_z} \mathrm{d}z_1 \int_0^{l_z} \mathrm{d}z_2 \frac{z_{12}^2}{r_{12}} \left(\frac{\partial u_{12}^{[\mathrm{a,a}]}}{\partial r_{12}}\right)$$

$$\times\, \rho_{\mathrm{T}}^{(\mathrm{a,a})}(x, y, z_1; \boldsymbol{r}_2). \qquad (3.126)$$

Exactly analogous results may be found for the x and y components of the pressure tensor. After multiplication of both sides of these expressions by the area element of the remaining two co-ordinates followed by integration, the results may be substituted in the hydrostatic equation for ϕ^{F} (cf Eqn (2.26))

$$\phi^{\mathrm{F}}\mathscr{A} = \int_V [\tfrac{1}{2}(p_{xx} + p_{yy}) - p]\,\mathrm{d}\boldsymbol{r} \qquad (3.127)$$

and the thermodynamic result (3.121) is recovered. This demonstrates the essential consistency at the molecular level of the hydrostatic and thermodynamic approaches (Navascués and Berry, 1977; Evans, 1974).

If we write p'_{zz} for p in Eqn (3.125) and integrate both sides over the area of the surface, the result can be expressed in the form of an ensemble average for p'_{zz}. Similar procedures result in analogous equations for the x and y components of $p'_{\alpha\alpha}$, thus for the α direction component,

$$\langle p'_{\alpha\alpha}\rangle V = \langle N\rangle kT - \left\langle \sum_i r_{\alpha i}(\partial u_i^{[1]}/\partial r_{\alpha i}) - \sum_{i>j} \frac{r_{\alpha ij}^2}{r_{ij}} \left(\frac{\partial u_{ij}^{(2)}}{\partial r_{ij}}\right)\right\rangle. \qquad (3.128)$$

When the interfacial system under consideration contains fluid phases only, the term containing the singlet potential will of course be absent and $p'_{\alpha\alpha} \to p_{\alpha\alpha}$.

Equation (3.128) can be substituted into (3.127) to give an ensemble average expression for $\phi^{\mathrm{F}}\mathscr{A}$ which could alternatively have been written directly from Eqn (3.121). This form of equation is particularly useful in machine simulations, since the sums can be calculated directly from the co-ordinates and potentials of individual molecules. It is interesting to note that Eqn (3.128) is very closely related to the pressure equations for a uniform fluid derived from the Clausius virial (Münster, 1969), but there appears to be no rigorous derivation of (3.128) from this starting point when separate components of the pressure tensor have to be considered (Lane and Spurling, 1976).

Some interesting consequences follow from (3.125) and (3.126). In the first place the pair interaction integrals in Eqn (3.125) vanish at the plane $z = 0$, giving

$$\beta p = \rho_{\mathrm{T}}^{(\mathrm{a})}(x, y, z = 0) - \int \mathrm{d}\boldsymbol{\omega}_1 \int_0^{l_z} \frac{\partial \beta u_1^{[\mathrm{a}]}}{\partial z_1} \rho_1^{(\mathrm{a})} \, \mathrm{d}z_1. \tag{3.129}$$

For the special case of an infinitely repulsive wall with no attractive forces, (3.129) gives the limiting density at the wall as

$$\rho_{\mathrm{T}}^{(\mathrm{a})}(x, y, z = 0) = \beta p \tag{3.130}$$

an equation attriouted to Fisher (1964).

Secondly the invariance in pressure normal to the surface can be exploited by setting $\partial \beta p / \partial z = 0$ in (3.125), which leads to

$$\frac{\partial \rho^{(\mathrm{a})}(\boldsymbol{r}_1)}{\partial z_1} = \int \rho^{(\mathrm{a})}(\boldsymbol{r}_1) \left(\frac{\partial \beta u_1^{[\mathrm{a}]}}{\partial z_1} \right) \mathrm{d}\boldsymbol{\omega}_1 - \iiint \frac{z_{12}}{r_{12}} \left(\frac{\partial \beta u_{12}^{[\mathrm{a,a}]}}{\partial r_{12}} \right) \rho_{12}^{(\mathrm{a,a})} \, \mathrm{d}\boldsymbol{r}_2 \, \mathrm{d}\boldsymbol{\omega}_2 \, \mathrm{d}\boldsymbol{\omega}_1. \tag{3.131}$$

This is the component in the z-direction of the first equation of the BBGKY (Bogoliubov, Born, Green, Kirkwood, Yvon) hierarchy; its implications are considered more fully in the next section.

5. Integral equations

(a) The BBGKY equations

The hierarchy of equations associated with the names of Bogoliubov (1946), Born and Green (1946), Kirkwood (1935) and Yvon (1935) is of particular importance for non-uniform systems since it has already produced some quite satisfactory and useful results in certain cases. We shall first develop the equations for a fairly general situation in which there are two species (a, b) in the presence of an external field due to a rigid solid. The extension to a larger number of species will be seen to be easily made by induction; alternatively the terms relating to the rigid solid can easily be removed and the remainder of the general equations applied to cases where only fluid phases are present. It is also worth noting that, although the derivation is carried out here by differentiation with respect to translational co-ordinates only, an analogous set of equations could be written, in which variation with respect to the orientational components of the co-ordinates is considered. A more general version of the equations, which includes time dependence, and is thus applicable to non-equilibrium systems, may also be derived.

We consider a system containing N_a molecules of type a, N_b of type b and specify the first n_a type a and the first n_b type b molecules in the distribution function, which, by an obvious extension of the notation introduced earlier, is therefore written as $\rho_{1,\ldots\sigma}^{(n_\alpha,n_b)}$, where $\sigma = n_a + n_b$. We write Eqn (3.14) for this particular system and differentiate with respect to $\boldsymbol{r}_1$, giving

$$\frac{\partial \rho_{1\ldots,\sigma}^{(n_a,n_b)}}{\partial \boldsymbol{r}_1} = \frac{1}{\Xi}\sum \frac{(\boldsymbol{\lambda}/\boldsymbol{\Lambda}^3)^N}{(\boldsymbol{N}-\boldsymbol{n})!}\int\cdots\int\left(\frac{-\beta\partial U_N}{\partial \boldsymbol{r}_1}\right)\exp(-\beta U_N)$$

$$\times\, d\boldsymbol{q}_{n_a+1},\ldots, d\boldsymbol{q}_{N_a};\, d\boldsymbol{q}_{N_a+1+n_b},\ldots, d\boldsymbol{q}_{N}. \tag{3.132}$$

Of the terms in the potential energy derivative $(\partial U_N/\partial \boldsymbol{r}_1)$, that from the singlet potential, is easily dealt with, since only one component in the first summation in Eqn (3.28) depends on $\boldsymbol{r}_1$. The treatment of the pair and higher-order interaction terms is more complicated, but it may be deduced with the aid of Fig 3.6 that the derivative up to the second-order terms in the potential energy can be written as

$$\frac{\partial \rho_{1\ldots\sigma}^{(n_a,n_b)}}{\partial \boldsymbol{r}_1} = -\rho_{1\ldots\sigma}^{(n_a,n_b)}\left\{\frac{\partial \beta u_1^{[a]}}{\partial \boldsymbol{r}_1} + \sum_{i=2}^{n_a}\frac{\partial \beta u_{1i}^{[a,a]}}{\partial \boldsymbol{r}_1}\right.$$

$$\left. + \sum_{i=N_a+1}^{N_a+N_b}\frac{\partial \beta u_{1i}^{[a,b]}}{\partial \boldsymbol{r}_1} + \cdots\right\} - \int \frac{\partial \beta u_{1,n_a+1}^{[a,a]}}{\partial \boldsymbol{r}_1}\rho^{(n_a+1,n_b)}\, d\boldsymbol{q}_{n_a+1}$$

$$- \int \frac{\partial \beta u_{1,N_a+n_b+1}^{[a,b]}}{\partial \boldsymbol{r}_1}\rho^{(n_a,n_b+1)}\, dq_{N_a+n_b+1}. \tag{3.133}$$

Equation (3.133) defines a hierarchy of integro-differential equations for the system, where each order of distribution function is expressed in terms

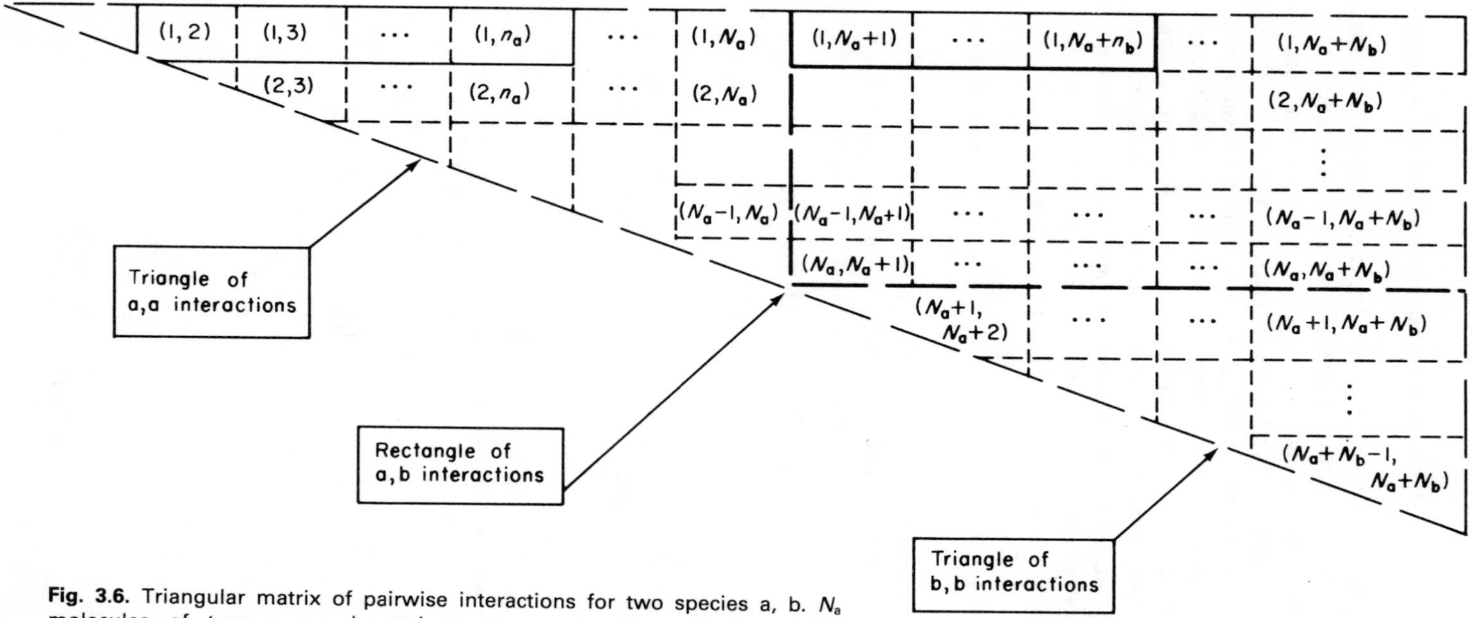

Fig. 3.6. Triangular matrix of pairwise interactions for two species a, b. N_a molecules of type a are located at $q_{1a}, \ldots, q_{N_a}$; those of type b at $q_{N_a+1}, \ldots, q_{N_a+N_b}$. Each bracketed pair (i, j) corresponds to a pairwise interaction term $u_{ij}^2(r_i, r_j)$. The pair terms enclosed in solid rectangles correspond to a pair potential term appearing in the derivatives outside the integral sign in Eqn (3.133).

of integrals over higher-order distributions. The first equation in this hierarchy relates the singlet density function, which is central to the understanding of adsorption systems, to the pair distribution function; it can be put into a more useful form by further rearrangement (Berry, 1974). In order to reduce the notation somewhat we now consider a simplified system in which the adsorbate fluid comprises a single, spherically symmetric species in contact at a planar interface with a rigid solid adsorbent. Multicomponent systems containing molecules of lower symmetry can easily be handled by methods indicated already, and likewise the vector notation encompasses the possibility of spatial curvature at the interface.

With the above restrictions, Eqn (3.133) can be written

$$\frac{\partial}{\partial \boldsymbol{r}_1}[\ln \rho^{(1)}(\boldsymbol{r}_1) + \beta u^{[1]}(\boldsymbol{r}_1)] = -\int \frac{\rho^{(2)}(\boldsymbol{r}_1, \boldsymbol{r}_2)}{\rho^{(1)}(\boldsymbol{r}_1)} \left(\frac{\partial \beta u^{[2]}_{12}}{\partial \boldsymbol{r}_1}\right) \mathrm{d}\boldsymbol{r}_2. \quad (3.134)$$

Considering the z-component only, and making use of Eqn (3.21), this equation may be integrated to give an expression for the singlet density at a position (x, y, z) in relation to that at a reference position (x, y, z_r)

$$\ln[\rho^{(1)}(x, y, z)/\rho^{(1)}(x, y, z_r)] + \beta[u^{[1]}(x, y, z) - u^{[1]}(x, y, z_r)]$$

$$= \int_z^{z_r} \mathrm{d}z_1 \int \rho^{(1)}(\boldsymbol{r}_2)\, g^{(2)}(\boldsymbol{r}_1, \boldsymbol{r}_2) \frac{z_{12}}{r_{12}} \left(\frac{\partial \beta u^{[2]}_{12}}{\partial r_{12}}\right) \mathrm{d}\boldsymbol{r}_2. \quad (3.135)$$

This may be further developed by transforming to a cylindrical co-ordinate system (Fig 3.7) centred on molecule 1 with axis normal to the interface

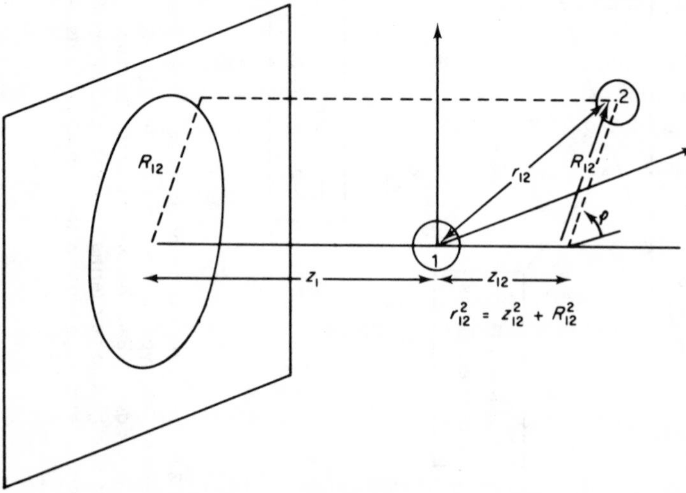

Fig. 3.7.

in which the vector locating 2 has the components $(R_{12}, \varphi, z_{12})$, where $R_{12} = (r_{12}^2 - z_{12}^2)^{1/2}$, is the projection of r_{12} onto the x–y plane. Under this transformation,

$$\left.\begin{aligned} x_2 &\rightarrow R_{12}\cos\varphi + x_1 \\ y_2 &\rightarrow R_{12}\sin\varphi + y_1 \\ z_2 &\rightarrow z_{12} + z_1 \\ \mathrm{d}\boldsymbol{r}_2 &\rightarrow R_{12}\,\mathrm{d}R_{12}\,\mathrm{d}\varphi\,\mathrm{d}z_{12} \rightarrow r_{12}\,\mathrm{d}r_{12}\,\mathrm{d}\varphi\,\mathrm{d}z_{12} \end{aligned}\right\} \tag{3.136}$$

and the integral $I(z)$ on the right-hand side of (3.135) becomes

$$I(z) = 2\pi \int_z^{z_r} \mathrm{d}z_1 \int_{-z_1}^{l_z - z_1} \mathrm{d}z_2 \int_\sigma^\infty \mathrm{d}r_{12} \times \rho^{(1)}(\boldsymbol{r}_2) g^{(2)}(z_1, z_{12}, r_{12}) z_{12} \left(\frac{\partial \beta u_{12}^{[2]}}{\partial r_{12}}\right) \tag{3.137}$$

where σ is a hard-sphere cutoff for $u^{[2]}$.

Equation (3.135) with (3.136) can provide a description of the adsorption between two planes separated by a distance l_z. This is an adequate model for the pore spaces in certain porous carbons (Stoeckli, 1974), however its use in this context still requires some auxiliary information relating to ρ, which will not in this case be equal to the density of the external uniform phase.

When the adsorbent is a single plane with no x–y periodicity in the adsorbent potential, z_r may be set at a large enough distance for the molecules to be no longer influenced by the external field and the density acquires a uniform value ρ. In practice the integrations must be carried out over some finite range l_z and it is necessary to allow for this by subtracting the uniform phase identity, obtained from (3.133):

$$O = \int_z^{l_z} \mathrm{d}z_1 \int_{-z_1}^{l_z - z_1} \mathrm{d}z_2 \int_\sigma^\infty \mathrm{d}r_{12} \rho \tilde{g}^{(2)}(r_{12}) z_{12} \left(\frac{\partial \beta u_{12}^{[2]}}{\partial r_{12}}\right) \tag{3.138}$$

to give

$$\ln[\rho^{(1)}(z)/\rho] + \beta u^{[1]}(z) = \int_z^{l_z} \mathrm{d}z_1 \int_{-z_1}^{l_z - z_1} \mathrm{d}z_2 \int_\sigma^\infty \mathrm{d}r_{12} \times [\rho_2^{(1)} g^{(2)}(z_1, z_{12}, r_{12}) - \rho \tilde{g}^{(2)}(r_{12})] z_{12} \left(\frac{\partial \beta u_{12}^{[2]}}{\partial r_{12}}\right). \tag{3.139}$$

In order to use this equation to determine $\rho^{(1)}(z)$ it is necessary to have an expression for $g^{(2)}$ which, even with the simplifications introduced at this stage, is a non-uniform three-dimensional function. The effectiveness of the approximations introduced to handle this problem determine the limitations inherent in the use of the BBGKY equations and will be discussed more fully in Chapter 7.

(b) The OZ equation

This equation was first introduced into statistical mechanical theory by Ornstein and Zernike in 1914 in connection with the theory of fluctuations. It may be regarded as essentially a defining equation for a new function, the pairwise direct correlation function $c^{(\mathrm{a,b})}(\boldsymbol{q}_1, \boldsymbol{q}_2)$. Subsequent developments, using functional differentiation have shown that a set of n-particle direct correlation functions may be defined in analogy to the n-particle distribution functions and that $c^{(2)}$ often behaves rather like the pair potential function. The extension of the OZ equation to non-uniform systems is of relatively recent origin (Perram and White, 1975; Percus, 1976; Henderson *et al.*, 1976).

It is self-evident that the correlation between two particles placed at $\boldsymbol{r}_1$ and $\boldsymbol{r}_2$ will become zero as $|\boldsymbol{r}_1 - \boldsymbol{r}_2| \to \infty$, and since this is the limiting behaviour of $(g^{(\mathrm{a,b})} - 1)$ it is convenient to define this quantity as the total correlation function $h^{(\mathrm{a,b})}(\boldsymbol{q}_1, \boldsymbol{q}_2)$ (see Eqn (3.24)). The total correlation can be thought of as arising from two contributions: (i) the direct correlation between $\boldsymbol{q}_1$ and $\boldsymbol{q}_2$ and (ii) the correlation transmitted indirectly via other particles in the fluid. For example, when a particle of type α is at $\boldsymbol{q}_3$ the direct correlation between $\boldsymbol{q}_1$ and $\boldsymbol{q}_3$ is "transmitted" to $\boldsymbol{q}_1$ and, since the total correlation between $\boldsymbol{q}_1$ and $\boldsymbol{q}_3$ is $h^{(\mathrm{a},\alpha)}(\boldsymbol{q}_1, \boldsymbol{q}_3)$, the contribution of this third particle to $h^{(\mathrm{a,b})}(\boldsymbol{q}_1, \boldsymbol{q}_2)$ is $h^{(\mathrm{a},\alpha)}(\boldsymbol{q}_1, \boldsymbol{q}_3)\ c^{(\mathrm{b},\alpha)}(\boldsymbol{q}_2, \boldsymbol{q}_3)$. It must also be taken into account that the third particle can be at any position in the fluid and that the number of particles in unit volume at $\boldsymbol{q}_3$ is $\rho^{(\alpha)}(\boldsymbol{q}_3)$. Finally we must incorporate the possibility that species α at $\boldsymbol{q}_3$ can be any among the various species present in the system, so that

$$h^{(\mathrm{a,b})}(\boldsymbol{q}_1, \boldsymbol{q}_2) = c^{(\mathrm{a,b})}(\boldsymbol{q}_1, \boldsymbol{q}_2) + \sum_{\alpha=\mathrm{a,b}} \int \rho^{(\alpha)}(\boldsymbol{q}_3) h^{(\mathrm{a},\alpha)}(\boldsymbol{q}_1, \boldsymbol{q}_3) c^{(\mathrm{b},\alpha)}(\boldsymbol{q}_2, \boldsymbol{q}_3)\, \mathrm{d}\boldsymbol{q}_3. \quad (3.140)$$

This is the Ornstein–Zernike equation generalized to a mixture of species some of which may be distributed in a non-uniform way through the system. We note that Eqn (3.140) as written, includes orientational as well as translational correlations. Of course the system of interdependent correlations just described is quite symmetrical and it would be therefore equally valid to transpose the pairs (a, b) and (1, 2) which would generate a second set of equations, similar in form to (3.140) with the correlations

$h_{23}^{(b,\alpha)}$, $c_{13}^{(a,\alpha)}$ under the integral sign. It should also be noted that the selected pair (a, b) can be any of the pairs of species, like or unlike, present in the system. Thus (3.140) defines a set of simultaneous integral equations for the vectors $\boldsymbol{h}$ or $\boldsymbol{c}$ which have the components $h_{ij}^{(\alpha,\beta)}$, $c_{ij}^{(\alpha,\beta)}$ in which $\alpha, \beta = \mathrm{a}, \mathrm{b}, \ldots$ and $i, j = 1, 2, \ldots$.

For a single-component non-uniform fluid, Eqn (3.140) reduces to

$$h_{12}^{(2)} = c_{12}^{(2)} + \int \rho_3^{(1)} h_{13}^{(2)} c_{32}^{(2)} \, \mathrm{d}\boldsymbol{q}_3. \tag{3.141}$$

This is an exact equation, but in order to make use of it a closure must be found in the form of a relationship between $c^{(2)}$ and $h^{(2)}$ (or $g^{(2)}$). For the uniform fluid case such a closure is sufficient to render (3.141) useful in the determination of $g^{(2)}$. However for non-uniform fluids it is clear that $\rho^{(1)}$ remains undetermined in such a scheme and the introduction of further approximations, for example for $g^{(2)}$ in terms of the correlation function for a uniform phase $\tilde{g}^{(2)}$, would be necessary. In the absence of any auxiliary information it would then become difficult to decide which approximation involved the more serious inaccuracies.

A method of establishing OZ equations in a form suitable for application to adsorption systems was proposed by Perram and co-workers (Perram and White, 1975; Perram and Smith, 1976), by Percus (1976) and by Henderson *et al.* (1976). The starting point is Eqn (3.140) written for a mixture in which each component is present with a uniform density. One component in the mixture is taken to a limit where it becomes infinitely dilute and its molecular radius can be considered to be very large in comparison to that of other species. In the limit of infinite radius this species becomes a plane adsorbent wall; it will be denoted by w. Specializing to a two-component mixture (a, w), where a now denotes fluid adsorbate, Eqn (3.140) with $\rho_{\mathrm{w}} \to 0$, yields the following four equations,

$$\tilde{h}_{12}^{(\mathrm{a,a})} = \tilde{c}_{12}^{(\mathrm{a,a})} + \rho_{\mathrm{a}} \int \tilde{h}_{13}^{(\mathrm{a,a})} \tilde{c}_{32}^{(\mathrm{a,a})} \, \mathrm{d}\boldsymbol{q}_{3\mathrm{a}} \tag{3.142a}$$

$$\tilde{h}_{12}^{(\mathrm{w,w})} = \tilde{c}_{12}^{(\mathrm{w,w})} + \rho_{\mathrm{a}} \int h^{(\mathrm{w,a})} c^{(\mathrm{a,w})} \, \mathrm{d}\boldsymbol{q}_{3\mathrm{a}} \tag{3.142b}$$

$$h_{12}^{(\mathrm{w,a})} = c_{12}^{(\mathrm{w,a})} + \rho_{\mathrm{a}} \int h_{13}^{(\mathrm{w,a})} \tilde{c}_{32}^{(\mathrm{a,a})} \, \mathrm{d}\boldsymbol{q}_{3\mathrm{a}} \tag{3.142c}$$

$$h_{12}^{(\mathrm{a,w})} = c_{12}^{(\mathrm{a,w})} + \rho_{\mathrm{a}} \int \tilde{h}_{13}^{(\mathrm{a,a})} c_{32}^{(\mathrm{a,w})} \, \mathrm{d}\boldsymbol{q}_{3\mathrm{a}} \tag{3.142d}$$

in which ρ_{a} is the density of the component a at a large distance from the component w, and a tilde denotes a function in a uniform phase having this density.

Of these equations, (3.142a) is simply the conventional OZ equation for a uniform fluid a, and equation (3.142b) is of no further interest in the present context. In the remaining equations $h_{12}^{(w,a)}$ expresses the density variation in a due to the presence of w and therefore has the significance of a singlet correlation function $h^{(a)}$, although it originates as a pairwise property. Under these conditions the pair distribution $\rho^{(w,a)}$ can be interpreted as a singlet distribution for the adsorbate fluid, $\rho^{(a)}$, with an appropriate translation of the origin of the z-co-ordinate to the surface of the w-particle. We may therefore write

$$\rho^{(a)} = \rho_a g^{(a)} = \rho_a(1 + h^{(a)}) \qquad \text{with} \qquad h^{(w,a)} \equiv h^{(a)} \tag{3.143}$$

and (3.142c, d) may be put in a form which applies to the non-uniform fluid only,

$$h^{(1)}(z) = c^{[1]}(z) + \rho_a \int h^{(1)}(z')\tilde{c}^{(2)}(|\mathbf{r} - \mathbf{r}'|)\, d\mathbf{r}' \tag{3.144a}$$

$$h^{(1)}(z) = c^{[1]}(z) + \rho_a \int \tilde{h}^{(2)}(|\mathbf{r} - \mathbf{r}'|)c^{[1]}(z')\, d\mathbf{r}'. \tag{3.144b}$$

These equations are complementary, but (3.144a) is the more useful in practice. Note that $c^{[1]}$ here has the interpretation $c_{12}^{(w,a)}$ and is to be distinguished from the true singlet direct correlation function defined below in Eqn (3.158a). Equations (3.144) may be transformed to a cylindrical co-ordinate system with its origin on a fixed particle at z in the same way as was shown for the BBGKY equations.

(c) Functional differentiation methods

The OZ equation is centred upon the direct correlation function and indeed may be regarded as a definition of this function. In order to make the equation useful, therefore, it is necessary to know more about c.

A powerful method for acquiring this knowledge is available in functional differentiation techniques. The wide ranging possibilities of this technique were very fully developed in an article by Percus (1964) and have been exploited by Lux and Münster (1967). The mathematical method and its relationship to diagram expansions have been described by Morita and Hiroike (1961), a general exposition is available in the standard work by Volterra (1959) and recent texts on statistical mechanics contain excellent summaries of the essentials (e.g. see, Münster, 1969, Baxter, 1970, McDonald and Hansen, 1976). We give a brief outline in Appendix 2 for the sake of completeness.

The grand partition function Ξ^A for the adsorbate phase, defined in (3.33), provides a convenient starting point. We have chosen a single-component system, the adsorbate fluid containing N^a molecules of type a, to simplify the notation at this stage. The extension to multicomponent systems follows along lines similar to those indicated for the BBGKY equations in § 5(a). The canonical partition function is defined in (3.32), the configurational integral in (3.31), and the potential energy of the adsorbate, U_{N_a} in (3.28). It is useful to introduce a new notation by defining the quantities $\zeta^{[1]}$, $\varepsilon^{[n]}_{ijk\ldots}$ where,

$$\frac{\lambda^{N_a}\exp(-\beta U_{N_a})}{\Lambda^{\nu N_a}} = \frac{\lambda^{N_a}\exp\left(-\beta\sum_{i=1}^{N_a} u_i^{[a]}\right)\exp\left(-\beta\sum_{i>j} u_{ij}^{[a,a]}\right)\ldots}{\Lambda^{\nu N_a}}$$

$$= \prod_i \zeta_i^{[a]} \prod_{i>j} \varepsilon_{ij}^{[a,a]} \prod_{i>j>k} \varepsilon_{ijk}^{[a,a,a]} \ldots \quad (3.145a)$$

with

$$\zeta_i^{[a]} = \lambda \exp(-\beta u_i^{[a]})/\Lambda^{\nu} \quad (3.145b)$$

$$\varepsilon_{ij}^{[a,a]} = \exp(-\beta u_{ij}^{[a,a]}) \quad (3.145c)$$

$$\varepsilon_{ijk}^{[a,a,a]} = \exp(-\beta u_{ijk}^{[a,a,a]}) \ldots \quad (3.145d)$$

and where, as before, the superscripts [a], [1]; [a, a], [2] are interchangeable if only a single component is being considered.

With this notation the grand partition function can be written,

$$\Xi^{A} = \sum_{N_a \geq 0} \frac{1}{N_a!} \int \cdots \int \prod_{i\geq 1} \zeta_i^{[1]} \prod_{i>j} \varepsilon_{ij}^{[2]} \ldots \, d\boldsymbol{q} \quad (3.146)$$

similarly the generic n-particle distribution function of (3.14) may be written, for the single-component adsorbate,

$$\rho^{(n)} = \frac{1}{\Xi^{A}} \sum_{N_a \geq n} \frac{1}{(N_a - n)!} \int_{(N_a - n)} \cdots \int \prod_{i\geq 1} \zeta_i^{[1]} \prod_{i>j} \varepsilon_{ij}^{[2]} \ldots \, d\boldsymbol{q}_{n+1}. \quad (3.147)$$

Thus the functional derivative of Ξ^{A} with respect to $\zeta^{[1]}$ is

$$\frac{\delta \Xi^{A}}{\delta \zeta_1^{[1]}} = \frac{\Xi^{A} \rho_1^{(1)}}{\zeta_1^{[1]}} \quad (3.148a)$$

or

$$\frac{\delta \ln \Xi^{A}}{\delta \zeta_1^{[1]}} = \frac{\rho_1^{(1)}}{\zeta_1^{[1]}}. \quad (3.148b)$$

(Note that $\delta \ln \bar{\Xi}^{A}/\delta \zeta_1^{[1]}$ is zero for a uniform phase so that subsequent equations relate directly to the excess free energies of Eqn (3.37b).

The second derivative of $\ln \Xi^{\mathrm{A}}$ is easily obtained from these equations using the methods of elementary calculus for the differentiation of a product to give

$$\frac{\delta^2 \ln \Xi^{\mathrm{A}}}{\delta \zeta_2^{[1]} \delta \zeta_1^{[1]}} = \frac{\delta(\rho_1^{(1)}/\zeta_1^{[1]})}{\delta \zeta_2^{[1]}} = \frac{\rho_{12}^{(2)} - \rho_1^{(1)}\rho_2^{(1)}}{\zeta_1^{[1]}\zeta_2^{[1]}} = \frac{\mathscr{F}_{12}^{(2)}}{\zeta_1^{[1]}\zeta_2^{[1]}} \tag{3.149}$$

in which $\mathscr{F}_{12}^{(2)}$ is the Ursell function defined in (3.23). By a straightforward extension of this procedure the higher-order Ursell functions may be defined and expressed in terms of generic distributions. Thus in general,

$$\frac{\delta^n \ln \Xi^{\mathrm{A}}}{\delta \zeta_1^{[1]} \delta \zeta_2^{[1]} \ldots \delta \zeta_n^{[1]}} = \frac{\mathscr{F}_{12\ldots n}^{(n)}}{\zeta_1^{[1]}\zeta_2^{[1]} \ldots \zeta_n^{[1]}}. \tag{3.150}$$

Equation (3.148b) can be written in the form of an expression for the singlet distribution function

$$\frac{\delta \ln \Xi^{\mathrm{A}}}{\delta \ln \zeta_1^{[1]}} = \frac{\zeta_1^{[1]}}{\Xi^{\mathrm{A}}} \frac{\delta \Xi^{\mathrm{A}}}{\delta \zeta_1^{[1]}} = \rho^{(1)} \tag{3.151}$$

from which elementary manipulations lead to the second derivative

$$\frac{\delta^2 \ln \Xi^{\mathrm{A}}}{\delta \ln \zeta_2^{[1]} \delta \ln \zeta_1^{[1]}} = \frac{\delta \rho_1^{(1)}}{\delta \ln \zeta_2^{[1]}} = \mathscr{F}_{12}^{(2)} + \rho_1^{(1)}\delta_{12} \tag{3.152}$$

where δ_{12} is a delta function.

Bearing in mind the definition of $\zeta_i^{[1]}$ (Eqn (3.145b)) it can be seen that (3.152) essentially expresses the effect of a variation in the external field at $\boldsymbol{q}_2$ on the shape of the singlet distribution function. This effect has two components, $\rho_1^{(1)}\delta_{12}$, the direct effect at $\boldsymbol{q}_1$, and a longer-range interaction expressed through $\mathscr{F}^{(2)}$ which tends to zero as the separation $|\boldsymbol{r}_1 - \boldsymbol{r}_2|$ between positions 1 and 2 increases. If the inverse derivative $\delta \ln \zeta_2^{[1]}/\delta \rho_1^{(1)}$ is considered, it is reasonable to postulate that this has similar components, and we provisionally designate the component corresponding to $\mathscr{F}_{12}^{(2)}$ as $\mathscr{C}_{12}^{(2)}$ such that

$$\frac{\delta \ln \zeta_1^{[1]}}{\delta \rho_2^{(1)}} = \frac{\delta_{21}}{\rho_2^{(1)}} - \mathscr{C}_{12}^{(1)}. \tag{3.153}$$

Equations (3.152) and (3.153) can be combined in the form

$$\int \frac{\delta \rho_1^{(1)}}{\delta \ln \zeta_3^{[1]}} \cdot \frac{\delta \ln \zeta_3^{[1]}}{\delta \rho_2^{(1)}} \mathrm{d}\boldsymbol{q}_3 = \int (\rho_1^{(1)}\rho_3^{(1)}h_{13}^{(2)} - \rho_1^{(1)}\delta_{13}) \left(\frac{\delta_{23}}{\rho_2^{(1)}} - \mathscr{C}_{32}^{(2)}\right) \mathrm{d}\boldsymbol{q}_3. \tag{3.154}$$

After expanding the brackets and using the properties of the δ function, Eqn (3.154) becomes

$$h_{12}^{(2)} = \mathscr{C}_{12}^{(2)} + \int \rho_3^{(1)} h_{13}^{(2)} \mathscr{C}_{32}^{(2)} \, \mathrm{d}\boldsymbol{q}_3 \tag{3.155}$$

which is the non-uniform OZ equation (3.141) if $\mathscr{C}_{12}^{(2)}$ is identified with $c_{12}^{(2)}$.

It is to be noted that the preceding development does not constitute a derivation of the OZ equation any more than the more intuitive approach outlined in the previous section. The latter might be regarded as a physical argument relating to the properties of the direct correlation function, whereas the development from Eqn (3.153) relies on a mathematical postulate and opens the way to a fuller understanding of the concept of a direct correlation function. For example, (3.153) may be rewritten

$$c_{12}^{(2)} = \frac{\delta}{\delta\rho_2^{(1)}} \ln\,(\rho_1^{(1)}/\zeta_1^{[1]}) \tag{3.156}$$

$$\left(\text{since} \frac{\delta \ln \rho_1^{(1)}}{\delta \rho_2^{(1)}} = \frac{\delta \ln \rho_1^{(1)}}{\delta \ln \rho_2^{(1)}} \cdot \frac{1}{\rho_2^{(1)}} = \frac{\delta_{21}}{\rho_2^{(1)}}\right)$$

which immediately implies that all orders of direct correlation function may be generated by successive differentiation. For a multicomponent system, the nth-order correlation function is

$$c_{12\ldots n}^{(\mathrm{a,b,\ldots})} = \frac{\delta^{n-1} \ln(\rho_1^{(\mathrm{a})}/\zeta_1^{[\mathrm{a}]})}{\delta\rho_2^{(\mathrm{b})}\delta\rho_3^{(\mathrm{c})} \ldots \delta\rho_n^{(\mathrm{a})}} = \frac{\delta c_{12\ldots,n-1}^{(\mathrm{a,b,\ldots})}}{\delta\rho_n^{(\alpha)}}. \tag{3.157}$$

The fundamental generating function is the singlet direct correlation function

$$c^{(1)} = \ln(\rho^{(1)}/\zeta^{[1]}). \tag{3.158a}$$

This can be rearranged in the form

$$\rho^{(1)} = \frac{\lambda}{\Lambda^3} \exp[-\beta u^{[1]} + c^{(1)}]. \tag{3.158b}$$

If $c^{(1)}$ is zero, Eqn (3.158b) is the well known Henry law expression giving the singlet density for a gas of non-interacting particles in an external field $u^{[1]}$. Thus $c^{(1)}$ expresses the effect on $\rho^{(1)}$ of mutual interactions amongst the adsorbate particles.

In an alternative approach (Mermin, 1965; Evans, 1979; Yang *et al.*, 1976), $c^{(1)}$ can be related to the thermodynamic potentials. We observe that Eqn (3.158a) can be put in yet another form as

$$\beta\mu = \ln\,(\rho^{(1)}\Lambda^3) - c^{(1)} + \beta u^{[1]}. \tag{3.158c}$$

The Helmholtz free energy A of the non-uniform fluid may be written with the aid of Eqn (3.79) for the internal energy as

$$A = U - TS = \int u_1^{[1]}\rho_1^{(1)}\,\mathrm{d}\boldsymbol{q}_1 + \tfrac{1}{2}\iint u_{12}^{[2]}\rho_{12}^{(2)}\,\mathrm{d}\boldsymbol{q}_1\,\mathrm{d}\boldsymbol{q}_2 + \ldots + U_{\mathrm{K}} - TS \quad (3.159)$$

and its functional derivative with respect to $\rho^{(1)}$ can be expressed as

$$\frac{\delta A}{\delta\rho^{(1)}} = u_1^{[1]} + \frac{\delta A_{\mathrm{int}}}{\delta\rho_1^{(1)}} \quad (3.160)$$

where A_{int} is the intrinsic Helmholtz energy for a uniform fluid of adsorbate particles having the local density $\rho_1^{(1)}$. The grand potential Ω of the system, given in Eqn (2.41), can be written as

$$\Omega = A - \mu\int_{V\mathrm{A}} \rho_1^{(1)}\,\mathrm{d}\boldsymbol{q}_1 \quad (3.161)$$

and is subject to the minimization condition $\delta\Omega/\delta\rho^{(1)} = 0$ (Mermin 1965). Employing this condition with (3.160) and comparing with (3.158c) shows that the intrinsic chemical potential is

$$\beta\mu_{\mathrm{int}} = \frac{\delta\beta A_{\mathrm{int}}}{\delta\rho^{(1)}} = \ln\,(\rho^{(1)}\Lambda^3) - c^{(1)}. \quad (3.162)$$

In an ideal gas with density $\rho^{(1)}$, the chemical potential is $\beta\mu_{\mathrm{id}} = \ln(\rho^{(1)}\Lambda^3)$, so that $c^{(1)}$ is identified once again as an expression of the adsorbate molecule interactions, since

$$\Delta\mu_{\mathrm{int}} = \mu_{\mathrm{int}} - \mu_{\mathrm{id}} = -c^{(1)}/\beta. \quad (3.163)$$

The definition of intrinsic chemical potential given in (3.162) can be used to put (3.158c) in the form

$$\mu = \mu_{\mathrm{int}} + u^{[1]} \quad (3.164)$$

which is a relationship of fundamental significance for non-uniform systems (Evans, 1979). It should be noted that since $\bar{\Xi}^{\mathrm{A}}$ for the uniform gas phase in Eqns (3.34–3.37) is invariant, the results following from the condition $\delta\Omega/\delta\rho^{(1)} = 0$, would also be obtained had we used Ω^{Σ} for the surface excess in place of Ω.

Having established these basic relationships we now proceed to consider some examples of the use of functional differentiation for the derivation of integral equations and density expansions.

(d) Integral equations with Ursell function kernels

From Eqn (3.149) we have

$$\frac{\delta(\rho_1^{(1)}/\zeta_1^{[1]})}{\delta\zeta_2^{[1]}} = \frac{\mathscr{F}_{12}^{(2)}}{\zeta_1^{[1]}\zeta_2^{[1]}}. \tag{3.165}$$

If $\rho_1^{(1)}/\zeta_1^{[1]}$ is Taylor expanded about a reference system value, denoted by superscript °, then

$$\frac{\rho_1^{(1)}}{\zeta_1^{[1]}} = \left(\frac{\rho_1^{(1)}}{\zeta_1^{[1]}}\right)^{\circ} + \int \left[\frac{\mathscr{F}^{(2)}}{\zeta_1^{[1]}\zeta_2^{[1]}}\right]^{\circ} (\zeta_2^{[1]} - \zeta_2^{[1]\circ})\, \mathrm{d}\boldsymbol{q}_2 + \ldots . \tag{3.166a}$$

Equation (3.166a) may be rewritten as an expansion for $\rho^{(1)}$, where the values of $\rho^{(1)}$ and $\mathscr{F}^{(2)}$ in the reference system are known.

If the reference system is non-uniform and not too far removed from the unknown system it is reasonable to truncate the expansion at the first term, leaving

$$\rho_1^{(1)} = \rho_1^{(1)\circ} \exp(-\beta u_1^{[1]})\Big\{1 + \int \rho_2^{(1)\circ} h_{12}^{(2)\circ} [\exp(-\beta u_1^{[1]})/\exp(-\beta u_1^{[1]\circ}) - 1]\, \mathrm{d}\boldsymbol{q}_2\Big\}. \tag{3.166b}$$

In the particular case where the reference phase is uniform with density ρ, Eqn (3.166b) simplifies to give

$$\rho_1^{(1)} = \rho \exp(-\beta u_1^{[1]}) \left\{1 + \rho \int \tilde{h}_{12}^{(2)} [\exp(-\beta u_2^{[1]}) - 1]\, \mathrm{d}\boldsymbol{q}_2 + \ldots\right\} \tag{3.166c}$$

in which the higher terms involve $\tilde{g}^{(3)}$ and higher-order correlation functions. It should be noted in connection with the application of (3.166b, c) that the actual boundary conditions for the integration can be of crucial significance in determining the structure of the reference phase. As pointed out by Findenegg and Fischer (1975), $\rho^{(1)\circ}$ may not be equal to ρ even when the attractive wall force is turned off. This is due to asymmetry in the intermolecular forces acting among the adsorbate molecules themselves.

An alternative form of (3.166) can be obtained using (3.152) in a Taylor expansion of the singlet density to give,

$$\rho_1^{(1)} = \rho\left\{1 - \beta u_1^{[1]} - \beta\rho \int h_{12}^{(2)} u_2^{[1]}\, \mathrm{d}\boldsymbol{q}_2\right\} \tag{3.167}$$

which is known as Yvon's equation. Equation (3.166) is more appealing in the context of adsorption since it gives the correct Henry's law limit when adsorbate–adsorbate interactions are absent.

The above type of procedure can be usefully generalized to give higher-order distribution functions (Percus, 1964). To see this we introduce a conditional distribution function in the following way. Suppose that a single fixed particle is placed at $\boldsymbol{q}_1$ in a fluid originally containing N particles numbered 2 to $N+1$ and that its interaction with the external field and with other particles is φ, where

$$\varphi = u_1^{[1]} + \sum_{i>1}^{N+1} u_{i1}^{[2]}. \tag{3.168}$$

The grand partition function for this new system of $(N+1)$ particles is

$$\Xi(\varphi) = \sum \frac{\lambda^{N+1}}{\Lambda^{\nu(N+1)}(N+1)!} \int \underset{N+1}{\cdots} \int \exp(-\beta U_N)\exp(-\beta\varphi)\,\mathrm{d}\boldsymbol{q}_1, \mathrm{d}\boldsymbol{q}_2 \ldots \mathrm{d}\boldsymbol{q}_{N+1} \tag{3.169}$$

and the pair distribution function for this system can be written

$$\Xi(\varphi)\rho_{12}^{(2)}(\varphi) = \sum \frac{\lambda^{N-1}}{\Lambda^{\nu(N-1)}(N-1)!} \int \cdots \int \exp(-\beta U_N)\exp(-\beta\varphi)\,\mathrm{d}\boldsymbol{q}_3 \ldots \mathrm{d}\boldsymbol{q}_{N+1}. \tag{3.170}$$

The singlet distribution function in the old system, before addition of the particle, can be put in a similar form:

$$\Xi(0)\rho_2^{(1)}(0) = \sum \frac{\lambda^{N-1}}{\Lambda^{\nu(N-1)}(N-1)!} \int \cdots \int \exp(-\beta U_N)\,\mathrm{d}\boldsymbol{q}_3 \ldots \mathrm{d}\boldsymbol{q}_{N+1}. \tag{3.171}$$

When both sides of this expression are multiplied by $\exp(-\beta\varphi)$, it is seen that

$$\exp(-\beta\varphi)\Xi(0)\rho_2^{(1)}(0) = \Xi(\varphi)\rho_{12}^{(2)}(\varphi) \tag{3.172}$$

whence the singlet distribution in the new system is given by

$$\rho^{(1)}(\varphi) = \exp(-\beta\varphi)\ \Xi(0)/\Xi(\varphi),$$

so that (3.172) can be put finally in the form

$$\rho_1^{(1)}(\varphi)\ \rho_2^{(1)}(0) = \rho_{12}^{(2)}(\varphi). \tag{3.173}$$

This result is to be interpreted as an expression giving the change in the singlet distribution function when the field due to the new particle is turned on; that is, the singlet distribution $\rho_2^{(1)}(0)$ in the old system changes to $\rho_{12}^{(2)}(\varphi)/\rho_1^{(1)}(\varphi)$ in the new one, and the "internal external potential" due to the particle at 1 is $u_{1i}^{[2]}$ which replaces $u_1^{[1]}$.

Equation (3.166a) written for particles 2 and 3, with Eqn (3.173) gives

$$\frac{\rho_{12}^{(2)}}{\rho_1^{(1)}\varepsilon_{12}^{[2]}} - \frac{\rho_2^{(1)}}{\varepsilon_{12}^{[2]\circ}} = \int \left[\frac{\mathcal{F}_{23}^{(2)}}{\varepsilon_{12}^{[2]}\varepsilon_{13}^{[2]}}\right]^0 \delta\varepsilon_{13}^{[2]}\,\mathrm{d}\boldsymbol{q}_3 \tag{3.174}$$

where the reference state is the old system with $\varepsilon_{12}^{[2]\circ} = 1$. This equation can be simplified to give

$$g_{12}^{(2)} = \varepsilon_{12}^{[2]}\left[1 + \int \rho_3^{(1)} h_{23}^{(2)} f_{13}^{(2)}\,\mathrm{d}\boldsymbol{q} + \ldots\right] \tag{3.175}$$

in which $f_{ij} = \varepsilon_{ij} - 1$ is the Mayer function, and it has been assumed that the addition of one extra molecule effects a negligible change in the pair correlation function. This is an integral equation for $g^{(2)}$ in the non-uniform system within the limits implied by the truncation of the expansion, but its solution requires the singlet distribution which could be obtained in principle by coupling with another equation such as the BBGKY or Eqn (3.166).

In a uniform system (3.175) reduces to an expression for $\tilde{g}^{(2)}$

$$\tilde{g}^{(2)} = \varepsilon_{12}^{(2)}\left[1 + \rho\int \tilde{h}_{12}^{(2)} f_{13}^{(2)}\,\mathrm{d}\boldsymbol{q}_3 + \ldots\right] \tag{3.176}$$

which is the initial equation in the Kirkwood–Salsburg hierarchy. (Kirkwood and Salsburg, 1953).

(e) Integral equations with direct correlation function kernels

The inverse equations which define $c^{(n)}$ can be treated in a similar manner to give a functional expansion in $\delta\rho^{(1)}$ instead of $\delta\zeta^{[1]}$. From (3.153) and (3.166) respectively we can obtain

$$\ln(\zeta_1^{[1]}/\zeta_1^{[1]\circ}) = \int \delta_{12}(\rho_2^{(1)}/\rho_2^{(1)\circ} - 1)\,\mathrm{d}\boldsymbol{q}_2 - \int c_{12}^{(2)\circ}(\rho_2^{(1)} - \rho_2^{(1)\circ})\,\mathrm{d}\boldsymbol{q}_2 + \ldots \tag{3.177a}$$

and

$$\ln(\rho_1^{(1)}/\zeta_1^{[1]}) = \ln(\rho_1^{(1)\circ}/\zeta_1^{[1]\circ}) + \int c_{12}^{(2)\circ}(\rho_2^{(1)} - \rho_2^{(1)\circ})\,\mathrm{d}\boldsymbol{q}_2 + \ldots \tag{3.177b}$$

where the omitted terms will contain higher-order integrals over $c^{(3)}$, $c^{(4)}$, etc., and the $^\circ$ superscripts again indicate the reference state.

The degree of approximation involved in truncating the Taylor series at the first term could be quite serious when the only known reference state lies at some distance from the system of interest.

A somewhat different method of approach is to introduce a coupling parameter ξ which can be changed infinitesimally slowly. We consider some general functional $\psi(\xi)$ which changes its shape as ξ changes, and impose limits between the reference state ψ° and the final state ψ such that

$$\psi(\xi) = \begin{cases} \psi^\circ & \xi = 0 \\ \psi & \xi = 1 \end{cases}. \tag{3.178a}$$

If $\psi(\xi)$ depends linearly on ξ we have

$$\psi(\xi) = \psi^\circ + \xi[\psi - \psi^\circ] \tag{3.178b}$$

and thus

$$\delta\psi = \int_0^1 \left(\frac{\partial\psi}{\partial\xi}\right) \mathrm{d}\xi. \tag{3.178c}$$

When $\ln \zeta$ and $\rho^{(1)}$ are coupled in this way, Eqn (3.177a) is replaced by

$$\begin{aligned} \ln(\zeta_1^{[1]}/\zeta_1^{[1]\circ}) = {} & \int \delta_{12}(\rho_2^{(1)}/\rho_2^{(1)\circ} - 1)\,\mathrm{d}\boldsymbol{q}_2 \\ & - \int_0^1 \int c_{12}^{(2)}(\xi)\,[\rho_2^{(1)} - \rho_2^{(1)\circ}]\,\mathrm{d}\xi\,\mathrm{d}\boldsymbol{q}_2 \end{aligned} \tag{3.179a}$$

and (3.177b) is replaced by

$$\ln(\rho_1^{(1)}/\zeta_1^{[1]}) = \ln(\rho_1^{(1)\circ}/\zeta_1^{[1]\circ}) + \int_0^1 \int c_{12}^{(2)}(\xi)\,[\rho_2^{(1)} - \rho_2^{(1)\circ}]\,\mathrm{d}\xi\,\mathrm{d}\boldsymbol{q}_2. \tag{3.179b}$$

Although these equations are exact they still present the problem that $c_{12}^{(2)}(\xi)$ must be determined for a range of values of the coupling parameter. In principle, at least, this problem can be solved by again introducing an extra particle at $\boldsymbol{q}_1$, which promotes an "internal external field". Here the modification to the system can sensibly be regarded as infinitesimal, and truncation of the Taylor series after the first term should not involve serious error, so that (3.177a) and (3.177b) become

$$-\beta u_{12}^{[2]} = h_{12}^{(2)} - \int \rho_3^{(1)} c_{23}^{(2)} h_{13}^{(2)}\,\mathrm{d}\boldsymbol{q}_3 \tag{3.180a}$$

and

$$\ln g_{12}^{(2)} + \beta u_{12}^{[2]} = \int \rho_3^{(1)} c_{23}^{(2)} h_{13}^{(2)}\,\mathrm{d}\boldsymbol{q}_3. \tag{3.180b}$$

If these results are compared with the non-uniform OZ equation (3.141), they produce two well known approximations to $c^{(2)}$. Thus from (3.180a),

$$c^{(2)} = -\beta u^{[2]} \qquad \text{(DH or MSA)}. \tag{3.181a}$$

This is the Debye–Hückel or mean spherical approximation. In its latter guise the approximation is applied in a specified region of the potential, usually where this is attractive. It is particularly useful when hard-core approximations are employed because in this case one has the two conditions

$$g^{(2)} = 0, \quad r < \sigma \qquad \text{and} \qquad c^{(2)} = -\beta u^{[2]}, \quad r > \sigma.$$

Equation (3.180b) produces the convoluted hypernetted chain equations of Morita and Hiroike (1960):

$$c^{(2)} = h^{(2)} - \ln g^{(2)} - \beta u^{[2]} \qquad \text{(CHNC)}. \tag{3.181b}$$

A third approximation to $c^{(2)}$ can be obtained by expanding $\rho_1^{(1)}/\zeta_1^{[1]}$, rather than its logarithm, and retaining the first two terms only from the exponential of the integral in (3.177b). This leads to

$$\frac{\rho_1^{(1)}}{\zeta_1^{[1]}} \frac{\zeta_1^{[1]\circ}}{\rho_1^{(1)\circ}} = 1 + \int c_{12}^{(2)\circ}[\rho_2^{(1)} - \rho_2^{(1)\circ}]\, \mathrm{d}\boldsymbol{q}_2 + \ldots \tag{3.177c}$$

which could also be put into the coupling parameter form of Eqn (3.179):

$$\frac{\rho_1^{(1)}}{\zeta_1^{[1]}} \frac{\zeta_1^{[1]\circ}}{\rho_1^{(1)\circ}} = 1 + \int_0^1 \int c_{12}^{(2)}(\xi)\, [\rho_2^{(1)} - \rho_2^{(1)\circ}]\, \mathrm{d}\xi\, \mathrm{d}\boldsymbol{q}_2. \tag{3.179c}$$

This again leads to an equation for the two-particle correlation function via the usual route of introducing an extra particle:

$$g_{12}^{(2)} \exp(\beta u_{12}^{[2]}) = 1 + \int \rho_3^{(1)} h_{13}^{(2)} c_{23}^{(2)}\, \mathrm{d}\boldsymbol{q}_3. \tag{3.180c}$$

In this case a comparison of Eqn (3.180c) with the non-uniform OZ equation gives the Percus–Yevick (PY) approximation to $c^{(2)}$:

$$c^{(2)} = g^{(2)}[1 - \exp(\beta u^{[2]})] \qquad \text{(PY)}. \tag{3.181c}$$

Any of the Eqns (3.181a, b, c) can be written as a function of the coupling parameter ξ and used as a closure to solve the non-uniform OZ equation. The solutions so obtained can be used in principle with any of the Eqns (3.179a, b, c) to give the singlet distribution functions. So far this system of equations has not been exhaustively explored, but it can be said that it will be subject to the well known limitations (see e.g. Barker and Henderson, 1976) which apply to the uniform versions of Eqn (3.181) in addition to those which may arise in non-uniform systems.

A less rigorous alternative is to assume some form for the dependence of $c^{(2)}(\xi)$ on ξ. For example if we choose the linear dependence of (3.178), the integrals in (3.179) become

$$\int^1 \int c_{12}^{(2)}(\xi)\, [\rho_2^{(1)} - \rho_2^{(1)\circ}]\, \mathrm{d}\xi\, \mathrm{d}\boldsymbol{q}_2 \to \int [c_{12}^{(2)\circ} + c_{12}^{(2)}]\, [\rho_2^{(1)} - \rho_2^{(1)\circ}]\, \mathrm{d}\boldsymbol{q}_2. \tag{3.182}$$

The far more drastic assumption, that $c^{(2)}(\xi)$ is independent of ξ, is equivalent to truncating Eqns (3.177) at the first term, but retains a greater flexibility in the choice of parameters ($\boldsymbol{q}_2$ or density) for which the constant $c^{(2)\circ}$ is chosen (Ebner *et al.*, 1976). An obvious possibility at this stage of approximation is to make $c^{(2)\circ}$ equivalent to the direct correlation function for a uniform fluid, $\bar{c}^{(2)}$, at some average position or density computed from the local properties of the fluid. Some examples of the application of this type of approximation are discussed in Chapter 7.

The added particle method which led to (3.181) can easily be extended to the case where a particle of type w is inserted into a fluid of type a particles. A rigorous development is given in Appendix 3, where it is shown that the singlet distribution in the fluid is changed according to

$$\rho_2^{(a)}(0) \to \rho_{12}^{(w,a)}(\varphi)/\rho_1^{(w)}(\varphi). \tag{3.183}$$

The case of particular interest here is where the added particle is inserted into a uniform fluid of type a molecules, and $\rho^{(a)} = \rho_a$, the uniform density. If we write the pair potential $u^{(w,a)}$ as an external field $u^{[a]}$, Eqns (3.180) now take the form

$$\beta u^{[a]} + h_{12}^{(w,a)} = \rho_a \int h_{13}^{(w,a)} \bar{c}_{32}^{(a,a)} \, d\boldsymbol{q}_{3a} \tag{3.184a}$$

$$\ln g_{12}^{(w,a)} + \beta u^{[a]} = \rho_a \int h_{13}^{(w,a)} \bar{c}_{32}^{(a,a)} \, d\boldsymbol{q}_{3a} \tag{3.184b}$$

$$g_{12}^{(w,a)} \exp(\beta u^{[a]}) - 1 = \rho^a \int h_{13}^{(w,a)} \bar{c}_{32}^{(a,a)} \, d\boldsymbol{q}_{3a}. \tag{3.184c}$$

If these equations are compared with the HAB equation (3.144a), it is clear that the three approximations of Eqn (3.181) will be obtained for $c^{(w,a)}$. When Eqns (3.184) are put in the form (3.144a), these approximations are respectively

$$c^{[1]} = -\beta u^{[1]} \qquad \text{(DH/MSA)} \tag{3.185a}$$

$$c^{[1]} = h^{(1)} - \ln g^{(1)} - \beta u^{[1]} \qquad \text{(CHNC)} \tag{3.185b}$$

$$c^{[1]} = g^{(1)}[1 - \exp(\beta u^{[1]})] \qquad \text{(PY).} \tag{3.185c}$$

Within the context of the HAB theory, therefore, any of the equations (3.184) can be written as expressions for the singlet distribution. If we now choose the reference state in Eqns (3.177) to be a uniform fluid of density ρ_a, then it is readily shown that these equations are identical with (3.184) at the level of approximation implied by the truncation at the first term of the Taylor expansion. The HAB theory is thus seen to be equivalent to the assumption that the higher-order integrals in the functional Taylor expansion involving $c^{(3)}$, $c^{(4)}$, etc., can be ignored. As already pointed out, this assumption is not likely to be valid when $\rho^{(a)}$ differs greatly from ρ_a

and the neglected higher-order terms may be more important in non-uniform systems than for the analogous determination of pair distributions in uniform fluids. Higher-order correction terms have been calculated for uniform fluids by Verlet and Levesque (1967).

(f) Analogues of the BBGKY equation

A minor modification of the functional Taylor expansion method can be used to develop analogues of the BBGKY equation. To illustrate this method we consider two functionals ψ and χ and write, with Δ here a small increment,

$$\psi(\boldsymbol{q}_1 + \Delta\boldsymbol{q}_1) = \psi(\boldsymbol{q}_1) + \int \frac{\delta\psi_1}{\delta\chi_2} \times [\chi(\boldsymbol{q}_2 + \Delta\boldsymbol{q}_2) - \chi(\boldsymbol{q}_2)]\,\mathrm{d}\boldsymbol{q}_2 + \ldots. \tag{3.186}$$

In the limit $\Delta\boldsymbol{q} \to 0$ this yields the exact result

$$\frac{\partial\psi_1}{\partial\boldsymbol{q}_1} = \int \left(\frac{\delta\psi_1}{\delta\chi_2}\right)\left(\frac{\partial\chi_2}{\partial\boldsymbol{q}_2}\right)\mathrm{d}\boldsymbol{q}_2. \tag{3.187}$$

If we now set $\psi = c^{(1)} = \ln(\rho^{(1)}/\zeta^{[1]})$, $\chi = \rho^{(1)}$ and use Eqn (3.157) for the pair correlation function, there results

$$\frac{\partial \ln \rho_1^{(1)}}{\partial\boldsymbol{q}_1} + \beta\frac{\partial u_1^{[1]}}{\partial\boldsymbol{q}_1} = \int c_{12}^{(2)}\left(\frac{\partial\rho_2^{(1)}}{\partial\boldsymbol{q}_2}\right)\mathrm{d}\boldsymbol{q}_2 \tag{3.188}$$

which may be compared with the BBGKY equation (Eqn (3.132)). A similar result is found if we set $\psi = \rho^{(1)}/\zeta^{[1]}$, $\chi = \zeta^{[1]}$ and use Eqn (3.188) for the Ursell function. This leads to the expression

$$\frac{\partial \ln \rho_1^{(1)}}{\partial\boldsymbol{q}_1} + \beta\frac{\partial u_1^{[1]}}{\partial\boldsymbol{q}_1} = -\int \rho_2^{(1)}h_{12}^{(2)}\left(\frac{\partial\beta u_2^{[1]}}{\partial\boldsymbol{q}_2}\right)\mathrm{d}\boldsymbol{q}_2. \tag{3.189}$$

By choosing χ to be the external field and ψ as $\rho^{(n)}$, the hierarchy of BBGKY equations themselves, including the initial singlet form of Eqn (3.134), may be obtained (Percus, 1964; Evans, 1979) by this method.

Since the left-hand sides of Eqns (3.134), (3.188) and (3.189) are identical it follows that the integrals (though not of course the integrands) from the right-hand sides of these equations can be equated, giving

$$-\int c_{12}^{(2)}\left(\frac{\partial\rho_2^{(1)}}{\partial\boldsymbol{q}_2}\right)\mathrm{d}\boldsymbol{q}_2 = \int \rho_2^{(1)}g_{12}^{(2)}\left(\frac{\partial\beta u_{12}^{[2]}}{\partial\boldsymbol{q}_2}\right)\mathrm{d}\boldsymbol{q}_2 = \int \rho_2^{(1)}h_{12}^{(2)}\left(\frac{\partial\beta u_2^{[1]}}{\partial\boldsymbol{q}_2}\right)\mathrm{d}\boldsymbol{q}_2. \tag{3.190}$$

These equations provide a set of self-consistency conditions for the non-uniform pair correlation functions.

Like the BBGKY equation itself, (3.188) (3.189) are exact but are unclosed since $c^{(2)}$, $h^{(2)}$ and $g^{(2)}$ must be obtained from higher-order equations in the hierarchy, or by the introduction of other approximations or by more intuitive methods.

Equation (3.186) can also be used in slightly modified form by incorporating the coupling parameter technique introduced earlier in the three Eqns (3.178) and applying (3.178c) to ψ or χ. Equation (3.187a) then becomes

$$\int \left(\frac{\partial \psi_1}{\partial \xi}\right) \mathrm{d}\xi = \iint \left(\frac{\delta \psi_1}{\delta \chi_2}\right)_\xi \left(\frac{\partial \chi_2}{\partial \xi}\right) \mathrm{d}\xi \, \mathrm{d}\boldsymbol{q}_2. \tag{3.187b}$$

With the identifications $\psi \equiv \rho^{(1)}/\zeta^{[1]}$ and $\chi \equiv \zeta^{[1]}$, this leads to a new equation

$$\ln(\rho_1^{(1)}/\rho_1^{(1)\circ}) - \ln(\zeta_1^{[1]}/\zeta_1^{[1]\circ}) = \int_0^1 \int \rho_2^{(1)}(\xi)\, h_{12}^{(2)}(\xi) \ln(\zeta_2^{[1]}/\zeta_2^{[1]\circ})\, \mathrm{d}\xi \, \mathrm{d}\boldsymbol{q}_2 \tag{3.191}$$

where the superscript $^\circ$ implies, as usual, the reference state, which is conveniently chosen to be a phase of uniform density with $u^{[1]} = 0$. This result was originally obtained (Findenegg and Fischer, 1975) by a different route in which the external field is coupled in and $\rho^{(1)}(\xi)$ differentiated with respect to the coupling parameter, a method which was introduced into statistical mechanics by Kirkwood (1935) and used to derive integral equations for higher-order correlation functions. The method described here is easily adapted to these cases by suitable choice of ψ and χ. When $\psi = c^{(1)}$ and $\chi \equiv \rho^{(1)}$, Eqn (3.179c) is recovered.

As with Eqn (3.179c), Eqn (3.191) can be rendered more tractable by introducing linear assumptions for $\rho^{(1)}(\xi)$ and $h^{(2)}(\xi)$ as in (3.178), which converts Eqn (3.191) into the form

$$\begin{aligned}\ln(\rho_1^{(1)}/\rho_1^{(1)\circ}) - \ln(\zeta_1^{[1]}/\zeta_1^{[1]\circ}) = \int \{ &\rho_2^{(1)\circ} h_{12}^{(2)\circ} + \tfrac{1}{2}[\rho_2^{(1)\circ} h_{12}^{(2)} - 2\rho_2^{(1)\circ} h_{12}^{(2)\circ} + \rho_2^{(1)} h_{12}^{(2)\circ}] \\ &+ \tfrac{1}{3}[\rho_2^{(1)} h_{12}^{(2)} - \rho_2^{(1)\circ} h_{12}^{(2)} - \rho_2^{(1)} h_{12}^{(2)\circ} + \rho_2^{(1)\circ} h_{12}^{(2)\circ}]\} \ln(\zeta_2^{[1]}/\zeta_2^{[1]\circ})\, \mathrm{d}\boldsymbol{q}_2.\end{aligned} \tag{3.192a}$$

If, on the other hand, $h^{(2)}(\xi)$ is set to $\bar{h}^{(2)}$ at some appropriately chosen density, the integral on the right-hand side of (3.192a) reduces to a much simpler form (Navascués, 1976), giving

$$\ln(\rho_1^{(1)}/\rho_1^{(1)\circ}) - \ln(\zeta_1^{[1]}/\zeta_1^{[1]\circ}) = \int \tfrac{1}{2}\bar{h}_{12}^{(2)}(\rho_2^{(1)} + \rho_2^{(1)\circ}) \ln(\zeta_2^{[1]}/\zeta_2^{[1]\circ})\, \mathrm{d}\boldsymbol{q}_2. \tag{3.192b}$$

An application of Eqn (3.191) which is particularly interesting in the context of multilayer physical adsorption arises if we choose the reference phase to be one in which a semi-infinite slab of adsorbate liquid replaces the semi-infinite slab of the solid adsorbent phase. We write, $u_1^{[1]\mathrm{P}} = (u_1^{[1]\mathrm{SOLID}} - u_1^{[1]\mathrm{LIQUID}})$, as the change in potential at $z_1 > z_0$ when the replacement is made. At high adsorbate coverage, it is reasonable to assume that the adsorbate phase is approximately uniform with density ρ_l appropriate to (p, T) of the system, and to replace $h^{(2)}(\xi)$ by the corresponding uniform function $\bar{h}^{(2)}$. Then (3.192) can be replaced by

$$\ln(\rho_1^{(1)}/\rho_l) = -\beta u^{[1]\mathrm{P}}\left[1 + \rho_l \int \bar{h}_{12}^{(2)}\, \mathrm{d}\boldsymbol{q}_2\right]. \tag{3.193}$$

According to Eqn (3.112c) the square bracket in this equation is equal to $kT\rho_l\kappa$, the compressibility equation for the uniform fluid, and therefore

$$\ln(\rho^{(1)}/\rho_l) = -u_1^{[1]\mathrm{P}}\rho_l\kappa. \tag{3.194}$$

If the above assumptions are applied to Eqn (3.188) or its analogues, this equation can be rearranged to read

$$\left[1 - \rho_l \int c_{12}^{(2)}\, \mathrm{d}\boldsymbol{q}_2\right]\frac{\partial \rho_1^{(1)}}{\partial z_1} = -\beta\rho_l\left(\frac{\partial u_1^{[1]\mathrm{P}}}{\partial z_1}\right) \tag{3.195}$$

where we have also assumed that $\rho^{(1)}$ and $u^{[1]}$ have z-dependence only. Again the square bracket is the standard expression, $(kT\rho_l\kappa)^{-1}$ and therefore (3.195) can be written

$$\left(\frac{\partial \rho_1^{(1)}}{\partial z_1}\right) = -\rho_l^2\kappa\left(\frac{\partial u_1^{[1]\mathrm{P}}}{\partial z_1}\right). \tag{3.196}$$

Bearing in mind that $\rho_1^{(1)} \to \rho_l$ and $u_1^{[1]\mathrm{P}} \to 0$ as $z_1 \to \infty$, and that $(\rho^{(1)} - \rho_l) \ll \rho_l$ everywhere, it is easily shown that (3.194) and (3.196) are equivalent results. Equation (3.196) was derived by Steele (1974) using a different route with similar assumptions.

(g) Perturbation theory and van der Waals theory

In a sense many of the equations discussed so far could be described as perturbation theory equations inasmuch as they are expansions about a reference state in which higher-order terms are neglected. A crucial requirement for the success of such theories, therefore, is that the reference state should be chosen in such a way as to ensure that higher-order terms are actually negligible. In uniform fluids this has been achieved for spherical molecules with considerable success, and several excellent accounts are available (Smith, 1973; Barker and Henderson, 1976; Andersen *et al.*, 1976). For non-spherical species the problem has proved to be more difficult—none of the possible candidates has a clear advantage at present (Gray, 1975).

In uniform fluids the most important part of the intermolecular potential energy comes from the pair interactions, and the form of the potential suggests a natural division into short-range repulsive and long-range attractive parts (Zwanzig, 1954), with the latter providing a relatively small contribution to the thermodynamic properties. In fact this idea underlies the original van der Waals theory of fluids (van der Waals, 1893; Rowlinson, 1979) in which the repulsive part of the potential is replaced by an infinitely repulsive hard sphere and a very simple statistical model is used for the thermodynamic properties of the hard-sphere reference system. Modern theories have improved on both these aspects of the choice of reference state.

A theory for non-uniform systems employing these principles was also proposed by van der Waals, and has enjoyed a revival in more recent times (Rowlinson, 1980). For reasons which will become clearer below, this type of theory is expected to be at its best when applied to systems with smoothly varying density profiles in contrast to those with strongly oscillating profiles such as are likely to be found in fluid–solid adsorption systems.

However, this limitation may not be too restrictive, since the application of van der Waals theory to fluid solid adsorption has shown promising results (Sullivan, 1979). Moreover, recent work using functional Taylor expansion has helped to clarify the underlying approximations and has led to the development of parallel theories using the direct correlation function. The latter have been shown to be capable of dealing with strongly oscillating profiles.

The most useful starting point here is a thermodynamic potential rather than the singlet density. The latter, which is often the object of prime interest, can be obtained, for example, by using a trial function and minimizing the grand potential Ω (Saam and Ebner, 1977).

Recalling that, $\beta\Omega = -\ln \Xi$, we functionally differentiate $\ln \Xi$ with respect to $\varepsilon_{12}^{[2]}$ to give a result analogous to (3.148b):

$$-\frac{-\delta\beta\Omega}{\delta\varepsilon_{12}^{[2]}} = \frac{\delta \ln \Xi}{\delta\varepsilon_{12}^{[2]}} = \frac{1}{2}\frac{\rho_{12}^{(2)}}{\varepsilon_{12}^{[2]}} \tag{3.197}$$

a functional Taylor expansion of Ω can therefore be written as

$$\Omega - \Omega^\circ = \tfrac{1}{2}\iint \rho_{12}^{(2)\circ}\delta u_{12}^{[2]}\,\mathrm{d}\boldsymbol{q}_1\,\mathrm{d}\boldsymbol{q}_2 \tag{3.198}$$

or, introducing a coupling parameter in the usual way via Eqn (3.187),

$$\Omega - \Omega^\circ = \tfrac{1}{2}\iiint \rho_{12}^{(2)}(\xi)\,[u_{12}^{[2]} - u_{12}^{[2]\circ}]\,\mathrm{d}\boldsymbol{q}_1\,\mathrm{d}\boldsymbol{q}_2\,\mathrm{d}\xi. \tag{3.199}$$

This equation is exact, and it is clear from Eqn (3.161) that the left-hand side in the last two expressions is in fact identically equal to $A - A^\circ$. The reference state is chosen to correspond to the short-range part of the potential, i.e. the repulsive part in simple fluids. The method chosen for subdividing the potential varies in different versions of this type of theory, but whatever the particular choice, the remaining attractive part of the potential is

$$u^{[2]A} = u^{[2]} - u^{[2]\circ}. \tag{3.200}$$

In applying van der Waals or perturbation theory to non-uniform systems it is now necessary to introduce the assumption that the free energy of the reference state can be expressed in terms of a free energy density, or free energy per unit volume $\omega^\circ(\rho_1^{(1)})$, of a uniform fluid having a density equal to the local density $\rho_1^{(1)}$. thus

$$\Omega^\circ = \int \omega^\circ(\rho_1^{(1)})\,\mathrm{d}q_1. \tag{3.201}$$

This amounts to a "local equilibrium" assumption and is expected to hold if the density profile varies slowly enough over the range of interaction of the repulsive forces (Bongiorno *et al* 1976). It is this assumption, therefore, which imposes a limitation on the applicability of these theories.

The final stage necessary to make the theory usable is to replace $\rho_{12}^{(2)}(\xi)$ in the integral by a known quantity. Although an assumption of linear dependence on ξ, such as (3.178b), would again be an obvious choice, it would, as before, fail to overcome the difficulty of obtaining $\rho^{(2)}$ for non-uniform systems. A more useful assumption, therefore, is once again to take $\rho^{(2)}(\xi)$ as being independent of ξ and replace this by an appropriate quantity for the uniform reference fluid,

$$\rho_{12}^{(2)}(\xi) \to \rho_1^{(1)}\rho_2^{(1)}\tilde{g}_{12}^{(2)\circ}(\xi;\bar{\rho}) \tag{3.202}$$

where $\bar{\rho}$ might typically be chosen as $\bar{\rho} = \frac{1}{2}(\rho_1^{(1)} + \rho_2^{(1)})$.

To develop the parallel, direct correlation function, equivalent to these equations, we choose the Helmholtz free energy as a convenient entity on which to base an expansion. For the adsorbate fluid, from Eqn (3.160), this is

$$\beta A = \int \rho_1^{(1)}\beta u_1^{[1]}\,\mathrm{d}\boldsymbol{q}_1 + \beta A_{\mathrm{int}}. \tag{3.203}$$

The intrinsic free energy A_{int} can be split, as in Eqn (3.163), into an ideal gas part A_{id} and an "excess Helmholtz free energy", $\mathcal{H}$, where from standard statistical mechanics

$$\beta A_{\mathrm{id}} = \int \rho_1^{(1)}[\ln \rho_1^{(1)}\Lambda^3 - 1]\,\mathrm{d}\boldsymbol{q}_1 \tag{3.204}$$

and,

$$\beta A = \int \rho_1^{(1)}[\beta u_1^{[1]} + \ln(\rho_1^{(1)}\Lambda^3) - 1]\, \mathrm{d}\boldsymbol{q}_1 - \mathcal{H}. \tag{3.205}$$

With aid of (3.169) and (3.162) we see that

$$\frac{\delta \mathcal{H}}{\delta \rho_1^{(1)}} = c_1^{(1)}. \tag{3.206}$$

We now introduce a linear coupling parameter, as usual via Eqn (3.178) with the reference density $\rho_1^{(1)}(0) = 0$, and note that at this limit, $\mathcal{H}^\circ = 0$, thus,

$$\mathcal{H} = \int_0^1 \int c_1^{(1)}(\xi)\rho_1^{(1)}\, \mathrm{d}\boldsymbol{q}_1\, \mathrm{d}\xi \tag{3.207}$$

this equation may be integrated a second time using $c_{12}^{(2)} = \delta c_1^{(1)}/\delta\rho_2^{(1)}$, to give,

$$\mathcal{H} = \int_0^1 \int_0^1 \int\int c_{12}^{(2)}(\xi, \xi')\rho_2^{(1)}\rho_1^{(1)}\, \mathrm{d}\boldsymbol{q}_1\, \mathrm{d}\boldsymbol{q}_2\, \xi\, \mathrm{d}\xi\, \mathrm{d}\xi'. \tag{3.208}$$

Equation (3.208) can now be substituted into (3.205) which can be written in terms of a free energy density $a(\boldsymbol{q}_1)$,

$$\beta A = \int a(\boldsymbol{q}_1)\, \mathrm{d}\boldsymbol{q}_1 = \int \rho_1^{(1)}\Big[\beta u_1^{[1]} + \ln(\rho_1^{(1)}\Lambda^3) - 1 - \tfrac{1}{2}\int \rho_2^{(1)} c_{12}^{(2)*}\, \mathrm{d}\boldsymbol{q}_2\Big]\, \mathrm{d}\boldsymbol{q}_1 \tag{3.209a}$$

where $c^{(2)*}$ is defined by

$$c^{(2)*} = 2\int_0^1\int_0^1 c^{(2)}(\xi, \xi')\, \xi\, \mathrm{d}\xi\, \mathrm{d}\xi' \tag{3.209b}$$

and the free energy density of a hypothetical Henry law fluid of density $\rho_1^{(1)}$, in which there are no intermolecular interactions, is

$$a_{\mathrm{H}}(\rho_1^{(1)}) = \rho_1^{(1)}[\beta u_1^{(1)} + \ln(\rho_1^{(1)}\Lambda^3) - 1] \tag{3.209c}$$

Equation (3.209) is exact; its analysis can be further developed (Yang *et al.*, 1976; Singh and Abraham, 1977) by expanding $c^{(2)}$, which is a functional of the singlet density $\rho^{(1)}$, about a set of uniform direct correlation functions for the local density $\rho_1^{(1)}$, and taking integrals over the coupling parameters ξ, ξ' in Eqn (3.209b) to give

$$c_{12}^{(2)*}(\rho_1^{(1)}) = \tilde{c}_{12}^{(2)*}(\rho_1^{(1)}) + \int \tilde{c}_{123}^{(3)*}(\rho_1^{(1)})\, [\rho_3^{(1)} - \rho_1^{(1)}]\, \mathrm{d}\boldsymbol{q}_1 + \tfrac{1}{2}\int\int \tilde{c}_{1234}^{(4)*}(\rho_1^{(1)})\, [\rho_3^{(1)} - \rho_1^{(1)}]\, [\rho_4^{(1)} - \rho_1^{(1)}]\, \mathrm{d}\boldsymbol{q}_3\, \mathrm{d}\boldsymbol{q}_4 + \ldots. \tag{3.210}$$

Substitution of this expression into (3.209a) yields

$$\beta A = \int a_0(\rho_1^{(1)})\, d\boldsymbol{q}_1 - \tfrac{1}{2}\int \rho_1^{(1)} \int \tilde{c}_{12}^{(2)*}[\rho_2^{(1)} - \rho_1^{(1)}]\, d\boldsymbol{q}_2\, d\boldsymbol{q}_1$$
$$- \tfrac{1}{2}\int \rho_1^{(1)} \int \rho_2^{(1)} \int \tilde{c}_{123}^{(3)*}[\rho_3^{(1)} - \rho_1^{(1)}]\, d\boldsymbol{q}_3\, d\boldsymbol{q}_2\, d\boldsymbol{q}_1$$
$$- \tfrac{1}{4}\int \rho_1^{(1)} \int \rho_2^{(1)} \iint \tilde{c}_{1234}^{(4)*}[\rho_4^{(1)} - \rho_1^{(1)}]\,[\rho_3^{(1)} - \rho_1^{(1)}]\, d\boldsymbol{q}_4\, d\boldsymbol{q}_3\, d\boldsymbol{q}_2\, d\boldsymbol{q}_1 \quad (3.211)$$

where $a_0(\rho_1^{(1)})$ is the free energy density of a uniform fluid having the density $\rho_1^{(1)}$:

$$a_0(\rho_1^{(1)}) = a_{\mathrm{H}}(\rho_1^{(1)}) - \tfrac{1}{2}\rho_1^{(1)2} \int \tilde{c}^{(2)*}\, d\boldsymbol{q}_2. \quad (3.212)$$

The second term in (3.211) gives the correction for non-uniformity assuming that the direct correlation function depends only on the local density $\rho_1^{(1)}$, and the higher-order terms provide additional corrections which account for the non-local character of the direct correlation function.

This equation provides the basis for the development of a perturbation theory. To take this route the free energy Ω° of the reference system in Eqn (3.198) is further expanded about a secondary hard-sphere reference fluid. Since the theory now applies specifically to molecules of spherical symmetry we replace $\boldsymbol{q}_i$ by the translational co-ordinate $\boldsymbol{r}_i$ and write

$$\Omega^\circ = \Omega^{\mathrm{HS}} + \iint \left[\frac{\delta\Omega}{\delta\varepsilon_{12}}\right]^{\mathrm{HS}} [\varepsilon_{12}^{[2]\circ} - \varepsilon_{12}^{[2]\mathrm{HS}}]\, d\boldsymbol{r}_1\, d\boldsymbol{r}_2 \quad (3.213a)$$

where, from (3.197),

$$\frac{\delta\Omega}{\delta\varepsilon_{12}^{[2]}} = \frac{\rho_1^{(1)}\rho_2^{(1)}g_{12}^{(2)}}{2\varepsilon_{12}^{[2]}} = \tfrac{1}{2}\rho_1^{(1)}\rho_2^{(1)}y_{12}^{(2)}. \quad (3.213b)$$

The Weeks, Chandler, Andersen (WCA) (1971) perturbation theory for uniform fluids is based on the observation that $(\varepsilon_{12}^{(2)\circ} - \varepsilon_{12}^{(2)\mathrm{HS}})$, the so-called blip function, is non-zero for only a small range of separations in the vicinity of the hard-sphere diameter d. The value of this diameter is obtained by truncating the expansion of Eqn (3.213a) after the first integral term and setting $\Omega^\circ = \Omega^{\mathrm{HS}}$. Analytical equations for the calculation of $d(\rho, T)$ have been provided by Verlet and Weiss (1972). The second ingredient in the theory is the subdivision of the potential $u_{12}^{(2)}$ in such a way that

$$\left.\begin{aligned} u^{(2)\circ}(r) &= u^{(2)}(r) + u_{\min} && r < r_{\min} \\ &= 0 && r > r_{\min} \\ u^{(2)\mathrm{A}}(r) &= -u_{\min} && r < r_{\min} \\ &= u^{(2)}(r) && r > r_{\min} \end{aligned}\right\} \quad (3.214)$$

where the subscript min indicates the minimum in the potential function.

To extend these ideas to non-uniform fluids, we note that, from Eqns (3.204) and (3.205), the Helholtz free energy difference is

$$\beta(A - A^{\mathrm{HS}}) = \beta(\Omega - \Omega^{\mathrm{HS}}) = \mathcal{H} - \mathcal{H}^{\mathrm{HS}} \tag{3.215a}$$

and from Eqns (3.157) and (3.206),

$$c_{12}^{(2)} = \frac{\delta^2 \mathcal{H}}{\delta\rho_1^{(1)}\delta\rho_2^{(1)}}. \tag{3.215b}$$

Combining these definitions with (3.197) and (3.213), there results

$$c_{12}^{(2)} = c_{12}^{(2)\mathrm{HS}} + \tfrac{1}{2}\frac{\delta^2}{\delta\rho_1^{(1)}\delta\rho_2^{(1)}}\int\int \rho_1^{(1)}\rho_2^{(1)}\{y_{12}^{(2)\mathrm{HS}}[\varepsilon_{12}^{[2]0}/\varepsilon_{12}^{[2]\mathrm{HS}} - 1] - g_{12}^{(2)0}\beta u_{12}^{[2]\mathrm{A}}\}\,\mathrm{d}\boldsymbol{r}_1\,\mathrm{d}\boldsymbol{r}_2. \tag{3.216}$$

In their perturbation theory for non-uniform fluids, Singh and Abraham (1977) solved this equation for a uniform fluid which led to the following approximation for $c_{12}^{(2)*}$:

$$c_{12}^{(2)*} = \tilde{c}_{12}^{(2)*\mathrm{HS}} + \tfrac{1}{2}\tilde{y}_{12}^{(2)\mathrm{HS}}[\varepsilon_{12}^{[2]0}/\varepsilon_{12}^{[2]\mathrm{HS}} - 1] - \tfrac{1}{2}\tilde{g}_{12}^{(2)0}\beta u_{12}^{[2]A}. \tag{3.217}$$

This was then substituted into Eqn (3.211) after the higher-order terms had been put into approximate forms containing only $c^{(2)}$ and its derivatives.

The idea incorporated in the ACW "blip" function was also applied by these authors to $\rho^{(1)}$ and $\zeta^{[1]}$ by writing

$$\Omega^{\circ} = \Omega^{\mathrm{HW}} + \int (\rho^{(1)}/\zeta^{[1]})^{\mathrm{HW}}[\zeta^{[1]\circ} - \zeta^{[1]\mathrm{HW}}]\,\mathrm{d}z + \ldots \tag{3.218}$$

and using the assumption $\Omega^0 = \Omega^{\mathrm{HW}}$ to find the position d_{w} of the hard wall by equating the integral to zero.

The expansion in Eqn (3.211) can be developed in another way by Taylor expanding the density about its local value. In this way the free energy density $a(\rho^{(1)})$ can be expressed as a series in density gradients, an idea which was originated by van der Waals (1873) and revived and formalized by Cahn and Hilliard (1958). It is convenient here again to specialize to spherically symmetric molecules with centres located by a vector $\boldsymbol{r}$. Using tensor notation the free energy density may be put into the generalized form,

$$a(\rho^{(1)}) = a_0(\rho^{(1)}) + a_\alpha\left(\frac{\partial\rho^{(1)}}{\partial r_\alpha}\right) + a_{(1)\alpha\beta}\left(\frac{\partial\rho^{(1)}}{\partial r_\alpha}\right)\left(\frac{\partial\rho^{(1)}}{\partial r_\beta}\right) + a_{(2)\alpha\beta}\left(\frac{\partial^2\rho^{(1)}}{\partial r_\alpha\,\partial r_\beta}\right) + \ldots. \tag{3.219a}$$

Since $a(\rho^{(1)})$, and consequently the coefficients in this expansion, are invariant under rotation in $\boldsymbol{r}$, the expansion may be reduced to the simpler form,

$$a(\rho^{(1)}) = a_0(\rho^{(1)}) + a_1(\rho^{(1)})\left(\frac{\partial^2\rho^{(1)}}{\partial r_\alpha \partial r_\alpha}\right) + a_2(\rho^{(1)})\left(\frac{\partial\rho^{(1)}}{\partial r_\alpha}\right)\left(\frac{\partial\rho^{(1)}}{\partial r_\alpha}\right) + \dots \quad (3.219b)$$

The second-order derivative may be eliminated by rewriting this as

$$a(\rho^{(1)}) = a_0(\rho^{(1)}) + \tfrac{1}{2}m\left(\frac{\partial\rho^{(1)}}{\partial r_\alpha}\right)\left(\frac{\partial\rho^{(1)}}{\partial r_\alpha}\right) - \frac{\partial}{\partial r_\alpha}\left(a_1\frac{\partial\rho^{(1)}}{\partial r_\alpha}\right).$$

where $\tfrac{1}{2}m = \mathrm{d}a_1/\mathrm{d}\rho^{(1)} - a_2$, and it is to be noted that if $\rho^{(1)}$ becomes uniform at the system boundaries, the last term will vanish when $a(\rho^{(1)})$ is integrated over V. This condition is of course fulfilled for liquid–vapour interfaces in systems of infinite extent, but would not generally apply to solid–fluid adsorption when the solid is treated as an external field.

The coefficient m was evaluated approximately by van der Waals who showed

$$m \cong -\tfrac{1}{6}\int r^2 u^{(2)\mathrm{A}}(r)\,\mathrm{d}\boldsymbol{r}. \quad (3.220)$$

An exact evaluation of m follows if the density is assumed to be slowly varying so that we may write

$$\rho_i^{(1)} = \rho_1^{(1)} + (r_{i\alpha} - r_{1\alpha})\left(\frac{\partial\rho^{(1)}}{\partial r_\alpha}\right)_1 + \tfrac{1}{2}(r_{i\alpha} - r_{1\alpha})^2\left(\frac{\partial^2\rho^{(1)}}{\partial r_\alpha^2}\right)_1 + \dots. \quad (3.221)$$

After substitution of this expansion into (3.211) and lengthy manipulation, a_1 and a_2 can be found from which an expression for m follows:

$$\beta m = \tfrac{1}{6}\int r^2 \bar{c}^{(2)}\,\mathrm{d}\boldsymbol{r}. \quad (3.222)$$

This equation is related to (3.220) by the MSM approximation introduced earlier in the context of the functional Taylor expansion of the singlet density (Eqn (3.181) *et seq.*).

The condition that $\rho^{(1)}$ varies slowly and smoothly with position imposes a limitation on the use of the gradient expansion which therefore finds its most immediate application to liquid–vapour interfaces close to the critical temperature, but it has also been used with success for this type of interface away from this temperature (Widom, 1977, 1979). A related approximation which does not appear to suffer from this limitation can be obtained from Eqns (3.209a) and (3.212) by using the symmetry property $c_{12}^{(2)} = c_{21}^{(2)}$ which follows from the equality of the second-order cross-derivatives of ln Ξ (cf. (3.157)),

$$\beta A = \int \left\{ a_0(\rho_1^{(1)}) + \tfrac{1}{4} \int c_{12}^{(2)*}[\rho_2^{(1)} - \rho_1^{(1)}]^2 \, \mathrm{d}\boldsymbol{q}_2 \right\} \mathrm{d}\boldsymbol{q}_1. \qquad (3.223)$$

This approximation may also be regarded as a partial summation of the gradient expansion (Evans 1979, Ebner *et al.*, 1976), or as a generalization of the low-density limit of (3.209a) (McCoy and Davis, 1979). Its applicability is considered in Chapter 7.

An analysis of the gradient approximation and of three local density models which could be used with the BBGKY and other integral equations have been made by Davis and co-workers (Carey *et al.*, 1978; Davis and Scriven, 1978). The three local density approximations which they considered for g_{12} are

$$g_{12}^{(2)} = \tilde{g}^{(2)}[r_{12}; \rho^{(1)}(\tfrac{1}{2}(\boldsymbol{r}_1 + \boldsymbol{r}_2))] \qquad (3.224a)$$

$$g_{12}^{(2)} = \tilde{g}^{(2)}[r_{12}; \tfrac{1}{2}(\rho_1^{(1)} + \rho_2^{(1)})] \qquad (3.224b)$$

$$g_{12}^{(2)} = \tfrac{1}{2}\{\tilde{g}^{(2)}[r_{12}; \rho_1^{(1)}] + \tilde{g}^{(2)}[r_{12}; \rho_2^{(1)}]\}. \qquad (3.224c)$$

The analysis can be made by substitution into the non-uniform OZ equation (Eqn (3.140)) of expansions, analogous to Eqn (3.210), for $c^{(2)}$ and $g^{(2)}$ in terms of uniform correlation functions $\tilde{c}^{(2)}$ and $\tilde{g}^{(2)}$:

$$c^{(2)}(\rho_1^{(1)}) = \tilde{c}^{(2)}(\rho_1^{(1)}) + \int \tilde{c}^{(3)}(\rho_1^{(1)}) \, \mathrm{d}\boldsymbol{r}_3 + \int \tilde{c}^{(3)}(\rho_1^{(1)})[\rho_3^{(1)} - \rho_1^{(1)}] \, \mathrm{d}\boldsymbol{r}_3 + \ldots \qquad (3.225a)$$

$$g^{(2)}(\rho_1^{(1)}) = \tilde{g}^{(2)}(\rho_1^{(1)}) + \int \tilde{g}^{(3)}(\rho_1^{(1)})[\rho_3^{(1)} - \rho_1^{(1)}] \, \mathrm{d}\boldsymbol{r}_3 + \ldots. \qquad (3.225b)$$

The densities in these equations are then Taylor expanded about their local values as in Eqn (3.221). The difference between the local density approximation and the exact equations can be traced to the treatment of the correlations within the same plane normal to the interface. In the approximate formulae these correlations are the same as in the homogeneous fluid having the local density, but in fact they depend also on density gradients. It may be concluded that the model of an inhomogeneous fluid as a stack of thin homogeneous fluid slabs is conceptually incorrect, although the approximation may be perfectly acceptable in suitably chosen circumstances.

References

Amos, A. T. and Crispin, R. J. (1976). *In* "Theoretical Chemistry Advances and Perspectives" (H. Eyring and D. Henderson, eds), Vol. 2. Academic Press, New York and London.

Andersen, H. C., Chandler, D. and Weeks, J. D. (1976). *Adv. Chem. Phys.* **34**, 105.

Barker, J. A. (1975). *In* "Rare Gas Solids" (J. A. Venables and M. L. Klein, eds), Vol. 1. Academic Press, London and New York.

Barker, J. A. and Henderson, D. (1976). *Rev. Mod. Phys.* **48**, 587.

Barker, J. A., Fisher, J. A. and Watts. R. O. (1971). *Molec. Phys.* **21**, 657.

Baxter, R. J. (1970). *In* "Physical Chemistry: An Advanced Treatise" (H. Eyring and D. Henderson, eds), Vol. 6. Academic Press, New York and London.

Bell, R. J. (1970). *J. Phys. B: Atomic and Molecular Physics* **3**, 751.

Bell, R. J. and Zucker I. (1975). *In* "Rare Gas Solids" (J. A. Venables and M. L. Klein, eds), Vol. 1. Academic Press, London and New York.

Ben Ephraim, A. and Folman, M. (1976). *J. Chem. Soc. Faraday Trans.* **72**, 671.

Bennet, A. J. (1974). *Phys. Rev.* **B9**, 741.

Benson, G. C. and Yun, K. S. (1967). *In* "The Solid–Gas Interface" (E. A. Flood, ed.), Vol. 1. Arnold, London.

Berry, M. V. (1974). *J. Phys. A: Mathematical and General* **7**, 231.

Bogoliubov, N. N. (1946). *J. Phys. USSR.* **10**, 257.

Bogoliubov, N. N. (1962). *In* "Studies in Statistical Mechanics" (J. de Boer and G. E. Uhlenbeck, eds), Vol. 1. North Holland, Amsterdam.

Bongiorno, V., Scriven, L. E. and Davis, H. T. (1976). *J. Colloid. Interf. Sci.* **57**, 462.

Born, M. and Green, H. S. (1946). *Proc. Roy. Soc.* **188A**, 10.

Böttcher, C. J. F. (1973). "Theory of Electric Polarization". Elsevier, Amsterdam.

Broekhoff, J. C. P. and de Boer, J. H. (1967). *J. Catalysis* **9**, 8.

Buckingham, A. D. (1965). *Disc. Faraday Soc.* **40**, 232.

Buckingham, A. D. (1967). *Adv. Chem. Phys.* **12**, 107.

Buckingham, A. D. and Utting, B. D. (1970). *Ann. Rev. Phys. Chem.* **21**, 287.

Buff, F. P. (1960). *In* "Handbuch der Physik" Vol. X, 281. Springer-Verlag Berlin.

Buff, F. P. (1949). *J. Chem. Phys.* **17**, 338.

Carey, B. S., Scriven, L. E. and Davis, H. T. (1978). *J. Chem. Phys.* **69**, 5040.

Cahn, J. W. and Hilliard, J. E. (1958). *J. Chem. Phys.* **28**, 258.

Carlos, W. E. and Cole, M. W. (1979). *Phys. Rev. Lett.* **43**, 697.

Cohen, J. S. and Pack, R. T. (1974). *J. Chem. Phys.* **61**, 2373.

Crowell, A. D. (1958). *J. Chem. Phys.* **29**, 446.

Crowell, A. D. and Steele, R. B. (1961). *J. Chem. Phys.* **34**, 1347.

Davis, H. T. and Scriven, L. E. (1978). *J. Chem. Phys.* **69**, 5215.

de Boer, J. H. and Heller, G. (1937). *Physica* **4**, 1045.

Deryaguin, B. V. (1957). *In* "Proceeding of the 2nd International Congress on Surface Activity" (J. Schulman, ed.), Vol. 2, p. 153. Butterworth, London.

Dzyaloshinskii, I. E., Lifshitz, E. M. and Pitaevski, L. P. (1961). *Adv. Phys.* **10**, 165.

Ebner, C., Saam, W. F. and Stroud, D. (1976). *Phys. Rev.* **A14**, 2264.

Evans, R. (1974). *J. Phys. C: Solid State Physics* **7**, 2808.
Evans, R. (1979). *Adv. Phys.* **28**, 143.
Findenegg, G. H. and Fisher, J. (1975). *Disc. Far. Soc.* **54**, 38.
Fisher, I. Z. (1964). "Statistical Physics of Liquids", p. 109. University of Chicago Press.
Freeman, D. L. (1975). *J. Chem. Phys.* **62**, 941.
Gordon, R. G. and Kim, Y. S. (1972). *J. Chem. Phys.* **56**, 3122.
Gray, C. G. (1975). *Chem. Soc. Specialist Periodical Reports*: "Statistical Mechanics", Vol. 2.
Gready, J. E., Backsay, G. B. and Hush, N. S. (1978). *J. Chem. Soc. Faraday Trans. II* **74**, 1430.
Harasima, A. (1953). *J. Phys. Soc. Japan.* **8**, 343.
Henderson, D. Abraham, F. F. and Barker, J. A. (1976). *Molec. Phys.* **31**, 1291.
Hill, T. L. (1960). "Introduction to Statistical Mechanics". Addison-Wesley, New York.
Hirschfelder, J. O., Curtiss, C. F. and Bird, R. B. (1961). "Molecular Properties of Gases and Liquids", p. 968. Wiley, New York.
Hobson, J. P. (1974). *Crit. Rev. Solid State* **4**, 221.
Hornig, J. F. and Hirschfelder, J. O. (1952). *J. Chem. Phys.* **20**, 1812.
Horton, G. K. (1968). *Amer. J. Phys.* **36**, 93.
House, W. A. and Jaycock, M. J. (1975). *J. Chem. Soc. Faraday Trans. II* **71**, 1597.
Irving, J. H. and Kirkwood, J. G. (1950). *J. Chem. Phys.* **18**, 817.
Kelly, B. T. and Duff, M. J. (1970). *Carbon* **8**, 77.
Kihara, T. and Midzuno, Y. (1956). *J. Phys. Soc. Japan* **11**, 1045.
Kirkwood, J. G. (1935). *J. Chem. Phys.* **3**, 300.
Kirkwood, J. G. and Salzburg, Z. W. (1953). *Disc. Faraday Soc.* **15**, 28.
Kiselev, A. V. (1965). *Disc. Faraday Soc.* **40**, 221.
Kiselev, A. V. and Poshkus, D. P. (1958). *Zh. Fiz. Khim.* **32**, 2824.
Koide, A., Meath, W. J. and Allnatt, A. R. (1980). *Molec. Phys.* **39**, 895.
Landman, U. and Kleiman, G. G. (1977). *Chem. Soc. Specialist Periodical Reports*: "Surface and Defect Properties of Solids" Vol. 6.
Lane, J. E. and Spurling, T. (1976). *Aust. J. Chem.* **29**, 2103.
Lekner, J. and Henderson, J. R. (1977). *Molec. Phys.* **34**, 333.
Lennard-Jones, J. E. and Dent, R. M. (1928). *Trans. Faraday Soc.* **24**, 92.
Leonard, P. J. and Barker, J. A. (1975). *In* "Theoretical Chemistry, Advances and Perspectives" (H. Eyring and D. Henderson, eds), Vol. 1. Academic Press, New York and London.
Lifshitz, E. M. (1955). *J. Exp. Theor. Phys. USSR.* **29**, 94.
Linder, B. (1967). *Adv. Chem. Phys.* **12**, 142.
Lux, E. and Münster, A. (1967). *Z. Physik.* **199**, 165.
McClachlan, A. D. (1964). *Molec. Phys.* **7**, 381.
McClellan, A. G. (1952). *Proc. Roy. Soc.* **213A**, 274.
McCoy, B. and Davis, H. T. (1979). *Phys. Rev. A.* **20**, 1201.
McDonald, I. R. and Hansen, J. P. (1976). "Theory of Simple Liquids". Academic Press, New York and London.
McRury, T. B. and Linder, B. (1972). *J. Chem. Phys.* **56**, 4368.
Mahanty, J. and Ninham, B. W. (1976). "Dispersion Forces", Academic Press, London and New York.
Maitland, G. C. and Wakeham, W. A. (1978). *Molec. Phys.* **35**, 1443.

Margenau, H. and Kestner, N. R. (1969). "Theory of Intermolecular Forces". Pergamon, Oxford.

Mavroyannis, C. and Stephen, M. J. (1962). *Molec. Phys.* **5**, 629.

Mermin, N. D. (1965). *Phys. Rev.* **137**, 1441.

Meyer, E. F. (1967). *J. Phys. Chem.* **71**, 4416.

Meyer, E. F. and Deitz, V. R. (1967). *J. Phys. Chem.* **68**, 2926.

Morita, T. and Hiroike, K. (1960). *Prog. Theor. Phys.* **23**, 1003.

Morita, T. and Hiroike, K. (1961). *Prog. Theor. Phys.* **25**, 537.

Münster, A. (1969). "Statistical Thermodynamics," Vol. 1, p. 48. Springer-Verlag, Berlin; Academic Press, London and New York.

Murrell, J. N. (1976). *In* "Rare Gas Solids" (M. L. Klein and J. A. Venables, eds), Vol. 1. Academic Press, New York and London.

Navascués, G. (1976). *J. Chem. Soc. Faraday Trans. II* **72**, 2035.

Navascués, G. and Berry, M. V. (1977). *Molec. Phys.* **34**, 649.

Nicholson, D. (1968). *Trans. Faraday Soc.* **64**, 3416.

Nicholson, D., Parsonage, N. G. and Rowley, L. A. (1981). *Molec. Phys.* **44**, 629.

Ornstein, L. S. and Zernike, F. (1914). *Proc. Akad. Sci.* (Amsterdam) **17**, 793.

Pauly, H. and Toennies, J. P. (1965). *In* "Advances in Atomic and Molecular Physics" (D. R. Bates and I. Estermann, eds), Vol. 1. Academic Press, London and New York.

Percus, J. K. (1964). *In* "Classical Theory of Equilibrium Fluids" (H. L. Frisch and J. L. Lebowitz, eds), II-33. Benjamin, New York.

Percus, J. K. (1976). *J. Stat. Phys.* **15**, 423.

Perram, J. W. and Smith, E. R. (1976). *Chem. Phys. Lett.* **39**, 328.

Perram, J. W. and White, L. R. (1975). *Disc. Faraday Soc.*, **54**, 29.

Pierotti, R. A. and Thomas, H. E. (1971). *In* "Surface and Colloid Science". (E. Matijevic, ed.), Vol. 4, p. 93. Wiley, New York.

Pisani, C., Ricca, F. and Roetti, C. (1973). *J. Phys. Chem.* **77**, 657.

Price, G. L. (1974). *Surface Sci.* **46**, 697.

Putnam, F. A. and Fort Jr., T. (1977). *J. Phys. Chem.* **81**, 2164.

Rae, A. I. M. (1975). *Molec. Phys.* **29**, 467.

Rae, A. I. M. (1973). *Chem. Phys. Lett.* **18**, 574.

Richmond, P. (1975). *Chem. Soc. Specialist Periodical Reports*: "Colloid Science", Vol. 2.

Rogowska, J. M. (1978). *J. Chem. Phys.* **68**, 3910.

Rowley, L. A., Nicholson, D. and Parsonage, N. G. (1976). *Molec. Phys.* **31**, 365.

Rowlinson, J. S. (1979). *J. Stat. Phys.* **20**, 197.

Rowlinson, J. S. (1980). *Chem. Brit.*, **16**, 32.

Rybolt, T. R. and Pierotti, R. A. (1979). *J. Chem. Phys.* **70**, 4413.

Saam, W. F. and Ebner, C. (1977). *Phys. Rev. A.* **15**, 2566.

Sams Jr., J. R. (1964). *Trans Faraday Soc.* **60**, 149.

Sams Jr., J. R., Constabaris, G. and Halsey Jr., G. D. (1961). *J. Phys. Chem.* **64**, 1689.

Sams Jr., J. R., Constabaris, G. and Halsey Jr., G. D. (1962) *J. Chem. Phys.* **36**, 1334.

Schmit, J. N. (1976). *Surf. Sci.* **55**, 589.

Sinanoğlu, O. and Pitzer, K. S. (1960). *J. Chem. Phys.* **32**, 1279.

Singh, Y. and Abraham, F. F. (1977). *J. Chem. Phys.* **67**, 537.

Smith, W. R. (1973). *Chem. Soc. Specialist Periodical Reports*: "Statistical Mechanics" Vol. 1.

Srivastava, K. P. (1958). *J. Chem. Phys.* **28**, 543.
Steele, W. A. (1973). *Surface Sci.* **36**, 317.
Steele, W. A. (1974). "Interaction of Gases with Solid Surfaces". Pergamon, Oxford.
Steele, W. A. (1978). *J. Phys. Chem.* **82**, 817.
Stevens, W. J., Wahl, A. C., Garden, M. A. and Karo, A. M. (1974). *J. Chem. Phys.* **60**, 2195.
Stoeckli, H. F. (1974). *Helv. Chim. Acta.* **57**, 2195.
Stogryn, D. E. and Stogryn, A. P. (1966). *Molec. Phys.* **11**, 371.
Sullivan, D. E. (1979). *Phys. Rev. B.* **20**, 3991.
Takaishi, T. (1975). *Prog. Surf. Sci.* **6**, 43.
van der Waals, J. H. (1873). *Verh K. Akad. Wet* (Sect. 1), no. 8.
Verlet, L. and Levesque, D. (1967). *Physica* **36**, 254.
Verlet, L. and Weis, J. J. (1972). *Phys. Rev.* **A5**, 939.
Volterra, V. (1959) "Theory of Functionals". Dover, New York.
Wagner, A. F., Das, G. and Wahl, A. C. (1974). *J. Chem. Phys.* **60**, 1885.
Widom, B. (1977). "Statistical Mechanical Methods in Theory and Application" (U. Landman, ed.), Plenum, New York.
Widom, B. (1979). *Physica* **95A**, 1.
Yang, A. J. M., Fleming, P. D. and Gibbs, J. H. (1976). *J. Chem. Phys.* **64**, 3732.
Yvon, J. (1935). "La Theorie Statistique des Fluides et l'Equation d'Etat" Actualities Scientifiques et Industrielles, Vol. 203. Herman, Paris.
Zaremba, E. and Kohn, W. (1976). *Phys. Rev. B.* **13**, 2270.
Zwanzig, R. W. (1954). *J. Chem. Phys.* **22**, 1420.

Chapter 4

Simulation Methods

1. Introduction

Use of any of these simulation techniques leads to what is, in principle, an exact evaluation of the properties of the model system with its particular set of input parameters defining the intermolecular forces. The limitations on the exactness depend upon the size of the system studied and the "length of time for which the system is observed", although the rate at which the output results converge on the final answer is also quite strongly dependent upon the nature of the system.

Simulation studies have been used in two ways. On the one hand, the results can be compared with those from an approximate theory based on statistical mechanics and using the same set of intermolecular forces. This provides a test of the approximations made in the latter theory which does not depend upon knowledge of the forces or on experimental artefacts. Simulations used in this way to furnish exact data for a model are often referred to as "machine experiments". On the other hand, the simulation results may be compared with real (physical) experiments, when they provide a test of the assumptions employed in the model and, in particular, those concerning the intermolecular forces. Machine experiments are more versatile than the physical kind in that the parameters and, indeed, the model itself can be varied at will. Laboratory experiments are often limited by the existence or otherwise of suitable materials for study. In both cases it is, of course, nature which is the source of inspiration and ultimate object of interest as in any other scientific work.

The information which can be obtained from simulation studies can be divided into thermodynamic, structural and transport (dynamic) properties. All such studies yield values for the thermodynamic internal energy (U) as a function of temperature. The heat capacity, isosteric heat of adsorption, compressibility, and other properties can also be obtained by use of fluctuation formulae, e.g. Eqns (3.12), (3.105), (3.117). The virial theorem (Eqn (3.128)) may also be used to provide a value for the pressure. It should be noted that values obtained from fluctuation equations or from the virial theorem are usually much less accurate than those obtained by other methods since convergence of the estimates of the former during simulations is generally very slow. The entropy and the free energies are still more difficult to obtain from simulations, although methods which yield these quantities are becoming available and will be described in some detail because of their importance in work on adsorption. In spite of the undoubted importance of the thermodynamic properties, simulation studies have been increasingly directed towards the structural quantities such as $\rho^{(1)}$ and $\rho^{(2)}$. Knowledge of these gives a very direct picture of the local environment in a fluid. They are also very suitable for discussions of the approximations involved in integral equation theories since the closure approximations which these theories require are usually expressed in terms of the distribution or correlation functions. The third class of information that can be obtained, that on transport properties, is exemplified by the diffusion coefficients and the velocity autocorrelation function ($\langle \boldsymbol{v}(0) \cdot \boldsymbol{v}(t) \rangle$). With greater difficulty, the viscosity and thermal conductivity can also be evaluated.

In most cases one is interested in a system which is infinite in at least one direction and the problem arises of deducing from the necessarily finite simulation system the properties of the infinite system. The normal procedure is to invoke periodic boundary conditions, which is almost invariably done in the simple form

$$x(\boldsymbol{r} + \boldsymbol{n}\boldsymbol{R}) = x(\boldsymbol{r}) \tag{4.1}$$

where $\boldsymbol{r}$ is a position in the simulation box, $\boldsymbol{R}$ is a vector with components equal to the sides of the box and $\boldsymbol{n}$ is a diagonal matrix with integer components. This device overcomes the problem of having the molecules at the edge of the simulation box interacting only with molecules to one side of themselves. It does, of course, have the objection of introducing a spurious periodicity, but this is not usually a serious problem, especially if the properties of interest are short-ranged. Nevertheless, to reduce the risk of this spurious periodicity it is common to use the so-called minimum image (MI) approximation. According to this, each particle in the simulation box only interacts with one of the infinite number of images of any other particle, that image being the one closest to the original particle. Thus particle i would interact with j' but not with j itself or any of the other images of j in Fig. 4.1. This device has, furthermore, the important

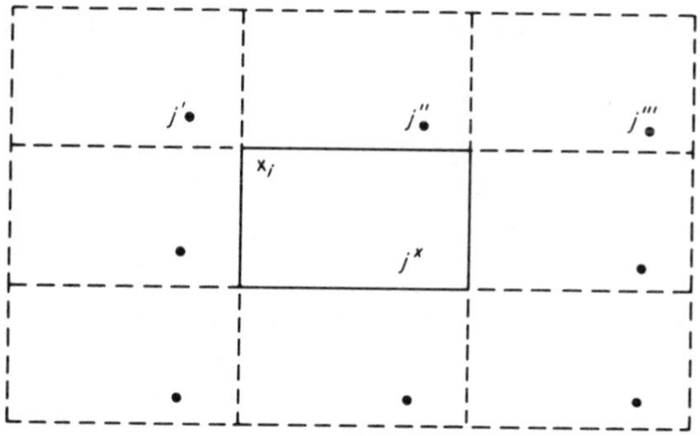

Fig. 4.1

practical advantage of reducing the amount of computing time needed to evaluate the energy of each configuration. An alternative method is one which takes account only of interactions between particles which are less than a certain distance apart, but uses the MI method to build up the cut-off sphere around each molecule. The distance is chosen to be not greater than one-half of the linear dimension of the box (l) and therefore prevents interactions with more than one image. In fact, to save computer time this radius (the spherical cut-off, SC) is often chosen to be much less than $l/2$.

Two main types of simulation procedure have been utilized: the Monte Carlo (MC) and molecular dynamics (MD) methods. The former is somewhat easier to apply but is limited to giving only static properties. The MD method is capable of providing information on time-dependent as well as time-independent properties.

2. Monte Carlo methods

(a) The canonical ensemble (*NVT*)

The possible configurations of the system are sampled with a known weighting and from these samples a variety of averaged quantities and histograms describing the structure can be formed. The extent of the phase space which is sampled can be varied, but by far the largest number of studies have employed the canonical ensemble (constant N, V and T); the isothermal-isobaric (constant N, p and T) and the grand canonical (constant T, V and λ) ensembles have also been used, and the latter has, indeed, been shown to have considerable advantages for certain types of adsorption problem. Here we shall describe the basis of the method in the *NVT* ensemble and then indicate the modifications which are required for the other ensembles.

The sampling of the *NVT* phase space is achieved by generating a Markov chain of states. In such a chain the probability of occurrence of a state at a particular point in the sequence of events is determined only by the immediately preceding state (but not by those at "earlier times"). The sequential development of the chain is governed by a probability matrix ($\boldsymbol{p}$) in which the element p_{ij} is the conditional probability that the $(t+1)$th state of the system will be j given that the tth state is i. There is, of course, a normalization condition $\Sigma_j p_{ij} = 1$, which indicates that the sum of the elements of each column of the matrix must be unity. For any given point (t) on the chain we can write the probability of occurrence of all the individual states at the $(t+1)$th point as a row vector $\boldsymbol{u}$. Then

$$\boldsymbol{u}(t+1) = \boldsymbol{u}(t)\,\boldsymbol{p} \tag{4.2}$$

By repeated application of Eqn (4.2) we obtain

$$\boldsymbol{u}(t+n) = \boldsymbol{u}(t)\,\boldsymbol{p}^n \tag{4.3}$$

Expanding $\boldsymbol{u}(t)$ in terms of the eigenvectors $\boldsymbol{\phi}_r$ of $\boldsymbol{p}$,

$$\boldsymbol{\phi}_r \boldsymbol{p} = \lambda_r \boldsymbol{\phi}_r \tag{4.4}$$

we get

$$\boldsymbol{u}(t) = \sum_r k_r^t \boldsymbol{\phi}_r \tag{4.5}$$

where λ_r is the rth eigenvalue of $\boldsymbol{p}$.

Substitution of Eqn (4.5) into Eqn (4.3) gives

$$\sum_r k_r^{t+n} \boldsymbol{\phi}_r = \sum_r k_r^t \lambda_r^n \boldsymbol{\phi}_r \tag{4.6}$$

One, and only one, of the eigenvalues must be unity, the others being less than unity. Therefore as $n \rightarrow \infty$

$$\boldsymbol{u}(t+n) = k_{r'}^t \phi_{r'} \tag{4.7}$$

where r' is the index number of the unit eigenvalue. Thus, in the limit as $n \rightarrow \infty$, $\boldsymbol{u}$ becomes independent of n. The remaining eigenvectors in Eqn (4.6) affect the approach to equilibrium but not the ultimate result. The steady-state value of $\boldsymbol{u}$ is then such that

$$\boldsymbol{u}\boldsymbol{p} = \boldsymbol{u} \tag{4.8}$$

The usefulness of the Markov chain depends upon the identification of an average taken over the chain with the ensemble average. The problem reduces to that of finding a suitable matrix $\boldsymbol{p}$, given that $\boldsymbol{u}$ must correspond to the natural (Boltzmann) probabilities of the states. There are, of course, many possible solutions to this problem. However, by far the largest number of MC simulations have used the scheme employed in the first paper of this type, that of Metropolis *et al.* (1953). They set

$$p_{ij} = M_i^{-1}\min\{1, \exp[-\beta(U_{N,j} - U_{N,i})]\}$$
$$p_{ii} = 1 - \sum_{j\neq i} p_{ij} \tag{4.9}$$

where M_i is the number of states directly accessible from state i. This satisfies Eqn (4.8) provided $M_i = M_j$, since

$$\sum_i u_i p_{ij} = \sum_i u_j p_{ji} = u_j \sum_i p_{ji} = u_j \tag{4.10}$$

where we have used the "microscopic reversibility" relation which follows from Eqn. (4.8):

$$\begin{aligned} u_i p_{ij} &= AM_i^{-1}\exp(-\beta U_{N,i})\min\{1, \exp[-\beta(U_{N,j} - U_{N,i})]\} \\ &= AM_i^{-1}\exp(-\beta U_{N,j}) \\ &= u_j M_j^{-1}\min\{1, -\exp[-\beta(U_{N,i} - U_{N,j})]\} \\ &= u_j p_{ji} \qquad \text{if } U_{N,j} \geqslant U_{N,i} \\ &= AM_i^{-1}\exp(-\beta U_{N,i}) \\ &= AM_i^{-1}\exp(-\beta U_{N,j})\min\{1, \exp[-\beta U_{N,i} - U_{N,j})]\} \\ &= u_j p_{ji} \qquad \text{if } U_{N,j} \leqslant U_{N,i}. \end{aligned} \tag{4.11}$$

An alternative scheme, which is symmetrical in i and j, has been proposed and used by A. A. Barker (1965). In this algorithm

$$p_{ij} = M_i^{-1}u_j/(u_i + u_j) \qquad p_{ii} = 1 - \sum_{j\neq i} p_{ij}. \tag{4.12}$$

Again, the "microscopic reversibility" relation is clearly satisfied

$$u_i p_{ij} = u_i u_j/(u_i + u_j) = u_j p_{ji} \tag{4.13}$$

and, in consequence, so is Eqn (4.8).

For a comparison of the efficiency of these transition matrices we are interested in the variance of the estimate of the mean of the quantity of interest (this is to be distinguished from the variance of the quantity itself, which is asymptotically related to physical properties, and so must be a constant for all valid simulation schemes). It has been shown that the algorithm of Metropolis *et al.* leads to a lower variance of the mean than the symmetrical scheme of Barker (Peskun, in Valleau and Whittington (1977). This is not to say that the rate of convergence for the former scheme is higher, since the results referred to only concern the behaviour in the limit of infinite chain length, but such tests as have been made suggest that the Metropolis scheme may have the advantage in this respect also (Barker, in Valleau and Whittington (1977)).

We turn now to the range of possible modifications which can be allowed between successive points on the Markov chain. In order to obtain correct reversibility between states it is necessary for j to be directly accessible from i if i is directly accessible from j, but not otherwise, i.e.

$$i \in \eta(j) \quad \text{if and only if} \quad j \in \eta(i) \tag{4.14}$$

where $\eta(i)$ is the domain of i.

Studies concerning translational motion are normally carried out in cartesian coordinates with a limit set to the maximum change in each of the x-, y- and z-coordinates of a particle (this maximum change which is usually the same for each direction, is often known as the "maximum step-length" (a), although, of course, the particle could be translated by up to $\sqrt{3}a$ in a single move). In anisotropic systems it may be advantageous for a_x, a_y and a_z to be different, and there is no objection to this. The actual move proposed is determined by the values of three random numbers (ξ_x, ξ_y, ξ_z) in the range 0 to 1:

$$\Delta x = a_x(1 - 2\xi_x) \qquad \Delta y = a_y(1 - 2\xi_y) \qquad \Delta z = a_z(1 - 2\xi_z) \tag{4.15}$$

Thus Δx must lie between $-a_x$ and $+a_x$, and likewise for Δy and Δz. It is clear that this arrangement satisfies Eqn (4.14). There is no reason why several particles should not be moved at once, but this freedom of choice has been rarely exercised. Indeed, it is not difficult to envisage configurations for which movement of single particles would be faced by large energy barriers whilst a concerted movement of several particles could avoid such barriers. An example is the lattice system shown in Fig. 4.2, in which the placing of two particles on adjacent sites gives rise to a large energy of repulsion. Motion from the configuration shown (1, 3, 5) to the (2, 4, 6) configuration would be difficult if only one particle could be moved at a time because it would be necessary to pass through states with two adjacent sites occupied. On the other hand, a concerted movement $1 \to 2$, $3 \to 4$, $5 \to 6$ would not encounter this difficulty. The objection to allowing concerted movement is that in most cases one does not know which of all the vast number of such motions to permit. Clearly, if concerted

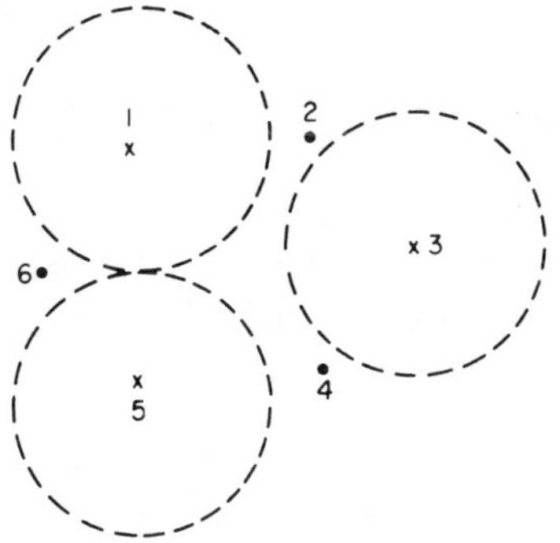

Fig. 4.2

motions are to be employed it is essential to have some precognition of which states are most likely to be occupied in the system under investigation. Otherwise, the choice of many improbable trial states would lead to unacceptably high rejection rates.

In 1977–8 two new, and closely related, methods were introduced for choosing trial modifications. Both of these, the "smart" and "force bias" methods, were designed so as to direct the Markov chain towards those trial states which were likely to have a good chance of being accepted. Thus, by examining the slope of the potential energy surface (the force) at the particle which it is proposed to move, one can discriminate against those directions which would involve a large increase in energy for a small displacement. The bias introduced by this procedure must be compensated by a corresponding adjustment to the probability of acceptance. Pangali *et al* (1978) noted that if the *underlying* transition probability from configuration R to configuration R' (the chance that R' would be chosen as the trial state if the current state were R) were taken to be $\exp[-\beta U_{\mathrm{N}}, (R')]$ then the required acceptance probability would be unity. However, to obtain the normalizing factor for that probability requires that the partition function be already known! Nevertheless, a considerable computational advantage can be obtained by making a simple expansion of $U_N(R')$ about R, giving an underlying probability for the transition of $c\exp[-\beta\{\nabla_j U_N \,.\, \Delta \boldsymbol{R}_j\}]$. (For di- and poly-atomic particles this becomes $c\exp[\beta\{\boldsymbol{F}_j \cdot \Delta \boldsymbol{r}_j + \boldsymbol{N}_j \cdot \Delta \boldsymbol{\omega}_j\}]$, where $\Delta \boldsymbol{R}_j$ has been broken down into a translational ($\Delta \boldsymbol{r}_j$) and a rotational ($\Delta \boldsymbol{\omega}_j$) displacement, with $\boldsymbol{F}_j = -\nabla_j U_N$ and $\boldsymbol{N}_j$ as the torque on molecule j.) The chance of acceptance must now be taken as

$$\min\left\{1, \frac{\exp[-\beta\{U_N(R') - U_N(R)\}]}{\exp[-\beta\{\nabla_j U_N \cdot \Delta R_j]}\right\} \tag{4.16}$$

The "smart" MC method, which was proposed almost simultaneously with the former by Rossky *et al.* (1978) is based on expressing a displacement as the sum of a term proportional to the force in that direction and a second term which is chosen randomly:

$$\Delta \boldsymbol{r} = \beta A \boldsymbol{F} + \boldsymbol{s} \tag{4.17}$$

For a particular distribution of the random variable $W(\boldsymbol{s})$ the probability that a trial displacement $\Delta \boldsymbol{r}$, taking the system from state i to state j, will be chosen is obtained by integration over $\boldsymbol{s}$ as

$$p_{ij}^* = \frac{\int \mathrm{d}\boldsymbol{s}\, W(\boldsymbol{s})\, \delta(\Delta \boldsymbol{r} - \beta A \boldsymbol{F}_i - \boldsymbol{s})}{\int \mathrm{d}(\Delta \boldsymbol{r}) \int \mathrm{d}\boldsymbol{s}\, W(\boldsymbol{r})\, \delta(\Delta \boldsymbol{r} - \beta A \boldsymbol{F}_i - \boldsymbol{s})} \tag{4.18}$$

A legitimate asymmetric algorithm may then be obtained by setting the probability of acceptance of the trial state as

$$\min(1, p_{ji}^* u_j / p_{ij}^* u_i) \tag{4.19}$$

Rossky *et al.* have given particular attention to the case in which $\boldsymbol{s}$ is a Gaussian random variable,

$$W(\boldsymbol{s}) = (4A\pi)^{-3/2} \exp(-\boldsymbol{s}^2/4A) \tag{4.20a}$$

for which Eqn (4.18) becomes

$$p_{ij}^* = C \exp[-(\Delta \boldsymbol{r} - \beta A \boldsymbol{F}_i)^2/4A] \tag{4.20b}$$

They have shown that the method shows much more rapid convergence (expressed in terms of configurations) than the normal Metropolis scheme for two model systems.

Both methods require the evaluation of forces at every step and so, although these techniques may have an advantage over conventional MC methods in overcoming ergodic problems, it is not obvious that they are superior to MD, which also requires the evaluation of forces but yields in addition, information on time-dependent properties.

So far, we have only examined translational movement and for many adsorption problems there is still much work to be done on the behaviour of structureless particles for which, of course, only the one type of motion is possible. If, however, the molecules are not structureless it is necessary to permit changes in orientation of the particles. These orientational changes are usually combined with translational movements but they may be separate from them, each trial move then being either a pure translation or a pure rotation. These matters are discussed later in this chapter (§6).

As the actual Markov chains are non-infinite, any quantities obtained by averaging over them would be affected by the (arbitrarily chosen) starting configuration. In an attempt to eliminate this effect it is usual to monitor certain quantities and to reject all data pertaining to the chain before these quantities have converged to within specified limits. For example, the mean energy $\langle U_N \rangle$ over successive blocks of states is computed and steadiness of this quantity is taken as an indication of convergence. Even when the system has "equilibrated" there will still be fluctuations of $\langle U_N \rangle$ about the mean value, but there should not be any discernible drift. Quantities such as the heat capacity, which are determined from fluctuation terms $f(U_N, U_N)$, etc. (Ch. 3, §2), require a much longer equilibration time than is necessary for the computation of $\langle U_N \rangle$. The determination of the components of the pressure tensor through the ensemble averages given in Eqn (3.128) is unfortunately another slowly converging process, so that the quantities derived from it usually have a high error associated with them. For anything other than simple spherical molecules the computation of the forces which are needed for the evaluation of the virial components is expensive in computer time. However, since this quantity is not required for the decision-making process of the chain, it is not necessary to evaluate it at every step. For systems having hard core potentials the calculation of the virial requires the value of the appropriate density function ($\rho^{(2)}$ for bulk systems; $\rho^{(1)}$ and $\rho^{(2)}$ for adsorption systems in which both adsorbate–adsorbate and adsorbate–substrate potentials have hard cores) at the cut-off distance. This necessitates extrapolation of the density function(s) to the cut-off distance(s), which may introduce considerable error.

In uniform systems the only structural quantity which is commonly calculated (apart from its Fourier transform) is $\rho^{(2)}$; for adsorption systems $\rho^{(1)}$, which describes the variation in number density as a function of distance from the interface, is often of greater interest and has certainly been more often evaluated. Indeed, if one follows the argument pursued by some integral equation theories which generate the interface by "enlarging" a single foreign molecule (Ch. 3, §5(b)) then $\rho^{(1)}$ is seen to correspond to $\rho^{(2)}$ for a uniform system. In an interfacial system $\rho^{(2)}$ depends upon the length and orientation of the inter-particle vector and also on the position of the centre of mass. Even for structureless particles, therefore, $\rho^{(2)}(\boldsymbol{r}_1, \boldsymbol{r}_2)$ is a six-dimensional quantity, although this may be reduced by the symmetry of the surface, e.g. to three-dimensional for a continuum surface. It is clear, however, that it is unreasonable to attempt a sufficiently full tabulation of this quantity to enable smooth interpolation and extrapolation to be achieved. It is therefore necessary to compromise. If the data are spread over too large a number of histogram boxes the number of "events" in each category will be small and the statistics will be poor. Where $\rho^{(1)}$ shows that the system has fairly clearly defined layers a number of workers have considered $\rho^{(2)}$ as a function of r_{ij} for each layer, averaging over the orientation of the inter-particle vector (Rowley *et al.* 1976a, b; Lane and Spurling, 1976). When this is done a decision has to be taken about the treatment of the many cases in which the two molecules concerned in a pairing are in different layers, and there is no uniformity in how this choice has been made. One possibility is to let such a pair contribute equally to the histograms for the two layers concerned (but with only one-half of the weight of a pair which was entirely in one layer). Another possibility is to ignore such pairs and consider only pairs in which both particles are within the layer. Far better than neglecting such pairs is to employ a coarse division of the angle between the inter-particle vector and the normal to the surface; division into just two or three parts greatly increases the information which can be conveyed. Another problem with regard to $\rho^{(2)}$ is that variations may occur which are trivially related to the variations in $\rho^{(1)}$ in the regions where the two particles are situated. In principle, this difficulty could be avoided by considering $g^{(2)}$ rather than $\rho^{(2)}$, since it is usually the correlations within the fluid that are of prime interest. Although this has been done in some instances, the computation of $g^{(2)}$, instead of $\rho^{(2)}$, presents increased difficulties. Since the histograms for $\rho^{(2)}$ are based on coarse divisions of the relevant variables it is not altogether satisfactory to divide the values of $\rho^{(2)}$ obtained by $\{\overline{\rho^{(1)}}\}^2$, where $\overline{\rho^{(1)}}$ is the coarse-grained average. A better procedure is to obtain first $\rho^{(1)}$ at small intervals of the variable(s) in the usual way and then to use the factor $1/\{\rho^{(1)}\}^2$ as weighting for each point in compiling the histograms for $g^{(2)}$.

(b) The canonical ensemble: computations of entropy and free energy

A severe weakness of *NVT* MC simulations (and also of the *NPT* MC and all the MD simulations which will be described later) is that they do not yield values for the entropy or free energy. For adsorption systems the free energy is an even more crucial quantity than it is for uniform systems. This is because interest is often centred on the equilibrium between a surface region and the bulk phase.

All MC simulations enable ratios of integrals over configuration space to be evaluated, but they do not yield the values of the integrals themselves. Thus, in an *NVT* ensemble,

$$\langle X \rangle = \int X \exp(-\beta U_N)\, \mathrm{d}\boldsymbol{q} \Big/ \int \exp(-\beta U_N)\, \mathrm{d}\boldsymbol{q} \qquad (4.21)$$

can be determined, but not the numerator or denominator separately. However, it is necessary to have the latter integral in order to obtain the Helmholtz free energy. In principle $\int \exp(-\beta U_N)\,\mathrm{d}\boldsymbol{q}$ could be obtained by setting $X = \exp(+\beta U_N)$ since

$$\langle \exp(+\beta U_N) \rangle = \int 1\,\mathrm{d}\boldsymbol{q} \Big/ \int \exp(-\beta U_N)\,\mathrm{d}\boldsymbol{q} = V \Big/ \int \exp(-\beta U_N)\,\mathrm{d}\boldsymbol{q}. \quad (4.22)$$

However, the statistics of the evaluation of $\langle \exp(+\beta U_N) \rangle$ by averaging over the Markov chain are extremely poor, since almost all of the contribution to the average is made by a very small number (maybe two or three) of configurations which have large values of U_N and are consequently infrequent. The estimate of $\langle \exp(+U_N) \rangle$, therefore, depends very much on these few configurations (their number and their energy); all the other configurations sampled together make only a tiny contribution. In fact, a method of this type exactly counteracts the major advantage of the Metropolis importance sampling method.

Another approach is to relate the integral to that of a similar, but simpler, system for which the integral is known. This can be done by writing the energy of a configuration as the sum of an expression equal to that which the simpler system would have and a difference term. The difference term may then be "switched on" or "off" by means of a coupling parameter (ξ) which ranges between 0 and 1:

$$\begin{aligned} U_N(\xi) &= U_N(\xi=0) + \xi\{U_N(\xi=1) - U_N(\xi=0)\} \\ &= U_N(\xi=0) + \xi\,\delta U_N \end{aligned} \quad (4.23)$$

where $\xi = 0$ and 1 lead to the reference system and the system of interest, respectively. Now,

$$\begin{aligned} &\mathrm{d}\left\{\ln \int \exp(-\beta U_N(\xi)\,\mathrm{d}\boldsymbol{q}\right\} \Big/ \mathrm{d}\xi \\ &= \int \exp[-\beta\{U_N(\xi=0) + \xi\delta U_N\}](-\beta\delta U_N)\,\mathrm{d}\boldsymbol{q} \Big/ \int \exp[-\beta U_N(\xi)]\,\mathrm{d}\boldsymbol{q} \\ &= -\beta\langle \delta U_N \rangle_\xi \end{aligned} \quad (4.24)$$

where $\langle \delta U_N \rangle_\xi$ is the average value of δU_N when the distribution is that for the energies of interaction with the value ξ.

Integrating Eqn (4.24) from $\xi = 0$ to $\xi = 1$ we get

$$\ln Z_N(\xi = 1) - \ln Z_N(\xi = 0) = -\beta \int_0^1 \langle \delta U_N \rangle_\xi \, d\xi. \tag{4.25}$$

If $\langle \delta U_N \rangle_\xi$ is evaluated by ordinary MC means at 5–10 well chosen values of ξ then the integral can, in most cases, be determined with sufficient accuracy. Combining this with the known value of $\ln Z_N(\xi = 0)$ (the reference system) $\ln Z_N(\xi = 1)$ is obtained. Hence the Helmholtz free energy ($\beta A = \ln \Lambda^3 + \ln(N!) - \ln Z_N$) and the entropy ($S = (U - A)/T$) are obtained.

The choice of reference system is very important for the success of this procedure. Apart from the obvious requirement that its configuration integral must be known, there is another more subtle condition, which should be satisfied: each of the interaction potentials which are to be "turned on" should have a hard core. If the reference potentials are less hard than the real potentials then the distribution factor $\exp[-\beta\{U_N(\xi = 0) + \xi\delta U_N\}]$ will permit close approaches of the particles for the smaller values of ξ, and these configurations lead to large positive contributions to $\langle \delta U_N \rangle_\xi$. For large values of ξ (say 0·25–1·0) this will not happen, as the Boltzmann factor will then make large positive values of δU_N very improbable. Nevertheless, on account of the uncertainties in $\langle \delta U_N \rangle_\xi$ for low ξ, the value of the integral in Eqn (4.25) will be unreliable. If, on the other hand, the reference system has hard cores, these very close approaches will be prevented for all values of ξ. Simple hard-sphere or hard-disc systems come immediately to mind as possible reference systems, especially so as it is often possible to obtain reliable values for the configuration integrals. There still remains to be decided the position of the infinite rise in potential energy, the collision diameter of the hard spheres or discs (Fig. 4.3). This must be chosen so as to prevent the situation described above from occurring: the particles must not be able to approach close enough to give very large values of δU_N. If, at the other extreme, the diameters are too large, then a part of the real potential which exerts a non-negligible influence on the properties of the system will not be sampled at all. In an early application of this technique to adsorption systems, Stroud and Parsonage (1971), in calculating the configuration integral for various numbers (1–8) of spherical Lennard-Jones molecules in a zeolite cavity, represented the potential energy as

$$\sum_i u_i^{(1)\mathrm{HS}} + \sum_{i>j} u_{ij}^{(2)\mathrm{HS}} + \xi\left\{\sum_i u_i^{(1)\mathrm{LJ}} + \sum_{i>j} u_{ij}^{(2)\mathrm{LJ}}\right\} \tag{4.26}$$

They found that the hard-sphere cut-off should be $0{\cdot}90$–$0{\cdot}95\sigma$, where σ is the distance at which the Lennard-Jones potential is zero. The evaluation

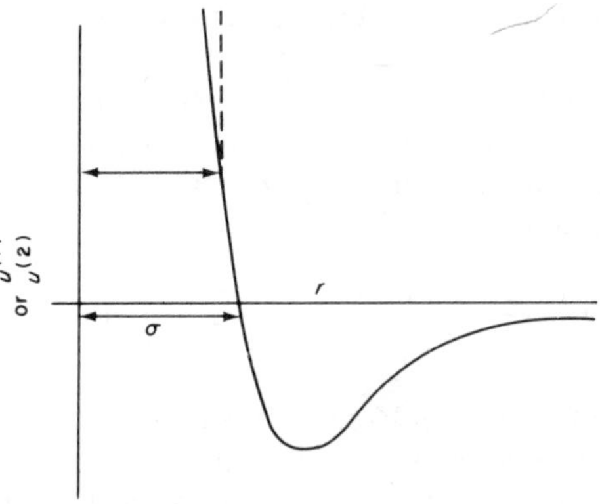

Fig. 4.3. Potential energy curves. Full curve, true potential energy; broken curve, part of hard-sphere (disc) reference potential.

of $Z_N(\xi = 0)$ was carried out by determining the fraction of randomly generated configurations which would not cause overlap of the spheres. Even with the use of modifications of the simplest procedures for achieving this, the estimate of $Z_N(\xi = 0)$ rapidly decreased in accuracy as the "free volume" decreased. Coldwell (1973, 1974) has suggested that this kind of method is practicable up to ~70% of the close-packed density. Thus, for a given number density of particle, the estimate would become less reliable as the diameter of the hard spheres increased. The choice of hard cut-off distance was made so as to minimize the sum of the errors from the integral term and the second term on the left-hand side of Eqn (4.25). Patey and Valleau (1973) have used a similar procedure for a fluid of dipolar hard spheres, but in this case there was a natural choice for the cut-off.

This technique (the coupling parameter method) may, alternatively, be considered as being an integration with respect to β from $\beta = 0 (T = \infty)$ to the temperature of interest. Looking at the coupling parameter method in this way, another solution to the problem of determining free energies suggests itself. This is to perform an integration over density from a value for which the configuration integral is known up to the density of interest.

$$(\partial A/\partial \rho)_{T,N} = (\partial A/\partial V)_{T,N}(\partial V/\partial \rho)_{T,N} = Np/\rho^2 \tag{4.27}$$

$$A(\rho, T) = A(\rho_0, T) + \int_{\rho_0}^{\rho} Np/\rho^2 \, d\rho \tag{4.28}$$

$$\ln Q_N(\rho, T) = \ln Q_N(\rho_0, T) - (N/kT) \int_{\rho_0}^{\rho} (p/\rho^2) \, d\rho \tag{4.29}$$

The obvious choice for the reference density is one that is so low that the system behaves ideally. For a uniform system this is a perfect gas; for a gas–solid adsorption system it would be a system having no interactions between the adsorbed particles. In the latter case, since there is no single density for such a system, the integration would be carried out at constant area ($\mathscr{A}$) using the alternative equation:

$$A(V, T) = A(V_0, T) - \int_{V_0}^{V} p \, \mathrm{d}V. \tag{4.29'}$$

In all cases the integration is likely to be a long one and may involve a phase change. If the latter would occur it is necessary to impose some constraint, which may not in fact be physically realizable, to prevent the transition. For uniform systems this has been done by Hoover and Ree (1967) and Hansen and Verlet (1969), who divided the volume into a number of imaginary regions and restricted the number of particles which could be present in each of these. These restrictions prevented the condensation of particles into some regions, leaving others empty, which would be the manifestation of the phase transition. Valleau and Whittington (1977) have pointed out that the intermediate simulations on the path used to avoid the transition are of no intrinsic interest and the process is to that extent wasteful. In contrast, each value of ξ in the coupling parameter integration method gives information about a particular temperature above the one of prime interest. Little or no use has been made of the density integration techique for interfacial problems, although there does not appear to be any reason why it should not be used.

Methods have also been found for relating the configuration integral required, to that of a reference system in a succession of stages. The number of stages needed is greater, the greater the dissimilarity between the real and the reference systems. For a single intermediate reference system we have

$$\begin{aligned}
\exp\{-\beta(A - A_0)\} &= \frac{\int \exp(-\beta U_N) \, \mathrm{d}\boldsymbol{q}}{\int \exp(-\beta U_{N_0}) \, \mathrm{d}\boldsymbol{q}} \\
&= \frac{\int \exp\{-\beta(U_N - U'_N)\} \exp(-\beta U'_N) \, \mathrm{d}\boldsymbol{q}}{\int \exp(-\beta U'_N) \, \mathrm{d}\boldsymbol{q}} \\
&\quad \times \frac{\int \exp\{-\beta(U'_N - U_{N_0})\} \exp(-\beta U_{N_0}) \, \mathrm{d}\boldsymbol{q}}{\int \exp(-U_{N_0}) \, \mathrm{d}\boldsymbol{q}} \\
&= \langle \exp\{-\beta(U_N - U'_N)\} \rangle_{0'} \langle \exp\{-\beta(U'_N - U_{N_0})\} \rangle_0
\end{aligned} \tag{4.30}$$

where $\langle \ \rangle_{0'}$ and $\langle \ \rangle_0$ refer to averages taken over the distributions for the intermediate and reference systems, respectively. Equation (4.30) can be generalized to allow a larger number of intermediate systems (″, ‴, $\overset{n'}{\cdot}$. .):

$$\exp\{-\beta(A - A_0)\} = \langle\exp\{-\beta(U_N - U'_N)\}\rangle_{0'}\langle\exp\{-\beta(U'_N - U''_N)\}\rangle_{0''} \ldots \langle\exp\{-\beta(U^{n'}_N - U_{N_0})\}\rangle_0 \quad (4.31)$$

Valleau and co-workers have made further progress by showing how non-Boltzmann distributions can be employed to obtain more reliable results from a given number of intermediate stages, or equally good results from a smaller number of such stages. The basic equation used refers to a modification of the distribution function by the inclusion of an extra factor w in addition to the Boltzmann factor. Then

$$\langle X\rangle = \frac{\int X\exp(-\beta U_N)\,\mathrm{d}\boldsymbol{q}}{\int \exp(-\beta U_N)\,\mathrm{d}\boldsymbol{q}} = \frac{\int (X/w)w\exp(-\beta U_N)}{\int w\exp(-\beta U_N)\,\mathrm{d}\boldsymbol{q}}\,\frac{\int w\exp(-\beta U_N)\,\mathrm{d}\boldsymbol{q}}{\int \exp(-\beta U_N)\,\mathrm{d}\boldsymbol{q}} = \langle X/w\rangle_w/\langle 1/w\rangle_w = \langle X/w\rangle_w\langle w\rangle \quad (4.32)$$

where $\langle\ \rangle_w$ and $\langle\ \rangle$ refer to averages over the modified and unmodified distributions, respectively (Torrie and Valleau, 1977). It is possible, with a suitable choice of w, to sample a broader range of states in a single stage of the type discussed than would be the case for the unmodified Boltzmann distribution, hence the name "umbrella sampling" which has been given to the device represented by Eqn (4.32). Unfortunately, there does not appear to be any systematic way of selecting the function w, so trial and error has been used so far. Bennett (1976) has observed that for some "bridging" ensembles, largely made up from discrete ensemble contributions, the simulation was very slow to equilibrate. This was because the intermediate ensemble was slow to sample all the regions, although it would eventually have done so (Torrie *et al.*, 1973). Rather better results were obtained by using a combination of a continuous set of ensembles.

In a related piece of work, Bennett (1976) has used the relation

$$\exp\{-\beta(A - A_0)\} = \langle W\exp(-\beta U_N)\rangle_0/\langle W\exp(-\beta U_{N_0})\rangle \quad (4.33)$$

which can be derived from Eqn (4.32) with $w = W\exp(-\beta U_N)$ together with the form of Eqn (4.30) in which the intermediate and reference systems are identical. Configurations which have infinite potential energy in either the real or the reference systems make no contribution to either numerator or denominator. This makes clear the necessity for the two systems (real and reference) to have sufficient overlap of their distributions.

The method requires the running of two Markov chains, one each for the real and the reference distributions, and Bennett's results suggest that approximately equal amounts of computer time should be devoted to each, the number of configurations in each chain being then inversely proportional to the time per step. Bennett has shown that no other division of computer time can be more than twice as efficient (measured by (time $\times\ \sigma^2)^{-1}$, where σ is the standard deviation) as the equal-time division. The improvement claimed over the previous method is that, whereas in using Eqn (4.31) *every* important region of configuration space of the real system should be well sampled in the reference ensemble, in Bennett's scheme the runs in each ensemble need only sample *some* of the important regions of the other system.

(c) The isothermal-isobaric ensemble (*NpT*) (McDonald, 1972)

Study of this ensemble arose from the hope that the greater flexibility it offered as compared with the *NVT* ensemble would lead to greater speed in sampling the various regions of configuration space. In particular, it was felt that by allowing V to fluctuate the likelihood of the fairly small systems used in the simulation becoming locked in a limited region of configuration space would be much reduced. This hope has not been entirely fulfilled, though with regard to obtaining p–V–T data it appears that evaluating $\langle V(N,p,T)\rangle$ in the *NpT* ensemble is more satisfactory than obtaining $p(N,V,T)$ from the average of the virial over the canonical ensemble. A further reason for performing *NpT* simulation is that one is often interested in the system at a particular pressure (near atmospheric pressure) rather than at a particular volume.

For systems with intermolecular potentials which depend only upon distance, the formulation of the *NpT* simulation is not greatly more difficult than for the *NVT* ensemble, although now both simple moves of a particle (or a few particles) and change of the simulation volume are permitted. For a change in the volume from V to $V + \mathrm{d}V$ the cartesian coordinates of each particle are scaled as $V^{1/3}$, e.g.

$$x' = x\{(V + \mathrm{d}V)/V\}^{1/3}$$

$$y' = y\{(V + \mathrm{d}V)/V\}^{1/3} \qquad z' = z\{(V + \mathrm{d}V)/V\}^{1/3} \tag{4.36}$$

Since all lengths are scaled as $V^{1/3}$ the r^{-6} and r^{-12} terms of Lennard-Jones interactions must scale as V^{-2} and V^{-4}, respectively, which renders the evaluation of the energy change arising from a volume change computationally quick. For potential terms which are not of the form ar^{-n} this simple procedure is not applicable, an example being the $A\exp(-\alpha r)$ repulsion term.

The probability of a particular configuration and volume in this ensemble is

$$p(V, \boldsymbol{q}) = \exp[-\beta\{U_N(\boldsymbol{q}) + pV\}]/\Pi \qquad (4.37)$$

where Π is the isothermal-isobaric partition function (Hill, 1956a). The asymmetric criterion for acceptance of a proposed modification, whether a move, a volume change, or a combination of the two, is expressed in terms of the ratio of the probabilities of the initial and trial states. If the ratio is $\leqslant 1$ the trial state is accepted; if the ratio is >1 a chance of acceptance of p_j/p_i is allowed and decided in the usual way by means of a random number.

This ensemble has not been exploited in interfacial studies, largely because the advantages it has are shared by the grand canonical ensemble, which, as will be seen in the next section, also has further qualities which are especially valuable in dealing with systems containing density gradients.

(d) The grand canonical ensemble ($TV\lambda$)

As compared with the canonical ensemble method this has flexibility with respect to N. Thus the density of the system is allowed to fluctuate and, it would be hoped, this would help to overcome ergodic difficulties. As compared with the NpT ensemble MC, for which the density changes produced by fluctuations in V are spread uniformly over the whole simulation box, in the $TV\lambda$ ensemble MC simulations the fluctuations in density, produced by creation or destruction of particles, are very localized. This ensemble also has the attraction that λ (or μ) is one of the independent variables, as it is in many real situations. Of particular relevance in the present context is the study of adsorption isotherms, where one is concerned with the value of $\langle N \rangle$ for fixed values of T and λ.

Early studies using this ensemble were confined to lattice models. Thus, for a two-dimensional adsorption system a configuration is defined by a set of values $c_1, c_2, \ldots$, each element of which may be 0 or 1, signifying an empty or occupied site, respectively. The probability of each configuration is

$$p(N, \boldsymbol{c}) = \exp\{-\beta U_N(\boldsymbol{c})\} \qquad (4.38)$$
$$\exp(\lambda \sum_i c_i)/\sum_c \exp\{-\beta U_N(\boldsymbol{c})\}\exp(\lambda \sum_i c_i)$$

where U_N includes the energies of interaction between admolecules and substrate and between pairs of admolecules. This is analogous to an Ising ferromagnet in an applied magnetic field and, indeed, transformations between these models are well known (Ch. 5, § 5). In the simplest approach to the simulation of such an adsorption model the trial move permitted is the reversal of the occupancy of a randomly chosen site. That is, if site i has been chosen, then

$$c_i \Rightarrow 1 - c_i \qquad (0 \Rightarrow 1 \text{ or } 1 \Rightarrow 0) \tag{4.39}$$

The quantity

$$u_i = \exp\{-\beta U_n(\boldsymbol{q})\} \exp\left(\lambda \sum_i c_i\right) \tag{4.40}$$

is then compared with the corresponding value u_j for the modified configuration and the probability of it being accepted may be set equal to

$$u_j/(u_i + u_j) \tag{4.41}$$

as in the symmetrical criterion of A. A. Barker (p. 139), or may be determined by an unsymmetrical procedure similar to that employed in the Metropolis algorithm for the canonical ensemble MC. Other modification schemes could, of course, be used in which the number of molecules added or subtracted could be more than one.

The adaptation of grand canonical ensemble MC methods to continuous (non-lattice) systems has proved to be considerably more difficult. A way in which the problem could be solved was first shown by Norman and Filinov (1969). Subsequently, Rowley *et al.* (1975) and Adams (1974, 1975) each reported satisfactory schemes. That of Rowley *et al.* (the "ghost" particle method) was based on the procedure for lattice models described above. "Real" and "ghost" molecules were moved throughout the system by a method similar to that used in the canonical ensemble MC method. However, since the "ghost" molecules do not interact with any other molecules, "real" or "ghost", their movements always corresponded to no change of potential energy and so were always accepted. Movement of "real" molecules, which interacted with other "real" molecules, was governed by the usual Boltzmann probability in the manner introduced by Metropolis *et al.* Apart from these translational moves there were also creation and destruction trials in which a "ghost" molecule was changed to a "real" one or vice versa. It is this type of trial which corresponds with that used in the grand canonical ensemble MC studies on lattices described above. In their paper describing this method, Rowley *et al.*, who applied it to bulk liquid–vapour equilibrium as an illustration, made an error with respect to the long-range corrections. As a consequence of this error a discrepancy was found with the previous results of Hansen and Verlet (1969) for the vapour pressure of the same system using a modified canonical ensemble method. It should be emphasized that the procedure of Rowley *et al.* is a correct one in all respects other than the way in which the long-range corrections are made. The error, first noticed as a result of a communication by J. A. Barker, does, however, have important implications for both this method and that of Adams, and also even for non-uniform systems in the canonical ensemble. These points will be discussed later (Ch. 6, § 2(a)).

Adams, following Norman and Filinov, used a method which differs primarily in that when a molecule is "destroyed" no record of its previous position is kept and there is no tendency for a "creation" event occurring soon after the "destruction" to take place at, or near to, the same position. In contrast, in the "ghost" particle method "creation" events can only occur at "ghost" sites and these move fairly slowly in the simulation. The latter method has the disadvantage that a larger amount of storage is required, since the coordinates of the "ghost" as well as the "real" particles must be preserved. For this reason, Rowley *et al.* in their subsequent work and all other workers have adopted Adams' procedure or methods closely related to it. All the algorithms allow translational moves, which are governed by the Metropolis formula and about which nothing further needs to be said, and the creation/destruction trials. For the derivation of a suitable algorithm the formula for the grand partition function can be written as

$$\Xi = \sum_N \frac{\lambda^N}{\Lambda^{3N} N!} \int \exp(-\beta U_N)\,\mathrm{d}\boldsymbol{q} \tag{4.42}$$

The fractional contribution of a particular region of configuration space (i) involving N molecules at positions $\boldsymbol{q}_1, \boldsymbol{q}_2, \ldots \boldsymbol{q}_N$ and of "volume" δ^N is then

$$\begin{aligned} u_i &= \Xi^{-1}(\lambda\delta/\Lambda^3)^N (N!)^{-1} \exp(-\beta U_N) \\ &= \Xi^{-1}(zV)^N (N!)^{-1} \exp(-\beta U_N)(\delta/V)^N \end{aligned} \tag{4.43}$$

Using a modified symmetric algorithm in which the factor $(\delta/V)^N$ is suppressed the probability of acceptance of an already chosen creation step $(N, i) \Rightarrow (N+1, j)$ may be taken as

$$\begin{aligned} &\frac{\{zV/(N+1)\}\exp(-\beta U_{N+1,j})}{\exp(-\beta U_{N,i}) + \{zV/(N+1)\}\exp(-\beta U_{N+1,j})} \\ &= \frac{\{zV/(N+1)\}\exp\{-\beta(U_{N+1,j} - U_{N,i})\}}{1 + zV/(N+1)\exp\{-\beta(U_{N+1,j} - U_{N,i})\}} \end{aligned} \tag{4.44}$$

and for a destruction step $(N, i) \Rightarrow (N-1, j)$ as

$$\frac{\{N/(zV)\}\exp\{-\beta(U_{N-1,j} - U_{N,i})\}}{1 + [N/(zV)\}\exp\{-\beta(U_{N-1,j} - U_{N,i})\}} \tag{4.45}$$

In a creation step the number of configuration regions which are accessible is $(N+1)V/\delta$, where V/δ is the number of regions of volume δ into which the new particle could be placed and the factor $(N+1)$ arises because the particle could be given any index number in the sequence $1, 2, \ldots, (N+1)$, i.e. it could be placed at the beginning, the end, or at any intermediate point in the sequence. Therefore the chance of that modification being chosen is $\alpha^{(+)}/\{(N+1)V/\delta\}$, where $\alpha^{(+)}$ is a preset number which is the fraction of all trials which are of the creation type. From this we see that the left-hand side of Eqn (4.11) is

$$\frac{\alpha^{(+)}}{(N+1)V/\delta}\frac{\{zV/(N+1)\}\exp[-\beta(U_{N+1,j}-U_{N,i})]}{1+\{(zV)/(N+1)\}\exp[-\beta(U_{N+1,j}-U_{N,i})]}$$

$$\times\frac{(zV)^N\exp(-\beta U_{N,i})(\delta/V)^N}{N!}$$

$$=\frac{\alpha^{(+)}z^{N+1}\exp(-\beta U_{N+1,j})\delta^{N+1}}{(N+1)(N+1)!\,[1+\{(zV)/(N+1)\}\exp\{-\beta(U_{N+1,j}-U_{N,i})\}]} \tag{4.46}$$

For the reverse (destruction) process $((N+1, j) \Rightarrow (N, i))$, the number of regions accessible is $N+1$, since any one of the $N+1$, particles may be chosen for destruction. The right-hand side of Eqn. (4.11) therefore becomes

$$\frac{\alpha^{(-)}}{N+1}\frac{\{(N+1)/(zV)\}\exp\{\beta(U_{N+1,j}-U_{N,i})\}}{1+\{(N+1)/(zV)\}\exp\{\beta(U_{N+1,j}-U_{N,i})\}}$$

$$\times\frac{(zV)^{N+1}\exp(-\beta U_{N+1,j})(\delta/V)^{N+1}}{(N+1)!}$$

$$=\frac{\alpha^{(-)}z^{N+1}\exp(-\beta U_{N+1,j})\,\delta^{N+1}}{(N+1)(N+1)!\,[1+\{(zV)/(N+1)\}\exp\{-\beta(U_{N+1,j}-U_{N,i})\}]} \tag{4.47}$$

showing that the criterion of microscopic reversibility (Eqn (4.11)) is satisfied provided that $\alpha^{(+)}=\alpha^{(-)}$. No thorough investigation of the effect on the convergence of varying $\alpha=\alpha^{(+)}=\alpha^{(-)}$ has been carried out, a value of about one-third being commonly employed. This corresponds to equal numbers of translational moves, creation trials and destruction trials.

An alternative scheme, which corresponds to the asymmetric procedure of Metropolis *et al.*, can be devised by replacing Eqns (4.44) and (4.45) by

$$\min\{1, \{zV/(N+1)\}\exp[-\beta(U_{N+1,j}-U_{N,i})]\} \tag{4.48}$$

for a creation step (4.48) and

$$\min\{1, \{N/(zV)\} \exp[\beta(U_{N+1,j} - U_{N,i})]\} \quad (4.49)$$

for a destruction step (4.49), respectively. This is the method adopted by Norman and Filinov (1969) and Adams (1974, 1975) and would seem to have the advantage previously associated with the ordinary Metropolis algorithm, namely that it gives the largest possible acceptance ratio whilst, of course, remaining consistent with microscopic reversibility.

The quantities most frequently evaluated in grand canonical ensemble MC simulations are $\langle N\rangle$, $\langle U_N\rangle$, and, for uniform systems, κ (from $\langle N^2\rangle - \langle N\rangle^2$). The configurational heat capacity most directly obtainable is $C_{\mu,V}$:

$$\begin{aligned} C_{\mu,V}/Nk &= (\partial\langle U_N - \mu N\rangle/\partial T)_{\mu,V}/(Nk) \\ &= \{\langle (U_N - \mu N)^2\rangle - \langle U_N - \mu N\rangle^2\}/Nk^2T^2 \end{aligned} \quad (4.50a)$$

(Hill, 1956b), which is not a directly measurable quantity but can be related to experimental quantities by thermodynamic equations. The definition of $C_{\mu,V}$ in the first part of Eqn (4.50a) is based on the rate of change of entropy with temperature, although this is partly due to the change in $\langle N\rangle$.

Schofield (1966) has shown how C_V itself may be derived from fluctuations in the grand canonical ensemble, the configurational contribution being

$$C_V = \frac{\langle U_N^2\rangle - \langle U_N\rangle^2 - \{\langle U_N\rangle\langle N\rangle\}^2/\{\langle N^2\rangle - \langle N\rangle^2\}}{\langle U_N N\rangle - \langle N\rangle kT^2} \quad (4.50b)$$

Of great interest is the fact that, provided the pressure can be obtained from a computation of the mean virial, then it is possible to obtain the entropy. For structureless particles the expression is

$$TS = 1{\cdot}5\langle N\rangle kT + \langle U_N\rangle - \mu\langle N\rangle + pV \quad (4.51)$$

with a corresponding increase in the kinetic energy term for more complicated particles. This equation is applicable to uniform systems. When the system has an interface, and is therefore anisotropic and non-uniform, p is not a simple scalar quantity and the separate components of the pressure tensor (Ch. 2, § 2) must be determined. Since the estimate of the mean virial converges slowly, the entropy values obtained in this way will be correspondingly slow to converge.

The same distribution functions as are obtained from the NVT and NpT ensembles can also be computed from the λVT ensemble, in particular $\rho^{(2)}$ and, for an adsorption system, $\rho^{(1)}$.

The grand canonical ensemble MC method has been disappointing for studies on dense fluids as a result of the low acceptance ratios which are found for the creation and destruction attempts. For the creation trials this is due to the very small chance of finding a hole large enough to accommodate a new particle without causing severe overlap repulsions. The reason for the low acceptance ratio for destruction attempts, although necessary in order to maintain a fixed ratio with that for creations, is more difficult to visualize. It arises because the removal of a particle without permitting relaxation of the remaining fluid leaves the system in an improbable state. As a consequence, the simulation samples very imperfectly the regions of phase space which involve variations of N from its most probable value. At lower densities this disadvantage disappears and is replaced by an important advantage. For dilute systems it is necessary to move the particles large distances in order to go from one important configuration to another. This movement over considerable distances can be more rapidly achieved by annihilation of a particle at one position followed by a creation at the new position. In the NVT and NpT MC procedures the movement would have to be accomplished by a long succession of short translations. For this reason the grand canonical ensemble MC method has been especially effective in the study of gas–solid adsorption, for which it is reasonable to suppose that adjustments in the dense regions near the wall occur primarily as a result of translational trials, whereas for the less dense regions further from the wall the creation/destruction parts of the algorithm are the more important.

Mezei (1980) has found a way of extending to higher density the region which can be efficiently examined by the grand canonical MC method. This involves the location in the system of voids or cavities of greater than a previously chosen size, creation attempts then being made preferentially at these places. Clearly the chance of acceptance of the creation step will be increased, but for there to be a net gain this increase in efficiency must more than counterbalance the loss arising from the need to locate the voids. An adjustable parameter in this method is the minimum radius of a cavity, within which a creation should be attempted. It turns out to be advantageous to decrease this radius as the density is increased. Mezei found that for a 2·2% acceptance the maximum density of a Lennard-Jones fluid at $T^* = 2{\cdot}0$ which could be studied rose from $\rho^* = 0{\cdot}65$ to 0·88, an important increase in the density range. The price of this improvement was an increase in the usage of computer time by a factor of 2·5. For a system with greatly varying density, as near the gas–solid interface, it would certainly be wasteful to carry out the preliminary search for cavities in the dilute region. In order for this cavity-biased method to maintain its advantage over the unbiased scheme it would be necessary to restrict the modification to the high-density regions.

3. Molecular dynamics methods

These methods originated in the work of Alder and Wainwright (1959) on hard-disc and hard-sphere systems. The method was soon extended so as to allow other types of interaction potential to be treated (Rahman, 1964; Verlet, 1967). Various "technical" improvements, mainly concerned with the integration of differential equations and with the "book-keeping" methods of handling the data, have since been developed. The extension to systems of molecules having angularly dependent intermolecular forces will be treated at the end of this section.

For structureless particles the equations of motion may be written as

$$d\boldsymbol{r}_i/\mathrm{d}t = \boldsymbol{v}_i \qquad \text{and} \qquad \mathrm{d}\boldsymbol{v}_i/\mathrm{d}t = -(1/m_i)(\partial U_N(\boldsymbol{r})/\partial \boldsymbol{r}_i) \tag{4.52}$$

where $\boldsymbol{r}_i$, $\boldsymbol{v}_i$ and m_i are the position, velocity and mass of the ith particle. These equations have to be solved in conjunction with the known initial data ($t = 0$) for the system together with any boundary conditions which are imposed. The linear momentum of the system $\left(\sum_i m_i \boldsymbol{v}_i\right)$ will, of course, be conserved during the time development of the system, and so the simulation is in fact a study at constant linear momentum as well as constant N, V and U. Thus when comparing the results of MD with other forms of theoretical treatment it should be remembered that it corresponds almost, but not quite, to the microcanonical ensemble of statistical mechanics.

Solution of Eqns (4.52) proceeds by a series of small time-steps (Δt). For example, the method of Verlet (1967) is based on a Taylor expansion of the position coordinates up to second order in the time,

$$\boldsymbol{r}_i(t \pm \Delta t) = \boldsymbol{r}_i(t) \pm \Delta t(\partial \boldsymbol{r}_i/\partial t) + \tfrac{1}{2}\Delta t^2(\partial^2 \boldsymbol{r}_i/\partial t^2) \tag{4.53}$$

from which

$$\begin{aligned} \boldsymbol{r}_i(t + \Delta t) + \boldsymbol{r}_i(t - \Delta t) &= 2\boldsymbol{r}_i(t) + \Delta t^2(\partial^2 \boldsymbol{r}_i/\partial t^2) = 2\boldsymbol{r}_i + \Delta t^2(\partial \boldsymbol{v}_i/\partial t) \\ &= 2\boldsymbol{r}_i - \Delta t^2(1/m_i)(\partial U_N/\partial \boldsymbol{r}_i) \end{aligned} \tag{4.54}$$

Equation (4.54) may be rearranged to give the form used by Verlet:

$$\boldsymbol{r}_i(t + \Delta t) = -\boldsymbol{r}_i(t - \Delta t) + 2\boldsymbol{r}_i - \Delta t^2(1/m_i)(\partial U_N/\partial \boldsymbol{r}_i). \tag{4.55}$$

Whether Eqn (4.55) is adequate or not depends upon the size of Δt: the larger Δt, the larger the errors introduced by truncating the Taylor expansion at the second-order term. Equation (4.55) is usually used with $\Delta t \approx 10^{-15}$ s, but occasionally with time-steps up to 10^{-14} s. The velocity of the particle is computed from the expression

$$\boldsymbol{v}_i(t) = [\boldsymbol{r}_i(t + \Delta t) - \boldsymbol{r}_i(t - \Delta t)]/2\Delta t. \tag{4.56}$$

In the so-called predictor-corrector method (Rahman, 1964) the starting information required is the position $\boldsymbol{r}_i$ at time $t - \Delta t$ and the positions, velocities and accelerations at time t. The prediction is then made that

$$\boldsymbol{r}_i(t + \Delta t) = \boldsymbol{r}_i(t - \Delta t) + 2\Delta t \boldsymbol{v}_i(t) \tag{4.57}$$

and

$$\boldsymbol{a}_i(t + \Delta t) = \boldsymbol{F}_i(t + \Delta t)/m_i \tag{4.58}$$

where the force $\boldsymbol{F}_i$ is a function of $\boldsymbol{r}_i(t + \Delta t)$, and $\boldsymbol{a}_i$ is the acceleration. These values of $\boldsymbol{r}_i$ and $\boldsymbol{a}_i$ are then used to obtain improved values of the velocities and positions

$$\boldsymbol{v}_i(t + \Delta t) = \boldsymbol{v}_i(t) + \tfrac{1}{2}\Delta t\{\boldsymbol{a}_i(t + \Delta t) + \boldsymbol{a}_i(t)\} \tag{4.59}$$

and

$$\boldsymbol{r}_i(t + \Delta t) = \boldsymbol{r}_i(t) + \tfrac{1}{2}\Delta t\{\boldsymbol{v}_i(t + \Delta t) + \boldsymbol{v}_i(t)\} \tag{4.60}$$

It will be noticed that both of these methods require information on the system at more than one time. For such algorithms, which are said to be non-self-starting, it is necessary to make some special provision for starting-up, although, of course, the outcome of the simulation should not be dependent upon this. A very simple choice is to assign zero velocities to all the particles at zero time.

The most time-consuming part of the simulation is the calculation of the forces, which has to be done at every step. However, most of the inter-molecular force contributions turn out to be entirely negligible and so a large saving in computer time can be made by devising a scheme for recognizing these contributions and rejecting them without calculation. This is usually done by preparing, for each molecule, a list of the particles which are within a sphere of radius R centred on the molecule of interest, where R is a distance somewhat greater than the distance at which the inter-molecular force would become non-negligible (R_c in Fig. 4.4). In

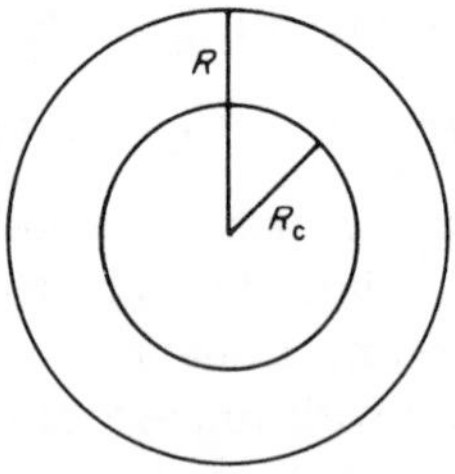

Fig. 4.4

evaluating the force on the central particle, only the interactions with molecules on the list need be computed. After about ten steps the lists should be up-dated to take account of the movement of molecules into and out of the range of the forces. The radius R must exceed the range of the forces by such an amount that a molecule outside the sphere at the zeroth step could not come within force "range" during ten steps. In forming the table the number of distances which must be examined is $N(N-1)/2 = O(N^2)$.

For simulations containing a larger number of molecules Quentrec and Brot (1973) have proposed an alternative "book-keeping" scheme. They divided the total space into boxes of side l, so that if $l = \frac{1}{2}\sigma$ the chance of finding two molecules in the same cell is very small indeed. The assignment of each molecule to its cell can be done very efficiently by dividing each of the x, y, z coordinates by l and retaining only the integral part of the resultant. For short-range forces it may then be adequate to compute the forces only for pairs of molecules which are in adjacent cells, the decision as to how many shells of adjacent cells need be considered being determined by the range of the forces and the accuracy required. In any case, the storage capacity required is $O(\mathcal{N})$, where $\mathcal{N}$ is the number of cells, and so is $O(N)$ at fixed volume per molecule, as against $O(N^2)$ for the list method described above.

The first quantity which needs to be determined is the temperature, which is calculated from the mean value of the translational kinetic energy

$$\tfrac{3}{2}k\langle T\rangle = \tfrac{1}{2}\langle p^2\rangle/m \qquad (4.61)$$

In MD simulations it almost invariably turns out that, after the system has been given its starting conditions and has settled down, $\langle T\rangle$ is found not to be the desired value. Adjustment is then made by appropriate scaling of all the velocities. $\langle T\rangle$ is evaluated again after a further simulation period and, if necessary, further adjustments are made. Another problem is that the integration algorithms are not exact and as a consequence $\langle T\rangle$ drifts away from the desired value. This difficulty can be reduced by using shorter time-steps, but this means that for a given amount of computational time the simulated time is shorter. Another way of improving the scheme is to use a more complicated, higher-order algorithm, but this increases the computational time needed per time step. With regard to the fluctuations in the mean value of a phase function it is worth noting that these are $O(N^{-1})$, whereas for a canonical ensemble MC simulation there are fluctuations of $O(\ln N/N)$ as well as of $O(N^{-1})$ (Hoover and Alder, 1967; Lado and Wood, 1968; Valleau and Card, 1972).

Quantities such as U and p can be obtained from $\langle U_N\rangle$ and the mean virial, as with MC simulations, but C_V is calculated from fluctuations in T

in MD as against fluctuations in U in canonical ensemble MC.

The autocorrelation functions, which are of the form

$$C(t) = \lim_{t' \to \infty} t'^{-1} \int_0^{t'} A(\tau)A(t+\tau)\,\mathrm{d}\tau \tag{4.62}$$

have proved to be some of the most interesting quantities to come out of MD simulations. Although they require separate storage for each particle, this burden is reduced since it is not necessary to note $A(t)$ at every time-step: if $A(t)$ is changing smoothly and not too rapidly, values for $A(\tau)$ and $A(t+\tau)$ may be recorded at less frequent intervals. An alternative method of obtaining $C(t)$ has been proposed by Futrelle and McGinty (1971): they accumulate the Fourier transforms of $A(t)$ for a large number of frequencies and obtain $C(t)$ at the end of the run from the inverse transform of $|\hat{A}(\omega)|^2$. The most widely calculated autocorrelation function is that for the velocity $\langle \boldsymbol{v}(0).\boldsymbol{v}(t)\rangle$. This quantity is primarily of interest in its own right, but it can also be related to the self-diffusion coefficient

$$D = d^{-1} \int_0^{\infty} \langle \boldsymbol{v}(0).\boldsymbol{v}(t)\rangle\,\mathrm{d}t \tag{4.63}$$

where d is the dimensionality of the system. The difficulty in using Eqn (4.63) arises from the very slow decay of $\langle \boldsymbol{v}(0).\boldsymbol{v}(t)\rangle$, as $\sim t^{-1/2}$. In almost every case it is better to obtain D from the linear part of a plot of $\{(\boldsymbol{r}_i(t) - \boldsymbol{r}_i(0)\}^2$ versus time, the first part of the plot being curved:

$$D = \frac{\{\boldsymbol{r}_i(t) - \boldsymbol{r}_i(0)\}^2}{2dt} \quad \text{or} \quad = \frac{\mathrm{d}}{\mathrm{d}t}\frac{\{\boldsymbol{r}_i(t) - \boldsymbol{r}_i(0)\}^2}{2d} \tag{4.64}$$

The computation of the viscosity and thermal conductivity coefficients are far more difficult than the self-diffusion coefficient. Most success has been achieved by the use of non-equilibrium methods, e.g. the calculation of the stress when a known rate of strain occurs (Ashurst and Hoover, 1973). A further improvement has been made by using a differential method introduced by Cicotti and Jacucci (1975) in which parallel MD trajectories, one with and one without the imposed stress, are employed. The change arising from the stress is obtained by difference and this has the important effect of greatly reducing the noise, at least at times which are not too long (Singer *et al*. 1980). No attention seems to have been given yet to the effect of substrate forces on the results.

Some consideration has been given to improving the efficiency of MD simulations by distinguishing between interactions at short and long distances on account of their very different correlation times (Finney, 1978; Streett, *et al.*, 1978). Since the long-distance interactions fluctuate much more slowly than those at shorter distances, and as they are also weaker, it is not really necessary to re-calculate as frequently the forces arising from them. This idea has been implemented by drawing around each molecule two cut-off spheres. Interactions with molecules in the innermost zone are re-calculated at every step, whereas those with molecules between the spherical surfaces are only completely re-calculated every n steps, where n is determined by trial and error.

Two tests of this modification have been made. Finney (1978) studied water, using the ST2 potential (Ch. 8, § 2(b)), and took the simplest view with regard to the force from the region between the spherical surfaces, namely that it was unchanged during the n steps. Streett *et al.*, in their work on fluid methane, assumed an mth-order Taylor expansion (with $m = 1$–3) for the force during the period between complete re-calculations of the force. In both pieces of work it appeared that $n = 10$ was quite tolerable. Thus Streett *et al.* found a percentage error in this secondary force region of only $\sim 1\%$ if $m = 2$ and $\sim 0\%$ if $m = 3$. The radii of the outer spheres were approximately those normally chosen for the (single) cut-off distance of a normal run (0·931nm for H_2O by Finney and 2·50σ for methane by Streett *et al.*). The inner spheres were, however, astonishingly small (0·338 or 0·578 nm for H_2O by Finney and 1·1σ for CH_4 by Streett *et al.*). Thus Finney's inner sphere contained on average only four molecules for the smaller radius and 21 for the larger, whilst Streett *et al.* had 1·4–3·1 and 24–29 according to the density of the fluid. Two queries remain about the usefulness of this method. One is the question of the amount of computational effort which must be expended in finding and justifying suitable choices for n, m and the radius of the inner sphere. The second concerns the applicability of the technique to systems, such as those in gas–solid adsorption, which have great variations in density.

A development which is in some measure related to the multiple time-step method and which was contemporaneous with it is that of so-called Brownian dynamics. As with the multiple time-step method, the problem to be faced is the existence of two time-scales and the virtual impossibility of carrying on the simulation long enough to take account of the slower process when the time-step must be small enough to deal with the faster one. The basic principle involved is that it is possible to perform calculations on some of the variables whilst representing the effect of the rest as a random noise. In Brownian dynamics the equation of motion of the solute particles of a solution are represented by the strict Langevin equation,

$$\mathrm{d}^2\boldsymbol{r}_i/\mathrm{d}t^2 = -A\,\mathrm{d}\boldsymbol{r}_i/\mathrm{d}t + \Big(-\sum_j \partial u_{ij}^{(2)}/\partial\boldsymbol{r} + \boldsymbol{F}\Big)/m_i \tag{4.67}$$

where $\boldsymbol{F}$ is a random force and A is the frictional constant, rather than by the Newtonian equations.

In the generalized Langevin equation the first term on the right-hand side of Eqn (4.67) is replaced by one including an integral over a memory function $f(t)$

$$-A\frac{\mathrm{d}\boldsymbol{r}_i}{\mathrm{d}t} \Rightarrow -\int_0^t F(t-s)\boldsymbol{p}(s)\,\mathrm{d}s \tag{4.68}$$

Thus in the generalized form the frictional term depends directly upon the state of the system at earlier times, whereas in the "strict" form it is only dependent upon the current state of the system. In this equation the interactions with the solvent are represented by the frictional and the random force terms. The solute–solute interactions $u_{ij}^{(2)}$ include solvent-mediated interactions, such as the hydrophobic interactions (Chapter 8), as well as direct interactions. Turq *et al.* (1977) have noted that for the use of the "strict", rather than the generalized, Langevin equation the correlation time for the random force $\boldsymbol{F}$ must be much less than that for the single-particle frictional constant. In their simulation Turq *et al.* chose $\boldsymbol{F}$ at regular intervals of t_s from a Gaussian distribution, the value then being held constant until the next choice was made. The success of this method could have important implications for surface chemistry simulations, in which the problem of having processes with several widely different time-scales is a frequent one.

Another extension of the MD approach is to other ensembles than the microcanonical (NVU) (Andersen, 1980). Runs corresponding to evaluations in the NpH ensemble, an ensemble that had previously been neglected, can be carried out by enclosing the system in a membrane of mass M so that volume changes can occur in which the scaled co-ordinates of all particles ($\rho_i = r_i/V^{1/3}$) are unchanged. This is similar to the scaling of co-ordinates used when implementing volume changes in the NpT MC simulations (§ 2(c)). The Hamiltonian of the new system becomes

$$H(\boldsymbol{r},\boldsymbol{p},V,\dot{V}) = \tfrac{1}{2}\sum_{i=1}^{N} m_i^{-1}\boldsymbol{p}_i\cdot\boldsymbol{p}_i + U_N(\boldsymbol{r}) + \tfrac{1}{2}M\dot{V}^2(t) + p_{\mathrm{E}}V(t) \tag{4.69}$$

where p_{E} is the pressure on the membrane, and the possibility of a surface term has been ignored. In terms of the scaled co-ordinates the Hamiltonian becomes

$$H(\boldsymbol{\rho},\dot{\boldsymbol{\rho}},V,\dot{V}) = \tfrac{1}{2}V^2\sum_i m_i\dot{\rho}_i\cdot\dot{\rho}_i + U_N(\rho_i V^{1/3},\ldots,\dot{\rho}_N V^{1/3}) + \tfrac{1}{2}M\dot{V}(t)^2 + p_{\mathrm{E}}V(t) \tag{4.70}$$

Andersen was able to show that a time average of any function $F(\boldsymbol{r},\boldsymbol{p};V)$

over the trajectory which can be derived from Eqn (4.70) and its associated equations of motion was equal to the Np_EH ensemble average of the function, apart from negligible errors,

$$\langle F\rangle = \langle F\rangle_{NpH}(N, p, U - \overline{\tfrac{1}{2}\Pi^2/M}) + O(N^{-2}) \tag{4.71}$$

where $\overline{\tfrac{1}{2}\Pi^2/M}$ is the mean kinetic energy associated with the motion of V, Π being the momentum conjugate to V. The static properties of the system should not depend upon the choice of M, though the latter does affect the rate of equilibration of the simulation (Haile and Graben, 1980). Andersen has suggested that it would be appropriate to choose M so that the time-scale for the volume fluctuation is approximately equal to the time taken for a sound wave to travel from one end of the sample to the other.

Andersen has proposed corresponding modifications to the normal MD procedure with a view to yielding averages equivalent to those over the NVT and NpT ensembles. To obtain NVT ensemble averages he suggested that stochastic collisions, random in timing, choice of particle involved, and effect, should be used so that an otherwise normal MD trajectory would be modified so as to sample all energies. The collisions would be regarded as being instantaneous, and between them the system would evolve according to the usual Newtonian laws. At each collision the momentum of the single particle affected would be chosen at random from a Boltzmann distribution, the particle thereby losing all memory of its former momentum. The choice of the frequency of the stochastic collisions should not, of course, affect the average values of the static quantities. However, it may be expected to alter greatly the rate of equilibration, since it governs the rate at which the trajectory can move from one energy to another. Andersen has suggested that this collision rate (ν) should be chosen so that the rate of decay of energy fluctuations along the MD trajectory should be approximately equal to that for small volumes of the real liquid (surrounded by larger volumes of the same liquid). On this basis he recommended $\nu = \tfrac{2}{3}a\lambda V^{1/3}$, where λ is the coefficient of thermal conductivity and a depends on the shape of the sample. To obtain averages equivalent to those over the NpT ensemble, he proposed a scheme which combined the new features of the NVT and NpH schemes. The three prescriptions are valid for the evaluation of position distribution functions as well as for static thermodynamic properties. It is not yet clear whether time-dependent correlation functions can be obtained from these modified trajectories.

The NVT and NpT ensemble schemes may offer considerable advantages over the normal NVU simulation when adsorption systems are being considered. The advantage these ensembles have over the NVU ensemble for uniform systems, namely that the temperature can be chosen with certainty, is more important still for adsorption. Especially if the energy of adsorption is large, it would be difficult to choose starting conditions which would yield the desired temperature when equilibrium was reached. NpT should, likewise, hold an advantage over the NVU and NVT schemes since, because the packing of molecules near the surface is very different from that in the bulk, the resultant pressure may be far from the desired one. If, however, the price of using these modified MD techniques is that no time-dependent correlation functions can be obtained a major advantage of MD over MC would be lost. Now consider the possibility that these developments can be carried further and an MD scheme devised giving averages over the grand canonical ensemble. Firstly, this would probably entail a system of stochastic collisions in order to be able to sample all energies, and thus get averages for constant T. Secondly, there would presumably be a need for a stochastic creation/destruction of particles in order to sample a range of N values. But, to have an advantage over the grand canonical MC technique it would have to yield extra information and/or converge more readily. With regard to the former, the possibility of obtaining values for the time-dependent correlation functions would be the main attraction, but the computation of $\langle \boldsymbol{v}(0) . \boldsymbol{v}(t) \rangle$, for example, would be put onto a less secure basis by the creation/destruction of particles.

An important development of Andersen's work on the removal of the constant volume restriction would appear to be the demonstration by Parrinello and Rahman (1980) that it is possible to let the *shape* of the simulation box vary also. In many simulations of dense systems in two or three dimensions it is likely that some possible structures are prevented from occurring because of their incompatibility with the dimensions of the simulation box. In other cases, it is felt that although two structures may both be compatible with the simulation box, transition between them may be rendered virtually impossible because of incompatibility of some of the transition states through which the system would have to pass. Flexibility of box shape (as well as size) may obviate both of these difficulties.

4. Long-range corrections: general

Much of the very early simulation work was on systems which, from the definition of the model, involved only inter-particle interactions which were completely and readily taken into account during the computation. Such systems were the Ising lattice with only nearest-neighbour interactions, and assemblies of hard-sphere and hard-disc particles. When more realistic potentials, such as the Lennard-Jones 12–6 potential for dense fluids, were introduced it soon became evident that a problem had arisen in the form of the infinite number of inter-particle interactions which now contributed to the overall energy of each configuration. It was clear that the number of interactions which could be exactly computed must be kept within the range of the computing power available. To this end two main procedures were adopted, both of these also having the effect of eliminating the spurious periodicity which could be introduced by the use of periodic boundary conditions.

In the case of the spherical cut-off the radius (R_c) can be chosen fairly freely. Although, if R_c is more than one-half of the linear dimension of the simulation box each particle may interact both with another particle and with the periodic images of that particle. However, the cut-off is often set at a much smaller value than this, typically 2·5–3·5σ for systems of Lennard-Jones molecules. The effect of interactions beyond this range is taken into account by applying long-range corrections. For a dense fluid of Lennard-Jones molecules McDonald and Singer (1969) showed that this correction could be applied with sufficient accuracy by means of a mean-field or van der Waals approach in which $g^{(2)} = 1$. This has the great advantage that, because the value is independent of the instantaneous positions of the particles, it does not vary during the simulation and so can be omitted at that stage, since it cannot affect the course which the simulation takes. The correction can be made at the end of the simulation, which is, of course, much more economic in computer time. The correction to the internal energy, for example, is

$$U^{\mathrm{LR}} = \langle U_N^{\mathrm{LR}} \rangle = \iint_{|r_1 - r_2| > R_c} \rho^{(1)}(\mathbf{r}_1)\rho^{(1)}(\mathbf{r}_2)u^{(2)}(|\mathbf{r}_1 - \mathbf{r}_2|)\,\mathrm{d}\mathbf{r}_1\,\mathbf{d}\mathbf{r}_2 \tag{4.72}$$

which in a uniform fluid reduces to

$$U^{\mathrm{LR}} = \tfrac{1}{2}V\rho^2 \int_{R_c}^{\infty} u^{(2)}(r)4\pi r^2\,\mathrm{d}r \tag{4.73}$$

For their system with a cut-off at 2.5σ, they were able to show that U^{LR}, itself only ~7% of U, could be estimated with an accuracy of ~1% in this way. In view of what has been said above, the distribution functions $\rho^{(1)}$ and $\rho^{(2)}$ would be unaffected by the long-range corrections in this approximation.

For simulations in the NpT and $TV\lambda$ ensembles the density varies along the Markov chain and so the correction should, in general, be applied to each configuration. If this is not done, then, bearing in mind that the correction is attractive, the more dense configurations suffer unfair discrimination since they are subject to the loss of larger attractive terms than are the less dense configurations. Such an omission led to the erroneously high vapour pressures found for liquid argon by Rowley *et al.* (1975). In that case it turned out that because of the very simple mathematical form of the expression for the correction to the chemical potential the correction could be applied easily at the end of the simulation without introducing any further error beyond that of the mean-field approximation (Rowley *et al.*, 1976b).

The effect of long-range corrections on thermodynamic properties can be conveniently treated by means of a coupling parameter ξ which turns on the long-range corrections. The effect on the Helmholtz free energy can be deduced as follows:

$$\beta A(\xi) = -\ln\left\{\int \exp\{-\beta(U_{N_0} + \xi U_N^{\mathrm{LR}})\}\,\mathrm{d}q\right\} \tag{4.74}$$

$$\begin{aligned}\partial(\beta A)/\partial\xi &= \int \beta U_N^{\mathrm{LR}} \exp\{-\beta(U_{N_0} + \xi U_N^{\mathrm{LR}})\}\,d\boldsymbol{q}\Big/\int \exp\{-\beta(\mathrm{U}_{N_0} \\ &\quad + \xi \mathrm{U}_N^{\mathrm{LR}})\}\,\mathrm{d}\boldsymbol{q} \\ &= \langle \mathrm{U}_N^{\mathrm{LR}}\rangle_\xi \end{aligned} \tag{4.75}$$

where U_{N_0} is the potential energy of a configuration ignoring long-range corrections; U_N^{LR} is the correction to the energy of a configuration when the long-range interactions are fully activated; $\langle U_N^{\mathrm{LR}}\rangle_\xi$ is the mean value of the long-range correction to the energy with the particle distribution appropriate to the energy with the coupling parameter at value ξ.

The correction to A is

$$A^{\mathrm{LR}} = \int_0^1 \mathrm{d}\xi\,\langle U_N^{\mathrm{LR}}\rangle_\xi \tag{4.76}$$

and to the chemical potential is

$$\mu^{\mathrm{LR}} = (\partial A^{\mathrm{LR}}/\partial N)_{T,V} \tag{4.77}$$

This yields an exact evaluation of μ^{LR}, and so, although it has been discussed in terms of the *NVT* ensemble the value for the *NpT* or $TV\lambda$ ensembles should be the same, at least within the usual tiny differences $\mathrm{O}(\mathrm{N}^{-1})$ found between ensembles. In fact, we only know $\langle U_N^{\mathrm{LR}}\rangle_\xi$ for $\xi = 0$, and so it is necessary to make the reasonable assumption that $\langle U_N^{\mathrm{LR}}\rangle_\xi \approx \langle U_N^{\mathrm{LR}}\rangle_{\xi=0}$. For the argon simulation mentioned above, Rowley *et al.* (1975) then found that $A^{\mathrm{LR}} \propto \rho^2$ at constant T (or $\propto N^2$ at constant T and V), so that $\mu^{\mathrm{LR}} = 2\langle U_N^{\mathrm{LR}}\rangle_{\xi=0}/N$. In general there would not be such a simple relationship between A^{LR} and N, and μ^{LR} would be obtained by numerical differentiation of A^{LR} with respect to N.

For the inhomogeneous systems with which one is dealing in adsorption the situation is far more complicated and it is not normally possible to apply the correction at the end of the run without introducing additional approximations even in the *NVT* ensemble. Consider a proposed move of a particle to a more dense region in an *NVT* simulation. As before, the stability of the dense region will be underestimated and there will be undue discrimination against the move. Furthermore the computation of U_N^{LR} for each configuration would be much more protracted than for a single configuration of a homogeneous system. Because of this, Rowley *et al.* (1976a, b) found it necessary to perform their simulation of gas–solid adsorption whilst omitting the long-range corrections and subsequently to apply a correction which involved approximations beyond those usually made in van der Waals theory. This procedure is described in Chapter 6, §2(a) where it is discussed in terms of the coupling parameter approach used above.

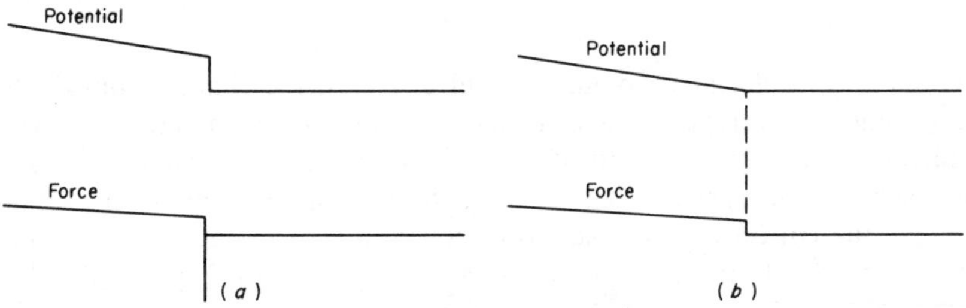

Fig. 4.5

Where MD simulations are being used there is an additional effect arising from the imposition of a cut-off: the existence of an impulsive force at the cut-off distance (Chapela *et al.*, 1977). This is frequently removed by making a shift in the potential so as to force it to be continuous at the cut-off (Fig. 4.5). Care must be exercised in comparing results from simulations using the modified cut-off scheme ((*b*) of Fig. 4.5) with those from MC studies, which use the unshifted potentials (*a*) (Murad *et al.*, 1979). For non-spherical molecules the shift involved is dependent upon their orientations. Adams *et al.* (1979) have shown how this can lead to non-constancy of the energy in MD runs on such substances. Suppose the molecule is at a distance just less than the cut-off. Its orientation will then be correlated with that of the central molecule in such a way as to favour a lower energy. If the molecule moves out of range its orientation will become less well correlated and if, subsequently, it returns to within the cut-off distance it will, on average, give a higher interaction energy with the central molecule than was the case initially. A somewhat similar effect occurs in the MI convention (Pangali *et al.*, 1980). In Fig. 4.6 the molecule (2′) which replaces (2) as it moves out of the box has the opposite orientation with respect to molecule (1) ((1) and (2) have substituents on adjacent parts; (1) and (2′) have them on remote parts). Since the orientations of

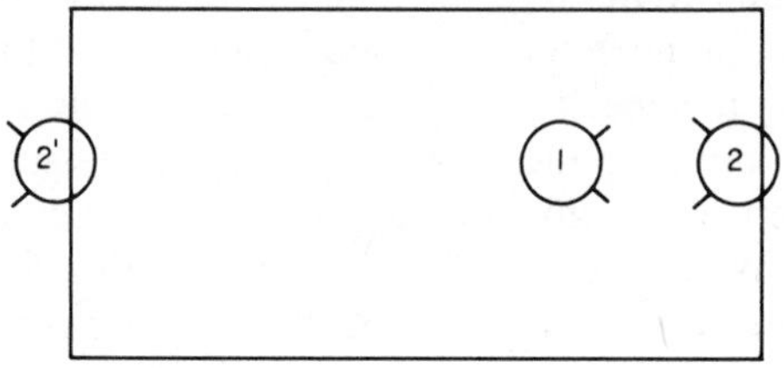

Fig. 4.6

(1) and (2) would have favoured lower energies, the change from (2) to (2′) would, on average, involve an increase in energy. Both of these effects, which lead to an upward drift of the temperature in MD simulations, have no exact counterpart in MC. Instead, the loss of correlation on passing outside the cut-off region leads to a reduction in $\rho^{(2)}$, since the molecule is no longer able to lower its potential energy by adopting a favourable orientation with respect to the central molecule.

5. Long-range corrections: electrostatic effects

When the inter-particle forces are of electrostatic origin the problem of correcting for long-range effects is much more difficult. The extra difficulty is derived from the slowness with which most such potentials fall off as the inter-particle distance is increased. For example, inter-ionic and inter-dipolar potentials fall off as r^{-1} and r^{-3}, respectively. That an ionic system has a definite value for the inter-ionic potential is dependent upon its electroneutrality, which causes a great deal of cancellation among terms which otherwise would lead to divergence. Likewise, convergence of the total dipole–dipole interaction energy is dependent upon the fact that the interaction is angle- as well as distance-dependent. For non-angle-dependent potentials of the form ar^{-n}, convergence requires $n > 3$ for three-dimensional systems; $n = 3$ would lead to divergence. All systems of interest should, of course, provided their Hamiltonians are correctly represented, have convergent potentials. However, the convergence may be only very slow, and herein lies the problem for simulations which are limited in the region over which the potentials can be summed and the amount of computing time which can be expended on energy evaluations. Several methods of dealing with this problem have been devised, none of which is by any means completely satisfactory.

The reaction-field method (Barker and Watts, 1973) depends upon a result obtained by Onsager (1936) for the interaction of a single dipole with a surrounding dielectric, a similar method having been earlier employed by Born (1920; see also Bell, 1931) for ions in a dielectric medium. In this approach a spherical cavity is formed in the medium with the dipole of interest at its centre. The dipole induces polarization in the medium (outside the sphere) which in turn produces a field in the cavity, and this "reacts" with the dipole at the centre of the sphere, giving a contribution to the energy of the configuration of

$$-2(\varepsilon_r - 1)(2\varepsilon_r + 1)^{-1}\boldsymbol{\mu}^2 R^{-3}(4\pi\varepsilon_0)^{-1} \tag{4.78}$$

where R is the radius of the sphere. For a simple, axially symmetric quadrupolar molecule, e.g. N_2 or CO_2, the corresponding expression is

$$-1{\cdot}5(\varepsilon_r - 1)(3\varepsilon_r + 2)^{-1}\boldsymbol{\Theta}^2 R^{-5}(4\pi\varepsilon_0)^{-1}. \tag{4.79}$$

When this type of calculation is applied in simulations the cut-off sphere is much larger than that proposed by Onsager and other early workers, who assumed the sphere to be just large enough to accommodate a single polar molecule. In a typical simulation the number of molecules in the sphere would be ~50. Equation (4.78) is accordingly modified by replacing $\boldsymbol{\mu}$ by $\boldsymbol{P}(=\Sigma_i \boldsymbol{\mu}_i)$, the resultant moment of the molecules in the sphere. Since $\boldsymbol{P}$ depends upon the configuration of the dipoles within the cut-off sphere it will vary during the simulation and so must be re-calculated at each step. Nevertheless, the computation involved is fairly inexpensive and application of the reaction-field correction for isotropic uniform systems, to which the above discussion refers, is always worthy of consideration. For anisotropic and/or non-uniform systems, into which class adsorption systems fall, more complicated, and as yet underived, equations would be needed. Furthermore, even when these equations are available it must be doubtful whether re-evaluation of the correction at each step would be feasible. The reaction field formulae are obtained by an imaginary "switching on" process of the dipoles or quadrupoles and therefore refer to the free energy contributions (A_{rf}). It follows that they lead to a correction factor of $\exp(-\beta A_{rf})$ being applied to the normal Boltzmann factor $\exp(-\beta U_N)$, where U_N includes only the interactions contained within the cut-off. The reaction field approach also involves an averaging over the fluctuating states of the dielectric.

For pure substances, as distinct from solutions of the polar substance in a non-polar solvent, the relative permittivity (ε_r) is not known, and, indeed, it is one of the quantities which one would hope to be able to obtain from the study. In principle, ε_r could be obtained by an iterative process: a value of ε_r could be guessed, a simulation carried out yielding a new value of ε_r, and so on. However, for a number of reasons this is not feasible. The procedure for obtaining ε_r from a simulation is uncertain, as will be explained below. Furthermore, the idea of using a simulation as one part of an iteration is unattractive. Fortunately, the reaction field correction to the energy is insensitive to the value of ε_r provided that $\varepsilon_r \gg 1$. Thus a fairly crude guess at ε_r may be adequate. In the NpT and $TV\lambda$ ensembles ε_r will vary as the density changes, providing a further reason why the correction should be re-evaluated at each step. In order to make feasible the calculation of the correction at each instant, it is necessary that it should be represented by a very simple formula. For interfacial systems two extra problems arise: (i) ε_r now varies with position. Thus a modification in an NVT ensemble simulation in which a polar molecule moves towards a region of higher ε_r will be favoured by the inclusion of the reaction field, which again is a reason why the correction must be re-calculated for each step. (ii) ε_r is in fact a tensor whose components are all dependent upon position. The symmetry of the system will reduce the number of these components which are independent. Furthermore, the spherical cavity, which is an appropriate choice for a bulk system, no longer gives simple equations for the adsorption systems. It might appear that a non-spherical cavity would be more suitable for such a system, but, be that as it may, no successful applications of the reaction field method for anything other than a spherical cavity have been made.

In applying the reaction field method it is usual to ignore the fact that ε_r is frequency-dependent and to use the static value instead. In MD, but not MC, computations it would be conceivable that $\varepsilon_r(\omega)$ would be calculable, although in practice it would be scarcely feasible. For MC simulations it does not even seem to be conceivable that a self-consistent calculation could be set up. Friedman (1975) tentatively suggested a way of taking the frequency dependence of ε_r into account in MD calculations by means of a time lag τ, so that the images at time t corresponded to the configuration at time $t - \tau$. Van Gunsteren *et al.* (1978) have investigated this time-lag procedure more thoroughly, but it now appears that it faces serious, if not insuperable, obstacles on account of its failure to satisfy microscopic reversibility.

A modification of the reaction field approach has been suggested in which the full expression for the reaction potential at a point ($r, x = \cos\theta$), where r, θ are the polar coordinates, due to a charge $+e$ at (s, 1),

$$\phi(r,x) = \frac{1-\varepsilon_r}{4\pi\varepsilon_0}\frac{e}{R}\sum_{n=0}^{\infty}\frac{n+1}{n+\varepsilon_r(n+1)}\left(\frac{rs}{R}\right)_n P_n(x) \qquad 0 \leqslant r \leqslant R \quad (4.80)$$

(R being the radius of the spherical cavity) is replaced by

$$\phi = \phi^{(0)} + \phi^{(1)} + \phi^{(2)} + \ldots \quad (4.81)$$

with

$$\phi^{(k)}(r,x) = -\frac{\varepsilon_r^{-1}}{(\varepsilon_r+1)^{k+1}}(4\pi\varepsilon_0)^{-1}\frac{R}{s}\frac{e}{R^2/s}\sum_n (n+1)^{-k}\left(\frac{r}{R^2/s}\right)^n P_n(x) \quad (4.82)$$

and only $\phi^{(0)}$ is retained (Friedman, 1975). This truncation is only reasonable if $\varepsilon_r \gg 1$. The corresponding equation for a dipole or a quadrupole would be obtained by treating it as the limiting case of separated charges. A serious criticism of this approximation is that $\phi^{(0)}$ for a molecule has a singularity at $s = R$, for then the charge (at s, 1) touches its image (at R^2/s, 1). It is therefore necessary to add a suitable repulsive term to overcome this. Very little use has been made of this method. However, it may be more suitable for studying adsorption because the approximation is exact for slab systems and the restriction to $\varepsilon_r \gg 1$ does not apply. Thus the anisotropy of these systems is here an advantage, although for simulations one could not take a cavity that was infinite in two dimensions, as the use of the results for a slab would require. The other feature of adsorption studies, the non-uniformity, would again be a disadvantage. Friedman has also pointed out that when the usual reaction-field equation (4.78) is employed one is only taking into account the dipole moment of the sample, the higher moments being rejected. As the size of the cavity is increased the relative contributions from the higher moments becomes steadily less, but for the size of system (and cavity) normally studied it is far from negligible. Friedman claims that in such circumstances his image reaction-field method would give better results.

An alternative method of correction, originated by Ewald (1921) and developed by Kornfeld (1924), depends upon the assumption that the configuration in the simulation box (of side l) is repeated in all other replicated boxes. Thus it assumes that the corresponding ordering extends to infinity. Ewald then showed that the pairwise electrostatic energy terms could be written as the sum of two series, one in ordinary space and the other in reciprocal space.

$$
\begin{aligned}
(4\pi\varepsilon_0)U_N^{\text{el}} = \frac{2\pi}{3}\sum_i\sum_j e_ie_j \\
\sum_{k\neq 0} \boldsymbol{k}^{-2}\cos[\boldsymbol{k}\,.\,(\boldsymbol{r}_j-\boldsymbol{r}_i)] \\
\exp[-k^2/(4\alpha^2)] + \tfrac{1}{2}\sum_i\sum_j{}' e_ie_j\sum_{\boldsymbol{k}} \\
\operatorname{erfc}(\alpha|\boldsymbol{r}_j-\boldsymbol{r}_i+l\boldsymbol{n}|)\,.\,|\boldsymbol{r}_j-\boldsymbol{r}_i+l\boldsymbol{n}|^{-1} \\
-\tfrac{1}{2}\sum_i e_i^2\alpha/\pi^{\frac{1}{2}} \qquad (4.83)
\end{aligned}
$$

$\boldsymbol{n} = n_x, n_y, n_z$, where n_x, n_y, n_z are integers and $\boldsymbol{n} = 0$, is included and $\boldsymbol{k} = (2\pi/l)\boldsymbol{n}$, where $n = 0$, is omitted.

The parameter α, which may be freely chosen, is adjusted so that both series converge. Changes in α have opposite effects on the convergence of the two series, so an optimum value of α is sought. With a suitable choice of α (Woodcock and Singer (1971) used $\alpha = 5{\cdot}714/l$) it was found that only three Fourier terms were needed for the coulombic interactions in a study of molten salts in which thermodynamic and structural, but not dielectric, properties were calculated.

For interactions between point dipoles the formula corresponding to Eqn (4.83) is

$$
\begin{aligned}
(4\pi\varepsilon_0)U_N^{\text{el}} = \sum_{i>j}\sum [\boldsymbol{\mu}_i\,.\,\boldsymbol{\mu}_jB(r_{ij}) - (\boldsymbol{\mu}_i\cdot\boldsymbol{r}_{ij})(\boldsymbol{\mu}_j\,.\,\boldsymbol{r}_{ij})C(r_{ij})] \\
-2\alpha^3\mu^2/(3\pi^{1/2}) + \frac{2\pi}{l^3}\sum_{k\neq 0}\boldsymbol{k}^{-2}\exp[-k^2/(4\alpha^2)]\Big\{\sum_j \boldsymbol{\mu}_j\,.\,\boldsymbol{k}\cos(\boldsymbol{k}\,.\,\boldsymbol{r}_{ij})/l)^2 \\
+\Big[\sum_j(\boldsymbol{\mu}_j\,.\,\boldsymbol{k})\sin(\boldsymbol{k}\,.\,\boldsymbol{r}_{ij})/l\Big]^2\Big\} \qquad (4.84)
\end{aligned}
$$

where

$$B(r) = \operatorname{erfc}(\alpha r)\,r^{-3} + (2\alpha/\pi^{1/2})\exp(-\alpha^2r^2)r^{-2}$$

and

$$C(r) = 3\operatorname{erfc}(\alpha r)r^{-5} + (2\alpha/\pi^{1/2})\exp(-\alpha^2r^2)(2\alpha^2r^{-2} + 3r^{-4})$$

(Adams and Mcdonald, 1976). Adams (1980) found that for the evaluation of ε_r for a liquid of point dipoles as many as 125 Fourier vectors were needed, in spite of the sum over the reciprocal lattice being $\leqslant 1\%$ of the total coulombic sum. This sensitivity is typical of the problems encountered in calculating dielectric, as distinct from thermodynamic, quantities.

Ladd (1978) has used another method which, like the Ewald method, assumes that there is long-range order of the configurations of the simulation box. However, he employed a multipole representation for dealing with the interactions between boxes, there being a set of multipoles at the centre of each box. The values of these multipoles depend upon the current configuration within the simulation box. The total interaction energy can then be written as a sum of terms, each of which is made up of a product of two of these configuration-dependent multipoles and a configuration-independent lattice sum.

$$U'_N = -(1/30)(4\pi\varepsilon_0)^{-1}\langle\mu_\alpha\Omega_{\beta\gamma\delta}\rangle T_{\alpha\beta\gamma\delta} -(1/1890)(4\pi\varepsilon_0)^{-1}\langle\mu_\alpha\psi_{\beta\gamma\delta\varepsilon\phi}\rangle T_{\alpha\beta\gamma\delta\varepsilon\phi}\ldots \tag{4.85}$$

where

$$T_{\alpha\beta\gamma\delta} = \sum_{n\neq0} T^{(n)}_{\alpha\beta\gamma\delta} = \sum_{n\neq0} \nabla_\alpha\nabla_\beta\nabla_\gamma\nabla_\delta(1/|\boldsymbol{R}^{(n)}|)$$

$$T_{\alpha\beta\gamma\delta\varepsilon\phi} = \sum_{n\neq0} T^{(n)}_{a\ldots\phi} = \sum_{n\neq0} \nabla_\alpha\ldots\nabla_\phi(1/|\boldsymbol{R}^{(n)}|)$$

$\boldsymbol{R}^{(n)}$ is the centre of the nth cell, and $\Omega_{\beta\gamma\delta}$ and $\psi_{\beta\gamma\delta\varepsilon\phi}$ are the octopole and 32-pole moments of the contents of the basic cell referred to its centre as origin. Ladd gives values for $T_{\alpha\beta\gamma\delta}$ and $T_{\alpha\beta\gamma\delta\varepsilon\phi}$. In a simulation of a liquid of dipolar hard spheres he found that the two terms shown in the expansion were sufficient for the evaluation of the mean energy. A corresponding expression may be derived for the effect of the outer cells on the force on the outer molecule.

The dependence of the properties on the method of correction used is least for those thermodynamic and distribution functions which are not, or are only slightly, angularly dependent. When the quantity desired is the permittivity or an angular correlation function, more careful consideration is needed (Pangali *et al.*, 1980). Just such a careful treatment has been made by de Leeuw *et al.* (1980) for the evaluation of the permittivity from simulations. For a system of dipoles they showed that when a large sphere composed of the replicated simulation boxes of permittivity ε_r is encased in a much larger sphere of permittivity ε'_r, then

$$\frac{(\varepsilon_r - 1)(2\varepsilon'_r + 1)}{3(\varepsilon_r + 2\varepsilon'_r)} = yg(\varepsilon'_r) \tag{4.86}$$

where

$$y = 4\pi\rho\mu^2(4\pi\varepsilon_0)^{-1}(9kT)^{-1}$$

and

$$g(\varepsilon'_r) = \{\langle M^2\rangle - \langle M\rangle^2\}/(N\mu^2)$$

Thus ε_r can be evaluated if ε_r' and $g(\varepsilon_r')$ are known. $g(\varepsilon_r')$ can be determined from a simulation run in which a correction is applied on the basis of there being a reaction field from the medium of the outer sphere. Three cases are of interest. The first is that in which the inner and outer region have the same permittivity. Equation (4.86) then becomes

$$(\varepsilon_r - 1)(2\varepsilon_r + 1)/9 = yg(\varepsilon_r) \tag{4.87}$$

This corresponds to the condition which one is *thought* to be imposing in most simulations. The second situation is that in which the outer region is a vacuum ($\varepsilon_r' = 1$), giving

$$(\varepsilon_r - 1)/(\varepsilon_r + 2) = yg(1) \tag{4.88}$$

the Clausius–Mosotti relation. Runs with $\varepsilon_r' = \varepsilon_r$ and $\varepsilon_r' = 1$ would give very different values for $g(\varepsilon_r')$, but both when combined with the appropriate equation ((4.87) or (4.88)) should give the same value for ε_r. However, de Leeuw *et al.* point out that the uncertainty in the value of ε_r obtained in the two cases may be very different. In practice, it is not possible to apply the condition $\varepsilon_r' = \varepsilon_r$, since ε_r would then need to be known at the start of the simulation. The third choice, and one which it is possible to implement, is $\varepsilon_r' = \infty$, for which

$$(\varepsilon_r - 1)/3 = yg(\infty) \tag{4.89}$$

De Leeuw *et al.* found that this choice for ε_r', which corresponds to a conducting outer region, gave more consistent values for ε_r and $g(\varepsilon_r')$ than did $\varepsilon_r' = 1$. It turns out that the Ewald–Kornfeld summation, as normally used, is equivalent to the $\varepsilon_r' = \infty$ case rather than to $\varepsilon_r' = \varepsilon_r$ as was previously thought (Adams, 1980). Hoskins and Smith (1980) have made a corresponding MC study for ionic melts, examining the effect on the distribution functions (g_{++}, g_{+-}, g_{--}) and energies of the value of ε', the permittivity outside the large sphere. They found that the results were quite sensitive to ε' for systems with high densities of ions. Figure 4.7 shows results for $\varepsilon' = 1$ (a vacuum) and $\varepsilon' = \infty$ (a conducting medium), the results for the simple MI method being indistinguishable from those for periodic boundary conditions with $\varepsilon' = 1$. Although Valleau and Whittington (1977) have criticized those methods, and in particular the Ewald method, which introduce a spurious periodicity into the system, it seems that the considerations introduced by de Leeuw *et al.* provide a more correct explanation of the differences found between the results using the various treatments of long-range corrections.

Adams (1980) has found that for highly polar fluids, e.g.

$$\mu^* = \mu/(4\pi\varepsilon_0\sigma^3 kT)^{\frac{1}{2}} = (2.75)^{\frac{1}{2}}$$

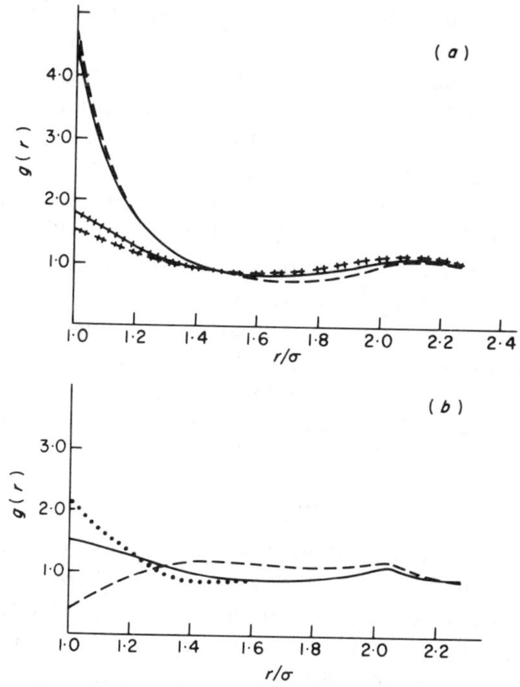

Fig. 4.7. Radial distribution functions for ionic melts at two values of the plasma parameter $\Gamma = 3q^2\eta^{\frac{1}{3}}/(4\pi\varepsilon_0 kT\sigma)$. Reduced density, $\eta = \pi N\sigma^3/(6L^3) = 0{\cdot}3503$.

(*a*) $\Gamma = 5$. MI: $g^{(2)}_{+-}$(——), $g^{(2)}_{++} = g^{(2)}_{--}$(——). PBC with $\varepsilon' = 1$: indistinguishable from MI on this scale. PBC with $\varepsilon' \to \infty$: $g^{(2)}_{+-}$(– – –), $g^{(2)}_{++} = g^{(2)}_{--}$ (++ ++ ++).

(*b*) $\Gamma = 20$. MI: $g^{(2)}_{++} = g^{(2)}_{--}$ (——). PBC with $\varepsilon' = 1$: $g^{(2)}_{++} = g^{(2)}_{--}$(······). PBC with $\varepsilon' \to \infty$: $g^{(2)}_{++} = g^{(2)}_{--}$ (– – – –).

(After Hoskins and Smith, 1980).

the most important factor limiting the accuracy of determinations of ε_r is the inability to include long-range angular correlations. These correlations are weak, but are, nevertheless, of crucial importance for systems of high permittivity. It has been suggested that one is concerned here with correlations over 10–20σ (Levesque *et al.*, 1977). If this is so, then simulation boxes containing more than 1000 molecules would be necessary. Less polar fluids present much less of a problem. Adams found good agreement for a fluid with $\mu^* = 1{\cdot}0$ and $\varepsilon_r = 8$ between the Ewald–Kornfeld and spherical cut-off approaches (using simulation boxes of 500–864 particles) and the HNC theory.

In the face of all the difficulties which have been mentioned, two encouraging points may be made. The first is that the simulations of Rahman and

Stillinger (1971) and Watts (1974) have shown that the structure (atom–atom distribution functions) of water can be reproduced with surprising success even when the long-range corrections are omitted altogether. The second point, which may provide a partial explanation of that success, is that, where a molecule possesses both dipole and quadrupole moments, the effects of these tend to cancel (Watts, 1977). It is well known that in water the quadrupole moment often acts in opposition to the dipole and, indeed, is often the more important (Buckingham, 1959). As would be gathered from what has been said previously, this consolation is not extendable to the predictions of the permittivity.

So far, nearly all the attempts to obtain ε_r have relied upon relationships with the fluctuation of polarization. Some attention has, however, also been devoted to an alternative method based on the response to a small applied electric field (E). From the mean polarization ($\langle P \rangle_E$) produced, the permittivity may be obtained as

$$\varepsilon_r = \lim_{E \to 0} \{1 + \langle P \rangle_E / E\}. \tag{4.90}$$

The extrapolation is a difficult one because for those values of E for which $\langle P \rangle_E$ may be accurately determined the ratio $\langle P \rangle_E / E$ is far from being constant; at lower E it is difficult to obtain reliable values of $\langle P \rangle_E$. Adams (1980) has suggested that the non-constancy of $\langle P \rangle_E / E$ is a consequence of the use of the Ewald–Kornfeld summation, rather than an intrinsic property of the system. If this is correct, then it leaves open the possibility that this response method, if coupled with a different method for calculating the configurational energy, may become important.

In view of the problems encountered in uniform systems, the prospects for corresponding calculations on non-uniform systems do not seem good. Here ε_r is a non-uniform tensor. That is, for adsorption onto a planar interface it will have four independent components (ε_{xx}, ε_{zz}, ε_{xz} and ε_{xy}, where the surface is normal to the z-direction). Each of these components is at least z-dependent (x, y-dependence might also be considered if the adsorbent field is periodic). Indeed, one of the points of interest in doing simulation studies on such systems would be to see how ε_r varied with z, the distance from the interface. Quite apart from difficulties in the physics of the problem, it is clear that the collection of data for the evaluation of the four independent ε_r components which are essentially fluctuation terms will be plagued by poor statistics, since each configuration or time-step will only contribute a very limited amount of information for each layer. In uniform systems every molecule is contributing towards the evaluation of a single quantity, the isotropic and uniform permittivity. Perhaps it would be better to ask for what purpose these permittivity data were needed. If for use in more conventional theories concerned with ionic and dipolar solutions, then it is probably preferable to obtain from the simulation a quantity more directly related to the property of the solution which has been measured. It may well transpire that this property itself is less difficult to study than the intermediate quantity, the permittivity. Indeed, the problem with regard to the permittivity is an example of a situation in which large-scale thermodynamic concepts become increasingly obscure in small non-uniform systems. Another example of this type which has been worked out in some detail and discussed elsewhere in this book is that of adsorption into cavities and pores. However, the need to quantify long-range effects in terms of a mean-field approach does require an assessment of "global" properties of this type.

Another approach to the problem of long-range interactions has been proposed by Hockney but so far used only by his group (Hockney *et al* 1973). The force between two particles is split into a rapidly varying part, which is subject to a spherical cut-off of radius R_c, and a slowly varying part, which is obtained by differencing the potential given for a regular set of mesh points. The computation of the first part of the force is carried out in the usual way, with the cut-off radius being typically 2 or 4 times the mesh separation used for the long-range force. The evaluation of the potential at the mesh points (256×256) was achieved by a fast Fourier transform solution of Poisson's equation, this being repeated at each cycle. No cut-off of any kind is involved in this process, and the method is therefore free from the ambiguities which arise for most other methods. An important advantage is that the number of computing operations required for each cycle increases as N, rather than N^2, where N is the number of particles in the simulation box. Thus Hockney *et al* were able to study a two-dimensional representation of KCl containing 10000 ions. Although no results have been reported for three-dimensional systems, it would appear that the method would still be practicable. As for the feasibility of its use for studying adsorption this would appear to be more doubtful. For such non-uniform systems the possibility of representing the coulomb potential as a sum of Fourier terms seems more questionable. A weakness of the method in this case would appear to be that it uses a uniform grid, whereas a non-uniform mesh chosen to suit the problem should have an advantage.

6. Polyatomic substances

Apart from the consideration given to the long-range correction for systems composed of polar molecules, discussion in this chapter has been concentrated on systems composed of monatomic molecules. When it is desired to study di- and polyatomic molecules, the procedure becomes more complicated, although the principles remain the same. There is, for instance, a very large increase in the amount of computer time required to evaluate the energies and forces for each configuration. Thus for a non-linear triatomic molecule for which a potential based on three point centres of force situated at the atomic positions is assumed, there would be nine atom–atom contributions to be calculated. Even for quite simple polyatomic molecules, this aspect of the simulation soon becomes the most time-consuming part. Because of this, it is advisable to keep down the number of force centres assumed, even at the expense of some loss of realism. The disposition of the centres of force will also have some effect on the computing required, a more symmetrical arrangement leading to economies in the number of independent distances which must be evaluated. So far, all that has been said applies to both MC and MD simulations. We turn now to the effects which apply specifically to one or the other of these.

(a) Monte Carlo

In choosing trial moves, orientational as well as translational changes now have to be considered. A single proposed change could be purely translational, purely rotational or a hybrid of these. It is, for example, legitimate to proceed through a cycle of m translational moves followed by n reorientational moves, although such a scheme would be uncommon. The choice of cycle would be made with a view to minimizing the convergence time of the simulation. If equilibrium with respect to reorientation were thought to be the rate-determining factor, then a cycle having a large proportion of purely (or nearly purely) rotational moves would be a reasonable choice. If a large proportion of translational moves were used many of these might be redundant.

With regard to a proposed reorientation, either alone or combined with a translation, both the axis of the reorientation and the angular displacement must be decided. A procedure which has been successfully adopted is to make a rotation about an axis through the molecule parallel to one of the space-fixed cartesian axes (x, y or z) chosen at random. The centre of the molecule for rotation purposes could, or course, be the centre of mass, but in some circumstances it may be preferable to choose the electrical centre: this would be likely to be so if the intermolecular forces were dominated by the electrostatic interactions and a translation of the electrical centre could lead to an unduly large energy increment. The choice of the maximum angular displacement would usually be made so as to lead to an ~50% acceptance rate for rotational moves. However, if the reorientational motion is subject to an approximately periodic rotational energy barrier, it may be advantageous to adopt a much larger value so as to include the possibility of a jump from one potential well to the next. The use of Euler angles for the choice of rotational trial states is more complicated and is usually found to be wasteful.

(b) Molecular dynamics

The equations of motion for the centre of mass are as for monatomic particles, but those for the rotational motion are more complicated. For the latter, two methods of representation have been used.

The first of these employs the set of principal axes of the molecule together with the Euler angles (θ, ϕ, ψ) required to define the orientation of the molecule-fixed axes with respect to the space-fixed axes. The torque (N_p) on the molecule in terms of the molecule-fixed axes is first computed from the known forces acting on the molecule, and this then leads by way of the Newton–Euler equations to the value of the derivative of the angular velocity:

$$\begin{aligned} I_{px}(\mathrm{d}\omega_{px}/\mathrm{d}t) - \omega_{py}\omega_{pz}(I_{py} - I_{pz}) &= N_{px} \\ I_{py}(\mathrm{d}\omega_{py}/\mathrm{d}t) - \omega_{pz}\omega_{px}(I_{pz} - I_{px}) &= N_{py} \\ I_{pz}(\mathrm{d}\omega_{pz}/\mathrm{d}t) - \omega_{px}\omega_{py}(I_{px} - I_{py}) &= N_{pz} \end{aligned} \tag{4.91}$$

where I_{px}, I_{py}, I_{pz} are the moments of inertia about the three molecule-fixed axes, and ω_{px}, ω_{py}, ω_{pz} are the corresponding angular velocities (Goldstein, 1950).

Since the Euler angles themselves also change with time their values must be recalculated at each step from the equations:

$$\begin{aligned} \omega_{px} &= (\mathrm{d}\phi/\mathrm{d}t)\sin\theta\sin\psi + (\mathrm{d}\theta/\mathrm{d}t)\cos\psi \\ \omega_{py} &= (\mathrm{d}\phi/\mathrm{d}t)\sin\theta\cos\psi - (\mathrm{d}\theta/\mathrm{d}t)\sin\psi \\ \omega_{pz} &= (\mathrm{d}\phi/\mathrm{d}t)\cos\theta + \mathrm{d}\psi/\mathrm{d}t. \end{aligned} \tag{4.92}$$

Rahman and Stillinger (1971), in their classical work on water, evaluated five time-derivatives for the centre of mass coordinates and four for the angular motion, which they then used to obtain values for the quantities x, y, z, θ, ϕ, ψ and ω_{px}, ω_{py}, ω_{pz} at the end of each time step.

The second scheme (Evans, 1977; Evans and Murad, 1977) employs quaternions, these being a type of four-dimensional complex number

$$\bar{u} = \chi\bar{1} - \eta\bar{i} + \xi\bar{j} - \zeta\bar{k} \tag{4.93}$$

in which the last three unit vectors are such that

$$\bar{i}^2 = \bar{j}^2 = \bar{k}^2 = -1 \tag{4.94}$$

and they have the cyclic property

$$\overline{ij} = \bar{k} = -\overline{ji} \tag{4.95}$$

the quaternion parameters are related to the Eulerian angles by the equation

$$\begin{aligned} \chi &= \cos\tfrac{1}{2}\theta\cos\tfrac{1}{2}(\psi + \phi) \\ \eta &= \sin\tfrac{1}{2}\theta\cos\tfrac{1}{2}(\psi - \phi) \\ \xi &= \sin\tfrac{1}{2}\theta\sin\tfrac{1}{2}(\psi - \phi) \\ \zeta &= \cos\tfrac{1}{2}\theta\sin\tfrac{1}{2}(\psi + \phi). \end{aligned} \tag{4.96}$$

The torque on each molecule is first evaluated in the laboratory frame ($\boldsymbol{N}$) and this is then converted to the molecule-fixed frame ($\boldsymbol{N}_p$) by means of the rotational matrix

$$\begin{pmatrix} -\xi^2 + \eta^2 - \zeta^2 + \chi^2 & 2(\zeta\chi - \xi\eta) & 2(\eta\zeta + \xi\chi) \\ -2(\xi\eta + \zeta\chi) & \xi^2 - \eta^2 - \zeta^2 + \chi^2 & 2(\eta\chi - \xi\zeta) \\ 2(\eta\zeta - \xi\chi) & -2(\xi\zeta + \eta\chi) & -\xi^2 - \eta^2 + \zeta^2 + \chi^2 \end{pmatrix} \tag{4.97}$$

(This matrix has the important computational advantage over the corresponding Eulerian matrix that it does not require the evaluation of any trigonometric functions.) $\boldsymbol{N}_p$ may then be used with the equation connecting $\boldsymbol{\omega}_p$ with the derivatives of the quaternions,

$$\begin{pmatrix} \omega_{px} \\ \omega_{py} \\ \omega_{pz} \\ 0 \end{pmatrix} = 2 \begin{pmatrix} -\zeta & \chi & \xi & -\eta \\ -\chi & -\zeta & \eta & \xi \\ \eta & -\xi & \chi & -\zeta \\ \xi & \eta & \zeta & \chi \end{pmatrix} \begin{pmatrix} \dot{\xi} \\ \dot{\eta} \\ \dot{\zeta} \\ \dot{\chi} \end{pmatrix} \tag{4.98}$$

for the integration of the equations of motion (Eqn. (4.91)). An important advantage of the quaternion scheme is that it is free from the singularities which arise when Eulerian angles are used. Efficiencies improved by a factor of ~10 have been reported for dense fluid systems.

References

Adams, D. J. (1974). *Mol. Phys.* **28**, 1241.
Adams, D. J. (1975). *Mol. Phys.* **29**, 307.
Adams, D. J. (1980). *Mol. Phys.* **40**, 1261.

Adams, D. J. and McDonald, I. R. (1976). *Mol. Phys.* **32**, 931.
Adams, D. J., Adams, E. M. and Hills, G. J. (1979). *Mol. Phys.* **38**, 387.
Alder, B. J. and Wainwright, T. E. (1959). *J. Chem. Phys.* **31**, 459.
Alder, B. J. and Wainwright, T. E. (1970). *Phys. Rev.* **A1**, 18.
Andersen, H. C. (1980). *J. Chem. Phys.* **72**, 2384.
Ashurst, W. T. and Hoover, W. G. (1973). *Phys. Rev. Lett.* **31**, 206.
Barker, A. A. (1965). *Aust. J. Phys.* **18**, 119.
Barker, J. A. and Watts, R. O. (1973). *Mol. Phys.* **26**, 789.
Bennett, C. H. (1976). *J. Comput. Phys.* **22**, 245.
Bell, R. P. (1931). *Trans. Faraday Soc.* **27**, 797.
Born, M. (1920). *Z. Physik* **1**, 45.
Buckingham, A. D. (1959). *Q. Rev.* **13**, 183.
Chapela, G. A., Saville, G., Thompson, S. M. and Rowlinson, J. S. (1977). *J. Chem. Soc. Faraday Trans. II* **73**, 1133.
Cicotti, G. and Jacucci, G. (1975). *Phys. Rev. Lett.* **35**, 789.
Coldwell, R. L. (1973). *Phys. Rev.* **A7**, 270.
Coldwell, R. L. (1974). *Phys. Rev.* **A10**, 897.
de Leeuw, S. W., Perram, J. W. and Smith, E. R. (1980). *Proc. R. Soc. London, Ser. A* **373**, 27, 57.
Evans, D. J. (1977). *Mol. Phys.* **34**, 317.
Evans, D. J. and Murad, S. (1977). *Mol. Phys.* **34**, 327.
Ewald, P. P. (1921). *Annln. Phys.* **64**, 253.
Finney, J. L. (1978). *J. Comput. Phys.* **28**, 92.
Friedman, H. L. (1975). *Mol. Phys.* **29**, 1533.
Futrelle, R. P. and McGinty, D. J. (1971). *Chem. Phys. Lett.* **12**, 285.
Goldstein, H. (1950). "Classical Mechanics", p. 158. Addison-Wesley, Reading, Mass.
Haile, J. M. and Graben, H. W. (1980). *J. Chem. Phys.* **73**, 2412.
Hansen, J. P. and Verlet, L. (1969). *Phys. Rev.* **184**, 151.
Hill, T. L. (1956). "Statistical Mechanics" (a) p. 67, (b) p. 106. McGraw-Hill, New York.
Hockney, R. W., Goel, S. P. and Eastwood, J. W. (1973). *Chem. Phys. Lett.* **21**, 589.
Hoover, W. G. and Alder, B. J. (1967). *J. Chem. Phys.* **46**, 686.
Hoover, W. G. and Ree, F. H. (1967). *J. Chem. Phys.* **47**, 4873.
Hoskins, C. S. and Smith, E. R. (1980). *Mol. Phys.* **41**, 243.
Kornfeld, H. (1924). *Z. Phys.* **22**, 27.
Ladd, A. J. C. (1977). *Mol. Phys.* **33**, 1039.
Ladd, A .J. C. (1978). *Mol. Phys.* **36**, 463.
Lado, F. and Wood, W. W. (1968). *J. Chem. Phys.* **49**, 4244.
Lane, J. E. and Spurling, T. H. (1976). *Aust. J. Chem.* **29**, 2103.
Levesque, D., Patey, G. N. and Weis, J. J. (1977). *Mol. Phys.* **34**, 1077.
McDonald, I. R. (1972). *Mol. Phys.* **23**, 41.
McDonald, I. R. and Singer, K. (1969). *J. Chem. Phys.* **50**, 2308.
Metropolis, N., Rosebluth, A. W., Rosenbluth, M. N., Teller, A. H. and Teller, E. (1963). *J. Chem. Phys.* **21**, 1087.
Mezei, M. (1980). *Mol. Phys.* **40**, 901.
Murad, S., Evans, D. J., Gubbins, K. E., Streett, W. B. and Tildesley, D. J. (1979). *Mol. Phys.* **37**, 725.
Norman, G. E. and Filinov, V. S. (1969). *High Temp. Res.* **7**, 216.

Onsager, L. (1936). *J. Am. Chem. Soc.* **58**, 1486.
Pangali, C., Rao, M. and Berne, B. J. (1978). *Chem. Phys. Lett.* **55**, 413.
Pangali, C., Rao, M. and Berne, B. J. (1980). *Mol. Phys.* **40**, 661.
Parrinello, M. and Rahman, A. (1980). *Phys. Rev. Lett.* **45**, 1196.
Patey, G. N. and Valleau, J. P. (1973). *Chem. Phys. Lett.* **21**, 297.
Quentrec, B. and Brot, C. (1973). *J. Comput. Phys.* **13**, 430.
Rahman, A. (1964). *Phys. Rev.* **136**, A405.
Rahman, A. and Stillinger, F. H. (1971). *J. Chem. Phys.* **55**, 3336.
Rossky, P. J., Doll, J. D. and Friedman, H. L. (1978). *J. Chem. Phys.* **69**, 4628.
Rowley, L. A., Nicholson, D. and Parsonage, N. G. (1975). *J. Comput. Phys.* **17**, 401.
Rowley, L. A., Nicholson, D. and Parsonage, N. G. (1976a). *Mol. Phys.* **31**, 365.
Rowley, L. A., Nicholson, D. and Parsonage, N. G. (1976b). *Mol. Phys.* **31**, 389.
Schofield, P. (1966). *Proc. Phys. Soc.* (*London*) **88**, 149.
Singer, K., Singer, J. V. L. and Fincham, D. (1980). *Mol. Phys.* **40**, 515.
Stillinger, F. H. and Rahman, A. (1974). *J. Chem. Phys.* **60**, 1545.
Streett, W. B., Tildesley, D. J. and Saville, G. (1978). *Mol.. Phys.* **35**, 639.
Stroud, H. J. F. and Parsonage, N. G. (1971). *"2nd International Conference on Molecular Sieve Zeolites, Worcester, Mass"*, Advances in Chemistry Series No. 102, p. 138. American Chemical Society, Washington, D.C.
Torrie, G. M. and Valleau, J. P. (1977). *J. Comput. Phys.* **23**, 187.
Torrie, G. M., Valleau, J. P. and Bain, A. (1973). *J. Chem. Phys.* **58**, 5479.
Turq, P., Lantelme, F. and Friedman, H. L. (1977). *J. Chem. Phys.* **66**, 3039.
Valleau, J. P. and Card, D. N. (1972). *J. Chem. Phys.* **57**, 5457.
Valleau, J. P. and Whittington, S. G. (1977). *In* "Modern Theoretical Chemistry, Vol. 5, Statistical Mechanics. Part A: Equilibrium Techniques". Ch. 4. Plenum, New York and London.
Van Gunsteren, W. F., Berendsen, H. J. C. and Rullman, J. A. C. (1978). *Faraday Disc.* **66**, 58.
Verlet, L. (1967). *Phys. Rev.* **159**, 98.
Watts, R. O. (1974). *Mol. Phys.* **28**, 1069.
Watts, R. O. (1977). *Chem. Phys.* **26**, 367.
Woodcock, L. V. and Singer, K. (1971). *Trans. Faraday Soc.* **67**, 12.

Chapter 5

Monolayers

1. Dimensionality of adsorbed films

Although all real adsorption systems are three-dimensional, it is in multilayer physisorption that this is most clearly evident. However, it is also true even if no multilayers are formed, provided that the molecules of the adsorbed species can be positioned at various distances from the adsorbing surface. The system can utilize this extra distance parameter so as to minimize its free energy. Monolayers which are buckled in a regular or irregular fashion then become possible (Fig. 5.1). Regularly buckled monolayers may form, most obviously, if the adsorbate–adsorbent potential is periodic, so that the value of z at which the minimum potential energy occurs varies in a regular periodic fashion over the surface. If the minimum as a function of z is very deep and narrow it may well be adequate to consider only one value of z for any given pair of (x, y) coordinates. The

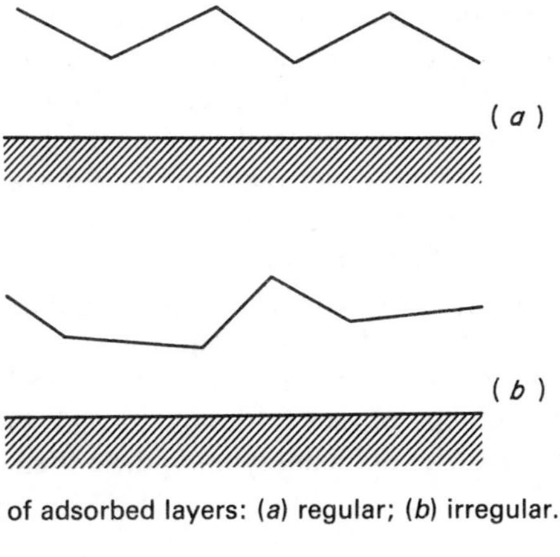

Fig. 5.1. Buckling of adsorbed layers: (*a*) regular; (*b*) irregular.

system then becomes effectively two-dimensional. Buckling could also occur if a larger number of adsorbate molecules were on the surface than could be packed into a flat layer. In chemisorption the binding of the monolayer to the surface will be very much stronger than the intermolecular adsorbate interactions and an increase in adsorbate concentration above the amount required for monolayer saturation would be unlikely, since the price to be paid for moving some of the molecules away from the position of minimum adsorbate–adsorbent energy would be too great. Furthermore, there is no experimental evidence for such buckling in chemisorbed layers. Thus, for chemisorption and many physisorption systems at coverages up to a monolayer, a theoretical understanding of the behaviour of ideal two-dimensional systems is clearly pertinent.

2. Epitaxy

Real adsorbent surfaces are structured, and it will be useful to distinguish between layers which are commensurate with the underlying surface (epitaxial adsorbates) and those that are not. If the periodic variation of the potential in the x, y-directions is sufficiently strong then it may be that effectively the adsorbed molecules only reside on the most-favoured sites, as determined by the adsorbent structure. This is an extreme example of a commensurate or epitaxial phase. At the other extreme, if the periodic potential does not depend sensitively on x and y, then the adsorbate may largely ignore this periodicity and form a structure which is determined by the packing of the adsorbate. This would be an incommensurate or non-epitaxial phase. Intermediate situations can, of course, occur in which many, but not all, of the adsorbed molecules are in the positions of minimum adsorbent potential.

The order of a one-dimensional structure may be simply defined so that nth order means that every nth adatom has an equivalent position in the adsorbent potential well to that of the zeroth. Unfortunately, this does not provide an unique definition unless the number of wavelengths of the periodic potential to be found between one adatom and the next to be found in an equivalent position is also given. Figure 5.2 shows two adlayer structures of each of the first three orders. For a two-dimensional surface this definition may be readily adapted by applying it separately along the two lattice vectors. In terms of this nomenclature a non-epitaxial phase is an $(\infty \times \infty)$ epitaxial phase. A great deal of theoretical consideration has been given to non-epitaxial phases (although in real systems they are less common than the low-order epitaxial type) and we shall start by examining them.

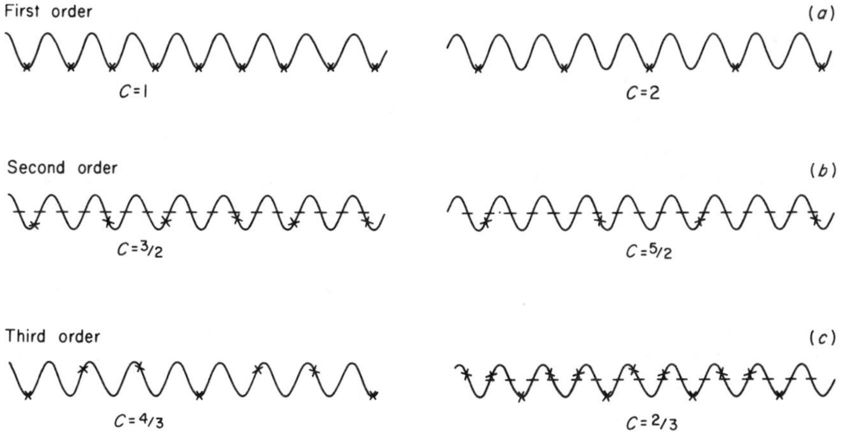

Fig. 5.2. Schematic diagram of one-dimensional commensurate structures of order 1–3: × denotes position of an adatom in external periodic potential. C = number of wavelengths of periodic potential between successive adatoms. (After Ying, S. C., 1971).

3. Systems in vanishing periodic substrate field

(a) Long-range positional order

Peierls (1934, 1935) was the first to present an argument which indicated that such long-range order could not exist at $T \neq 0$ for a system infinite in two dimensions because the amplitude of displacement of the molecules would diverge. The long wavelength phonons are the key to this argument. For these, $kT \gg \hbar\omega$ at any other than the lowest temperatures, and so the occupancy of each vibrational state is adequately given by the classical approximation as

$$n_{ks} = kT/\hbar\omega_{ks} \tag{5.1}$$

where n_{ks} and ω_{ks} are the number of phonons and the frequency, respectively, of the mode of the sth branch which has the wave vector k.

Equating the energy in that mode with the mean potential energy we have

$$\tfrac{1}{2}f_{ks}\langle x_{ks}^2\rangle = \frac{kT}{\hbar\omega_{ks}}\,\hbar\omega_{ks} \tag{5.2}$$

where f_{ks} is the force-constant of the mode. But

$$f_{ks} = \omega_{ks}^2 m \tag{5.3}$$

and hence

$$\langle x_{ks}^2 \rangle = 2kT/(\omega_{ks}^2 m) \tag{5.4}$$

where m is the reduced mass for the mode. The mean total displacement $\langle x^2 \rangle$ can be found by summation of $\langle x_{ks}^2 \rangle$ over all modes. Replacing the sum by an integral and using the Debye approximation, which gives a distribution of frequencies of $g(\omega)\,d\omega \propto \omega\,d\omega$, the expression for $\langle x^2 \rangle$ becomes

$$\langle x^2 \rangle = \frac{kT}{m} \int_{\omega_l}^{\omega_D} \frac{\omega\,d\omega}{\omega^2} = (kT/m)\ln(\omega_D/\omega_l) \tag{5.5}$$

where ω_D and ω_l are, respectively, the Debye frequency and the lowest frequency, as determined by the size of the sample. Clearly

$$\omega_l \propto 1/l \qquad \text{and} \qquad \omega_D \propto 1/a \tag{5.6}$$

where l and a are the linear dimension of the sample and the interparticle spacing, respectively. The lower limit of (5.5) diverges as $\ln(l/a)$ as $l \to \infty$. This is a very weak divergence, and its implications for real systems have been discussed by several workers, e.g. Imry (1979). The adsorbent particles can be of appreciable size before the divergence becomes a problem. Thus, if $l \approx 5 \times 10^8 a$, corresponding to a crystal of length ~10 cm, $\ln(l/a) \approx 20$. For the usual size of simulation system $l/a \approx 30$ and $\ln(l/a) \approx 3{\cdot}4$, and it will be seen that $\langle x^2 \rangle$ is far from "exploding". This accounts for the observation of solid–fluid transitions in two-dimensional systems (Alder and Wainwright, 1962; Woods, 1968; see also p. 193), a result which at first sight seems to be forbidden by theory. The simulations mentioned here mostly employed periodic boundary conditions, rather than the free boundary conditions as in a real finite system, and the imposition of these conditions eliminates modes with wavelengths greater than the size of the simulation box.

As has been seen, Peierls' theory employs the assumption that the motions may be broken down into harmonic vibrations. Subsequently, it has been shown how the non-existence of a solid phase may be proved rigorously for an infinite system (Mermin and Wagner, 1966; Mermin, 1968). Their proof is valid provided that the inter-particle potential $(u^{(2)}(r))$ satisfies the condition

$$\nabla^2 u^{(2)}(r) = r^{-4-|\varepsilon|} \qquad \text{as } r \to \infty \tag{5.7}$$

where ε is any increment, and

$$u^{(2)}(r) - \lambda r^2 |\nabla^2 \mathrm{u}^{(2)}(\mathrm{r})| > |\mathrm{A}|\mathrm{r}^{-2-|\varepsilon|} \qquad \text{for } r \approx 0 \tag{5.8}$$

for some value of $\lambda > 0$.

Mie potentials with $m > n > 2 + |\varepsilon|$ are included in the class of potentials for which Eqns (5.7) and (5.8) hold. Hard-disc potentials are, however, not covered by the theorem. This treatment, which is therefore much more general than that of Peierls, may also be applied to systems having magnetic, superfluid or superconducting order. In comparing the properties of these systems it is usual to denote the number of degrees of freedom in the parameter describing the order of the system by n, and the dimensionality of the system by d. Thus a three-dimensional Ising model has $n = 1$, $d = 3$, whilst the corresponding Heisenberg model would have $n = 3$, $d = 3$. The superfluid and superconducting systems each have $n = 2$, since the parameter is a complex number, which necessarily has two components. The arrays of molecules which are of interest in this book have $n = 2$, $d = 2$, since x- and y-coordinates are free to be fixed for each particle. The analogous magnetic system (with $n = 2$) is known as the planar XY or planar rotator model.

For the case of two-dimensional crystalline order, it is necessary to find a suitable definition of an order parameter. Mermin (1968) defined a set of positional order parameters in terms of the averaged Fourier components of the singlet density:

$$\rho_k = N^{-1}\langle\hat{\rho}_k\rangle \tag{5.9}$$

$$= N^{-1}\int \ldots \int \mathrm{d}\boldsymbol{r} \exp(-\beta U_N)\hat{\rho}_k(\boldsymbol{r})/Z_N$$

where the configuration integral for positional coordinates alone (cf. Eqn (3.7)) is

$$Z_N = \int \ldots \int \mathrm{d}\boldsymbol{r} \exp(-\beta U_N) \tag{5.10}$$

and

$$\hat{\rho}_k(\boldsymbol{r}) = \sum_{i=1}^{N} \exp(-i\boldsymbol{k}\cdot\boldsymbol{r}_i). \tag{5.11}$$

If $\rho_{\boldsymbol{k}} \neq 0$ for one or more $\boldsymbol{k}$ values then the system displays positional order. Mermin (1968) was able to show that $\rho_{\boldsymbol{k}} \leqslant 1/(\ln N)^{1/2}$ and $\leqslant 1/N^{1/2}$ for two- and one-dimensional systems, respectively. Clearly, $\rho_{\boldsymbol{k}} \rightarrow 0$ as $N \rightarrow \infty$ for both cases, but in the former the approach to zero is very gradual as was first noted by Peierls.

One way of appreciating the effect of dimensionality (d) on ordering is to evaluate the domain wall energy for various systems. For a system having a continuous symmetry parameter (ϕ) with the energy given by $E = J\int_0^\infty |\mathrm{d}\phi/\mathrm{d}r|^2$, the energy of a single domain of the opposite kind can be expressed in terms of the overall length (l) and lattice spacing (a) of the system. It is found to be

$$J(d-2)a^{d-2} \qquad \text{for } d>2 \tag{5.12}$$

$$J/\ln(l/a) \qquad \text{for } d=2 \tag{5.13}$$

and

$$J(2-d)l^{d-2} \qquad \text{for } d<2. \tag{5.14}$$

For $d \leqslant 2$, therefore, the wall energy $\rightarrow 0$ as $l \rightarrow \infty$ (Imry, 1979). However, in accord with what has been said above, for $d=2$ the approach to zero is very slow. The entropy term, omitted from the above expressions, favours the formation of domains. Thus, for $d \leqslant 2$ an infinite number of domains occur at $T \neq 0$; for $d>2$ the opposing energy and entropy terms lead to an equilibrium domain structure.

(b) Other kinds of ordering

The slowness with which $\rho_k \rightarrow 0$ as $N \rightarrow \infty$ suggested to Mermin that some other kind of order may be present in the two-dimensional system and he was able to show that such a quantity was the correlation function for the orientation of inter-particle vectors:

$$\mathscr{C}(\boldsymbol{a};\boldsymbol{R},\boldsymbol{R}') = \langle[\boldsymbol{r}(\boldsymbol{R}+\boldsymbol{a})-\boldsymbol{r}(\boldsymbol{R})]\cdot[\boldsymbol{r}(\boldsymbol{R}'+\boldsymbol{a})-\boldsymbol{r}(\boldsymbol{R}')]\rangle \tag{5.15}$$

where $\boldsymbol{r}(\boldsymbol{R}+\boldsymbol{a})$ is the displacement of the particle which would in the "solid" phase be at $\boldsymbol{R}+\boldsymbol{a}$ and $\boldsymbol{a}$ is a unit vector of the solid (Fig. 5.3). Thus Eqn (5.15) gives the correlation between the relative displacement of two neighbouring particles near $\boldsymbol{R}$ with the same quantity for corresponding particles near $\boldsymbol{R}'$, and is a measure of the directional long-range order. In

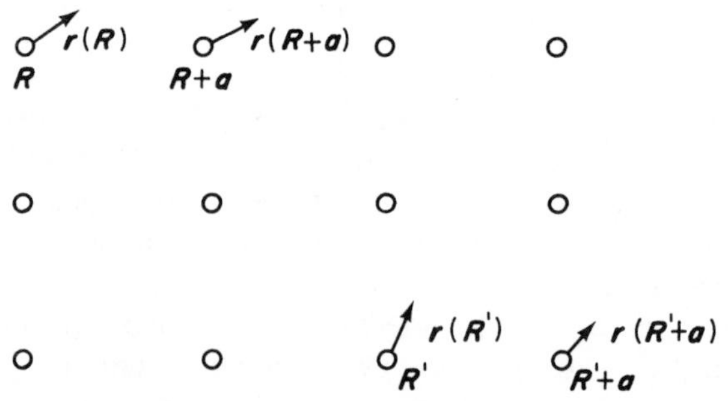

Fig. 5.3. Displacements (***r***) from the mean lattice positions.

the harmonic approximation it tends to the non-zero value a^2 as the separation $|\boldsymbol{R} - \boldsymbol{R}'|$ tends to infinity for a crystalline solid. Difficulties can arise with the evaluation of the function $\mathscr{C}(\boldsymbol{a}; \boldsymbol{R}, \boldsymbol{R}')$ when the atomic displacements are large because of uncertainty with regard to the undisplaced positions.

A question which now arises is whether a phase transition may occur at a non-zero temperature even though the long-range positional order is zero throughout the temperature region. Important evidence on this came first from the study of magnetic disorder: from calculations of the magnetic susceptibility using a high-temperature series expansion, Stanley and Kaplan (1966) concluded that a two-dimensional Heisenberg model ($n = 3, d = 2$) did display such a transition. It now seems that this conclusion is incorrect (Yamaji and Kondo, 1973). However, it is fairly certain that there is a transition of the Stanley–Kaplan type in the planar XY model ($n = 2$, $d = 2$) (Stanley, 1968). At all temperatures below the transition point the susceptibility was found to be infinite. (A distinction has to be drawn here between the true two-dimensional XY and the two-dimensional planar XY or planar rotator. In the former the ordering vectors have three degrees of freedom (although only two of these are involved in the interactions), whereas in the latter the "spins" are confined to the two-dimensional plane and so have only two components.)

The work of Stanley and Kaplan has led others to explore the possibility of types of ordering being concerned in these transitions other than those which are usually taken to represent the state of the sample. Kosterlitz and Thouless (1973) and Kosterlitz (1974) have considered the concept of a "topological order", which can exist even when the normal long-range order is absent. For simple two- and three-dimensional solids the order concerned and the processes which eventually lead to its becoming zero are centred around the behaviour of dislocations. At low temperatures dislocations occur, but they are present as closely bound pairs consisting of dislocations of opposite sign. In this condition the dislocations exert very little influence on the order within the solid. However, as the temperature is raised these pairs dissociate and further isolated dislocations are also created, and it is this build-up of the concentration of isolated dislocations which causes the disappearance of the order. The theory has also been applied to magnetic solids and superfluids. In the former each entity (spin) in the system has its own orientation. Thus, whereas for the melting solid we were concerned with displacements of the particles, here we are interested in their orientations. Corresponding to dislocations in the melting solid we have disclinations in the magnetic system (Nabarro, 1967). The distinction between these is illustrated in Fig. 5.4. For a dislocation a cut is made in the dislocation-free cylinder (a) and the two faces are shifted

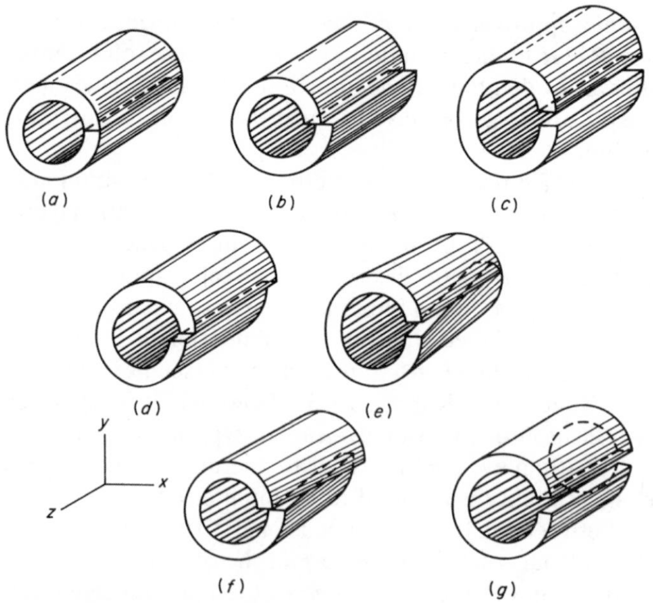

Fig. 5.4. Formation of dislocations and disclinations: (*a*) fault-free cylinder; (*b*) and (*c*) edge dislocations; (*d*) screw dislocation; (*e*) and (*f*) edge disclinations; (*g*) screw disclination. (After Nabarro, 1967).

with respect to each other in the *x*- or *y*-direction, giving an edge dislocation, or in the *z*-direction, giving a screw dislocation. To obtain disclinations the faces are rotated with respect to each other; (*e*) and (*f*), in which the axis of this rotation is perpendicular to the axis of the cylinder, are known as edge disclinations; likewise, (*g*), in which the axis of the rotation is parallel to the cylinder axis, leads to a screw disclination. A magnetic solid could, of course, have both dislocations and disclinations. The same is true of liquid crystals, an example being the smectic B phase, in which the centres of the molecules lie on a regular lattice. It appears likely that disclinations (concerned with the orientation of the inter-particle vector) also play a part in the melting of two-dimensional solids (p. 194).

In the development of their theory, Kosterlitz and Thouless had to make many approximations: some of these were of the mean field type, some took the form of retaining only leading terms, and others were based on rough estimates of the relative sizes of various quantities. Nevertheless, it is very probable that the ideas behind the theory are correct. Disagreement has, however, been expressed with respect to some of the details of the treatment. For example, the values of the critical indices they deduced

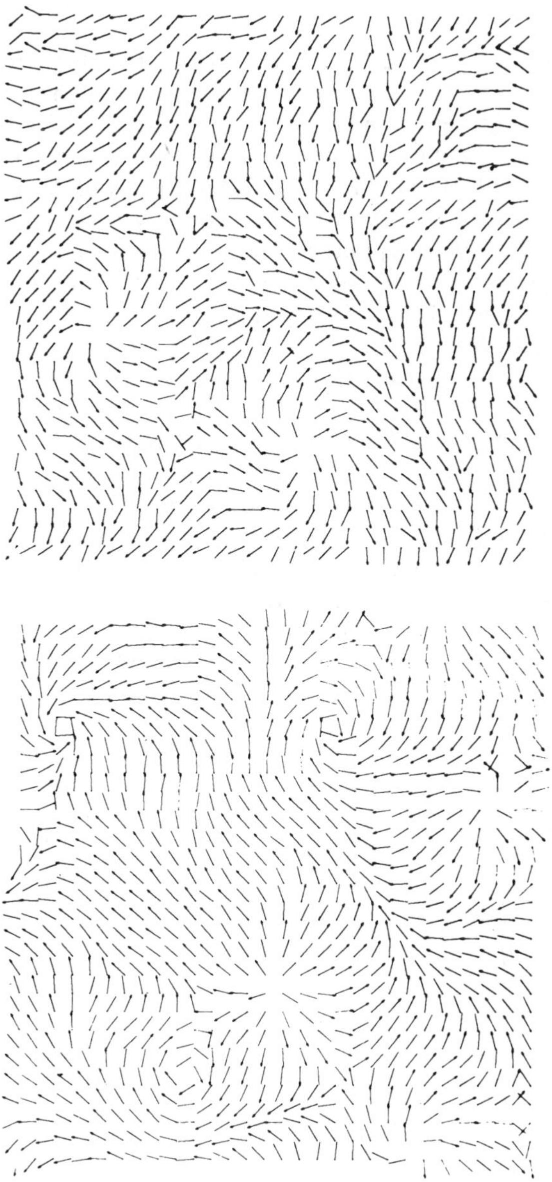

Fig. 5.5. Snapshot spin configurations of the classical *XY* model $H_N = -4J\sum_{i>j} \mathbf{s}_i \, . \, \mathbf{s}_j$, where the $\mathbf{s}_i$ are 3-vectors of modulus $\frac{1}{2}$. (*a*) at $kT/J = 0{\cdot}2$; (*b*) at $kT/J = 0{\cdot}01$. Tails of spins lie on a square lattice; spins with arrowheads are directed slightly above the plane; spins without arrowheads are directed below the plane. Lengths of spins shown indicate the magnitude of the spin when projected onto the plane. (After Suzuki *et al.*, 1977).

are now thought to be in error. Thus, for η for the planar rotator, defined by $\langle\cos(\phi_0 - \phi_r)\rangle \sim r^{-\alpha(T)}$, with $\eta = \alpha(T_{SK})$, where ϕ_0 and ϕ_r are the orientations at the origin and at position $\boldsymbol{r}$, and T_{SK} is the transition temperature (analogous to the Stanley–Kaplan temperature) they give $\frac{1}{4}$, whereas Luther and Scalapino (1977) found it to be $1/\sqrt{8}(0{\cdot}353)$, with Monte Carlo simulations ($\eta = 0{\cdot}25$–$0{\cdot}5$) being unable to decide between these values (Miyashita *et al.*, 1978). Some support for the general basis of the theory came from Monte Carlo simulations of the two-dimensional classical *XY*-model by Suzuki *et al.* (1977) (Fig. 5.5). The snapshots, which refer to temperatures below that of the Stanley–Kaplan transition ($kT/J \approx 1$) show clear evidence of the existence of vortices. However, Kawabata and Binder (1977) have observed that, especially at the low temperatures such as that represented by Fig. 5.5(*b*), these structures are "frozen-in" rather than true equilibrium, states; they conclude that any simulations of this kind at $kT/J < 0{\cdot}3$ are suspect for this reason. Indeed, at $kT/J = 0{\cdot}01$ any vortices should be present as bound pairs rather than as free entities, and this was part of the reason for suspecting that the simulations at the low temperatures had not converged. Nevertheless, large regions of nearly aligned spins can be seen even at the higher temperature of Fig. 5.5(*a*), and it seems reasonable to assume that it is the ability of a small applied field to rotate these regions, rather as in aligning domains of a ferromagnet or liquid crystal, which leads

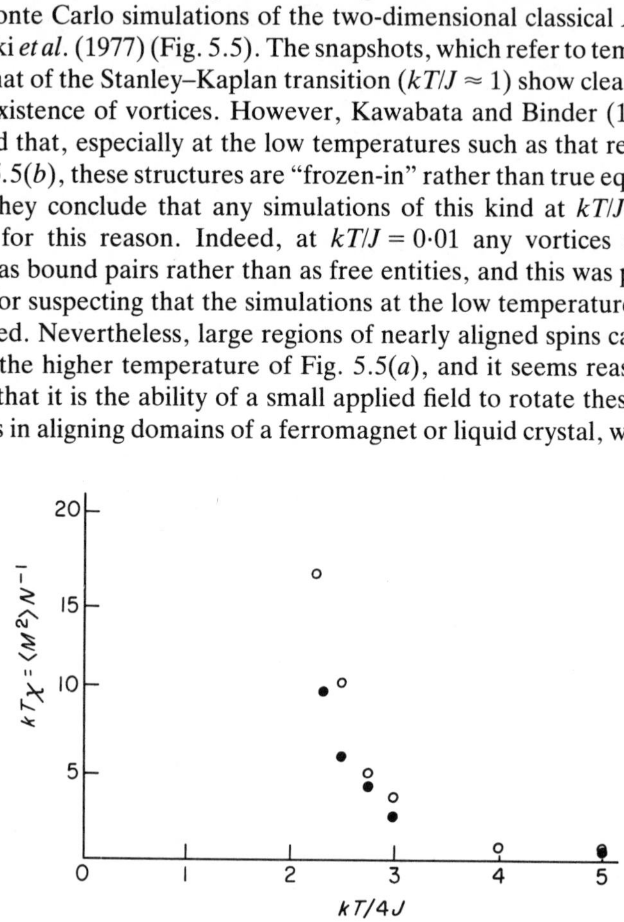

Fig. 5.6. Susceptibility in the *xy*-plane (χ) versus temperature for the quantal *XY*-model, $H_N = -4J\sum_{i>j}(\hat{s}_i^x . \hat{s}_j^x + \hat{s}_i^y . \hat{s}_j^y)$, where the $\hat{s}_i$ are operators with eigenvalues of $\frac{1}{2}$. Simulation data are shown for arrays of two sizes (15 × 15 (○) and 9 × 9 (●)). It is assumed that $\langle M^2\rangle$ has converged and that the true $\langle M\rangle = 0$, although the value found for $\langle M\rangle \neq 0$. Note that the transition temperature ($kT/J \approx 2$) is much higher than for the classical model. (After Suzuki *et al.*, 1977.)

to the divergent susceptibility in the x–y plane at these temperatures (Fig. 5.6). (Their work also shows no divergence in the susceptibility for the two-dimensional quantal Heisenberg model, in agreement with the view (Yamaji and Kondo, 1973; Kosterlitz and Thouless, 1978; Imry, 1979) that whereas the Stanley–Kaplan transition is real for the two-dimensional XY model it is not for the two-dimensional Heisenberg model.) However, of more relevance for adsorption studies is the behaviour of the planar rotator model. For this model it is found that a Stanley–Kaplan transition exists at $kT/J \approx 0{\cdot}9$–$1{\cdot}2$ (Fig. 5.7), below which the correlation function $\mathscr{C}(\boldsymbol{a}; \boldsymbol{R}, \boldsymbol{R}')$ (Eqn (5.15)) decays with distance as a power law, rather than in the exponential manner which is characteristic of a disordered phase (Miyashita *et al.*, 1978; Miyashita, 1980).

Using an entirely different technique (the renormalization group method—Ch. 2, §7(g)) Halperin and Nelson (1978) have obtained results which are in general agreement with those of Kosterlitz and Thouless. They found two transitions in the melting of a two-dimensional solid. At the lower of these (T_m) the first free dislocations appear and lead to an exponential decay of the translational correlation function $\mathscr{C}_k(\boldsymbol{R}) = \langle \hat{\rho}_k(\boldsymbol{R}) \hat{\rho}_k(\boldsymbol{0}) \rangle$ with a correlation length $\xi_+(T) \approx n_f^{-2}$ (ξ_+ diverges at T_m where

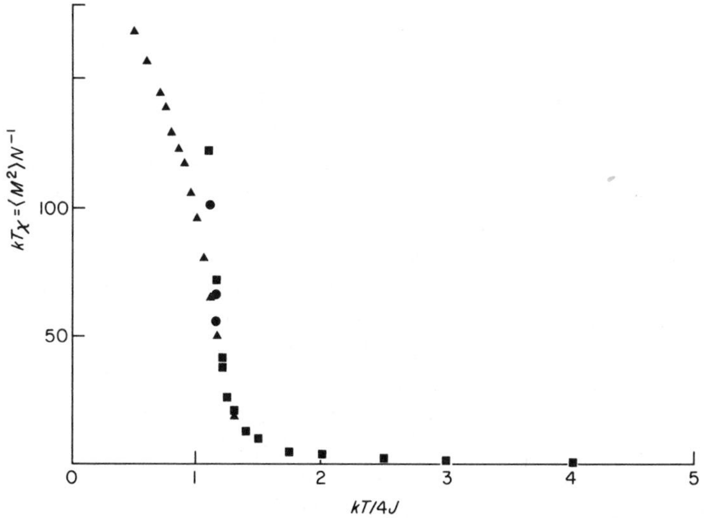

Fig. 5.7. Mean-square magnetization per spin ($\langle M^2 \rangle / N$) versus temperature for the planar rotator model, $H_N = -4J \sum_{i>j} \boldsymbol{s}_i . \boldsymbol{s}_j$, where the $\boldsymbol{s}_i$ are 2-vectors of modulus $\frac{1}{2}$. $\langle M^2 \rangle / N = \chi kT$ provided $\langle M \rangle = 0$, which should be so for a converged simulation. Points are shown for three simulation arrays (50 × 50(■), 30 × 30(●), 15 × 15(▲)). (After Miyashita *et al.*, 1978.)

n_f, the density of free dislocations, is zero). The elastic modulus vanishes at this temperature. However, the orientational order persists in the sense that the decay of the orientational correlation $\mathscr{C}(\boldsymbol{a}; \boldsymbol{R}, \boldsymbol{R}')$ has an inverse power form ($\propto r^{-\eta_6(T)}$). When the second transition temperature (T_i) is reached this order also disappears as a result of the increase of free disclinations, the correlation thereafter decaying exponentially with r. The intermediate phase, known as a "floating solid" or "hexatic" phase, is somewhat like a liquid crystal with six-fold orientations of nearest-neighbour inter-particle vectors. Both the transitions found in this treatment were gradual, rather than first-order. Toxvaerd (1975) had found van der Waals loops in the p–ρ curves for a two-dimensional Lennard-Jones fluid at $T^* = 0{\cdot}651$ and $0{\cdot}526$ which he obtained from a molecular dynamics (MD) study. In each case ρ^* was found to be $\approx 0{\cdot}80$–$0{\cdot}83$. The idea of a two-stage melting process has received support from the MD simulations of the Lennard-Jones system by Frenkel and McTague (1979). They observed a sharp drop in the correlation length of "hexatic" orientational order

$$\langle \psi_6^*(\mathbf{0})\, \psi_6(\boldsymbol{r}) \rangle / g^{(2)}(r) = A \exp(-r/\xi_6)$$

where

$$\psi_6(r) = \frac{1}{N} \sum_{i=1}^{N} \delta(\boldsymbol{r} - \boldsymbol{r}_i) \,.\, \frac{1}{6} \sum_{j=1}^{6} \exp(6\mathrm{i}\theta_{ij})$$

and $j = 1$–6 are the six nearest neighbours of atom i, at $T^* \approx 0{\cdot}57$. They did not find any hysteresis, which would have been expected if the transitions had been first-order, and the intermediate phase had a six-fold symmetry, which is exactly as would be expected for the "hexatic" phase. Subsequently Toxvaerd (1980) and Abraham (1980) have re-examined this system and have been unable to reproduce the results of Frenkel and McTague. Toxvaerd, using MD, observed (from snapshots) what appeared to be a phase separation and also found hysteresis. He commented that the MD results of Frenkel and McTague included a region of negative pressures, which it is natural to assume arose from a metastable one-phase system in the van der Waals' loop region. Abraham's results (Fig. 5.8), showing a hysteresis loop near $T^* \approx 0{\cdot}5$, $\rho^* \approx 0{\cdot}8$ when the system is reduced in density, is in reasonable accord with the earlier results of Toxvaerd, mentioned above, which had shown loops at $T^* = 0{\cdot}526$ and $0{\cdot}651$ with $\rho^* = 0{\cdot}80$–$0{\cdot}83$.

Barker *et al.* (1981) have subsequently determined the complete phase diagram for the system by two methods, one based on simulation, the other on perturbation theory for the fluid and the harmonic approximation for the solid. The results are in excellent agreement with each other, and do

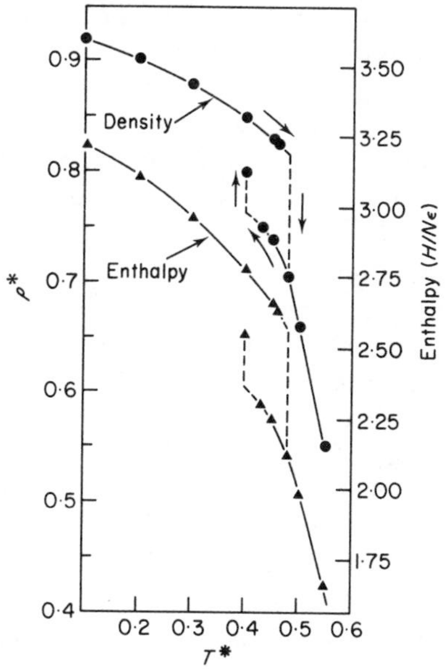

Fig. 5.8. Hysteresis loops in the density–temperature and enthalpy–temperature diagrams of the dense Lennard-Jones system at reduced pressure $p\sigma^2/\varepsilon = 0{\cdot}05$ from simulations in the isothermal–isobaric ensemble. (After Abraham, 1980.)

not show any hexatic region, the diagram being similar to that for a three-dimensional system. They suggest that the results of Frenkel and McTague *do* show incipient phase separation and that the persistence of directional order arises as an artefact of the simulation, because of the sluggishness of the simulation with regard to rotation of crystallites: melting and freezing of the solid essentially leaves the inter-particle directions unchanged. It is also suggested that the transitions proposed by Halperin and Nelson refer to mechanical instability (Abraham, 1981), and that thermodynamic melting may occur at a lower temperature. Both Zippelius *et al.* (1980) and Van Swol *et al.* (1981), who take opposing sides in this controversy agree that the transitions would give only weak static effects in two dimensions, certainly much weaker than those found in three-dimensional systems. Zippelius *et al.* (1980) point out that some of the *transport* properties should be more sensitive to the phase changes in two dimensions and so would be more suitable for their detection. An earlier suggestion by Halperin for reconciliation of the opposing views was that the melting may occur by a single first-order process at high temperatures and pressures, but by two successive second-order transitions at lower temperatures and pressures (Fig. 5.11). This proposal has, however, been contradicted by the results of Abraham (1980) (using an isobaric simulation to avoid isochoric effects), according to which the transition is first-order at all pressures, even some very low ones.

The situation with regard to a possible liquid–vapour transition is also not completely settled. Early workers (Tsien and Valleau, 1974; Henderson, 1977; Toxvaerd, 1975) all concluded that the situation was entirely analogous to that in three dimensions: there was a first-order transition up to a critical temperature. Tsien and Valleau found $T_c^* \approx 0{\cdot}625–0{\cdot}7$, $\rho_c^* = 0{\cdot}38$; Henderson found $T_c^* \approx 0{\cdot}56$, $\rho_c^* = 0{\cdot}32$; and Toxvaerd found $T_c^* \approx 0{\cdot}526–0{\cdot}651$. This uncertainty is associated with a large amount of scatter in the computed data in the region concerned. Frenkel and McTague, using *NpT* MC, also found wide density fluctuations and sluggish behaviour, but they concluded that it was not reasonable to classify these properties as being similar to those of the three-dimensional systems. Toxvaerd (1980) found similar behaviour, but his snapshot showed what appeared to be two distinct phases (liquid and vapour) in the simulation box. His final conclusion was that there is indeed a critical temperature at $T_c^* \approx 0{\cdot}56$, though some of the density variation observed could be attributed to the large fluctuations which would be expected even in a one-phase system close to a critical point.

Also of relevance to this discussion are some of the simulations carried out by Hanson *et al.* (1977) in their studies of the adsorption of argon on graphite. They made three types of run: two-dimensional without and with a periodic substrate potential, and three-dimensional without a periodic potential. The first of these is just the model considered above. The second, on which they did the greatest amount of work, was shown to give almost indistinguishable results at all temperatures for one particular density. If it may be assumed that the period potential would have similarly weak effects at the other coverages studied, then all the two-dimensional results could provide extra information on the behaviour of the two-dimensional Lennard-Jones system without the periodic field. They were, however, looking at systems which were below the density which would be associated with a solid phase. Figure 5.9 shows the energy and heat capacity curves obtained.

For all three coverages (or two-dimensional densities) there was a region centred at $T^* \approx 0{\cdot}38$ in which there was a large but gradual gain in energy. The heat capacity experiments of Chung (1979) confirm the size, shape and (approximately) the temperature of this feature for the real system. The value of $\rho_c^* = 0{\cdot}35$ taken from the previously discussed work corresponds to ~38% monolayer coverage and lies roughly mid-way between the 15% and 64% coverages for which values are presented, and it may

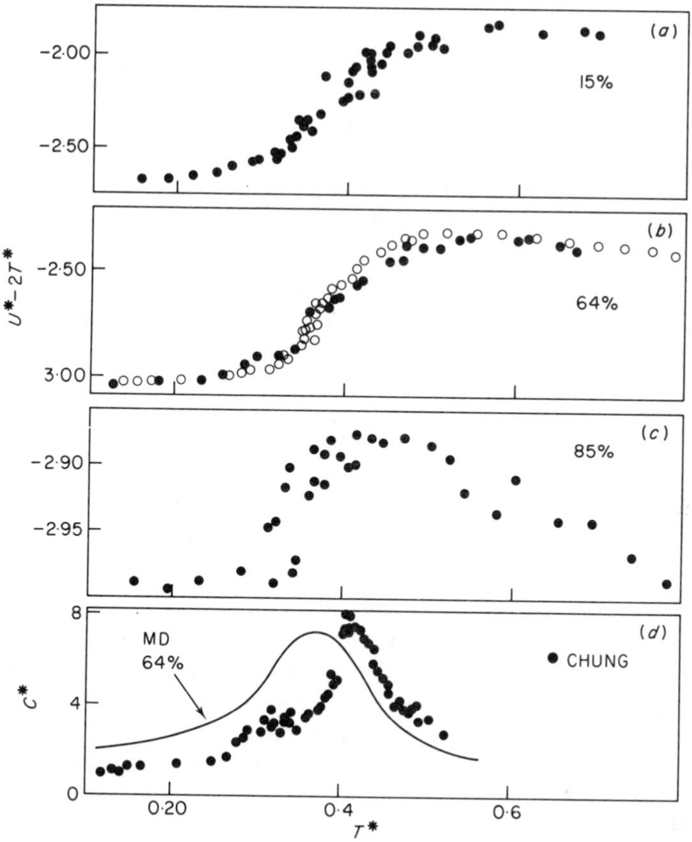

Fig. 5.9. Energy ($U^* = U/\varepsilon$) and heat capacity ($C^* = C/Nk$) versus reduced temperature ($T^* = kT/\varepsilon$) for a planar array of Lennard-Jones molecules at three densities (the value shown is the percentage of close-packed density). The energy of an array of classical two-dimensional harmonic oscillators has been deducted from the energy of the Lennard-Jones system. Also shown are the experimental results of Chung (1979) for C^* for 80% coverage. Open circles are for the model without periodic potential; closed circles are for a model with a periodic potential appropriate to the argon-graphite system. (After Hanson *et al.*, 1977.)

seem reasonable to assume that the maximum in C_v for that coverage would also occur at $T^* \approx 0{\cdot}38$. This value is much lower than all those quoted for the critical point of a two-dimensional fluid (0·526–0·7) (In fact the temperature of the heat capacity maximum seems to fall slowly as the coverage is increased.) This large (~50%) discrepancy between the temperature of a heat capacity maximum and a critical temperature obtained

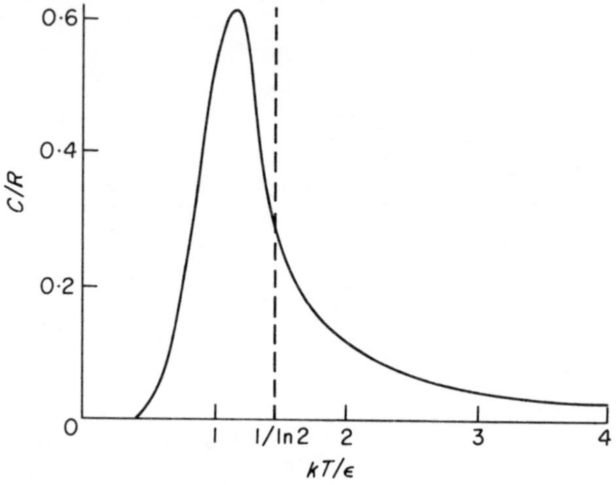

Fig. 5.10. Heat capacity versus temperature for the Rys F model. (After Lieb and Wu, 1972.)

from p–ρ isotherms recalls the situation for the two-dimensional Rys F model in zero electric field, in which the two temperatures are, respectively, $T^* \approx 1$ and $1/(\ln 2) = 1{\cdot}44$ (Fig 5.10) (Lieb and Wu, 1972). Snapshots of the simulation system of Hanson *et al.* suggested that in the runs concerned there was a single phase present, but with considerable clustering.

The phase diagram suggested by Zippelius *et al.* (1980) is represented

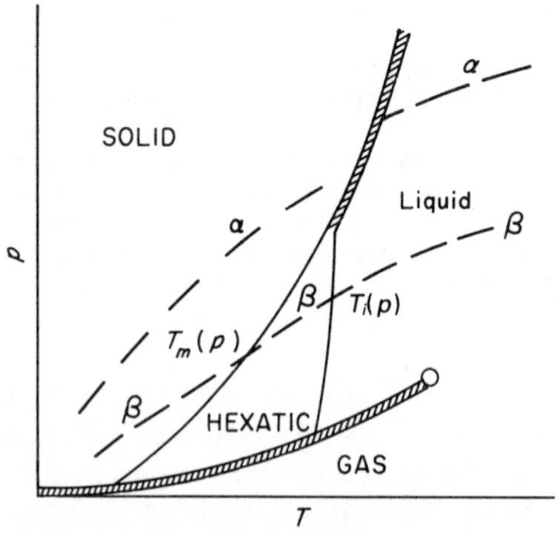

Fig. 5.11. Suggested phase diagram for a two-dimensional Lennard-Jones system. Broken lines are isochores. (Adapted from Zippelius *et al.*, 1980.)

schematically in Fig. 5.11. It can be seen that if the temperature is increased either at constant volume or constant pressure it is possible to obtain a single- or double-stage melting according to the region of the (T, p) diagram investigated.

Most of the work which has been described above, and also other simulation work which follows, is classical rather than quantum mechanical. In so far as the interest is primarily in critical behaviour this is probably not a serious deficiency, since it is now believed that for behaviour near critical points the classical approximation is a good one. This, it is held, is because in such regions it is the long-wavelength modes which are of primary importance and for these $kT_c \gg \hbar\omega$ (Hertz, 1976).

4. Systems in periodic substrate fields: epitaxial phases

It has been seen in the previous sections that there are still considerable uncertainties about the behaviour of the true two-dimensional system with no superimposed field. It is not surprising, therefore, that when a periodic substrate field is added the problem becomes one on which less progress has been made.

Following the classical papers on one-dimensional epitaxial and non-epitaxial systems by Frank and van der Merwe (1949), Ying (1971) examined the ground-state energy and the lattice-vibration spectrum for two-dimensional epitaxial phases of various kinds and compared these with the corresponding properties for a non-epitaxial phase. For a square array in which the substrate potential for an adatom position (x, y) is

$$W\{\cos(2\pi x/a) + \cos(2\pi y/a)\} \tag{5.16}$$

the total potential energy is written as:

$$\begin{aligned} U_N = {} & \tfrac{1}{2}\alpha \sum_{l,m} [\{x(l+1, m) - x(l, m) - \mu\}^2 \\ & + \{y(l, m+1) - y(l, m) - \mu\}^2] \\ & + \tfrac{1}{2}\beta \sum_{l,m} [\{x(l, m+1) - x(l, m)\}^2 \\ & + \{y(l+1, m) - y(l, m)\}^2] \\ & - W \sum_{l,m} [\cos(2\pi x(l, m)/a) + \cos(2\pi y(l, m)/a)] \end{aligned} \tag{5.17}$$

The first term arises from the change in spacing of adjacent molecules from the equilibrium spacing (μ) in the absence of the field; the second term is the potential energy arising from shearing of the adlayer; the third term is, of course, just the adsorbate–adsorbent potential (Eqn (5.16)). For the non-epitaxial layer the energy is found to be

$$E_\infty = 2t^2 \sin^2(c\pi) + t^4\pi^2 \sin^2(2\pi c) - \mathcal{U}J_1(2\pi t) \quad (5.18)$$

where J_1 is a first-order Bessel function of the first kind;

$$\mathcal{U} = 4W/\alpha a^2 \quad (5.19)$$

measures the ratio of the strength of the substrate potential to the elastic energy; and t is the smallest positive solution of the equations

$$t = (\pi U/2 \sin^2(c\pi))J_1'(2\pi t) \quad (5.20)$$

$$c = b - \pi t^2 \sin(2\pi c) \quad (5.21)$$

$b = \mu/a$ and J_1' is the derivative of J_1. A quantity $\bar{b}$, which is related to b by the equation

$$\bar{b} = \text{mod}(b; 1) = b - p \quad (5.22)$$

where p is the integer such that $\bar{b}$ lies between $-\frac{1}{2}$ and $\frac{1}{2}$, is the "mismatch" of the system.

Ying examined the ground-state energy for the non-epitaxial (incommensurate) and several low-order commensurate phases as a function of mismatch for several values of the substrate potential constant $\mathcal{U}$ (Fig. 5.12). A $\sqrt{\mathcal{U}} - \bar{b}$ "phase diagram" for 0 K can then be constructed (Fig. 5.13). Whatever the mismatch, sufficiently high values of $\mathcal{U}$ lead to an $n = 1$ layer. If $\bar{b} = 0{\cdot}5$, the $n = 2$ layer is the stable one provided $\mathcal{U}$ is not too large. When we turn to the lattice vibrations of the layer we find that the $n = 1$ structures have a gap from $\omega = 0$ to $\omega = \pi(\mathcal{U}\alpha/M)^{1/2}$; for $n = 2$ structures the gap is smaller, extending from $\omega = 0$ to $\omega = \pi\{(\mathcal{U}\alpha/M) \sin 2\eta t\}^{1/2}$, where M is the mass of the admolecule. The existence of a gap starting at $\omega = 0$ is a characteristic feature of commensurate layers which is an aid to recognition in neutron scattering. In contrast, the incommensurate layers show no such frequency gap.

One of the criticisms of Ying's work is that he did not consider interactions between domain walls which crossed. Van der Merwe (1970), Bak and Mukamel (1979) and Bak *et al.* (1979) have taken account of this wall crossing energy. Bak *et al.* wrote this energy contribution as λ/l^2, where l is the distance between adjacent parallel walls and λ is a parameter dependent upon the system. Hexagonal or one-dimensional domain wall structures were favoured according as $\lambda < 0$ or $\lambda > 0$. Applying the symmetry theory of transitions of Landau and Lifshitz (Ch. 2, §7(e)) they concluded that the transition would be first-order or continuous according as the

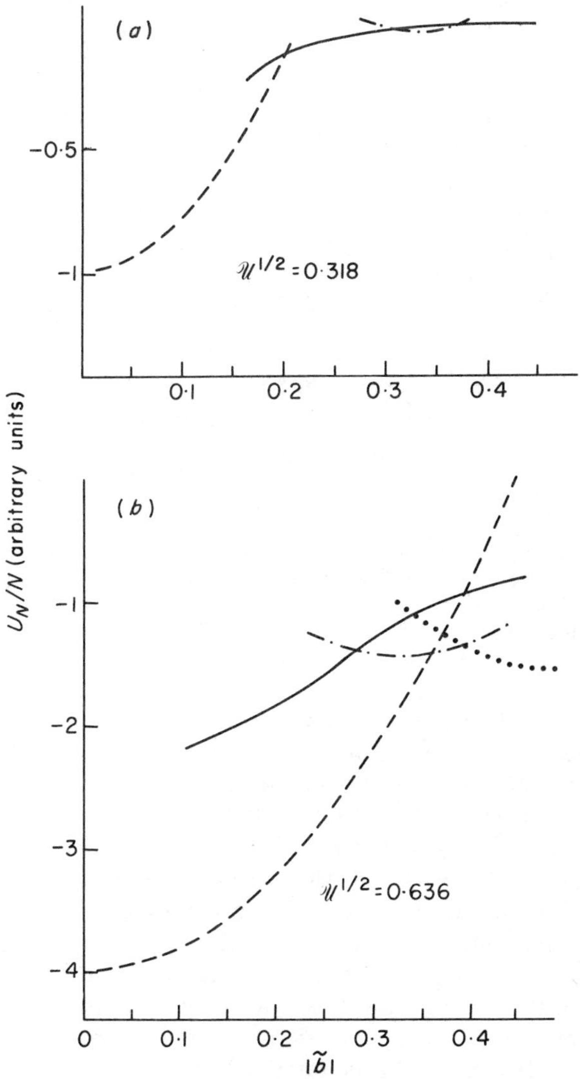

Fig. 5.12. Ground state energy per adatom versus mismatch ($|\tilde{b}|$) for incommensurate ($U_{N,\infty}/N$, ———) and three simple commensurate structures ($U_{N,1}/N$, – – – –; $U_{N,2}/N$, ······; $U_{N,3}/N$, –·–·–·–) with substrate potential constants (*a*) $\mathcal{U}^{1/2} = 0{\cdot}318$ and (*b*) $\mathcal{U}^{1/2} = 0{\cdot}636$. (After Ying, 1971.)

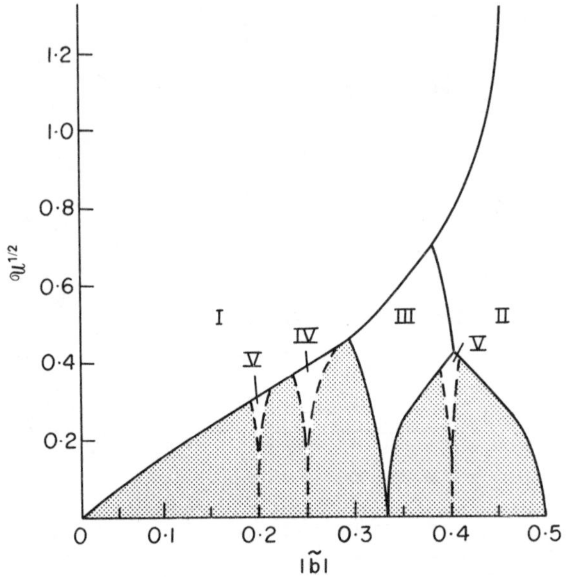

Fig. 5.13. Stability regions for five commensurate (I–V) and the incommensurate phases at 0 K: $\mathcal{U}^{1/2}$, where $\mathcal{U}$ is the substrate potential constant, versus the mismatch ($|\tilde{b}|$). Shaded areas indicate phases which are incommensurate or commensurate but of order >5. (After Ying, 1971.)

incommensurate domain structure was hexagonal or one-dimensional, and therefore according as $\lambda > 0$ or $\lambda < 0$. The domain walls of the incommensurate structure referred to here are the "misfit dislocations" of Venables and Schabes-Retchkiman (1977, 1978).

So far attention has been devoted to the balance between commensurate and incommensurate phases at 0 K, the variables being the strength and symmetry of the substrate field. Some work has recently been carried out with temperature and coverage as the variables, the field being regarded as fixed. The renormalization group treatment of Halperin and Nelson has already been mentioned with regard to its predictions for the system in zero substrate field. Their conclusions included a prediction of a phase diagram for a system with a periodic field (Fig. 5.14) (Halperin, 1979). Epitaxial (or commensurate) phases of the first three orders are shown. The results suggest that there should be two two-phase regions: vapour–liquid and liquid–epitaxial I. The diagram refers to a fixed not-too-large value of $\mathcal{U}$, the substrate field. It is thought that larger values of $\mathcal{U}$ would cause the floating-solid ("hexatic") region to be "squeezed out" between the fluid and epitaxial regions.

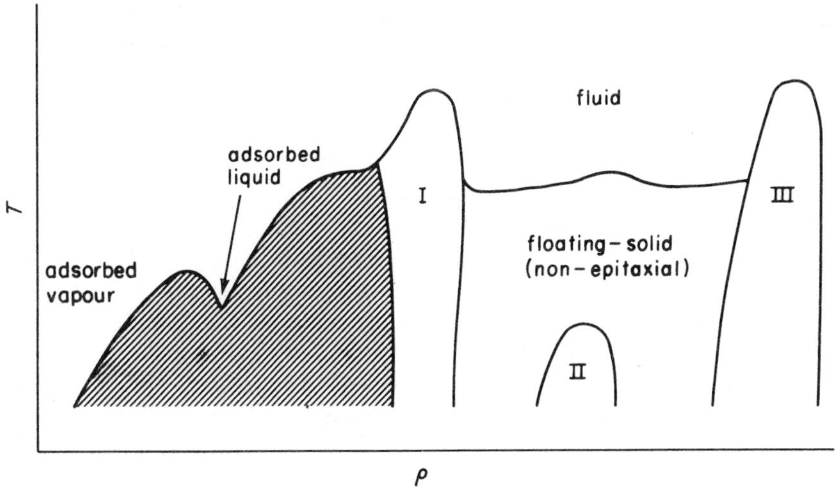

Fig. 5.14. T–ρ phase diagram for a two-dimensional system with periodic substrate potential. I, II and III are epitaxial phases of the first, second and third order, respectively. (After Halperin, 1979.)

Novaco and McTague (1977a, b) and Novaco (1979) have studied the problem of noble gas monolayers on graphite. An interesting result, which was subsequently supported by MD simulations, is that in the incommensurate phase, close to the transition there exist domains of commensurate "material", these growing as the transition is approached. They also predicted that remote from the transition the principal axes of the incommensurate phase would be rotated with respect to the axes of the substrate by amounts of the order of 3° for Ar, Kr or Xe, and 17° for Ne. These findings are in agreement with the low-energy electron diffraction (LEED) experiments of Chinn and Fain (1977) and Shaw *et al.* (1978) for Ar, though possibly not for Kr. Previous theoretical treatments of this type of problem had excluded from consideration this kind of adaptation between the lattices of the adsorbate and substrate. In the first two of their papers Novaco and McTague discussed the quasi-harmonic approximation to the problem. It is interesting that the orientation which they found occurred in spite of the energy advantage being tiny: for argon as adsorbate, with a lattice constant of 0·386 nm, the reduction in energy is only ~1 K, which is to be compared with a total adsorption energy of ~1100 K. Subsequently, the effect of zero-point and thermal vibration energy of the admolecules was examined (Novaco, 1979). The effect of these motions, which was incorporated as a renormalization of the force constants of the previous problem, turned out to be small for the orientational displacement (a change of ~0·2° on ~3·5° for argon on graphite). Novaco and McTague showed that the mean displacement of the adatom from its own monolayer site could be written as:

$$\langle \hat{r}_j^{\alpha} \rangle = \sum_{Gl} \varepsilon_l^{\alpha}(\boldsymbol{G})\, \eta_l(\boldsymbol{G}) \sin \boldsymbol{G} \,.\, (\boldsymbol{R}_j - \Delta) \tag{5.23}$$

where

$$\eta_l(\boldsymbol{G}) = \tilde{\phi}_G G^{\beta} \varepsilon_l^{\beta}(\boldsymbol{G}) / M\omega_l^2(\boldsymbol{G}). \tag{5.24}$$

Here $\varepsilon_l^{\alpha}(\boldsymbol{G})$ and $\eta_l(\boldsymbol{G})$ give the polarization and amplitude of the $(l\boldsymbol{G})$ distortion (or mass density) wave: $\boldsymbol{\Delta}$ is a centre of inversion symmetry of the graphite lattice; $\boldsymbol{G}$ is a reciprocal lattice vector of the graphite lattice; $\tilde{\phi}_G$, which is real and symmetric in $\boldsymbol{G}$, is related to the Fourier coefficients (ϕ_G) of the potential by $\phi_G = \tilde{\phi}_G \exp(-\mathrm{i}\boldsymbol{G} \,.\, \boldsymbol{\Delta})$. The internal energy of the phase at 0 K is then composed of three terms:

$$U = E_{\mathrm{L}} + \tfrac{1}{2} \sum_{l} \hbar\omega_l(\boldsymbol{q}) - \tfrac{1}{2} N \sum_{G,l} \tilde{\phi}_G G^{\alpha} \varepsilon_l^{\alpha}(\boldsymbol{G}) \eta_l(\boldsymbol{G}) \tag{5.25}$$

where E_{L} is the lattice energy of the monolayer in the absence of distortion waves. The second term, which is the zero-point energy of the vibrations, is unaffected by rotation of the monolayer lattice in this approximation, though not in that of the later paper (Novaco, 1979). Another term, not shown in Eqn (5.25), $\mathrm{N} \sum_{G} \sum_{\tau} \phi_G \delta_{G,\tau}$, which represents the interaction of the monolayer atoms at their undisplaced positions (reciprocal lattice $\boldsymbol{\tau}$) with the substrate, is zero. The final term in Eqn (5.25) is dependent upon the relative orientation of the lattices and determines the angle adopted. This (negative) term will be favoured by small mode frequencies $(\omega_l(\boldsymbol{q}))$, and consequently the transverse will be more important than the longitudinal modes. However, because of the form of $\eta_l(\boldsymbol{G})$ these modes cannot contribute unless $\theta \neq 0$. The orientational displacement causes ω to increase, opposing the previously mentioned effect and hence leading to an equilibrium value of θ. With regard to a possible incommensurate–commensurate transition, no quantitative information can be deduced from these theories. However, since the order parameter $(\eta_l(\boldsymbol{G}))$ can take any of a continuous range of values it is possible for such a transition to be a continuous one. If all the harmonics are included, rather than just a single "pure" distortion wave, the theory should tend towards the dislocation treatment of Kosterlitz and Thouless and the renormalization group approach of Halperin and Nelson. The picture of the adsorbate then tends to revert to one in which the adatoms occupy epitaxial positions with lines of dislocation also present (Novaco and McTague, 1977b).

One of the problems in dealing with the commensurate versus incommensurate question is that there is an infinite number of commensurate lattices which could exist. The first part of the problem may then be seen as finding which of the commensurate phases is the most stable, this being subsequently compared with the incommensurate phase. Fuselier *et al.* (1980) have carried out this procedure for argon on graphite using a method similar to that of Novaco and McTague. They found from their approximate statistical mechanical theory that the free energy could be written as

$$G = -kT\sum_{q\lambda} \ln[2\sinh\{\hbar\beta\omega_\lambda(\boldsymbol{q})/2\}] + \sum_i u_i^{(1)} - \tfrac{1}{2}\sum_{i,j} \boldsymbol{u}_i^{(1)'} \cdot \Phi_{ij}^{-1} \cdot \boldsymbol{u}_j^{(1)'} \quad (5.26)$$

where $u_i^{(1)}$ and $\boldsymbol{u}_i^{(1)'}$ are the energy and force arising from the interaction of the adatoms (on their own sites) with the substrate, Φ_{ij} is the force constant matrix for adatom–adatom interaction and $\omega_\lambda(\boldsymbol{q})$ is the frequency of the $(\lambda, \boldsymbol{q})$ mode of the monolayer. Replacing $u^{(1)}$ by the set of $u_G^{(1)}$, its Fourier transforms in terms of the lattice vectors of the graphite ($\boldsymbol{G}$), this becomes

$$\begin{aligned} G = {} & -kT\sum_{q\lambda} \ln[2\sinh\{\hbar\beta\omega_\lambda(\boldsymbol{q})/2\}] \\ & + N\sum_{G\tau} u_G^{(1)}\delta_{G\tau} - \tfrac{1}{2}N\sum_{G\lambda} u_G^{(1)}\boldsymbol{G}\cdot\boldsymbol{\varepsilon}_\lambda(\boldsymbol{G})\frac{1}{M\omega_\lambda^2(\boldsymbol{G})}\boldsymbol{G}\cdot\boldsymbol{\varepsilon}_\lambda(\boldsymbol{G}) \\ & + \tfrac{1}{2}N\sum\sum_{GG'\lambda\tau\neq 0} u_G^{(1)}\boldsymbol{G}\cdot\boldsymbol{\varepsilon}_\lambda(\boldsymbol{G})\frac{1}{M\omega_\lambda^2(\boldsymbol{G})}u_{G'}^{(1)}\boldsymbol{G}'\cdot\boldsymbol{\varepsilon}_\lambda(\boldsymbol{G})\,\delta_{G+G',\tau} \end{aligned} \quad (5.27)$$

where $\boldsymbol{\tau}$ is a reciprocal lattice vector of the monolayer and $\boldsymbol{\varepsilon}_\lambda(q)$ is an eigenvector of the matrix $\Phi_{ij}(\boldsymbol{q})$. The third and fourth terms are the static distortion contributions. The latter is zero for incommensurate lattices; the former is the term evaluated by Novaco and McTague. The second term is known as the lock-in term since it has a delta function form centred at the reciprocal lattice points of the substrate lattice. A graph of orientation (θ) against lattice constant is shown in Fig. 5.15 on which points for a large number of commensurate phases have been plotted. (The phase is defined by the integers p and q which are the smallest pair such that $\tan\theta = p/(\sqrt{3}q)$; the monolayer lattice constant is $a = a_c/N$, where N is an integer and a_c is the length of the smallest triangular cell which has its vertices at substrate site positions. In terms of these parameters the possible common phases can be represented as in Fig. 5.16.) The main feature of the plot is that the phases fall into several distinct branches. The free energies of these phases, omitting for the moment the distortion wave contribution, is such that only the main branch (2) phases have appreciable free energy advantages over the incommensurate phase; the subsidiary branch (1) has a small advantage. The phases of interest all have large unit

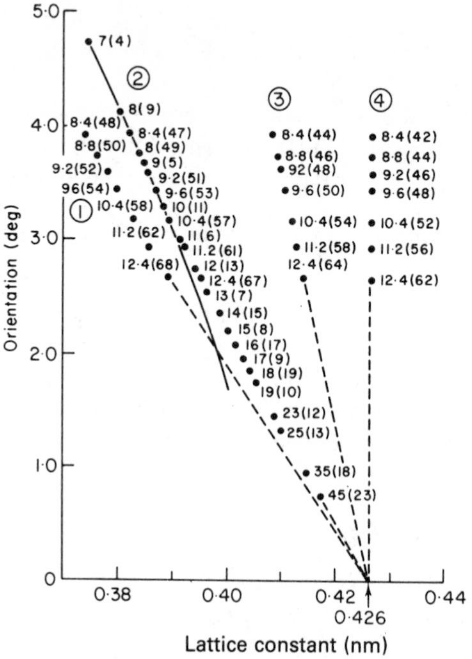

Fig. 5.15. Adsorbed monolayer orientation angle versus lattice constant for selected coincident lattices with small lattice constants. The commensurate structures are labelled by the numbers q/p and (N) (see Fig. 5.16). The $\sqrt{3} \times \sqrt{3}$ registered structure has a lattice constant of 0·426 nm. Full line: results of Novaco and McTague (1977) for the incommensurate adlayer. (After Fuselier *et al.*, 1980.)

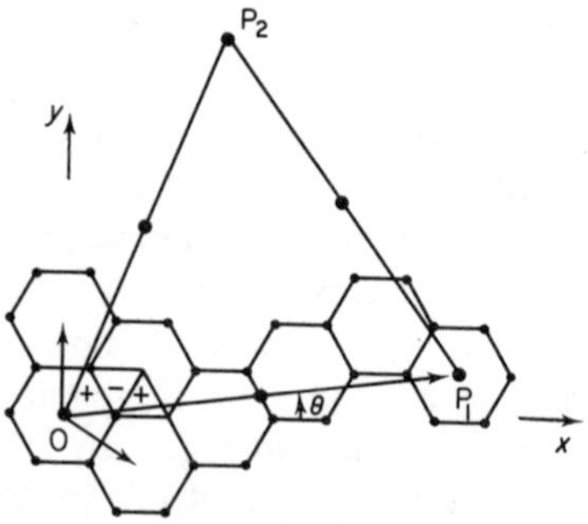

Fig. 5.16. An adsorbed monolayer with coincident lattice sites (CSL). The angle θ between the adsorbed monolayer and the substrate lattice is arctan$(1/5\sqrt{3})$ and so $p = 1$, $q = 5$ since $\theta = \arctan(p/q\sqrt{3})$. The ratio of OP_1 to the monolayer lattice constant $= N = 2$. (After Fuselier, 1980.)

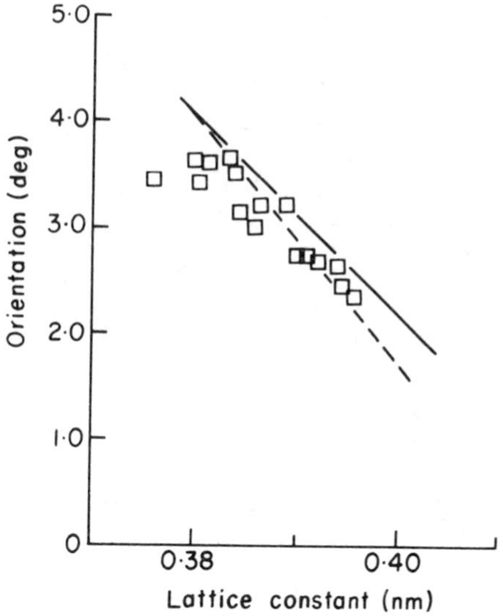

Fig. 5.17. Comparison of the predictions for orientation angle versus lattice constant of the adlayer by Fuselier *et al.* (full line) and Novaco and McTague (broken line) with the experimental points of Shaw *et al.* (□). Only the most stable of the "branches" of Fig. 5.15 (branch 2) is shown. (After Fuselier, 1980.)

cells, i.e. they are nearly incommensurate. this being so, it is not unreasonable to use the results of Novaco and McTague to evaluate the static distortion wave contribution. When this is done, the subsidiary branch may be ruled out as being now of higher free energy than the incommensurate phase. Retaining only the most stable (lowest G) phases the θ–a graph is found to differ only slightly from that deduced by Novaco and McTague, both of which are in fair agreement with the experimental results of Shaw *et al.* (1978) (Fig. 5.17). Neither these calculations nor those of Novaco and McTague, both of which depend upon linear response theory (ter Haar, 1977; Novaco, 1979), are suitable for dealing with the stronger perturbations from the substrate which occur in the vicinity of a commensurate–incommensurate transition. The work of Fuselier *et al.*, like the earlier papers of Novaco and McTague, deals only with the situation at 0 K.

The question of whether the non-epitaxial domains in the noble gas–graphite system are one-dimensional or hexagonal has been examined in detail by Shiba (1980) using a numerical solution of the sine-Gordon equation. He concluded that except for a small range of force and size parameters the hexagonal domains were the more stable. The generally

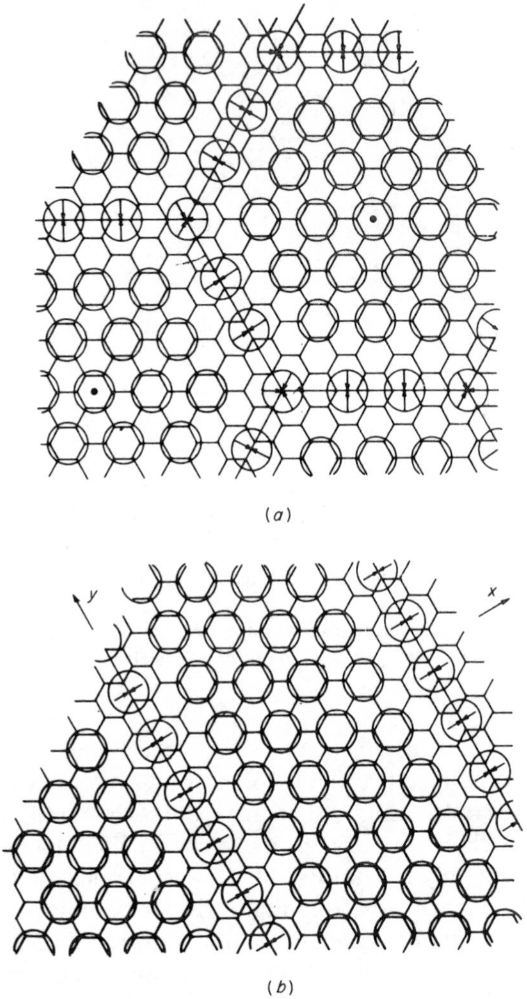

Fig. 5.18. Schematic diagrams showing (a) the hexagonal, (b) the uniaxial domain structures. Adatoms are represented by circles. (After Shiba, 1980.)

greater stability of the hexagonal domain system is achieved in spite of there being adatoms (at the meeting point of three domain walls) which are directly above carbon atoms and therefore in the position of maximum substrate–adatom energy (Fig. 5.18). The region, if any, where the one-dimensional domains were of lower free energy was close to that where the system was on the point of going over to a commensurate phase (Fig. 5.19). The figure refers to the non-rotated, non-epitaxial phase, but Shiba

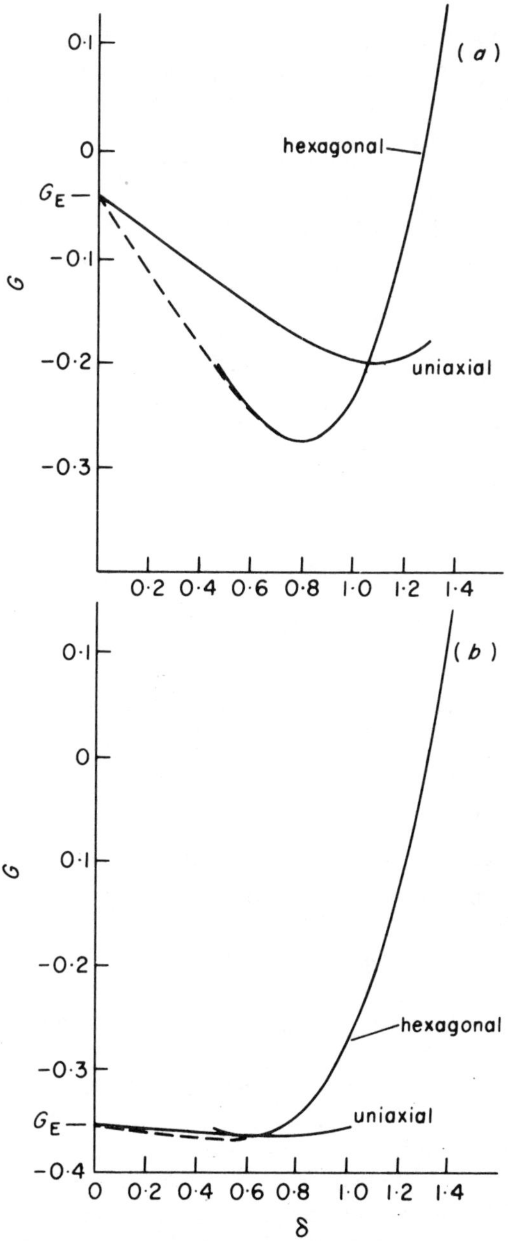

Fig. 5.19. Free energy (G) versus "actual mismatch" (δ) for non-rotated uniaxial and hexagonal domain structures at points in the non-epitaxial region (*a*) far from and (*b*) close to the epitaxial–non-epitaxial transition. G_E is the free energy of the epitaxial ($\delta = 0$) phase. (After Shiba, 1980.)

was able to show that the corresponding rotated phase gave way to the non-rotated phase before the transition was reached. It follows that, for a discussion of the transition itself, only the non-rotated phases are relevant. Within the regime of hexagonal domains Shiba found that the occurrence of a rotation, of the type described by Novaco and McTague, depended upon the value of c_T^2/c_L^2 where c_L and c_T are the longitudinal and transverse sound velocities, respectively. Low values of this ratio favoured relatively high values of θ, the displacement angle of Novaco and McTague.

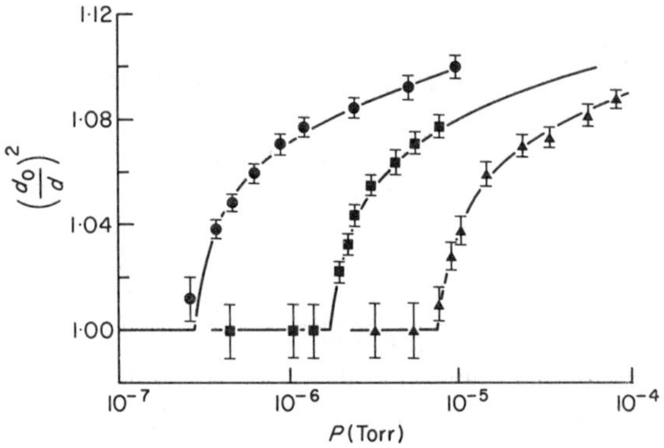

Fig. 5.20. Mean misfit (from LEED measurements) versus pressure at 54 K (circles) 57 K (squares) and 59 K (triangles) for the adsorption of krypton on graphite. d and d_0 are the mean observed and the epitaxial Kr–Kr distances, respectively. (After Fain and Chinn, 1977.)

Another question of interest is: what is the order of the commensurate–incommensurate transition? The best experimental data available indicates that the transition is continuous (Chinn and Fain, 1977; Stephens *et al.*, 1979; Vora *et al.*, 1979), but the possibility that it is partly first-order cannot be ruled out. Figure 5.20 shows a plot of mean misfit versus pressure illustrating the gradual nature of the transition. Accepting the transition as being of the critical type, such a plot enables a critical index to be fitted to the data. From similar data on krypton obtained from X-ray measurements the index β, defined by the equation $(d_s - d)/d_s \sim \ln(p/p_c)^\beta$, where d and d_s are the lattice parameters of the adsorbed layer and the substrate, respectively, was found to be $0{\cdot}32 \pm 0{\cdot}02$ and $0{\cdot}26 \pm 0{\cdot}06$ at 89·3 K and 80 K, respectively (Stephens *et al.*, 1979). However, there is a difficulty here in that, if both the phases concerned have hexagonal symmetry, the Landau theory states that the transition could not be continuous (Ch. 2, §7(e)). The conclusion to be drawn is that either both phases are truly hexagonal, in which case the transition must be at

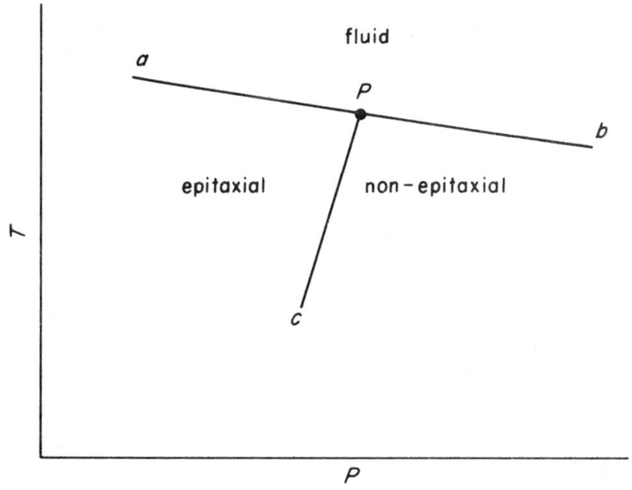

Fig. 5.21. Suggested phase diagram in the transition region. (After Bak and Mukamel, 1979.)

least partly first-order, or the transition is continuous and one or both of the phases has a distortion from true hexagonal symmetry (Bak *et al.*, 1979). Bak and Mukamel (1979) concluded that close to the point P (Fig. 5.21) the transition is first-order, enabling both phases to be of undistorted hexagonal structure, but that far from P the transition becomes continuous. This would mean that there must be a tricritical point somewhere on the boundary Pc, and suggests that there should be two kinds of incommensurate phase with a boundary between them, with one having a distortion from the pure hexagonal structure. This recalls the proposal of Halperin and Nelson that for zero or weak substrate field, and consequently the incommensurate adsorbate, there should be two kinds of phase. The order of the remaining phase boundaries of Fig. 5.21 are also somewhat uncertain. It is now believed that the commensurate–fluid transition (Pa) is of the kind described by a 3-state Potts model (Ch. 2, §7(f)) and is therefore continuous. (Domany *et al.*, 1978). The line Pb represents two-dimensional melting, which has been discussed earlier, and the nature of which is still unclear.

A two-dimensional study of the effect of a periodic substrate potential on $\rho^{(1)}(x, y)$ has been made by Prasad and Toxvaerd (1980). The periodic potential was that over the (111) face of an FCC lattice of the same particles. Their main interest was in the application of the BBGKY equations to this problem, and their starting point was Eqn (3.134) in two-dimensional form, in which they made the customary assumption that $\rho^{(2)}$ could be replaced by $\rho_1^{(1)}\rho_2^{(1)}\tilde{g}_{12}$. After this replacement, the equation could be integrated to give

$$\ln(C\rho_1^{(1)}) + \beta u_1^{(1)} = \int_{\mathscr{A}} \rho_2^{(1)} S(r_{12})\,\mathrm{d}r_{12} \tag{5.28}$$

where

$$S(r_{12}) = \int_{r_{12}}^{\infty} \tilde{g}(r)\,\beta(\partial u^{(2)}/\partial r)\,\mathrm{d}r \tag{5.29}$$

and C is an integration constant. For $u^{(2)}$ they used the 12–6 potential cut off at $2{\cdot}8\sigma$ and for $u^{(1)}$ the periodic part of the Steele potential, obtained from Eqn (3.82) when the inverse z-dependence is factored out. Only the first two terms in the summation over the wave vector $\boldsymbol{k}$ were significant and their contribution could be expressed in terms of the reciprocal lattice vectors $\boldsymbol{b}_1, \boldsymbol{b}_2$ by

$$\begin{aligned} u^{(1)} = {} & \alpha_1\{\cos(2\pi\boldsymbol{b}_1 \,.\, \boldsymbol{r}) + \cos(2\pi\boldsymbol{b}_2 \,.\, \boldsymbol{r}) + \cos(2\pi(\boldsymbol{b}_1 + \boldsymbol{b}_2) \,.\, \boldsymbol{r})\} \\ & + \alpha_2\{\cos 2\pi(\boldsymbol{b}_1 - \boldsymbol{b}_2) \,.\, \boldsymbol{r} + \cos 2\pi(2\boldsymbol{b}_1 + \boldsymbol{b}_2) \,.\, \boldsymbol{r} \\ & + \cos 2\pi(\boldsymbol{b}_1 + 2\boldsymbol{b}_2) \,.\, \boldsymbol{r}\} \end{aligned} \tag{5.30}$$

where $\boldsymbol{r}$ locates the adsorbate in a plane above the surface at a distance equal to the spacing between (111) layers. The coefficients α_1, α_2 were $0{\cdot}6236\varepsilon$ and $0{\cdot}0653\varepsilon$, respectively, in this plane, where ε is the parameter of the 12–6 potential. $\tilde{g}$ was obtained from MD simulation either as the radial distribution function in the absence of the external potential or as its mean value when the potential was turned on and calculated values of $\rho^{(1)}(x, y)$ were compared with those from the simulations. The agreement was found to be very close for both approximations even though small systematic differences exist between the two approximations to $\tilde{g}$.

5. Dilute epitaxial adsorbed phases

(a) Relationship to other model systems

The previous discussion of epitaxial (commensurate) phases has dealt almost exclusively with phases in which nearly all of the sites are occupied. In many cases of the chemisorption of small molecules or atoms onto metallic surfaces, regular overlayers with much lower density are often found. In order to explain these structures it is necessary to postulate adatom–adatom interactions which act over a distance comparable with the observed inter-atom spacing.

In physisorption, on the other hand, it is often reasonable to consider that only interactions of adatoms adsorbed on neighbouring sites need to be taken into account. The statistical mechanics of such systems are equivalent to two-dimensional lattice gas problems, the more simple of which may have known analytic solutions. Other examples may usually be simulated comparatively cheaply.

Table 5.1. Transformation between the magnetic Ising problem and the adsorption problem

Magnetic	Adsorption
$kT \ln Q + \mathcal{B}(\mathcal{H} - \frac{1}{2}zJ)$	$kT \ln \Xi$
$4J$	$\varepsilon^{(a,a)}$
$2\mathcal{H}$	$\mu + \varepsilon^{(a,s)} + \frac{1}{2}z\varepsilon^{(a,a)}$
$\frac{1}{2}(M + \mathcal{B})$	N
$2zJ - 2(\partial[\langle U_N\rangle + M\mathcal{H}]/\partial M)_{T,B}$	$q_{st}/N_A - \varepsilon^{(a,s)} = kT - (\langle U_N\rangle/\langle N\rangle)_{T,B} - \varepsilon^{(a,s)}$

If all the sites have adsorption energies which are equal ($-\varepsilon^{(s,a)}$) and adatoms on nearest-neighbour sites interact with an energy $-\varepsilon^{(a,a)}$ (all more remote interactions being assumed to be zero), then the problem reduces to that of the two-dimensional Ising lattice. The corresponding zero-field Ising problem, for which Onsager's famous work provided the solution, is equivalent to adsorption with a half-filled lattice ($\theta = 0{\cdot}5$). The more general non-zero field problem, which corresponds to $\theta \neq 0{\cdot}5$, has not been solved analytically, but several simulation studies have been carried out on it (Fosdick, 1963; Binder and Landau, 1976). In converting these data to the adsorption case the transformations used are derived from a comparison of the canonical ensemble for the magnetic case in field $\mathcal{H}$ with the grand canonical ensemble for the adsorption system with chemical potential μ. This leads to the equivalences shown in Table 5.1, based on the configurational energies

$$U_N = -4J \sum_{\substack{ij \\ nns}} s_i s_j - 2\mathcal{H} \sum_i s_i$$

and

$$U_N = -\varepsilon^{(a,a)} \sum_{\substack{ij \\ nns}} (s_i + \tfrac{1}{2})(s_j + \tfrac{1}{2}) - 2\varepsilon^{(a,s)} \sum_i (s_i + \tfrac{1}{2})$$

for the magnetic and adsorption systems, respectively. z is the coordination number of each lattice site and s_i may take the values $\pm\frac{1}{2}$. M (the magnetization) $= 2\sum_i s_i$. $\mathcal{B}$ is the number of sites. Similar transformations, especially between the magnetic and lattice gas systems, have been used for many years (Fisher, 1967). This model shows a gas–liquid type condensation of the adsorbate which on changing T at constant θ is gradual if $\theta = 0{\cdot}5$, but first-order for $\theta \neq 0.5$.

In moving on to more complicated models we shall first examine some results which have been obtained for lattices of the quadratic topology and then consider those for the triangular lattice, which corresponds to the most-studied adsorption systems involving the (0001) face of graphite.

(b) Quadratic and related lattices

A model of this class which has yielded somewhat surprising results is one in which two kinds of adsorption site (α and β) are recognized, with adsorption energies of $-\varepsilon_\alpha$ and $-\varepsilon_\beta$, respectively. It is further supposed that the α- and β-sites are arranged alternately on the lattice and that occupation of neighbouring sites leads to an extra energy contribution of $-4J$ ($J < 0$, a repulsion). This corresponds to a magnetic Ising model in which there is an alternating field which takes values of $\mathcal{H}_\alpha$ $(=\frac{1}{2}\varepsilon_\alpha)$ and $\mathcal{H}_\beta$ $(=\frac{1}{2}\varepsilon_\beta)$ at α- and β-sites respectively. This bears some resemblance to the staggered field ($\pm\mathcal{H}$ at neighbouring sites) which is often invoked in dealing with antiferromagnetic phases.

Thus, the energy of a configuration of the adsorption system may be written as

$$U_N = -2\sum_i \{(s_i^\alpha + \tfrac{1}{2})\mathcal{H}_\alpha + (s_i^\beta + \tfrac{1}{2})\mathcal{H}_\beta\} - 4J\sum_{\substack{ij\\ nns}} (s_i^\alpha + \tfrac{1}{2})\,(s_j^\beta + \tfrac{1}{2}) \tag{5.31}$$

where $s_i^\alpha, s_j^\beta = \pm\frac{1}{2}$. Putting

$$\mathcal{H} = \tfrac{1}{2}(\mathcal{H}_\alpha - \mathcal{H}_\beta) \tag{5.32}$$

and

$$\mathcal{H}^\circ = \tfrac{1}{2}(\mathcal{H}_\alpha + \mathcal{H}_\beta) \tag{5.33}$$

Eqn (5.31) becomes

$$U_N = -2B\mathcal{H}^\circ - JBz - \sum_i [2\{\mathcal{H}^\circ + \mathcal{H} + Jz\}s_\alpha^i + 2\{\mathcal{H}^\circ - \mathcal{H} + Jz\}s_\beta^i] - 4J\sum_{\substack{ij\\ nns}} s_i^\alpha s_j^\beta \tag{5.34}$$

When the grand partition function

$$\Xi(T, B, \mathcal{H}^\circ, \mathcal{H}, \mu) = \sum \exp\left[+\frac{1}{kT}\left\{2B\mathcal{H}^\circ - B\mu + JBz) + 4J\sum s_i^\alpha s_j^\beta + \sum_i (s_i^\alpha + s_i^\beta)\,(2Jz + 2\mathcal{H}^\circ - \mu) + 2\mathcal{H}\sum_i (s_i^\alpha - s_i^\beta)\right\}\right] \tag{5.35}$$

is formed, it is seen that the problem is equivalent to an Ising model in which both a uniform field ($\mathcal{H}^\circ$) and a staggered field ($\mathcal{H}$) are applied. It seems that no general set of results have been reported for such systems.

Some work has, however, been reported on the variation of the isosteric heat of adsorption with coverage for a model of this kind (Parsonage, 1970a, b). Figure 5.22 shows results for systems in which the nearest-neighbour interaction is repulsive ($J < 0$). The most notable feature is that

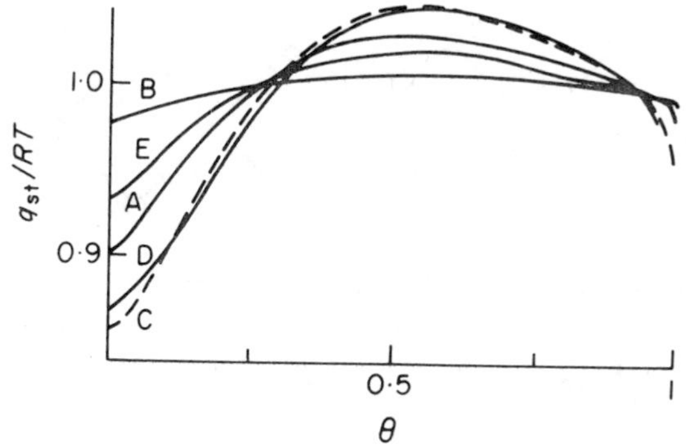

Fig. 5.22. Isosteric heat (q_{st}) against coverage (θ), expressed as a fraction of the total number of α-sites, for adsorption on a planar lattice. A, $x = 0{\cdot}1$, $y = 0{\cdot}01$; B, $x = 0{\cdot}01$, $y = 0{\cdot}01$; C, $x = 0{\cdot}3$, $y = 0{\cdot}01$; D, $x = 0{\cdot}5$, $y = 0{\cdot}01$; E, $x = 0{\cdot}05$, $y = 0{\cdot}01$; $x = \exp(-4\mathcal{H}/kT)$; $y = \exp(4J/kT)$. (After Parsonage, 1970a.)

q_{st} passes through a maximum as θ is increased. It is common in the discussion of experimental data to attribute such behaviour to the existence of attractive interactions between the adsorbed particles. The fact that the model studied here has only repulsive interactions is therefore a refutation of the argument that such behaviour *necessarily* implies attractive interactions. Indeed, bearing in mind that the distance between nearest-neighbour adsorption sites on real substrates will often be less than the sum of the radii of a pair of adsorbate particles, repulsions rather than attractions between admolecules should be fairly commonplace. It will be noted, however, that the magnitude of the maximum predicted by the model is never very large ($\lesssim 0{\cdot}3RT$) (The very pronounced maxima sometimes found experimentally are almost certainly due to attractions between the particles combined with a different model for the adsorption.) The molecular explanation of the maximum observed is as follows. At any temperature above 0 K and at low coverage there will be a proportion of the admolecules on the less good sites, this being approximately given by a Boltzmann function. As more molecules are added to the surface those on β-sites will be further handicapped by repulsive interactions with molecules on neighbouring sites. The molecules will therefore tend to revert to α-sites. q_{st} $(= -(\partial H/\partial n)_T)$, since it is a differential heat, will receive a large positive contribution from this source. Accompanying this process will be a reduction in the entropy, arising from the increased order, as the molecules are forced off the β-sites.

Some adsorbents may have a site structure which is quasi-one-dimensional and for these it is often possible to carry out the statistical mechanics

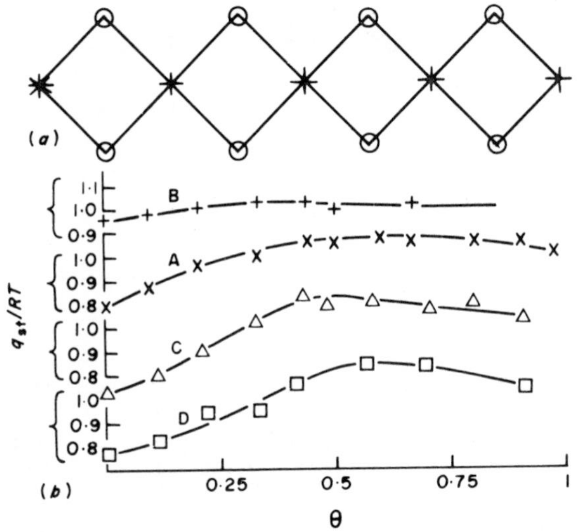

Fig. 5.23. Quasi-one-dimensional adsorption. (*a*) Site array: +, α-site; ○, β-site. (*b*) Isosteric heat (q_{st}) versus coverage (θ). Symbols as for Fig. 5.22. (After Parsonage, 1970a.)

without resort to a simulation. Thus, adsorption on the site array shown in Fig. 5.23(*a*) has been subjected to study by a transfer matrix approach (Hill, 1956). This led to the partition function being obtained as the larger root of a quadratic equation. Figure 5.23(*b*) shows results for a model in which nearest-neighbour adsorbed molecules are assumed to have a repulsive interaction. Again, a maximum in the graph of q_{st} against θ was found (Parsonage, 1970a).

Passing to the extreme case in which there are small groups of sites which are sufficiently far from each other that they behave quite independently, it is found that if each group consists of two α- and two β-sites, as in Fig. 5.24(*a*) then the maximum in q_{st} versus θ is again present although usually at a higher value of θ than for the previous examples (Fig. 5.24(*b*))

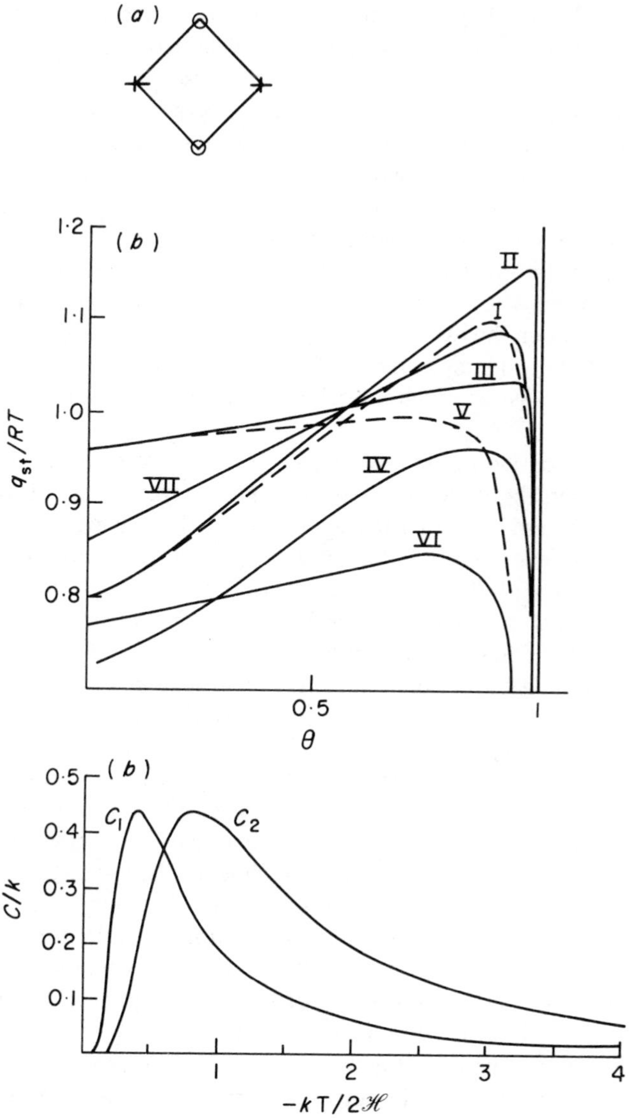

Fig. 5.24. Adsorption in separate cells. (*a*) Site array for a single cell: +, α-site; ○, β-site. (*b*) Isosteric heat (q_{st}) versus coverage (θ). I, $x = 0{\cdot}1$, $y = 0{\cdot}01$; II, $x = 0{\cdot}1$, $y = 0{\cdot}001$; III, $x = 0{\cdot}01$, $y = 0{\cdot}01$; IV, $x = 0{\cdot}3$, $y = 0{\cdot}01$; VI, $x = 0{\cdot}5$, $y = 0{\cdot}01$; VII, $x = 0{\cdot}05$, $y = 0{\cdot}01$; $x = \exp(-4\mathcal{H}/kT)$; $y = \exp(4J/kT)$. (After Parsonage, 1970b.) (*c*) Heat capacity versus temperature for one (C_1) and two (C_2) particles per cell. (After Richards *et al.*, 1976.)

(Parsonage, 1970b). It is also possible for the heat capacity at constant coverage to be less for a single adsorbed particle than for two. This occurs, for example, if the repulsive interaction for adsorption on neighbouring sites is so large that there is effective exclusion of such occupation. For a cell containing a single adsorbed particle the ground-state is doubly degenerate and has zero energy; the first excited state, which has energy $4\mathcal{H}$ is also doubly degenerate. For the two-particle system the ground-state again has zero energy but is now non-degenerate, whilst the excited state has energy $8\mathcal{H}$ and is also non-degenerate. In both cases on raising the temperature a Schottky anomaly in $C_{N(e)}$ will be found arising from the equalization of population between the two levels (Fig. 5.24(*c*)). At fairly low temperatures it is seen that $C_{N(e)}$ (1 molecule) $>$ $C_{N(e)}$ (2 molecules), even though the latter refers to twice as much adsorbed material as the former (Richards *et al.*, 1976).

So far our discussion of site-wise adsorption on surfaces has dealt with systems in which the structure of the adsorbed layer is fairly simple, and with the equilibrium between such layers and the three-dimensional gaseous adsorptive. However, it is possible for there to be several different adsorbed layer structures and their symmetry and the nature of the transitions between them has attracted considerable attention in recent years. In order to obtain a complicated field of phases it is necessary for there to be several competing interactions, rather than having just nearest-neighbour interactions and those of equal strength. We note that as compared with the models discussed on pp. 199–212 we are replacing complexity with respect to the adsorption energy of the sites by complexity with respect to the inter-molecular energy of adsorbate molecules. Once again, since we are interested in examining the system over a range of occupancies (θ), we have systems which are equivalent to Ising or Ising-related models in non-zero field. This being so, progress has largely come from MC calculations, with some valuable support from renormalization group studies. One of the attractive features of MC simulations of this kind is that, provided any ordered structures involved are fairly simple, quite small arrays, e.g. 30×30, are often adequate.

An early piece of work of this kind was that of Doyen *et al.* (1975) who included only a nearest-neighbour repulsion when studying adsorption on a square lattice. (This system is analogous to the antiferromagnet in a magnetic field studied earlier by D. P. Landau (1972).) If the number of adatoms was close to one-half of the total number of sites available then, at low temperatures, the $c(2 \times 2)$ structure (Fig. 5.25) was observed. However, at much lower coverages the adatoms became spread fairly evenly over the surface and no vestiges of the $c(2 \times 2)$ structure could be seen. If attractions between next-nearest-neighbour atoms were added,

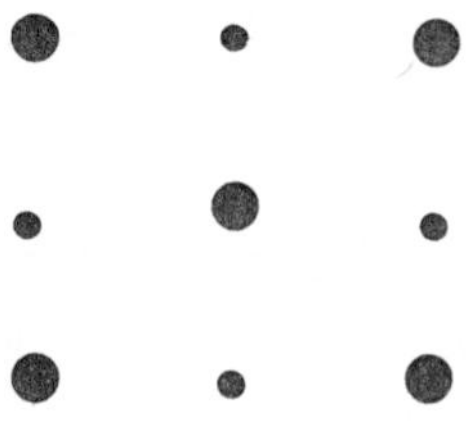

Fig. 5.25. The $c(2 \times 2)$ structure.

"islands" of the $c(2 \times 2)$ structure could still be seen at these low coverages. The latter behaviour is in line with the experimental results from structural studies on chemisorption systems (Somorjai and Szalkowski, 1971).

Much more extensive calculations have been made by Binder, Landau and co-workers. Representing the Hamiltonian of a configuration of adatoms in terms of the occupation numbers ($c_i = 0, 1$) of the sites as

$$H_N = \sum_{i>j} c_i c_j u^{(2)}(\boldsymbol{r}_i - \boldsymbol{r}_j) + \sum_i u_i^{(1)} c_i + \mathcal{H}_0 \tag{5.36}$$

where the first and second terms represent the intermolecular and the adsorbate–adsorbent interactions, respectively, it is easy to show that this is equivalent to the Ising model with Hamiltonian

$$H_N = -4 \sum_{i>j} J_{nn} s_i s_j - 4 \sum_{i>j} J_{nnn} s_i s_j - 2\mathcal{H} \sum_i s_i + \mathcal{H}_0 \tag{5.37}$$

where

$$4J(\boldsymbol{r}_i - \boldsymbol{r}_j) = -u^{(2)}(\boldsymbol{r}_i - \boldsymbol{r}_j)$$

and

$$2\mathcal{H}(\boldsymbol{r}_i) = -\left[u_i^{(1)} + \tfrac{1}{2} \sum_{j \neq i} u^{(2)}(\boldsymbol{r}_i - \boldsymbol{r}_j)\right]$$

and $s_i = \pm\frac{1}{2}$. For $J_{nn} < 0$ and $J_{nnn} = 0$, corresponding to an anti-ferromagnet in a field, the phase diagram is simple, with a (2×2) structure near $\theta = 0{\cdot}5$ and below a critical temperature. Elsewhere in the diagram there is a disordered array (lattice gas or fluid) (Fig. 5.26) (Binder and Landau, 1976). If $J_{nnn} > 0$, the (2×2) structure is stabilized by comparison with the lattice gas/fluid so that the phase region I is wider and extends to higher temperatures than was the case for $J_{nnn} = 0$. However, at lower temperatures two-phase systems occur. This set of parameters is in fact the one most likely to be appropriate for physisorbed layers. The final set of conditions was one in which both *nn* and *nnn* interactions were repulsive, this being more likely in chemisorption than in physisorption. A more

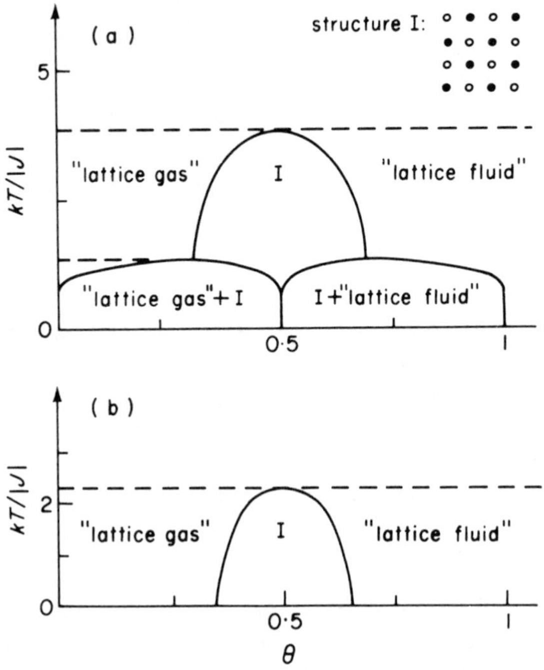

Fig. 5.26. Phase diagrams for adsorbed layers. (*a*) $J_{nn} < 0$, $J_{nnn}/J_{nn} = -0{\cdot}5$, (*b*) $J_{nn} < 0$, $J_{nnn} = 0$. (After Binder and Landau, 1976.)

complicated phase diagram was obtained showing three ordered structures as well as lattice–gas and lattice–liquid phases (Fig. 5.27(*a*)). The structure of the layer near $\theta = 0{\cdot}5$ is $c(2 \times 2)$, as found in previous examples. However, near $\theta = 0{\cdot}25$ the structure has rows which are alternately either completely empty or have every other site empty (antiferromagnetic rows). The $\theta = 0{\cdot}75$ phase is the complementary phase to that found near $\theta = 0{\cdot}25$, i.e. alternate rows are either completely filled or have every other site empty. In the absence of longer-range interactions these structures are degenerate, there being two positions for each antiferromagnetic row. (The resultant structures are not, therefore, (2×2) since the antiferromagnetic rows are not ordered with respect to each other.) Figure 5.27(*b*) show the corresponding diagram in terms of the temperature and the conjugate field. As for the quantitative reliability of the phase diagrams so obtained, it can be said that the alternatives to Monte Carlo, i.e. mean-field and renormalization group methods, also give results which are somewhat uncertain. Mean-field theories, as usual, give transition temperatures which are much too high (often about twice).

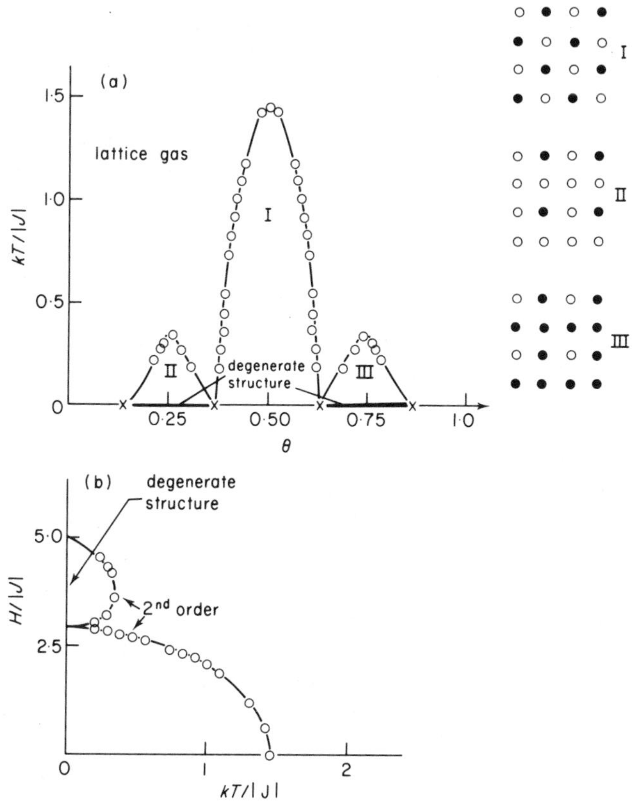

Fig. 5.27. Phase diagrams for adsorbed layers. $J_{nn} < 0$, $J_{nnn}/J_{nn} = 0{\cdot}25$. (*a*) T–θ diagram; (*b*) $\mathcal{H}$–T diagram. $\mathcal{H}$ is related to the chemical potential μ as in Table 8.1. The compositions or fields at which the structure is degenerate with respect to the antiferromagnetic layers are indicated. (After Binder and Landau, 1980.)

In chemisorption, observations of the structures of adsorbed layers are often used in conjunction with other data to determine the nature and strength of the interactions between adatoms. For example, for oxygen chemisorbed on the W(110) surface the formation of the $p(2 \times 1)$ structure demands three or more different types of interactions, with the attractive interactions being much weaker than the repulsive one. Since the LEED data are affected by the size of the "islands" of chemisorbed atoms, this size also has to be fitted to the data. It is found that the interaction along the ***a*** direction must be attractive ($-\varepsilon_a$), as must the *nnn* interaction

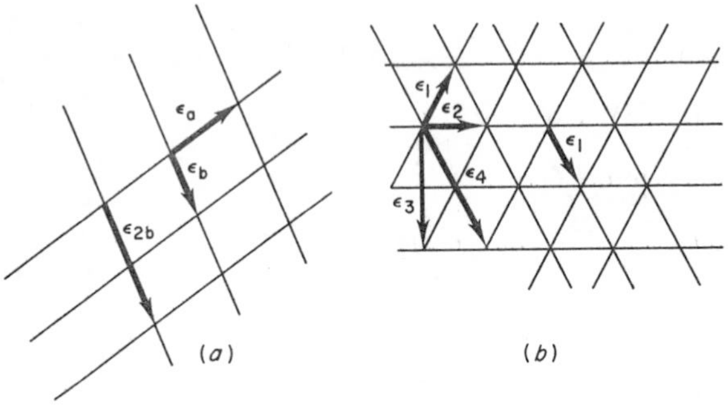

Fig. 5.28. Chemisorption on W(110). Interaction energy schemes: (*a*) Lu *et al.* (1977); (*b*) Williams *et al.* (1978).

($-\varepsilon_{2b}$) but the other *nn* interaction must be repulsive (ε_b) (Fig. 5.28(*a*)). Fitting has been done with $\varepsilon_a = \varepsilon_{2b}$, the values obtained being $\varepsilon_a = \varepsilon_{2b} = -6{\cdot}3 \pm 0{\cdot}5$ kJ mol^{-1} and $\varepsilon_b = 14 \pm 1$ kJ mol^{-1} (Lu *et al.*, 1977). However, these values are very sensitive (to the extent of a factor of 2) to the assumptions made and in particular to that concerning the equality of ε_a and ε_{2b}. Thus Williams *et al.* (1978) have subsequently fitted the transition temperature data obtained from LEED measurements to the interaction energy scheme shown in Fig. 5.28(*b*) with the values $\varepsilon_1 = -8{\cdot}8$, $\varepsilon_2 = \varepsilon_3 = +7{\cdot}1$ and $\varepsilon_4 = -2{\cdot}9$ kJ mol^{-1}. In making their fit they carried out MC simulations on a 30×30 lattice for 23 different sets of parameters. For the same system Ertl and Schillinger (1977) concluded that the inter-relationship between the parameters of Lu *et al.* (1977) should be $\varepsilon_a = \varepsilon_{2b} = -0{\cdot}9\varepsilon_b$ with all the parameters being $\sim$4 kJ mol^{-1}. Again a MC simulation with a 30×30 lattice was used for the fitting. In physisorption a better idea of the strengths of the interactions is usually obtainable from independent sources than is the case with chemisorption, so that most commonly one is trying to obtain predictions of structure which fit the observations rather than using the observations to fix the interactions. It should be remarked that the anisotropy represented by $\varepsilon_a \neq \varepsilon_b$ is attributed to the chemisorption of the oxygen: no intrinsic anisotropy of the surface is implied.

A more extensive examination of the possible phase diagram types for the anisotropic triangular lattice, to which the adsorption on W(110) corresponds, has been made by Binder *et al.* (1982) using a renormalization group method. For suitable values of the parameters (4×1) and (3×1) structures, as well as ($\sqrt{3} \times \sqrt{3}$), (2×2) and (2×1) structures, can occur. There is some evidence also that structures having modulations which are incommensurate with the underlying lattice can also occur.

A comprehensive survey of work of this kind up to 1979, including the experimental evidence from chemisorption studies, has been given by Bauer (1979).

(c) Triangular lattices: the graphite (0001) face

Stimulated by the many interesting and surprising experimental results obtained for adsorption on the basal plane of graphite (Ch. 1, §§1, 2), theoretical workers have also directed their attention towards these systems. The immediate reaction to the adsorption isotherm data for the noble gases obtained by Thomy and Duval (1969, 1970) was to suppose that the phase diagram was entirely analogous to that for a simple uniform three-dimensional system (Fig. 5.29), with a region of coexistence of adsorbed "gas" and "liquid" culminating in a critical point and also a "solid"–"liquid"–"gas" triple point. The existence of the triple point in these systems now seems less likely. First of all, Larher's work (1974) suggested that the liquid phase region between the two coexistence fields was very much smaller than was at first thought, and subsequently the experimental studies of Putnam and Fort (1975) have indicated that there is indeed no triple point at all.

Schick *et al.* (1977) used the renormalization group method in its position space form for systems of this kind. Their interest was primarily in helium as adsorbate, and for this a suitable model was one in which there was a

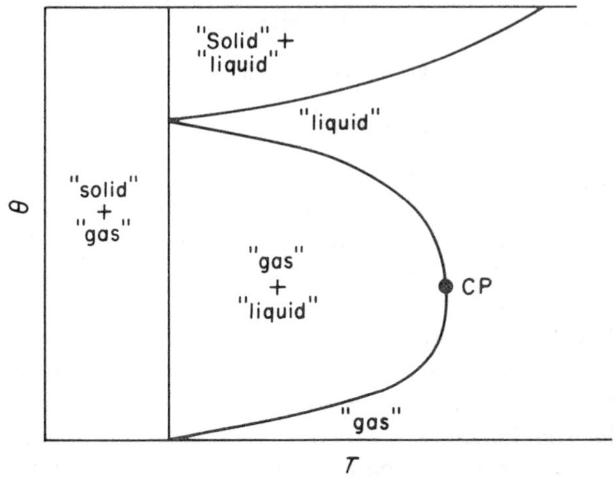

Fig. 5.29. Phase diagram for the adsorption layer proposed by Thomy and Duval (schematic).

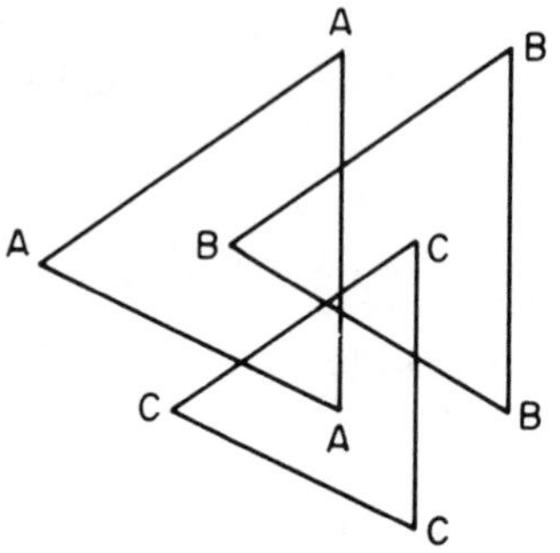

Fig. 5.30

repulsion between adjacent adsorbed atoms but in which every site could nevertheless be occupied. The system was therefore equivalent to an antiferromagnetic system in a magnetic field (Green and Hurst, 1964). The renormalization procedure involved the reduction of three interlocking triangular cells to a single triangular super-cell (Fig. 5.30). In the course of this process 3-spin interactions between mutually adjacent neighbours (*mnn*) were introduced, the Hamiltonian being

$$-\beta H_N = K \sum_{\substack{i>j \\ nn}} c_i c_j + \mathcal{H} \sum_i c_i + P \sum_{\substack{i>j>k \\ mnn}} c_i c_j c_k + NC \qquad (5.38)$$

with the spin-independent constant C given by

$$C = \tfrac{1}{2}(\beta\mu - \tfrac{3}{2}\beta u^{(2)} - \ln(\Lambda^2/a_c))$$

$$\Lambda = (2\pi\hbar^2\beta/m)^{1/2}$$

where $u^{(2)}$ = the interaction energy of a pair of adjacent adsorbed atoms; a_c = the area of the primitive cell of the substrate; $c_i = 0, 1$, and $K = J/kT$. By choosing the antiferromagnetic interaction parameter (K) to fit the data good results for the "gas"–"solid" transition curve and the heat capacity versus temperature curves (Fig. 5.31) were obtained.

A somewhat similar approach has been adopted by Berker *et al.* (1978). For krypton, which has been the most thoroughly studied adsorbate experimentally, the size of the atom is such that the occupation of one site precludes the simultaneous occupation of nearest-neighbour sites. The array of sites may, therefore, be broken down into three sub-lattices (A, B, C of Fig. 5.32(*a*)), such that the maximum coverage which can be achieved is that with all the sites of one kind, e.g. A, occupied, but none of the other sites. This suggests that a Potts lattice gas model would be appropriate. The sites were grouped into clusters of three, these being mutually adjacent, and two coordinates were assigned to describe the state of each cluster: t_i (=0 or 1) being the occupation number and s_i (=A, B or C) indicating the site, if any, which is occupied. This gives only an approximate representation of the real exclusion effect, since it allows simultaneous occupation of neighbouring sites if they are members of different clusters, e.g. *iB* and *jA* of Fig. 5.32(*a*). The resultant Hamiltonian is

$$-\beta H_N = -\beta \sum_{ij} V_{ij} t_i t_j - \Delta \sum_i t_i \tag{5.39}$$

where

$$-\beta V_{ij} = J(3\delta_{s_i s_j} - 1) + K. \tag{5.40}$$

Inspection of (5.39) shows that the interactions in the model Hamiltonian have both 3-state Potts and lattice gas character through the first and second terms, respectively. Δ is equivalent to the chemical potential, controlling the overall coverage. A type of Migdal "bond-moving" process (Ch. 2, §7(g)) was used to obtain a recursion formula, which was then applied to the system. They obtained the phase diagrams shown in Fig. 5.32(*b*). The absence of a triple point or a liquid–gas coexistence region is to be noted. Point B is a fourth-order point at which the three solid phases (symmetric with respect to ABC interchange) and the fluid phase become indistinguishable. The diagram shown is for the particular values of the interaction constants appropriate to krypton as adsorbate. The method also yields a "global" phase diagram, covering all ratios K/J (Fig. 5.32(*c*)). The $1/J$ and $-\Delta/J$ axes are equivalent to temperature and chemical potential axes for a given adsorbate; the K/J section which is relevant is determined by the interaction constants for the adsorbate + adsorbent system concerned. For $K/J = 0$, $P_3^{(0)}B^{(0)}$ is a line of Potts tricritical points (three equivalent solid phases being in equilibrium with a liquid phase) and giving continuous (second-order) transitions. Below the fourth-order point $B^{(0)}$, at which the three solid phases and the liquid become indistinguishable, the transition becomes first-order. At $K/J = \infty$ the behaviour is that found in bulk systems: a first-order liquid–gas transition terminating in an Ising critical point and a first-order (Potts tricritical) solid–liquid transition which is insensitive to change of chemical potential. Between these K/J sections (at $K/J = 4$) lies the 4-state Potts transition arising from the merging of the Ising gas–liquid critical line ($C^{(0)}P_4^{(4)}$) with the line of fourth-order transitions $P_4^{(4)}B^{(0)}$. The fourth-order Point B for the krypton system (Fig. 5.32(*b*)) is the intersection of this line $P_4^{(4)}B^{(0)}$ with the plane $K/J = 2{\cdot}10$, which corresponds to krypton.

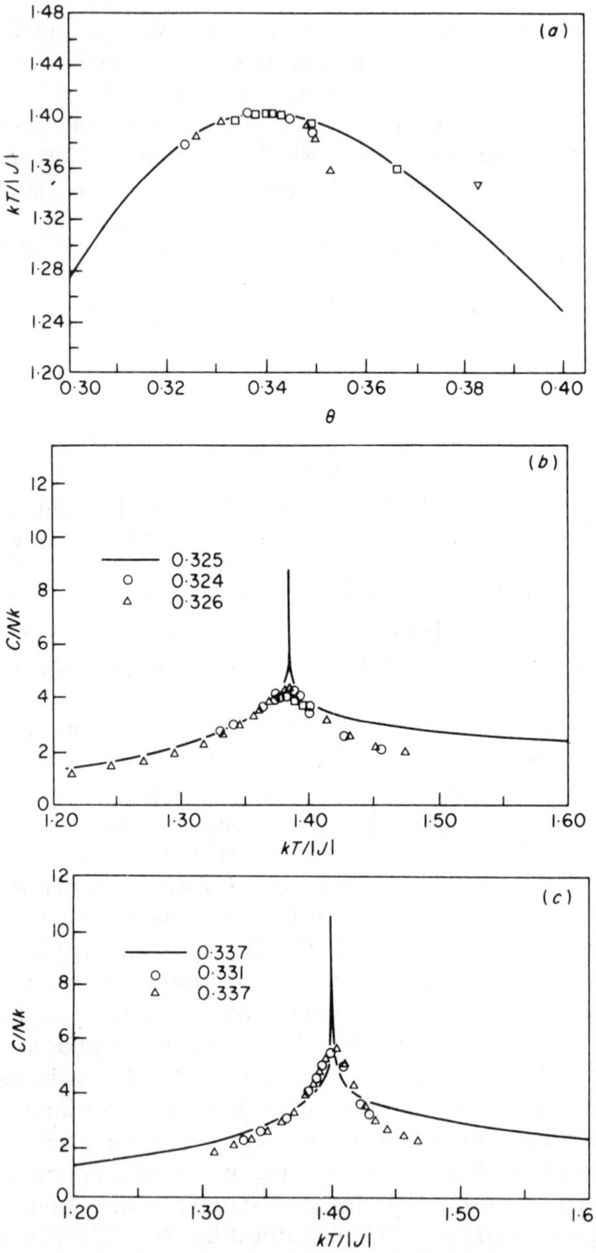

Fig. 5.31. Comparison of experimental and theoretical results for adsorption of He^4 on graphite. (*a*) "Liquid"–"gas" coexistence curve: solid line, theory; symbols, experiment. (*b*) to (*e*) Heat capacity versus temperature at four coverages:

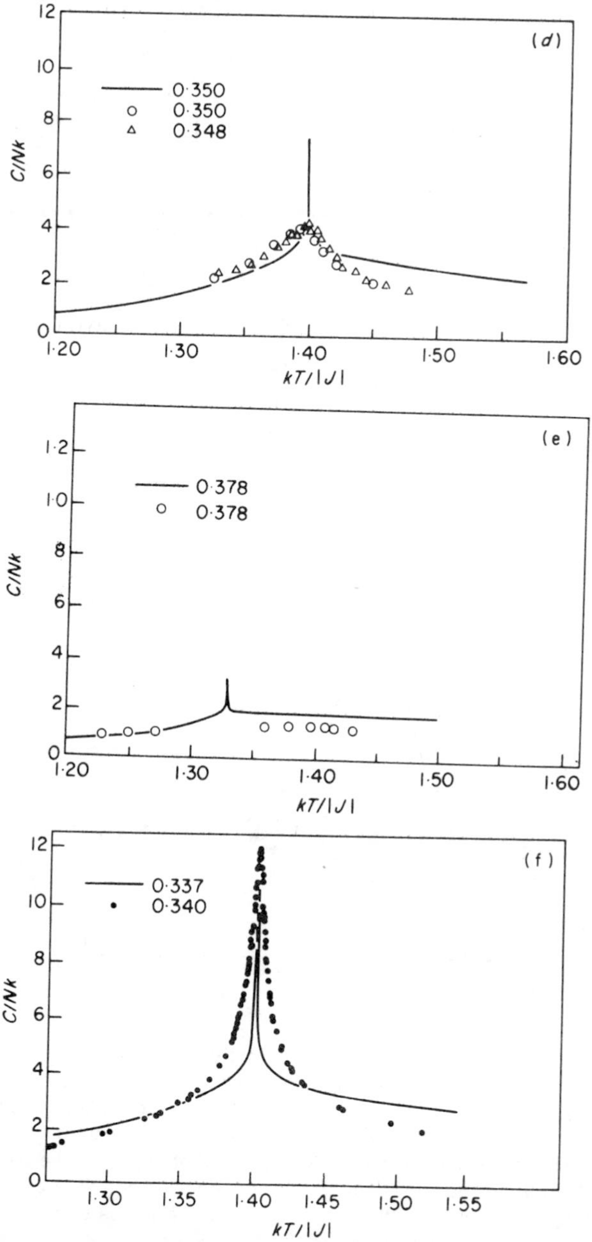

solid line, theory; symbols, experiment. (*f*) as (*b*) to (*e*) but experiments are for an improved substrate. (After Schick *et al.*, 1977.)

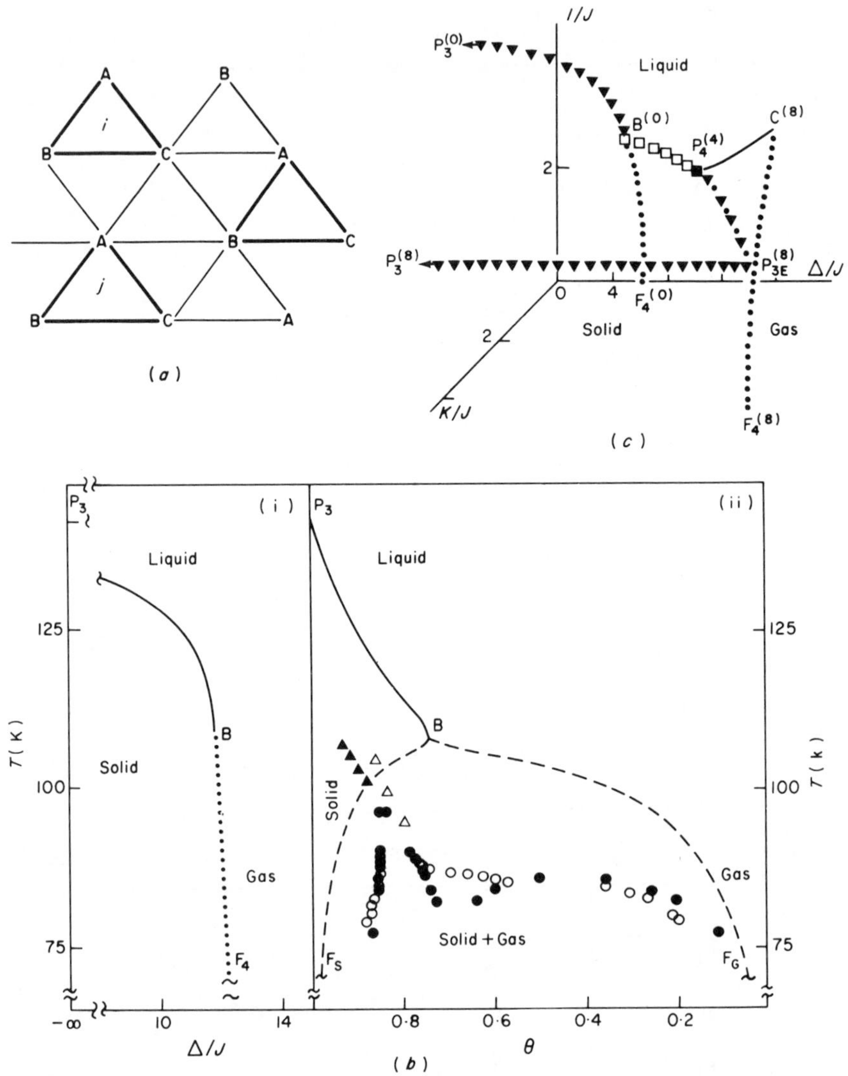

Fig. 5.32. Adsorption of krypton on graphite. (*a*) Sub-lattice and renormalization group elementary triangles. (*b*) Phase diagrams: (i) T versus Δ/J ($\propto$ chemical potential); (ii) T versus θ ($\theta = 1$) corresponds to complete filling of one-sub-lattice). RG results (– – –, first-order transition; ——, continuous transition). Experiment: ●, Thomy and Duval (1969, 1970) first-order; ○, Larher (1974) first-order; △, Putnam and Fort (1975, 1977) and Putnam *et al.* (1977), continuous; ▲, Thomy and Duval (1969, 1970) continuous. (*c*) Global phase diagram. Section shown is for $K/J = 0{\cdot}8$. Superscripts give other K/J values. ·····, first-order; ———, Ising critical; ▼, Potts tricritical; □, fourth-order; ▼·▼·, tricritical end-point. (After Berker *et al.*, 1978.)

Although it might appear that the global phase diagram described above could be used for the discussion of other adsorbates, it turns out that this is not necessarily so. Thus, Ostlund and Berker (1980), when wishing to consider the adsorption of xenon and methane on graphite, felt it necessary to adopt a different model and renormalization procedure in order to take account of the different relationship between interaction parameters which prevailed in those systems. For xenon, in which the most favourable interaction was that between third-nearest neighbours (as against second-nearest neighbours or *nnn* for krypton), Ostlund and Berker considered a cluster of 12 sites so arranged that both $\sqrt{3} \times \sqrt{3}$ and 2×2 symmetries could be preserved throughout the renormalization procedure. The state of each cluster (or supersite) was then described by a set of four numbers $(t, \tilde{t}, s, \tilde{s})$: t and $\tilde{t}$ are occupation variables (0 or 1) which indicate whether the supersite is vacant $(t = \tilde{t} = 0)$, predominantly in the $\sqrt{3} \times \sqrt{3}$ form $(t = 1, \tilde{t} = 0)$, or predominantly in the 2×2 form $(t = 0, \tilde{t} = 1)$; s and $\tilde{s}$ indicate the main occupied sub lattice for whichever form is predominant—s = A, B or C for $\sqrt{3} \times \sqrt{3}$, $\tilde{s}$ = A, B, C or D for 2×2. Because of these degeneracies, disordering of the $\sqrt{3} \times \sqrt{3}$ and 2×2 structures can be considered as being similar to the 3- and 4-state Potts transitions, respectively. The Hamiltonian describing the interaction between supersites had no fewer than eight interaction parameters, $H_N = -4 \sum_n J_n \sum_{i>j} c_i c_j$, where J_n refers to the interaction between nth-nearest neighbours and the summation is first over all nth-nearest-neighbours and then over all values of n. However, when the actual values calculated for these from the known Lennard-Jones equation are used, none of the phase diagrams is very complicated, though they do have a rich succession of forms (Fig. 5.33). Figure 5.33 is based on the use of all Lennard-Jones interactions out as far as the 20th-nearest-neighbours. A similar set of diagrams was derived by imposing nearest-neighbour exclusion ($J_1 = -\infty$), choosing J_2 and J_3 as for Fig. 5.33, setting $J_4 = 0{\cdot}3J_2$, and neglecting more distant interactions. For comparison, attempts were made to perform a Monte Carlo simulation on the system represented by the latter scheme, but these met with severe problems from hysteresis. The diagrams are very sensitive to the values chosen for the interaction constants and, since these values are somewhat uncertain, too much emphasis should not be placed on the quantitative comparison with real systems. Rather, interest is in the variety of features which can be found. Figure 5.33(*d*) displays a strangely shaped coexistence region, which can probably be best appreciated as being due to an incipient triple point, this being fully realized when σ/a reaches 1·646. Figures 5.33(*e*) and (*f*) show a simple triple point, of the kind originally proposed by Thomy and Duval to explain their experimental

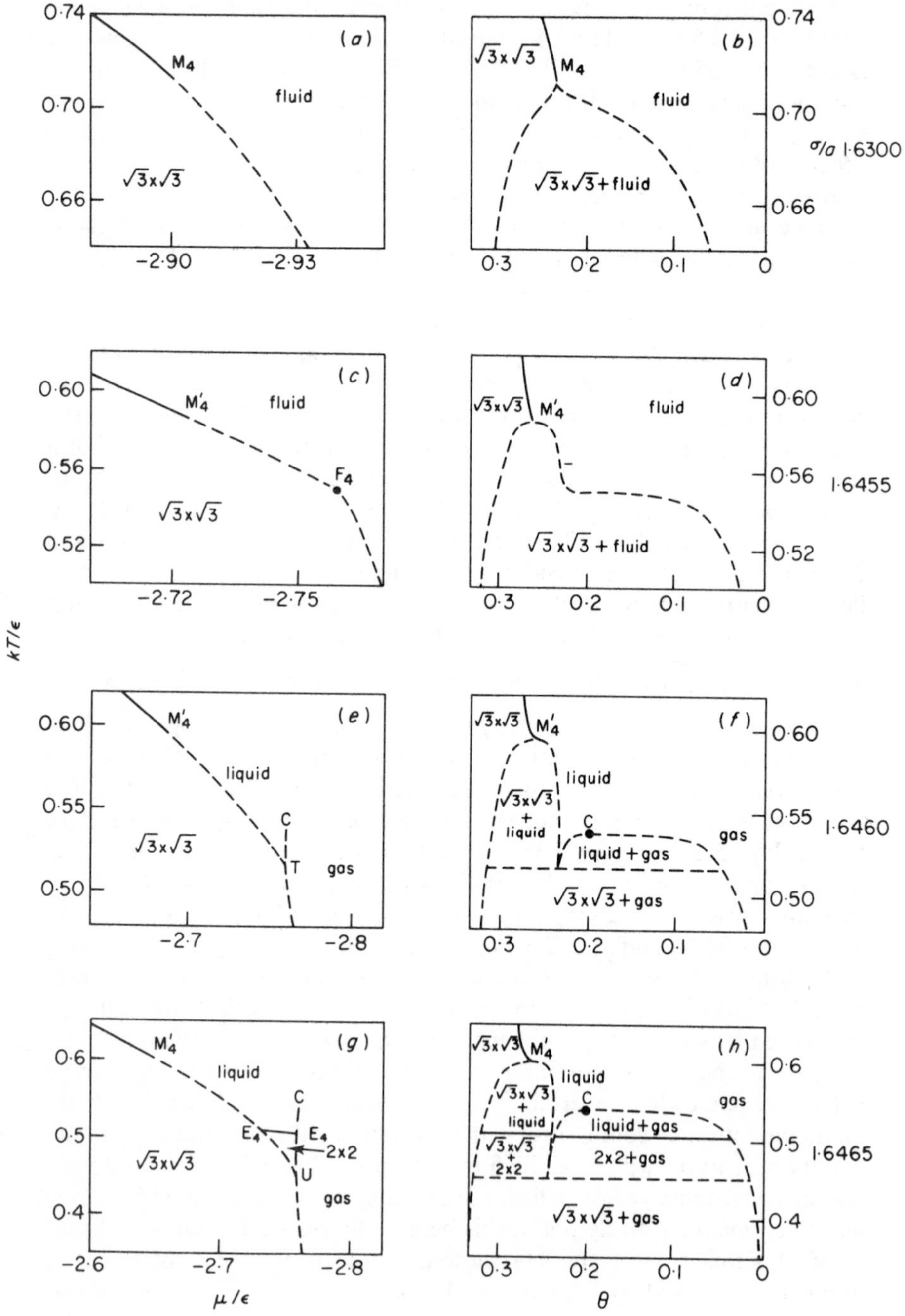

(a)
M4
fluid
√3x√3
0·74
0·70
0·66
−2·90
−2·93
(b)
√3x√3
M4
fluid
√3x√3+fluid
0·74
0·70
0·66
σ/a 1·6300
0·3
0·2
0·1
0
(c)
M′4
fluid
F4
√3x√3
0·60
0·56
0·52
−2·72
−2·75
(d)
√3x√3
M′4
fluid
√3x√3 + fluid
0·60
0·56
0·52
1·6455
kT/ε
(e)
M′4
liquid
C
T
√3x√3
gas
0·60
0·55
0·50
−2·7
−2·8
(f)
√3x√3
M′4
liquid
√3x√3 + liquid
C
gas
liquid + gas
√3x√3 + gas
0·60
0·55
0·50
1·6460
(g)
M′4
liquid
C
E4
E4
2x2
U
√3x√3
gas
0·6
0·5
0·4
−2·6
−2·7
−2·8
μ/ε
(h)
√3x√3
M′4
liquid
√3x√3 + liquid
C
gas
liquid+gas
√3x√3 + 2x2
2x2+gas
√3x√3+gas
0·6
0·5
0·4
1·6465
θ

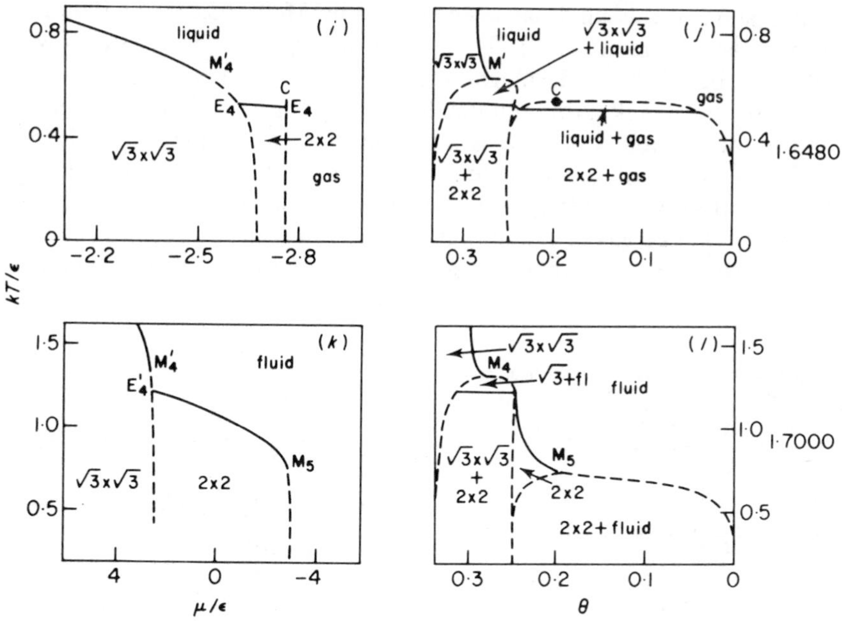

Fig. 5.33. Adsorption of Lennard-Jones gases on graphite: a = lattice constant of graphite; ε, σ are the Lennard-Jones parameters; $\theta = 1$ corresponds to all of the sites being occupied. (Ostlund and Berker, 1980.)

results. Figures 5.33(g) and (h) are interesting in that the 2×2 structure is present at intermediate temperatures only, being replaced by the $\sqrt{3} \times \sqrt{3}$ solid at lower temperatures and by fluid phases at higher temperatures. In the real systems which the calculation was originally intended to treat (physisorption of noble gases) there is an extra possibility which is not considered here: incommensurate adsorption. Indeed, it appears that for both xenon and methane the 2×2 solid, predicted by this work, is pre-empted by the formation of an incommensurate layer. Apart from the appearance of the incommensurate rather than the 2×2 phase, the experimentally determined phase diagram for methane is similar to the low-θ part of Fig. 5.33(h), which suggests that adsorbed methane has a larger effective radius (~0·405 nm) than is normally assigned to molecules in the gas (0·381 nm). Chemisorption, for which incommensurate layers are not possible, may provide examples which are closer still to these theoretically predicted diagrams. Thus, a LEED study of O on Ni(111) yielded a diagram which is very similar to Fig. 5.33(h) (Kortan *et al.*, 1979; Roelofs *et al.*, 1980).

A system of this kind has been studied by Monte Carlo simulation by Mihura and Landau (1977). The *nn* and *nnn* interaction constants were taken to be of the same magnitude and to be repulsive and attractive, respectively. Unlike the renormalization group calculations described above, the possibility of coverages greater than $\frac{1}{3}$ was allowed, and indeed the study covered the complete range of θ. Two ordered structures ($\theta = \frac{1}{3}$ and $\frac{2}{3}$) were found, there being continuous transitions to the disordered ("lattice gas or liquid") at higher temperatures for these compositions (Fig. 5.34). At low temperatures there is almost complete

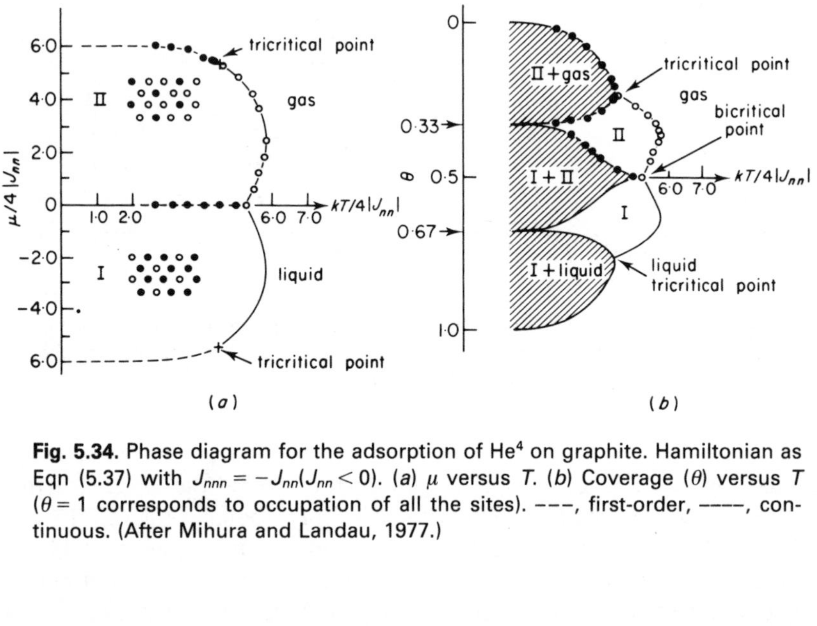

Fig. 5.34. Phase diagram for the adsorption of He[4] on graphite. Hamiltonian as Eqn (5.37) with $J_{nnn} = -J_{nn}(J_{nn} < 0)$. (*a*) μ versus T. (*b*) Coverage (θ) versus T ($\theta = 1$ corresponds to occupation of all the sites). ---, first-order, ——, continuous. (After Mihura and Landau, 1977.)

"immiscibility" of these two phases and the completely empty and completely full phases. As the temperature is raised the "miscibility" increases and there are symmetric tricritical points at which the first-order I (or II) → liquid (or gas) transition goes over to a continuous transition. The two continuous transitions (I → liquid and II → gas) meet the first-order transition (I → II) at a bicritical point, which is analogous to the "point" found for some antiferromagnets in a magnetic field (Ch. 2, §7(*d*)). No point of this kind was found in the studies on the square lattice. An oddity of these observations was that peak heat capacity values for $\theta = 0{\cdot}5$ were much smaller than for $\theta \neq 0{\cdot}5$. In comparing these results with those of Ostlund and Berker, it is important to note that the latter do not include any information for $\theta > \frac{1}{3}$, so that the (I + II) region and the bicritical point of Fig. 5.34 could not, of course, occur. In passing, it could be added that, in the region $\theta > \frac{1}{3}$, incommensurate phases supplant the commensurate in real systems, so that part of Fig. 5.34 is not of great practical importance. Ignoring the range $\theta > \frac{1}{3}$, Fig. 5.34 corresponds to curve (*b*) of Fig. 5.33. This is consistent with the fact that the condition $J_{nn} = -J_{nnn}$ corresponds to adsorbate molecules of size $\sigma/a = 1{\cdot}006$ interacting with the Lennard-Jones 12–6 potential, more distant neighbours than *nnn* being neglected.

All the results discussed so far in this section have been concerned with spherical or quasi-spherical adsorbate molecules. However, Southern and Lavis (1979, 1980) have successfully applied the renormalization group method to a system consisting of triangular molecules on the triangular lattice. For each molecule just two distinguishable orientations were permitted (Fig. 5.35(*a*)). If the intersite vectors of the adsorbent are numbered 1 to 6, the two directions for the adsorbate molecule are those with the vertices of the adsorbate molecule pointing in the 1, 3, 5 or 2, 4, 6 directions. The state of any site is written as $S = \pm 1, 0$, the latter indicating that it is vacant. The interaction energies for the four possible *nn* pairings were as in Fig. 5.35(*b*). The Hamiltonian for the complete array could then be written as the sum of contributions from each elementary triangle of the substrate lattice, S_A, S_B, S_C being the "spin coordinates" for the A, B and C sub-lattice members of that triangle:

$$H_N = \sum \left\{ \tfrac{1}{6}\mu(S_A^2 + S_B^2 + S_C^2) - \tfrac{1}{2}\left(\eta - \frac{w}{4}\right)(S_A S_B + S_B S_C + S_C S_A) - \tfrac{1}{2}\left(\varepsilon + \frac{w}{4}\right)\left(S_A^2 S_B^2 + S_B^2 S_C^2 + S_C^2 S_A^2\right) + \frac{w}{8}(S_A - S_B)(S_B - S_C)(S_C - S_A) \right\} \tag{5.41}$$

where the summation is to be taken over all elementary triangles. The renormalization procedure was of the block spin type used by Schick *et al.* (1977). Phase diagrams for a range of values of w, ε and η were obtained, some of these being shown in Fig. 5.35(*c*). Solids S and $\tilde{S}$ differ in having relative orientations of nearest neighbour pairs parallel and anti-parallel, respectively. Ideal S and $\tilde{S}$ also differ in that the former is close-packed whereas the latter has the honeycomb structure, only two-thirds of the sites being used. For both solids and also for all values of the parameters which were examined, the melting transition was second-order. This contrasts with the findings of Berker *et al.* (1978), which were that both first- and "second"-order melting was common.

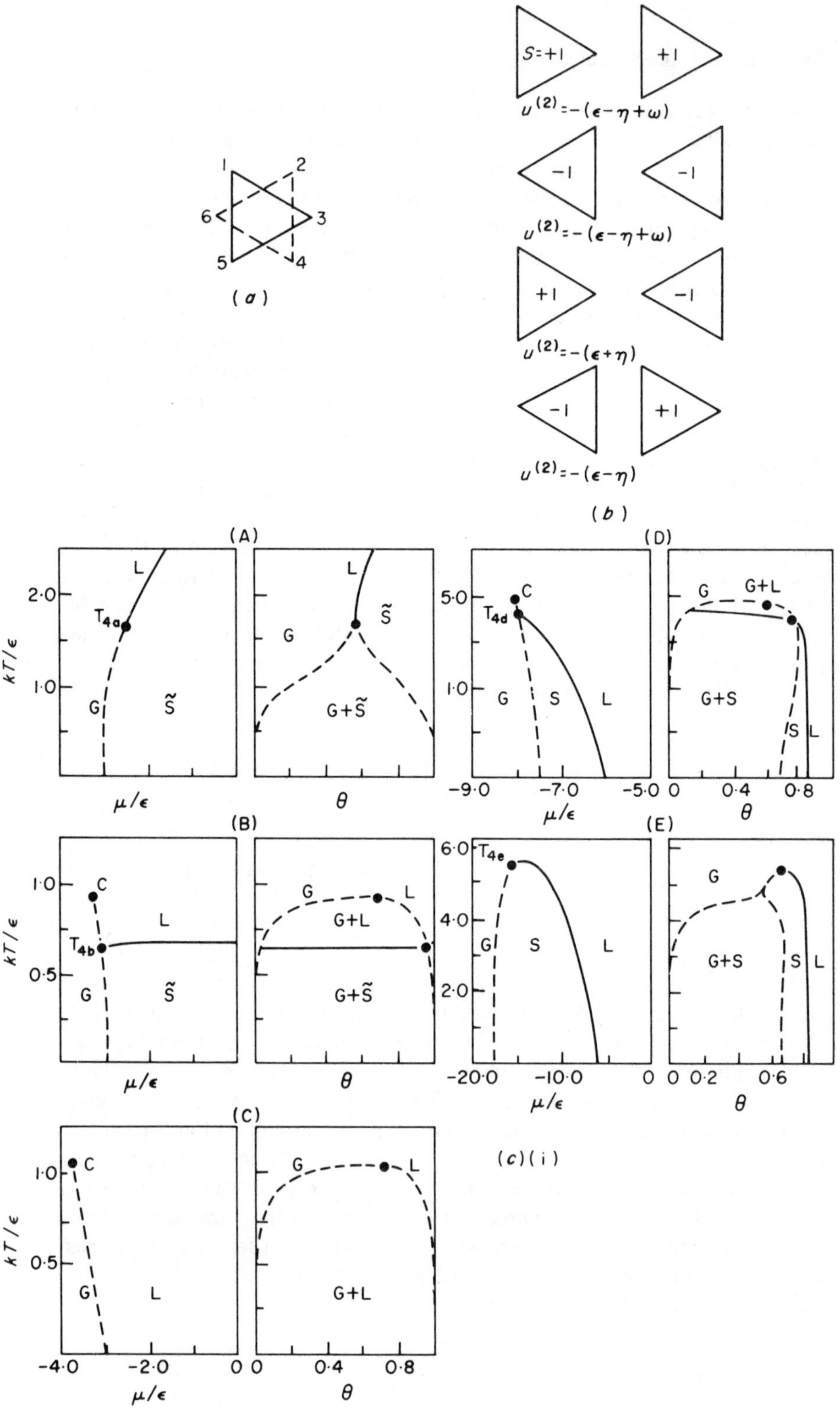

1
2
3
4
5
6
(a)
S=+1
+1
u(2)=-(ε-η+ω)
-1
-1
u(2)=-(ε-η+ω)
+1
-1
u(2)=-(ε+η)
-1
+1
u(2)=-(ε-η)
(b)
(A)
(B)
(C)
(D)
(E)
kT/ε
μ/ε
θ
L
G
S̃
G+S̃
T4a
C
T4b
G+L
T4d
S
G+S
T4e
2·0
1·0
0·5
5·0
6·0
4·0
-9·0
-7·0
-5·0
-20·0
-10·0
-4·0
-2·0
0
0·2
0·4
0·6
0·8
(c)(i)

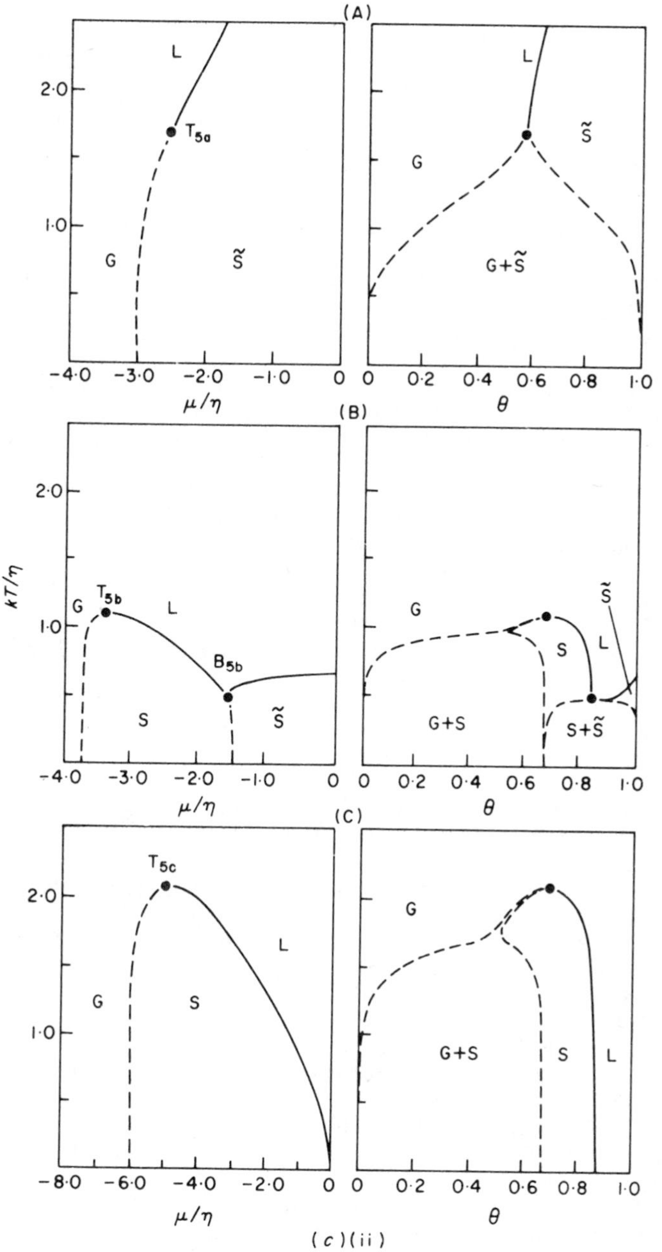

Fig. 5.35. Adsorption of triangular molecules on a triangular lattice. (*a*) Possible orientations of admolecules. ——, $S = +1$; ----, $S = -1$. (*b*) Interaction energy scheme. (*c*) Phase diagrams: temperature versus chemical potential (μ) and temperature versus coverage (θ).

(i) $\eta = 0$: (A) $\omega/\varepsilon = -5$; (B) $\omega/\varepsilon = -0{\cdot}5$; (C) $\omega/\varepsilon = 0$; (D) $\omega/\varepsilon = 4$; (E) $\omega/\varepsilon = 10$.

(ii) $\varepsilon = 0$: (A) $\omega/\eta = 0$; (B) $\omega/\eta = 3{\cdot}5$; (C) $\omega/\eta = 5$.

G = gas. L = liquid. The close-packed and open honeycomb solid phases are denoted by $\tilde{S}$ and S, respectively. ----, first-order; ——, continuous. (After Southern and Lavis, 1979.)

References

Abraham, F. F. (1980). *Phys. Rev. Lett.* **44**, 463.
Abraham, F. F. (1981). *Phys. Rep.* (to be published).
Alder, B. J. and Wainwright, T. E. (1962). *Phys. Rev.* **127**, 359.
Bak, P. and Mukamel, D. (1979). *Phys. Rev. B* **19**, 1604.
Bak, P., Mukamel, D., Villain, J. and Wentowska, K. (1979). *Phys. Rev. B* **19**, 1610.
Barker, J. A., Henderson, D. and Abraham, F. F. (1981). *Physica* **106A**, 226.
Bauer, E. (1979). Nato Advanced Study Institute for Phase Transitions in Surface Films (Dash, J. G. and Ruvalds, J., eds) p. 267. Plenum Press, New York and London.
Berker, A. N., Ostlund, S. and Putnam, F. A. (1978). *Phys. Rev. B* **17**, 3650.
Binder, K. and Landau, D. P. (1976). *Surface Sci.* **61**, 577.
Binder, K. and Landau, D. P. (1980). *Phys. Rev. B* **21**, 1941.
Binder, K., Kinzel, W. and Landau, D. P. (1982). *Surface Sci.* **117**, 232.
Chinn, M. D. and Fain, S. C. (1977). *Phys. Rev. Lett.* **39**, 146.
Chung, T. T. (1979). *Surface Sci.* **87**, 348.
Domany, E., Schick, M., Walker, J. S. and Griffiths, R. B. (1978). *Phys. Rev. B* **18**, 2209.
Doyen, G., Ertl, G. and Plancher, M. (1975). *J. Chem. Phys.* **62**, 2957.
Ertl, G. and Schillinger, D. (1977). *J. Chem. Phys.* **66**, 2569.
Fain, S. C. and Chinn, M. D. (1977). *J. de Phys.* **38**, C4–99.
Fisher, M. E. (1967). *Rep. Prog. Phys.* **30**, 615.
Fosdick, L. D. (1963). In "Methods in Computational Physics" (Alder, B. J., Fernbach, S. and Rotenberg, M., eds) Vol. 1, p. 245. Academic Press, New York and London.
Frank, F. C. and Van der Merwe, J. H. (1949). *Proc. Roy. Soc. A* **198**, 205, 216.
Frenkel, D. and McTague, J. P. (1979). *Phys. Rev. Lett.* **42**, 1632.
Fuselier, C. R., Raich, J. C. and Gillis, N. S. (1980). *Surface Sci.* **92**, 667.
Green, H. S. and Hurst, C. A. (1964). "Order–Disorder Phenomena," Ch. 2. Interscience, London, New York and Sydney.
Halperin, B. I. (1979). *Phys. Rev. B* **19**, 2457.
Halperin, B. I. and Nelson, D. R. (1978). *Phys. Rev. lett.* **41**, 121.
Hanson, F. E., Mandell, M. J. and McTague, J. P. (1977). *J. de Phys.* **38**, C4–76.
Henderson, D. (1977). *Mol Phys.* **34**, 301.
Hertz, J. A. (1976). *Phys. Rev. B* **14**, 1165.
Hill, T. L. (1956). "Statistical Mechanics," p. 309. McGraw-Hill, New York, Toronto and London.
Imry, Y. (1979). *CRC Crit. Revs. Solid St. Materials Sci.* **8**, 157.
Kawabata, C. and Binder, K. (1977). *Solid St. Commun.* **22**, 705.
Kortan, A. R., Cohen, P. I. and Park, R. L. (1979). *J. Vac. Sci. Technol.* **16**, 541.
Kosterlitz, J. M. (1974). *J. Phys. C: Solid State Physics* **7**, 1046.
Kosterlitz, J. M. and Thouless, D. J. (1973). *J. Phys. C: Solid State Physics* **6**, 1181.
Kosterlitz, J. M. and Thouless, D. J. (1978). *Prog. Low Temperature Phys.* **7**B, 371.
Landau, D. P. (1972). *Phys. Rev. Lett.* **28**, 449.

Landau, L. D. and Lifshitz, E. M. (1969). "Statistical Physics," 3rd edn, p. 433. Pergamon Press, Oxford, New York, Toronto, Sydney and Braunschweig.
Larher, Y. (1974). *J. Chem. Soc. Faraday Trans. I* **70**, 320.
Lieb, E. H. and Wu, F. Y. (1972). In "Phase Transitions and Critical Phenomena" (C. Domb and M. S. Green, eds.), Vol 1, p. 393. Academic Press, London and New York.
Lu, T-M., Wang, G-C and Lagally, M. G. (1977). *Phys. Rev. Lett.* **39**, 411.
Luther, A. and Scalapino, D. J. (1977). *Phys. Rev. B* **16**, 1153.
Mermin, N. D. (1968). *Phys. Rev.* **176**, 250.
Mermin, N. D. and Wagner, H. (1966). *Phys. Rev. Lett.* **17**, 1133.
Metropolis, N. A., Rosenbluth, A. W., Rosenbluth, M. N., Teller, A. H. and Teller, E. (1953). *J. Chem. Phys.* **21**, 1087.
Mihura, B. and Landau, D. P. (1977). *Phys. Rev. Lett.* **38**, 977.
Miyashita, S. (1980). *Prog. Theor. Phys.* **63**, 797.
Miyashita, S., Nishimori, H., Kuroda, A. and Suzuki, M. (1978). *Prog. Theor. Phys.* **60**, 1669.
Nabarro, F. R. N. (1967). "Theory of Crystal Dislocations." Clarendon Press, Oxford.
Novaco, A. D. (1979). *Phys. Rev. B* **19**, 6493.
Novaco, A. D. and McTague, J. P. (1977a) *Phys. Rev. Lett.* **38**, 1286.
Novaco, A. D. and McTague, J. P. (1977b) *J. de Phys.* **38**, C4–116.
Ostlund, S. and Berker, A. N. (1980). *Phys. Rev. B* **21**, 5410.
Parsonage, N. G. (1970a). *J. Chem. Soc., Ser. A*, 2859.
Parsonage, N. G. (1970b). *Trans. Faraday Soc.* **66**, 723.
Peierls, R. E. (1934). *Helv. Phys. Acta, Suppl. II*, 81.
Peierls, R. E. (1935). *Ann. Inst. Henri Poincaré* **5**, 177.
Prasad, S. D. and Toxvaerd, S. (1980). *J. Chem Phys.* **72**, 1689.
Putnam, F. A. and Fort, T. (1975). *J. Phys. Chem.* **79**, 459.
Putnam, F. A., Fort, T. and Griffiths, R. B. (1977). *J. Phys. Chem.* **81**, 2171.
Richards, E., Stroud, H. J. F. and Parsonage, N. G. (1976). *J. Chem. Soc. Faraday Trans. I.* **72**, 1759.
Roelofs, L. D., Einstein, T. L., Hunter, P. E., Kortan, A. R., Park, R. L. and Roberts, R. M. (1980). *J. Vac. Sci. Technol.* **17**, 231.
Schick, M., Walker, J. S. and Wortis, M. (1977). *Phys. Rev. B* **16**, 2205.
Shaw, G. C., Fain, S. C. and Chinn, M. D. (1978). *Phys. Rev. Lett.* **41**, 955.
Shiba, H. (1980). *J. Phys. Soc. Japan* **48**, 211.
Somorjai, G. A. and Szalkowski, F. Z. (1971). *J. Chem. Phys.* **54**, 389.
Southern, B. W. and Lavis, D. A. (1979). *J. Phys. C: Solid State Phys.* **12**, 5333.
Southern, B. W. and Lavis, D. A. (1980). *J. Phys. A: Mathematical and General* **13**, 251.
Stanley, H. E. (1968). *Phys. Rev. Lett.* **20**, 150.
Stanley, H. E. and Kaplan, T. A. (1966). *Phys. Rev. Lett.* **17**, 913.
Stephens, A. W., Heiney, P., Birgeneau, R. J. and Horn, P. M. (1979). *Phys. Rev. Lett.* **43**, 47.
Suzuki, M., Miyashita, S., Kuroda, A. and Kawabata, C. (1977). *Phys. Lett.* **60A**, 478.
Ter Haar, D. (1977). "Lectures on Selected Topics in Statistical Mechanics, " §8.2. Pergamon Press, Oxford, New York, Toronto, Sydney, Paris and Frankfurt.
Thomy, A. and Duval, X. (1969). *J. Chim. Phys.* **66**, 1966.
Thomy, A. and Duval, X. (1970). *J. Chim. Phys.* **67**, 1101.

Toxvaerd, S. (1975). *Mol. Phys.* **29**, 373.
Toxvaerd, S. (1980). *Phys. Rev. Lett.* **44**, 1002.
Tsien, F. and Valleau, J. P. (1974). *Mol. Phys.* **27**, 177.
Van der Merwe, J. H. (1970). *J. Appl. Phys.* **41**, 4725.
Van Swol, F., Woodcock, L. V. and Cape, J. N. (1980). *J. Chem. Phys.* **73**, 913.
Venables, J. A. and Schabes-Retchkiman, P. S. (1977). *J. de Phys.* **38**, C4–105.
Venables, J. A. and Schabes-Retchkiman, P. S. (1978). *Surface Sci.* **71**, 27.
Vora, P., Sinha, S. K. and Crawford, R. K. (1979). *Phys. Rev. Lett.* **43**, 704.
Williams, E. D., Cunningham, S. L. and Weinberg, W. H. (1978). *J. Chem. Phys.* **68**, 4688.
Wood, W. W. (1968). In "Physics of Simple Liquids." (Temperley, H. N. V., Rowlinson, J. S. and Rushbrooke, G. S., eds.), Ch. 5. North-Holland, Amsterdam.
Yamaji, K. and Kondo, J. (1973). *J. Phys. Soc. Japan* **35**, 25.
Ying, S. C. (1971). *Phys. Rev. B* **3**, 4160.
Zippelius, A., Halperin, B. I. and Nelson, D. R. (1980). *Phys. Rev. B* **22**, 2514.

Chapter 6

Three-dimensional Interfacial Systems

1. Introduction

The previous chapter dealt with adsorbed layers considered as two-dimensional systems and, as was seen, a considerable number of both exact and approximate results have been obtained, without, however, resolving all the problems which arise. In spite of this incompleteness of knowledge in the theory of truly two-dimensional systems, some progress has already been made on the next stage–the consideration of adsorbed phases as three-dimensional systems. Because of the extra difficulties involved virtually all the work has been based on computer simulation. The exceptions to this observation are mostly treatments using virial expansions (Pierotti and Thomas, 1971, 1974; Steele, 1974; Sams, 1974; Sokolowski, 1980). These expansions are clearly most suitable at high temperatures and low coverages, but the actual domain of convergence has not yet been established. A perturbation expansion, of the kind which has been so successful for uniform liquids, has also been examined, but it proved to be disappointing (Steele, 1977).

It is clear that a three-dimensional treatment is essential whenever multilayer adsorption is involved. However, even when the amount of adsorption is less than a statistical monolayer (the coverage which, if confined to the first layer, would give a complete monolayer), it is still questionable whether neglect of the third dimension can be justified (Ch. 5, § 1). Indeed, most of the three-dimensional simulation studies on gas–solid adsorption have been for such sub-monolayer coverages. There is, of course, a high price (in terms of computer time) to be paid for the inclusion of this extra dimension and such lengthy calculations only became feasible in the early 1970s with the advent of fast computers.

In this chapter we shall consider in turn gas–solid and solid–liquid interfaces and finally microclusters. The first of these suffers especially severely from the problems associated with non-uniformity. This affects simulation work mainly by the difficulties it raises in finding optimum conditions for the runs. In Monte Carlo studies on uniform systems, the choice of simulation parameters is primarily determined by the density of the medium and if this varies from point to point it becomes impossible to obtain a single set of conditions which optimizes the simulation everywhere in the system. Indeed, all the successful work on gas–solid adsorption has employed either the GCEMC or MD methods; the canonical ensemble Monte Carlo (CEMC) method has proved unsuitable for such systems.

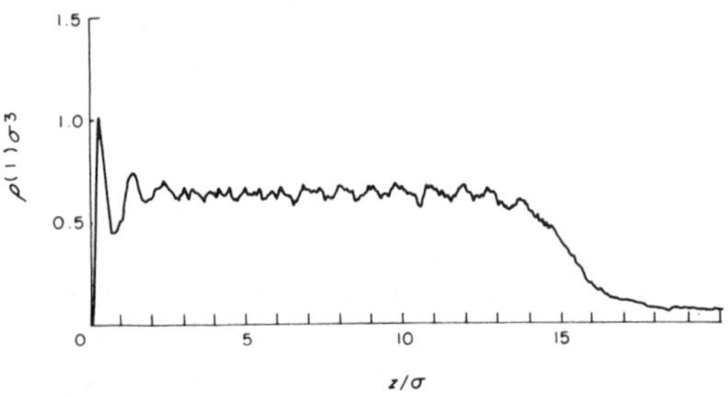

Fig. 6.1. Density profile for a gas–liquid interface at $T^* = 1{\cdot}127$ and $\rho\sigma^3 = 0{\cdot}622$. (After Chapela *et al.*, 1977.)

For liquid–solid studies the reverse is true: the CEMC method works well, but the GCEMC method is wasteful at these high densities. This has been a very active field of research, nearly all of it directed towards obtaining $\rho^{(1)}$ information and very little towards the evaluation of thermodynamic properties.

The gas–liquid interface has been the subject of much controversy. The main uncertainty has centred around the existence or otherwise of oscillations in $\rho^{(1)}$ close to the interface (Fig. 6.1). The weight of opinion is against the reality of these oscillations, but Osborn and Croxton (1977, 1980) and Rao *et al.* (1979) have shown how they may arise. Other quantities studied include the surface tension and, for a mixture, the Gibbs excess of one component at the surface. These systems are the subject of a recent book by Rowlinson and Widom (1982) and no further discussion will be given here.

The chapter finishes with a short description of the simulation work on microclusters. These are groups of, generally, less than 100 particles and interest in them is primarily centred on the ability of such small clusters to show melting and, if it occurs, its variation with size.

2. Gas–solid interfaces

(a) The structureless solid

The simplest type of system in this class is that studied by Rowley *et al.* (1976). For the graphite adsorbent, which it was intended to mimic, a semi-infinite three-dimensional continuum was assumed and the adsorptive was taken to be a noble gas. Lennard-Jones 12–6 potentials were assumed to be effective between the atoms of the adsorptive and also between each noble gas atom and each element of the wall. Integrating over the complete volume of the adsorbent the adsorbent–adsorbate potential assumes the well known 9–3 form (see Eqn (3.81c)):

$$u^{(1)}(z) = \frac{3^{3/2}}{2}\,\varepsilon_w\{(s_0/z)^9 - (s_0/z)^3\} \tag{6.1}$$

where ε_w and s_0 are energy and distance parameters characteristic of the system. For the adsorbate–adsorbate interactions the values chosen for $\varepsilon^{(Ar,Ar)}$ and σ_{Ar} were 119·8 K and 0·3405 nm, respectively, and ε_w and s_0 were chosen so as to give agreement with zero-coverage data for argon–graphite as, $\varepsilon_w/\varepsilon^{(Ar,Ar)} = 9{\cdot}24$ and $s_0/\sigma_{Ar} = 0{\cdot}5621$. In most of the simulations a single square adsorbent surface of area $46{\cdot}08\sigma_{Ar}^2$ (σ_{Ar} being the Lennard-Jones parameter for the argon–argon interaction) was used with a box length of $14{\cdot}03\sigma_{Ar}$. It turned out that at the higher coverages this length was only just sufficient to enable the gas at the far end of the box to be considered as being normal, bulk argon. Since the size of the box was fixed the number of particles in the GCEMC simulation depended upon the activity of the gas but was in the range 35–143. No fewer than 28 runs were made, these being spread over the coverage range $\theta = 0{\cdot}8$–$3{\cdot}2$ and the two temperatures $T^* = 0{\cdot}668$ (~80 K) and 1·002 (~120 K). Thus the work was directed towards the study of multilayer adsorption, and in this respect contrasts with most subsequent studies, which have been concerned with sub-monolayer coverages. Two main kinds of information were obtained. Firstly, tests of the previously developed, "traditional" theories, such as BET, "slab", and Polanyi potential theories, could be made without the intervention of heterogeneity and other experimental artefacts. These comparisons are discussed in Chapter 7, § 3. The second type of information was that provided by the distribution functions $\rho^{(1)}$ and $\rho^{(2)}$. Figure 6.2(*a*) shows the $\rho^{(1)}$ curves for the runs at the two temperatures in order of increasing coverage (θ). One characteristic, which is clearly shown in Fig. 6.2(*b*), is that the build-up of atoms in higher layers commences well before the lower layers are complete: the

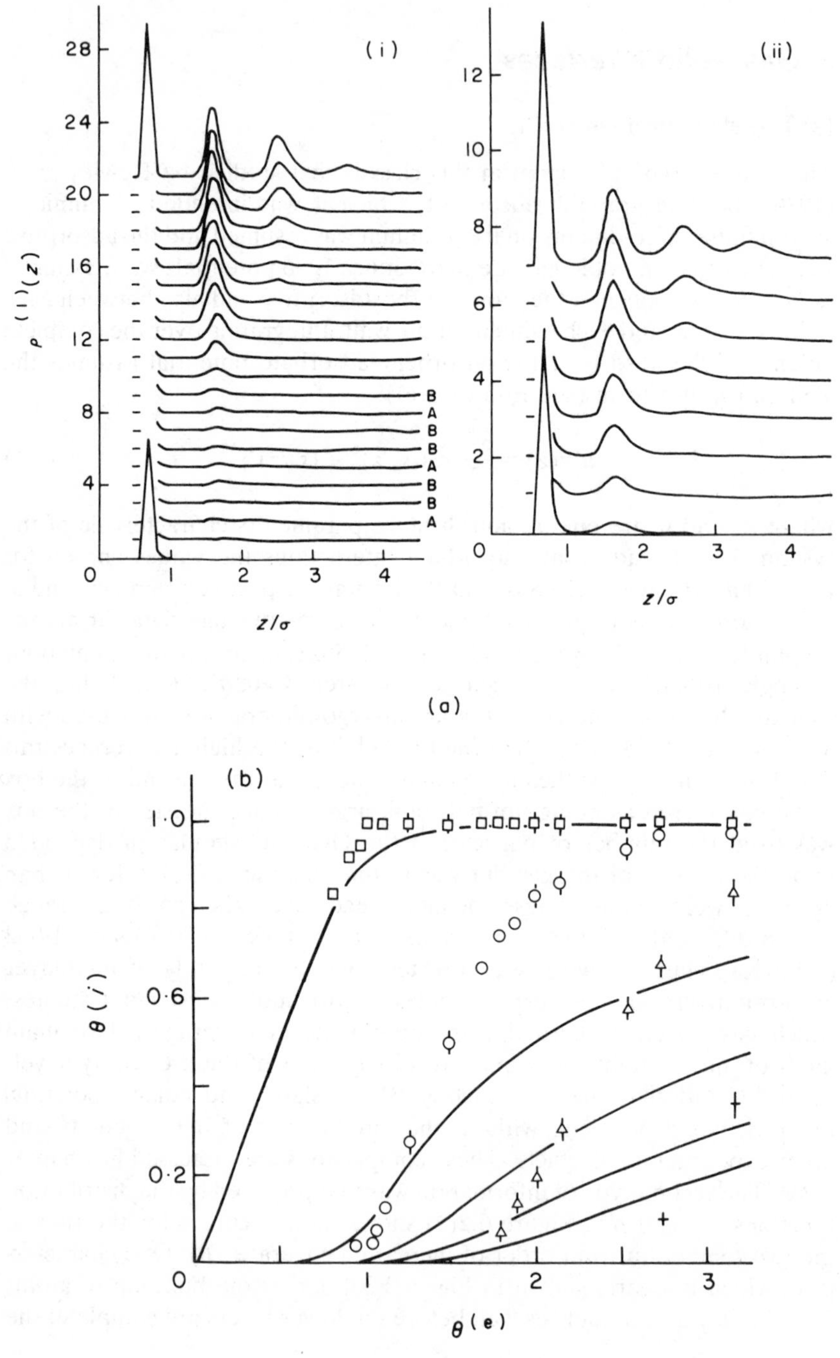
(i)
(ii)
$\rho^{(1)}(z)$
z/σ
B
A
B
A
B
B
A
(a)
(b)
$\theta(i)$
$\theta^{(e)}$

implications of this for the "slab" and other theories is discussed later (Ch. 7, § 3(b)). At the lower temperature the spacing of the maxima in $\rho^{(1)}$ are all $0{\cdot}92\sigma_{Ar}$, which corresponds to the spacing between close-packed layers in an FCC or CPH lattice. At the higher temperature, on the other hand, all the spacings are $\sim 1{\cdot}0\sigma_{Ar}$ with the exception of that between the first and second peaks which is $0{\cdot}92\sigma$. When, in subsequent analyses, it is desired to distinguish between the first, second, third, . . . layers of adsorption, a frequently used method is to take the minima in these $\rho^{(1)}$ curves as the boundaries between successive layers. The width of the peaks can be put onto a quantitative basis by calculating the width-at-half-height(H) for each layer (Fig. 6.3). This shows the expected broadening at the higher temperature, but also, surprisingly, that the outer peak ($i = 3$) is narrower than the inner ($i = 2$). The $\rho^{(2)}$ curves also reveal a number of interesting, although incompletely explained, features. At the lower temperature, a double peak is found near $2\sigma_{Ar}$ in most of the curves for the first layer (Fig. 6.4): this is usually associated with random dense packing of spheres (Finney, 1970). At coverages of just below $\theta = 1$, where $\rho^{(2)}$ curves both with and without this feature are found, the uncertainty seems to be associated with the existence of a "loop" in the corresponding adsorption isotherm (Fig. 6.5). The suggestion has been made that the two branches of the "loop", one of which must be metastable, correspond to different arrangements of the first layer. At the highest coverages studied the double peak near $2\sigma_{Ar}$ is present, which suggests, provided that the assignment of the double peak to dense random packing is correct, that the same type of packing is found in the first layer at those coverages. It would have been expected that the more ordered arrangement would have occurred when higher layers had been added. All the $\rho^{(2)}$ functions computed in this work were averages over orientations of the inter-particle vector, and this averaging no doubt hides a considerable amount of structural detail.

A very similar GCEMC study of the adsorption of argon on a continuum adsorbent has been subsequently carried out by Lane *et al.* (1980) in their

---

Fig. 6.2. The argon–"continuum graphite" interface. (*a*) Density profiles at (i) $T^* = 0{\cdot}668$, (ii) $T^* = 1{\cdot}002$. Read from the bottom, curves are for (i), $\theta^{(e)} = 0{\cdot}836$, 0·935, 1·036, 0·954, 0·970, 1·058, 0·973, 0·992, 1·073, 1·021, 1·121, 1·270, 1·505, 1·700, 1·802, 1·894, 1·912, 2·014, 2·156, 2·538, 2·737, 3·156; for (ii), $\theta^{(e)} = 0{\cdot}952$, 1·087, 1·279, 1·563, 1·765, 1·960, 2·406 and 2·948. A and B indicate the branch of the isotherm to which the points belong (see Fig. 6.5). Curves have been displaced vertically for clarity.

(*b*) Fractional occupancy of the *i*th layer ($\theta(i)$) versus total excess coverage ($\theta^{(e)}$) at $T^* = 0{\cdot}668$. MC points: $i = 1$ (□), 2 (○), 3 (△), 4 (+). Full lines are BET predictions. (After Parsonage and Nicholson, 1979.)

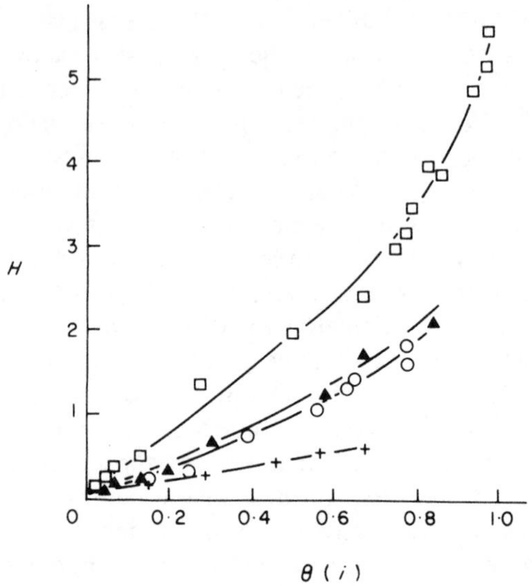

Fig. 6.3. The argon–"continuum graphite" interface. Variation of the shape of peaks in $\rho^{(1)}$ with occupancy. H is ratio of height to width-at-half-height for layer i against $\theta(i)$. $T^* = 0{\cdot}668$, $i = 2$ (□); $T^* = 0{\cdot}668$, $i = 3$ (△); $T^* = 1{\cdot}002$, $i = 2$ (○); $T^* = 1{\cdot}002$, $i = 3$ (+). (After Rowley *et al.*, 1976.)

critical discussion of the predictions made by Saam and Ebner (1978) from density functional theory. The same Lennard-Jones parameters were used for the argon–argon potential, but the parameters of the argon–wall potential, which was again of the 9–3 form, were chosen to represent CO_2 rather than graphite, as adsorbent. This causes the zero and minimum of the adsorbate–adsorbent potential to occur at $0{\cdot}78\sigma_{\mathrm{Ar}}$ and $0{\cdot}940\sigma_{\mathrm{Ar}}$, respectively, as compared with $0{\cdot}562\sigma_{\mathrm{Ar}}$ and $0{\cdot}675\sigma_{\mathrm{Ar}}$, for the graphite system. The results, which are for $T^* = 1{\cdot}1$, may be roughly compared with those of Rowley *et al.* (1976) for $T^* = 1{\cdot}002$ provided that an estimate of the temperature coefficient can be made and then applied to convert the latter data from $T^* = 1{\cdot}002$ to $T^* = 1{\cdot}1$. If the bulk gas is taken to be ideal, then a bulk density ρ corresponds to an activity of $\rho\sigma^{-3}$. Inserting the activity $z = \lambda/\Lambda^3 \cong \beta p$ into Eqn (2.76) gives the relationship

$$(\partial \ln z/\partial T)_\theta = (q_{\mathrm{st}} - RT)/RT^2 \tag{6.2}$$

from which the value of z at $T^* = 1{\cdot}002$ needed to give the same coverage as a given value of z at $T^* = 1{\cdot}1$ could be calculated. Using the values of q_{st} from Rowley *et al.*, it is found that $T^* = 1{\cdot}002$, $z \approx 0{\cdot}019\sigma^{-3}$

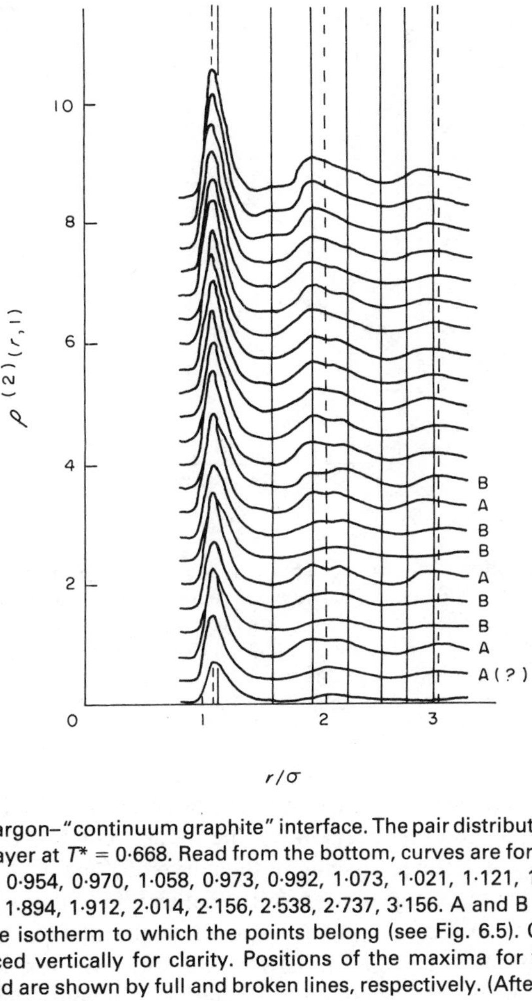

Fig. 6.4. The argon–"continuum graphite" interface. The pair distribution function for the first layer at $T^* = 0{\cdot}668$. Read from the bottom, curves are for $\theta^{(e)} = 0{\cdot}836$, 0·935, 1·036, 0·954, 0·970, 1·058, 0·973, 0·992, 1·073, 1·021, 1·121, 1·270, 1·505, 1·700, 1·802, 1·894, 1·912, 2·014, 2·156, 2·538, 2·737, 3·156. A and B indicate the branch of the isotherm to which the points belong (see Fig. 6.5). Curves have been displaced vertically for clarity. Positions of the maxima for the uniform solid and fluid are shown by full and broken lines, respectively. (After Parsonage and Nicholson, 1979.)

would give the same coverage as the system at $T^* = 1{\cdot}1$, $z = 0{\cdot}03$ studied by Saam and Ebner and by Lane *et al.* Comparing the $\rho^{(1)}$ curves for these corresponding conditions, it is found that the curve of Rowley *et al.* shows much more clearly defined second and third peaks. This may be partly due to the greater length of the runs used by Rowley *et al.* (2×10^6 against $0{\cdot}5 \times 10^6$), but probably mainly to the lower temperature (1·002 against 1·1). Bearing in mind that the position of the first peak is essentially determined by the minimum in the argon–wall potential, the positions of

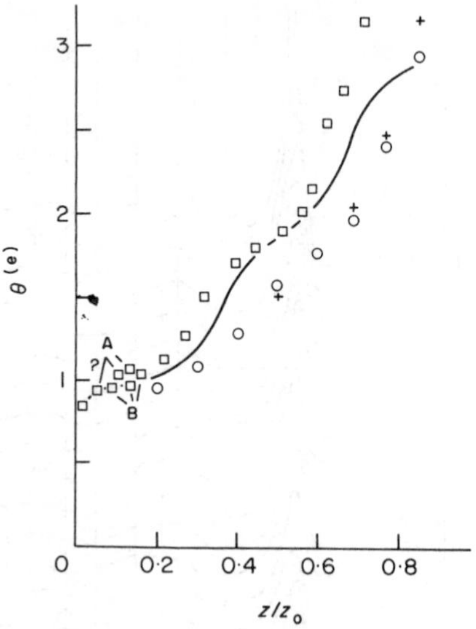

Fig. 6.5. Adsorption isotherms for argon on "continuum graphite". Excess coverage ($\theta^{(e)}$) versus activity (z). Activities are reduced by the saturation activity at T^* (z_0 values are taken from theory for MC points, from experiment for the experimental curve). $T^* = 0{\cdot}668$ (□), 1·002 (○). Full curve: experimental isotherm for 78 K. Crosses indicate positions to which nearby points move on using the "augmented box" method (p. 253). (After Parsonage and Nicholson, 1979.)

the respective second peaks are consistent: Rowley *et al.* found the spacing between the first and second peaks to be $0{\cdot}92\sigma$, which is consistent with the much less well defined spacing in the curve of Lane *et al.* As discussed below, the inclusion of corrections for the truncation of the potential causes the outer peaks to increase at the expense of the first peak (at constant coverage) (Rowley *et al.*, 1978). This would not, however, be in any way sufficient to obviate the argument of Lane *et al.* against the existence of "films" in adsorption (but note that "film formation" may be feasible under certain conditions, but not others—cf. the discussion in Ch. 7, § 2(b)).

A similar model had been previously used by Lane and Spurling (1976) in their attempt to simulate the adsorption of krypton on the basal plane of graphite. However, instead of the 9–3 adsorbent–adsorbate potential they used Eqn (3.81a) in the approximate form given in Eqn (8.2), the former being in turn an approximation to the "summed 10–4 potential". This potential (Eqn (8.2)) is not periodic in the x, y directions. A major difference from the work of Rowley *et al.* (1976) was that the object of the study was to obtain an understanding of the rather complex phase diagram found by Thomy and Duval (1970) for both Kr and Xe on graphite. For this reason their work was restricted to the submonolayer region. A simulation box with two adsorbent faces was used and in some cases the structure of the fluid near these was quite different. Whenever this occurred the information from the two faces was not averaged, but treated separately. Figure 6.6 shows some of the $\rho^{(2)}$ curves which were obtained. At 90·12 K

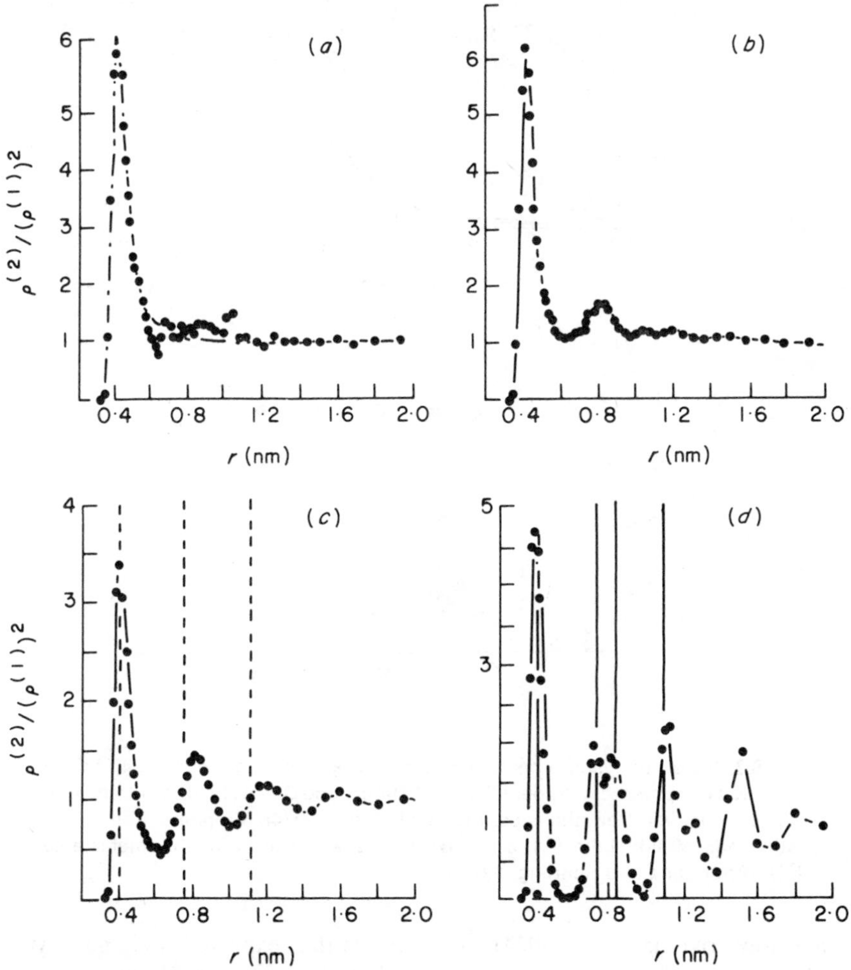

Fig. 6.6. Radial distribution functions for krypton on "continuum graphite" at 90·12 K. Positions of the maxima for the uniform solid and fluid are shown by full and broken lines, respectively. (*a*) $\theta = 0{\cdot}024$; broken curve, LJ gas at $p \to 0$. (*b*) $\theta = 0{\cdot}13$, gas-like. (*c*) $\theta = 0{\cdot}76$, liquid-like. (*d*) $\theta = 0{\cdot}98$. (After Lane and Spurling, Aust. J. Chem., 1976.)

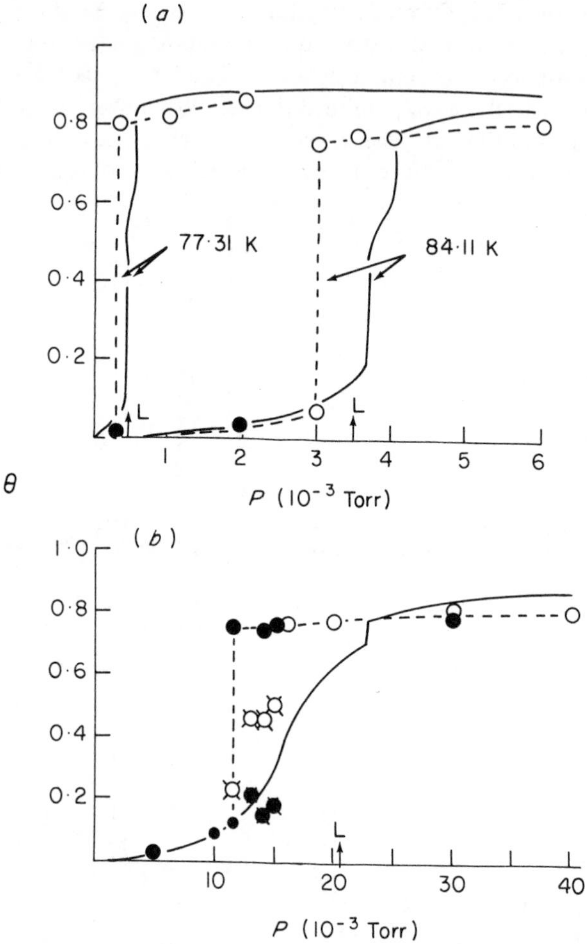

Fig. 6.7. Adsorption isotherms for krypton on graphite at (*a*) 77·31 K, 84.11 K (*b*) 90·12 K. Full curves, experimental data of Thomy and Duval (1970). L = transition pressure derived from the experiments of Larher (1974). MC points from simulation box of side 1 nm (○), 2 nm (●); points which are probably metastable (◌). (After Lane and Spurling, 1976.)

and low coverage ($\theta = 0{\cdot}024$), the curve is that expected for a gas. At a higher pressure gas-like ($\theta = 0{\cdot}13$) and liquid–like ($\theta = 0{\cdot}76$) phases were found to be in equilibrium. At still higher coverages ($\theta = 0{\cdot}98$) the structure became solid-like. Adsorption isotherms for three temperatures (77·31, 84·11 and 90·12 K) are shown in Fig. 6.7. Because of the absence of periodicity in the adsorbing potential it was not possible for an epitaxial phase to form in this simulation, although it is now known that the main condensed adsorbed phase is just such a phase. Spurling and Lane (1978) have subsequently incorporated a periodicity into the lattice potential (giving a potential barrier to translation of $\Delta\Psi \approx 40k$, see p. 88) but they did not find an epitaxial phase.

Lane and Spurling (1978) have also made more detailed studies of the pressure of the solid-like → liquid-like transition at 90·12 K, which they found to be ~0·1 Torr, as compared with the value of ~1·5 Torr found experimentally for the upper transition (Thomy and Duval, 1970). In this work they reverted to their original potential, which was without periodicity. Considerable care was taken to ensure that the outcome of the simulation was not affected by the ability of the structures to fit into the repeat unit of the simulation box: although the box was small ($\sim 2\sigma \times 2\sigma$) the dimensions could be varied. The location of the transition pressure was based on observations of the effect of increase of pressure on the radial distribution function $(\rho^{(2)}/(\rho^{(1)})^2)$ (Fig. 6.8). For $0{\cdot}015 \leqslant p$ (Torr) $\leqslant 0{\cdot}04$ no change in this function occurs. On increasing the pressure to 0·07 Torr a change in the shape of the function was observed and further modifications

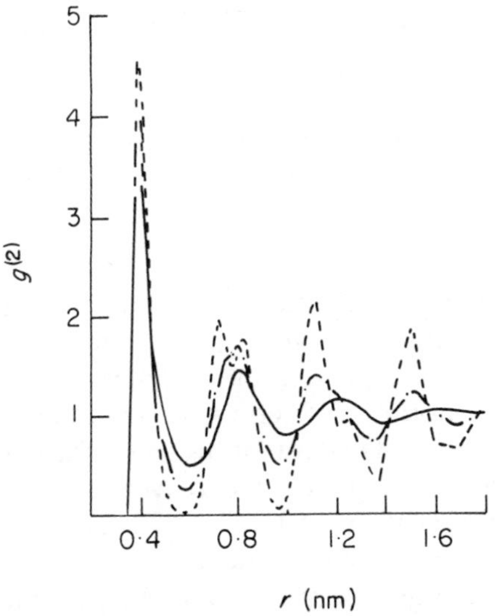

Fig. 6.8. Radial distribution functions for krypton on "continuum graphite" at 90·12 K. Pairs considered are those with inter-particle vector approximately parallel to the surface and with at least one particle having $z \approx 0{\cdot}36\sigma$. Full curve, $p = 0{\cdot}03$ Torr; chain curve, $p = 0{\cdot}5$ Torr; broken curve, $p = 0{\cdot}875$ Torr. (After Lane and Spurling, 1978.)

occurred on further increase of pressure. At 0·875 Torr the adsorbate was clearly solid-like, but a run at 0·5 Torr appeared to show a mixed structure. A run at 0·1 Torr resulted in one interfacial region being liquid-like, the other solid-like, and this was the pressure which they considered to be the best estimate for the transition pressure.

Hanson *et al.* (1977), whose two-dimensional MD simulations using an adsorbent–adsorbate potential of the Steele type with and without periodicity have already been described (Ch. 5, §4), made some runs with a

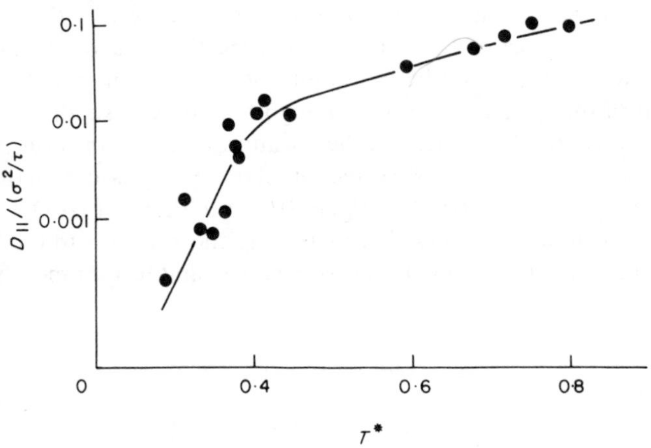

Fig. 6.9. Lateral diffusion coefficient ($D_{\parallel}$) for argon on "continuum graphite" at $\theta \approx 0{\cdot}73$ versus temperature, $\tau = 2 \times 10^{-12}$ s for argon. (After Hanson *et al.*, 1977.)

three-dimensional system. Their potential parameters were chosen so as to represent the adsorption of argon on graphite. As with Lane and Spurling, their attention was focused on coverages of less than a close-packed monolayer, but in their case effects at lower temperatures (down to ~15 K) were examined. The lateral diffusion coefficient for a coverage of ~73% of a statistical close-packed monolayer showed a sharp change in its temperature coefficient at ~50 K, ~7 K above the temperature at which an energy intake was found in the same simulation (Fig. 6.9). Similar, though more gradual, energy absorptions were found in the corresponding two-dimensional systems at about the same temperature (Fig. 5.9). An increase in the sharpness of transitions is normally observed on going from two to three dimensions, but in this case it may have been expected that the three-dimensional system would have behaved virtually as two-dimensional at such a low temperature and a coverage markedly less than the close-packed monolayer. As the potential used in this three-dimensional simulation was of the non-periodic 9–3 type, there is no possibility of epitaxy being involved in the transition. Since there are considerable similarities between this model and that of Rowley *et al.* (1976) it is of interest to compare the results as far as the limited range of overlap permits. In the region just below $\theta = 1$, where two branches were found for the isotherm by Rowley *et al.*, the results of Hanson *et al.* correspond to the *lower* of these (Fig. 6.5) (Parsonage and Nicholson, 1979). Rowley *et al.*, on the basis of their evaluation of the surface pressure, had concluded that the *upper* branch was the more stable.

Some attention has also been given to the adsorption of polyatomic molecules at solid surfaces. Thus Spurling and Lane (1980) have performed GCEMC runs on simulated NH_3 at $\theta \approx 0{\cdot}75$ on a non-periodic "graphite" surface at 251 K and 194 K. In evaluating the NH_3–NH_3 interactions the Lennard-Jones 12–6 potential was supplemented by terms arising from the dipole ($0{\cdot}49 \times 10^{-29}$ C m) and a cylindrically symmetric quadrupole ($-0{\cdot}774 \times 10^{-39}$ C m^2). The angle-averaged radial distribution functions ($g^{(2)}(z, r, \beta)$), where z is the distance of one particle from the wall and β is the angle between the inter-particle vector (of length r) and the normal to the surface, were liquid-like. The singlet distribution function $\rho^{(1)}(\cos \alpha)$, which is an average over all particles within the first layer of $\cos \alpha$, where α is the angle between the dipolar axis and the surface normal, has a maximum at $\cos \alpha = 0$. This indicates that the dipoles (and quadrupoles) tend to be parallel to the surface. One other feature is worthy of note: the admolecules were not uniformly distributed over the surface, but were gathered in large clusters. This observation is reminiscent of the results of the study of argon adsorption on a weakly periodic lattice at low temperatures by Hanson *et al.* (1977).

In all of the simulations in this section a cut-off was imposed on the adsorbate–adsorbate interactions. This was at $3{\cdot}39\sigma$ (Rowley *et al.*, 1976), $2{\cdot}7\sigma$ (Lane and Spurling, 1976), $5{\cdot}6\sigma$ (Spurling and Lane, 1980) and $2{\cdot}5\sigma$ (Hanson *et al.*, 1977). In the last three studies no correction was made to take account of this truncation, and, as will be seen, this can probably be justified because of the fairly low coverages which were under consideration. At the higher coverages which were the main concern of Rowley *et al.* it is clear that some correction must be made. As discussed previously (Ch. 4, § 4), when the correction is complicated to evaluate, it is almost essential that it should be applied at the end of the run, rather than at every step, yet it is not possible to make the correction accurately in this way for this type of system. For convenience, the correction equations were obtained by consideration of the corresponding canonical ensemble problem. A coupling parameter method (Ch. 4, § 4) was adopted leading to expressions for the long-range corrections (denoted by superscript LR) for the fixed amount of adsorbate. If in the equation

$$U_N^{\mathrm{LR}} = \tfrac{1}{2}\int \mathrm{d}\boldsymbol{r}_1\,\mathrm{d}\boldsymbol{r}_2\rho^{(1)}(\boldsymbol{r}_1;\xi)\rho^{(1)}(\boldsymbol{r}_2;\xi)g^{(2)}(\boldsymbol{r}_1,\boldsymbol{r}_2;\xi)u^{(2)}(r_{12}) \qquad (6.3)$$

the approximations are made that

$$g^{(2)}(\boldsymbol{r}_1,\boldsymbol{r}_2;\xi) = 1 \text{ and } \rho^{(1)}(\boldsymbol{r};\xi) = \rho^{(1)}(\boldsymbol{r};0),$$

then Eqn (4.76) becomes

$$A^{\mathrm{LR}} = \langle U_N^{\mathrm{LR}}\rangle_{\xi=0} \qquad (6.4)$$

$\langle U^{\mathrm{LR}}\rangle_{\xi=0}$ is, of course, evaluated using the uncorrected singlet distribution functions. An examination of the values of $\langle U^{\mathrm{LR}}\rangle_{\xi=0}$ as a function of N showed that a relation held of the form $\langle U^{\mathrm{LR}}\rangle_{\xi=0} \propto N^a$ with $a = 1{\cdot}996(1{\cdot}942)$ for $T^* = 0{\cdot}668(1{\cdot}002)$, which could be approximated by $\langle U^{\mathrm{LR}}\rangle_{\xi=0} \propto N^2$. Thus as for the homogeneous system (Ch. 4, § 4) $\mu^{\mathrm{LR}} = 2\langle U^{\mathrm{LR}}\rangle_{\xi=0}$, or expressing this result in terms of activities

$$z_{\xi=1} = z_{\xi=0}\exp(2\langle U^{\mathrm{LR}}\rangle/NkT). \qquad (6.5)$$

The corrections to z were ~4% at $\theta \approx 1$ rising to ~19% and ~12% at the highest coverages studied ($\theta \approx 3$) for $T^* = 0{\cdot}668$ and $1{\cdot}002$, respectively. Corrections to the distribution functions are far more difficult to deduce and apply. However, it can be shown that, with further approximations,

$$\rho^{(1)}(\boldsymbol{r};\xi=1) = \rho^{(1)}(\boldsymbol{r};\xi=0)[1 + \ln(z(1)/z(0)) - \beta\int \mathrm{d}\boldsymbol{s}_1\rho^{(1)}(\boldsymbol{s}_1;0)u^{\mathrm{LR}}(\boldsymbol{s}_1-\boldsymbol{r})]. \qquad (6.6)$$

By integration over all $\boldsymbol{r}$ it can be shown that Eqn (6.6) conserves N, as is necessary for a canonical approach. Figure 6.10 shows the uncorrected and corrected distribution functions for the runs at $T^* = 1{\cdot}002$ with θ in the range 1·7–3·4. It will be seen that there are considerable changes in the outer layers for the three highest coverages. These must be compensated by corresponding reductions in the inner layers so as to conserve N, but these latter alterations are not evident because of the much larger values of $\rho^{(1)}$ in the inner layers. Corrections to the isosteric heat of adsorption can also be applied:

$$q_{\mathrm{st},\xi=1} - q_{\mathrm{st},\xi=0} = \int_0^1 \mathrm{d}\xi(\partial q_{\mathrm{st}}/\partial\xi) \approx \int_0^1 \mathrm{d}\xi(\partial\{(\partial U/\partial N)_{T,V} + kT\}/\partial\xi)$$
$$\approx (\partial U^{\mathrm{LR}}/\partial N)_{T,V} \approx 2\langle U^{\mathrm{LR}}\rangle/N. \qquad (6.7)$$

The corrections are, again, most important at high coverages, but even there they do not exceed 4·5%. The corrections to the heat capacity at

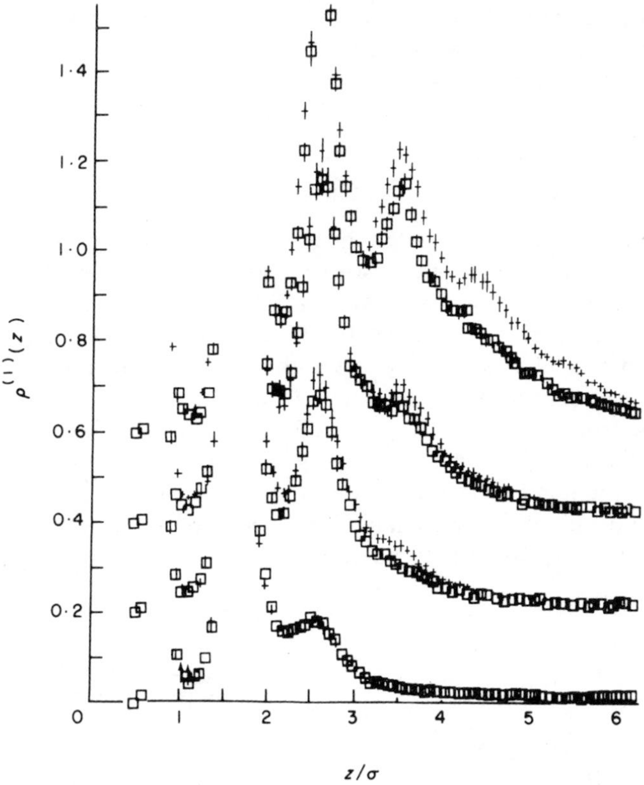

Fig. 6.10. Effect of cut-off on the density profiles of argon on "continuum graphite" at $T^* = 1{\cdot}002$. Cut-off: $3{\cdot}39\sigma$ (□), $10{\cdot}17\sigma$ (+) (using "augmented box" method). Excess coverages referred to for standard (large) cut-off are (from the bottom) 1·769 (1·724), 2·272 (2·358), 2·769 (2·820) and 3·381 (3·623). Graphs are displaced vertically for clarity. (After Rowley *et al.*, 1978.)

constant coverage $C_{\mathscr{A},N}$ are of even less importance: they are smaller than the sampling errors. Support for this method of making the correction was provided by further simulations (Rowley *et al.*, 1978) using the "augmented box" method of Wood and Parker (1957). In this method replicas are generated in all three directions so that the array now consists of the original box plus 26 replicas. A cut-off distance of one-half of the length of this "augmented box" (three times the original cut-off distance) was used and the minimum image criterion was abandoned: a molecule was able to interact with any molecule within its cell and also with any images within the cut-off. When long-range corrections to the new system were estimated they proved to be never more than comparable with the uncertainties in the data. Comparing the results with those obtained for the original system after correction showed good agreement for the run with $\theta \approx 1{\cdot}7$, becoming less satisfactory as θ is increased.

(b) Adsorbent with periodic potential

Nicholson *et al.* (1977) extended their work on the adsorption of argon on graphite by replacing the 9–3 adsorbent–adsorbate potential by the Steele potential with terms up to E_1 (see Eqn (3.82)). In order to avoid discriminating against certain structures by the choice of the dimensions of the simulation box, it was necessary to employ a larger-than-usual box. This included 972 carbon hexagons and was able to accommodate, with only slight distortion, either a close-packed or an epitaxial layer. Apart from a single point at $\theta = 1{\cdot}54$, all coverages were $<1{\cdot}2$, the object being to study the phenomena in the region near $\theta = 1$. The simulations suffered from a difficulty common to many studies with periodic potentials, that of passing from one favourable state to another. The solution employed was

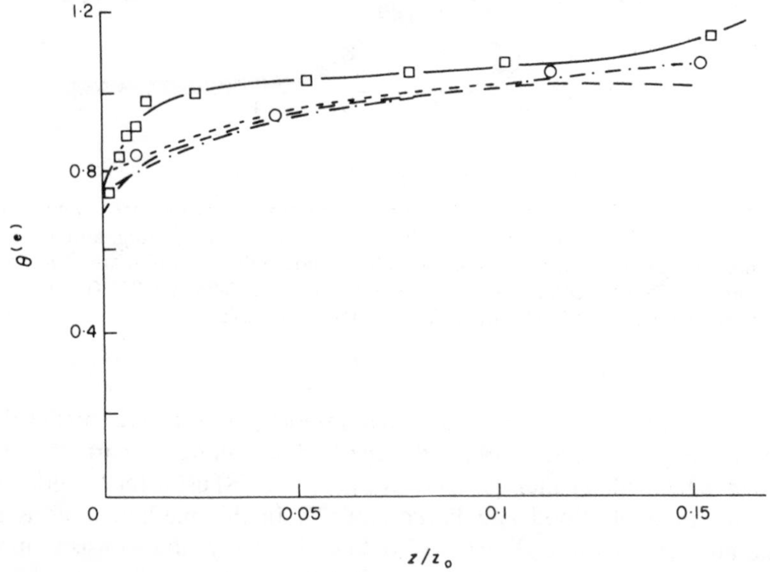

Fig. 6.11. Adsorption isotherms for argon on graphite at $T^* = 0{\cdot}668$ (77 K). Excess coverage ($\theta^{(e)}$) versus reduced activity (z/z_0), where z_0 is the saturation activity. MC: adsorbent with periodic potential (□), "continuum adsorbent" (○). Experimental curves: Grillet *et al.* (1979) (———); Goellner *et al.* (1975) (–·–·–), Hegde (unpublished results) (-----). (After Parsonage and Nicholson, 1979.)

to make runs from several different starting-points and to test for the relative stability of the resultant configurations by calculating $-\Omega/\mathscr{A} = \frac{1}{2}V(p_{xx}+p_{yy})/\mathscr{A}$ from the virial. The phase with the larger value of $-\Omega/\mathscr{A}$ ($=\phi + pV/\mathscr{A}$—Ch. 2, § 4) is the more stable, since this means that it makes the larger contribution to Ξ. Figure 6.11 shows the computed adsorption isotherm together with that previously obtained by the same authors for the continuum model of the adsorbent and also three sets of experimental results. The most striking, and disappointing, observation is that the results for the periodic adsorbent are in poorer agreement with experiment than those for the continuum adsorbent: the adsorbate seems to be attracted too strongly in spite of the extra effort made to provide a realistic adsorbent–adsorbate potential. Although careful consideration has been

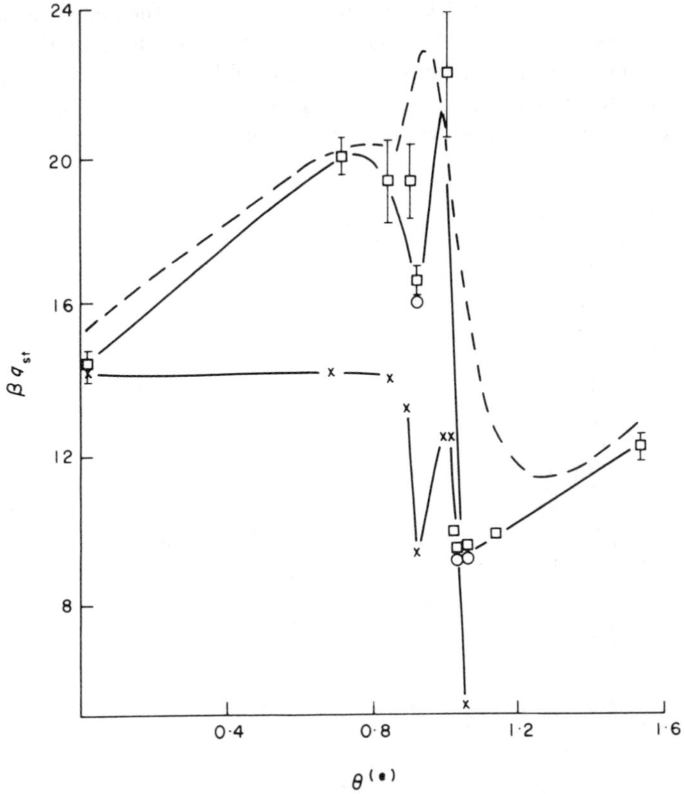

Fig. 6.12. Isosteric heat of adsorption (q_{st}) versus excess coverage ($\theta^{(e)}$) for argon on graphite at $T^* = 0{\cdot}668$ (77 K). Symbols as in Fig. 6.11. Also shown is the wall contribution to the isosteric heat ($\beta q_w + 1$) derived from the MC simulations (—×—).

given to the shapes and depths of the adsorbing potential wells, no completely satisfactory explanation has been produced. In spite of the big shift in the isotherm on going from the continuum to the periodic adsorbent, the transition region occurs at very nearly the same coverage ($\theta \approx 0{\cdot}85$) in both cases. The simulation results for the isosteric heat of adsorption (q_{st}) for the two models, periodic and continuum, are in good accord with each other and also with the experimental results of Grillet *et al.* (1979) (Fig. 6.12). The variation of q_{st} with increase in θ can be considered as being a gradual rise up to $\theta = 1$, followed by a sudden fall to lower values characteristic of second-layer adsorption. However, there is also some "structure" near the peak, with a sharp dip at $\theta \approx 0{\cdot}9$, which is found in both experimental and simulation results. Discussion of these results will be deferred until the simpler curves, obtained for systems with higher potential barriers, have been treated. Figure 6.13 shows the pair distribution function averaged over the first layer for four coverages in the range $\theta = 0{\cdot}675$–$1{\cdot}037$. The upper three show the double second peak which, as has been mentioned earlier, is often taken as a signature of random close-packing. In the earlier runs on the continuum model at the same

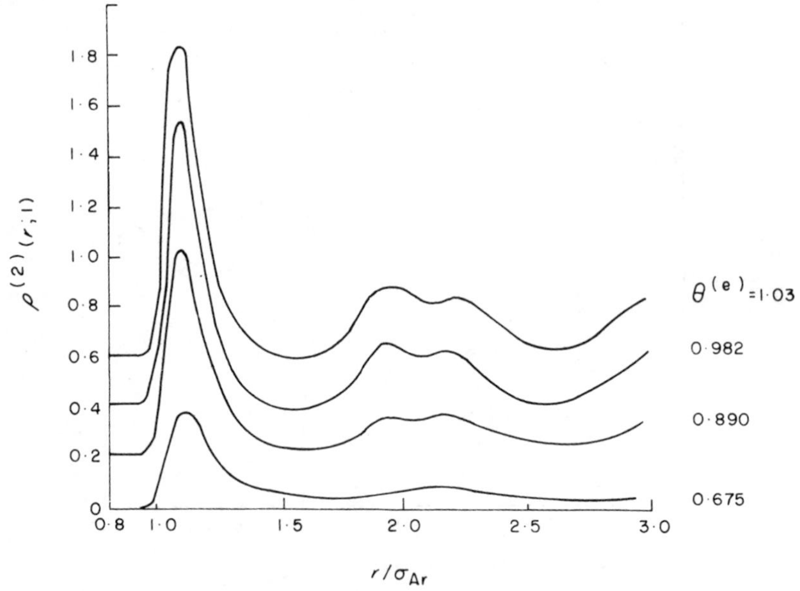

Fig. 6.13. First-layer, angle-averaged, pair distribution functions for argon on graphite at $T^* = 0{\cdot}668$. Curves are displaced vertically for clarity. (After Nicholson *et al.*, 1977.)

temperature and similar coverages, the upper, though not the lower, branch points showed this feature. A further observation which may be made from Fig. 6.13 is that there is a gradual change in the $\rho^{(2)}$ curves from the densely packed monolayer ($\theta = 1{\cdot}037$) to the liquid-like dilute layer ($\theta = 0{\cdot}675$). This is consistent with the fact that in two-dimensional systems it is possible to have a gradual decrease from regular, high-density packing to irregular lower density layers. (By contrast, in three dimensions there is a gap in the range of possible packings of hard spheres (Finney, 1970)). Another observation which can be made is that the position of the first maximum ($1{\cdot}10\sigma_{Ar}$) is very close to that for a "relaxed" two-dimensional crystal ($1{\cdot}11\sigma_{Ar}$) and somewhat higher than that for liquid argon ($\sim 1{\cdot}08\sigma_{Ar}$). Since the continuum adsorbent gave a value of $1{\cdot}08\sigma_{Ar}$, it seems that the periodicity of the lattice (spacing of the sites = $1{\cdot}25\sigma_{Ar}$) stretches the intermolecular separation somewhat, even though the energy barriers are very low ($\beta\Delta\psi = 0{\cdot}44$, where $\Delta\psi = \frac{1}{2}(u_C^{(1)} + u_{SP}^{(1)}) - u_S^{(1)}$ and $u_C^{(1)}$, $u_{SP}^{(1)}$ and $u_S^{(1)}$ are the potential minima over a carbon atom, a saddle-point and a hexagon site, respectively; or $\beta\Delta\psi' = 0{\cdot}51$ for $\Delta\psi' = u_C^{(1)} - u_S^{(1)}$).

One outcome of the work described in the preceding paragraphs is that the theory correctly predicts that there should not be an epitaxial phase for argon adsorbed on graphite at this temperature. The next question which may be asked is: for what value of the potential barrier does an epitaxial layer become stable? Some progress towards the answering of this question has been made by studies on systems which have high barriers (P-adsorbent, $\Delta\psi = 4{\cdot}89$, K-adsorbent, $\Delta\psi = 6{\cdot}70$) but which otherwise are the same as the argon–graphite system discussed above (Nicholson *et al.* 1981). The adsorbent–adsorbate potential used was

$$u^{(1)}(r) = \varepsilon_{gs}[E_0 + E_1 f_1 + bE_1(f_1 + 6)] \qquad (6.8)$$

where $\varepsilon_{gs}/k = 57{\cdot}8$ K for argon adsorption, and f_1, $E_0(z)$ and $E_1(z)$ have been given in Chapter 3, § 3(e). By varying b, the barrier height is changed without altering the potential minimum over a hexagon site position ($u^{(1)}/k = -1118$ K). Thus $b = -1$ and $b = 0$ correspond to the adsorbing potential with no periodicity and with that appropriate to Steele's (1973) model of graphite, respectively. Not surprisingly, the non-epitaxial condensed layer found for the J system ($\beta\Delta\psi = 0{\cdot}44$) was replaced by essentially epitaxial layers in which one-third of the hexagon centres could be occupied by adatoms (Fig. 6.14). The adsorption isotherms (Fig. 6.15) in the region of coverage up to $\theta = 0.805$ (the value for a complete epitaxial layer) correspond closely to what would be expected for a lattice gas condensation. For example, the coexistent phases are almost exactly symmetrical about the composition corresponding to one-half of the epitaxial

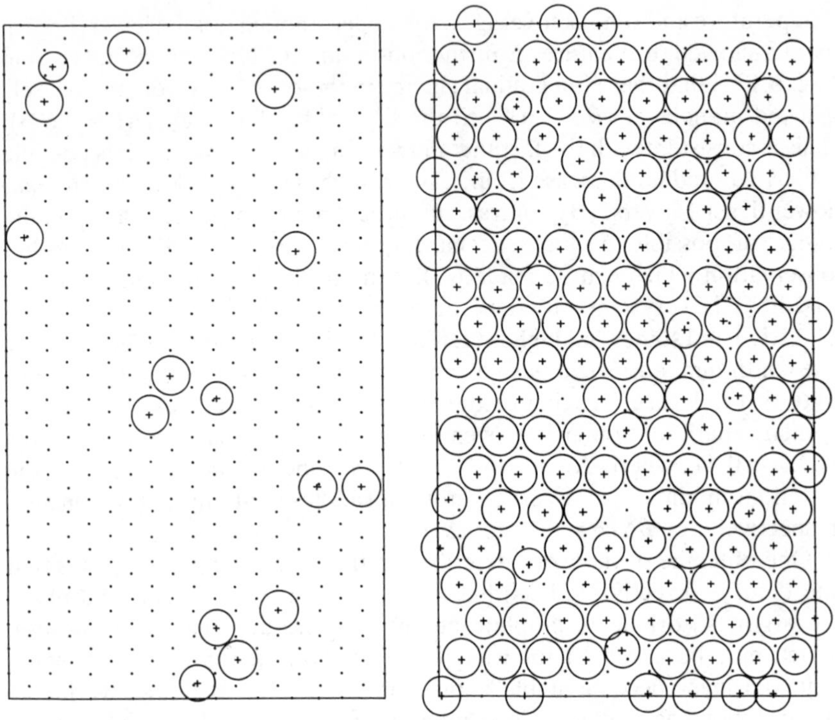
(a)

(b)

Fig. 6.14. Snapshots from the GCEMC simulation of argon on the high-barrier (K) adsorbent at $T^* = 0{\cdot}668$. The size of the circles indicates the distance of the adatoms from the wall: each circle is the intersection of a circle of diameter σ_{Ar} around the adatom and the plane through the most favoured sites. (*a*) $\theta^{(e)} = 0{\cdot}093$; (*b*) $\theta^{(e)} = 0{\cdot}732$. (After Nicholson *et al.*, 1981.)

sub-lattice positions being occupied. Snapshots of the phases on either side of the coexistence curve also suggest such a symmetry (Fig. 6.14). The gas pressure at which this condensation occurs rises as the barrier is increased, being ~0·03, 0·28 and 1·05 bar for the J-, P- and K-systems, respectively. This rise occurs because, whereas the potential minimum is the same for each series, for all other positions the energy is increased as the potential barrier is increased. The corresponding difference in the transition chemical potentials for the P- and K-series is $\sim 1{\cdot}8\varepsilon_{gs}$. For both of the high-barrier systems the more dilute of the coexistent phases has $\theta \approx 0{\cdot}1$. Figure 6.15 shows that though there is a large change in βq_{st} of ~4·5(6.3) for the K(P) adsorbent over the transition range, very little of this change (~0·7(0·3))

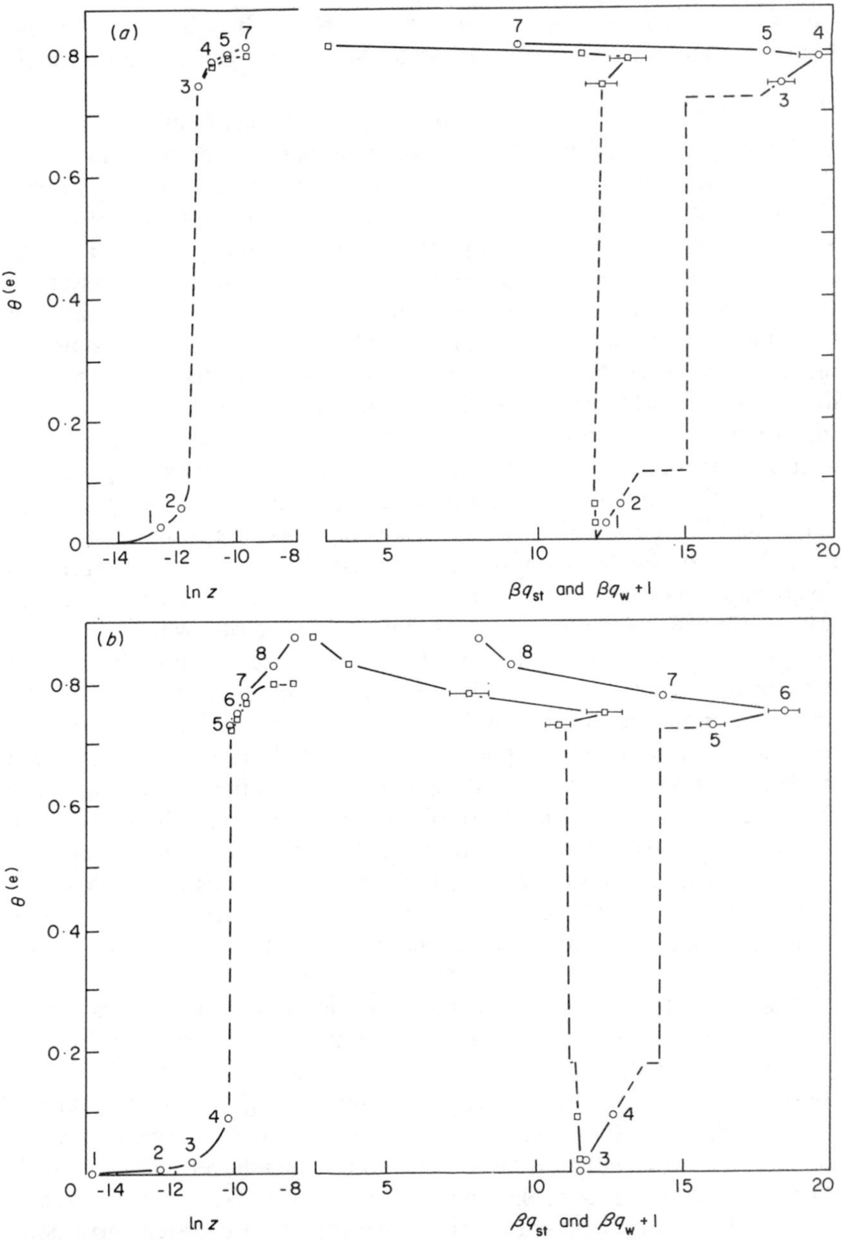

Fig. 6.15. The adsorption of argon on (*a*) the P- and (*b*) the K-adsorbent at $T^* = 0{\cdot}668$. Adsorption isotherms: excess coverage ($\theta^{(e)}$) versus activity. Isosteric heat (q_{st}) and the wall contribution ($\beta q_w + 1$) versus $\theta^{(e)}$. Numbers on the points indicate the run numbers. (After Nicholson *et al.*, 1981.)

can be attributed to changes in the adsorbent–adsorbate interaction. Thus both the high-barrier systems behave at these coverages in a fairly simple manner.

At coverages >0·805 characteristic sharp maxima are found in q_{st}. Figure 6.15 shows that a large part of the peak arises from the adsorbent–adsorbate (AS) potential, but that the contribution from the adsorbate–adsorbate (AA) potential is of the same sign. Two kinds of displacement of the adatoms were considered as the possible source of these effects. The first of these follows a suggestion previously made for two-dimensional systems (Parsonage, 1970). In this, it is assumed that at coverages less than that for which the peak occurs an appreciable number of the adatoms are off-site with respect to their x, y coordinates (although having the z-coordinate value which gives the minimum potential for those x, y values). As the completion of the epitaxial layer is approached, repulsion between adatoms causes the displaced atoms to revert to their epitaxial positions. However, this model would require that the adsorbate–adsorbate part of βq_{st} would decrease with increase of coverage because of the increase in the AA part of the energy (this being repulsive), and the simulation results do not support this prediction. The second kind of displacement considered was one in which the atom does not have the value of z which minimizes the potential for the given x, y position. The process responsible for the peak in q_{st} is then one in which the atoms revert to their site positions under the influence of attractive interactions from the increased number of adatoms on *nn* sub-lattice positions. This would cause both the AA and AS contributions to q_{st} to be of the same sign. Thus the predictions of this theory are in accord with the results of the simulation. Moreover, the second theory also seems the more reasonable since at the distance of nearest-neighbour sites on the sub-lattice the interactions would be attractive rather than repulsive. The fall in q_{st} with still further increase in coverage is almost certainly due to the build-up of the second adsorption layer.

The situation with regard to the isosteric heat of adsorption for the low-barrier, graphite-like J-system, which has been deferred until now, is more complicated. The positive-going peak and subsequent decline found for the high-barrier systems are again present, though now they occur near $\theta = 1$ rather than $\theta = 0{\cdot}805$. This difference is not unexpected, since the ideal condensed configuration towards which the adlayer tends is now the close-packed rather than the epitaxial monolayer. However, the peak is preceded by a negative "cusp" at $\theta \approx 0{\cdot}9$ for which the area corresponds to $\sim\frac{1}{2}\Delta\psi$. Analysis of this spike shows that it is almost entirely attributable to the AS interactions, but that the AA interactions make a more gradual negative contribution at a somewhat lower coverage ($\theta \approx 0{\cdot}81$). This suggests that as the epitaxial monolayer coverage is approached repulsions between some of the admolecules cause the layer to change to a fully incommensurate form, with loss of some AS attractive interactions and removal of some AA repulsions. Further increase of coverage causes the AS contribution to revert to its former value and the AA contribution to increase rapidly.

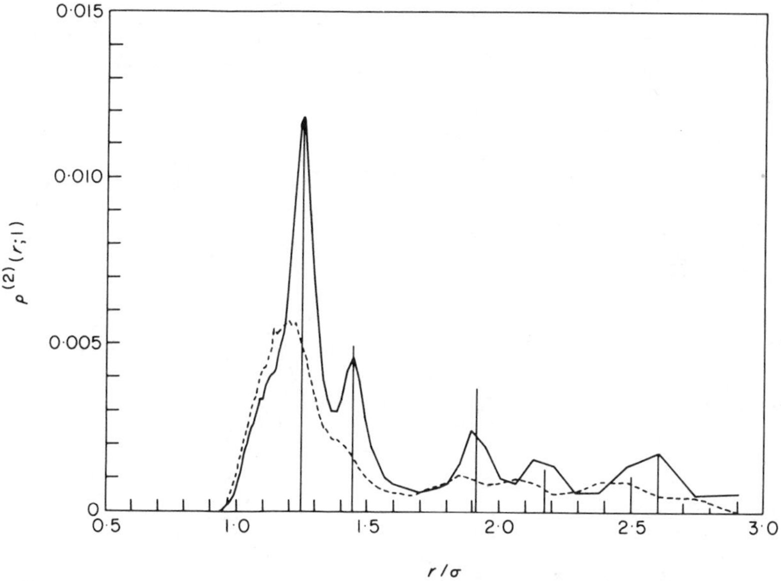

Fig. 6.16. Pair distribution functions for the adsorption of argon on the P- and K-adsorbents in the first layer at $T^* = 0{\cdot}668$. Vertical lines indicate the positions and relative intensities to be expected for the random lattice gas. ---, P2($\theta^{(e)} = 0.058$);——, K4($\theta^{(e)} = 0.093$). (After Nicholson *et al.*, 1981.)

Returning to the consideration of the high-barrier adsorbents, examination of the first-layer pair distribution function for a coverage of $\theta = 0{\cdot}093$ on the highest barrier (K) adsorbent shows that it corresponds well to what would be expected for a random lattice gas (RLG), i.e. a random distribution of adatoms on the hexagon centre site positions subject to exclusion of occupation of adjacent sites (Fig. 6.16). A shoulder on the first peak near $1{\cdot}12\sigma_{Ar}$ probably arises from clustering with, typically, two atoms on sites and a third touching them but off-site. For the low-coverage P-system ($\theta = 0{\cdot}058$) the conclusions are similar, but are modified by the greater ease with which displacements from site can occur: the peaks are therefore broader. At coverages corresponding to the condensed phase,

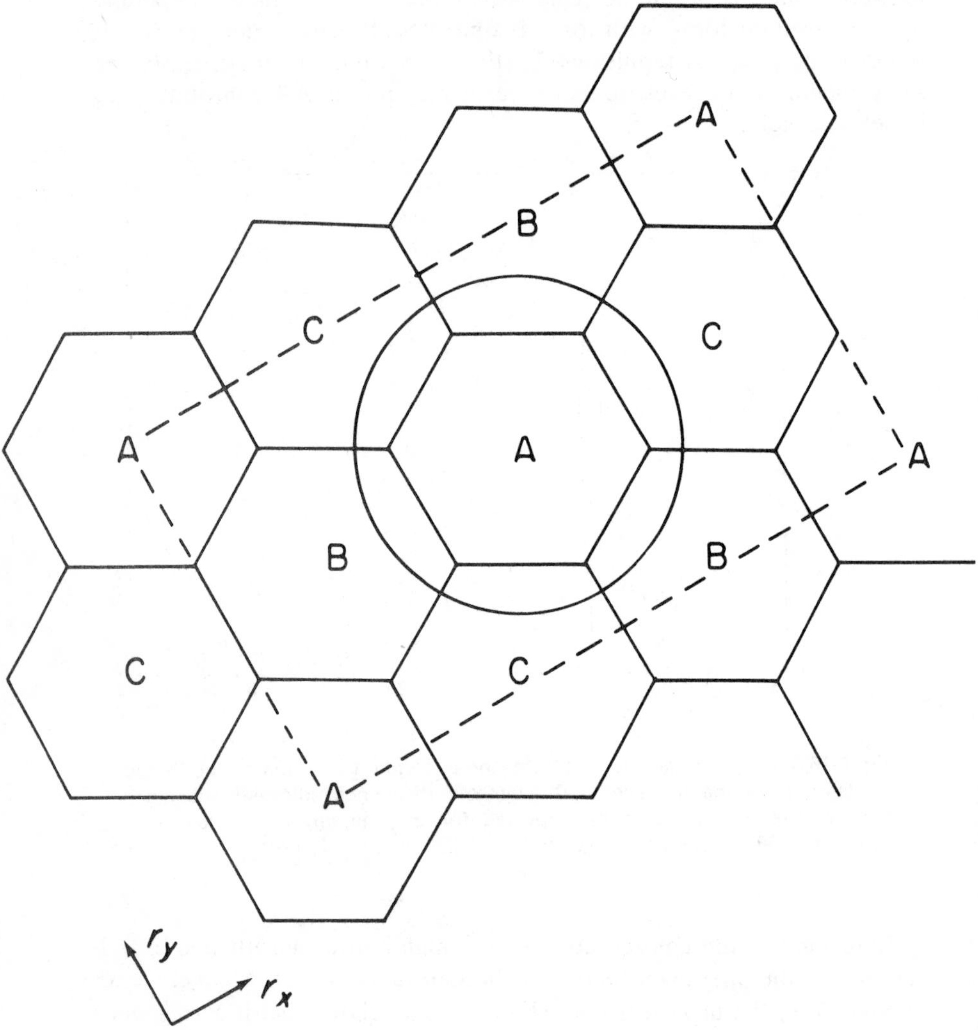

Fig. 6.17. The sub-lattices (A, B, C) of best adsorption sites for argon or krypton on graphite. The circle indicates the size of an argon atom ($r = 0{\cdot}3405$ nm). The unit cell of the graphite lattice is delimited by the broken lines.

maxima only occur at distances which would be possible with use of a single sub-lattice (say A of Fig. 6.17). The shoulder at $1{\cdot}12\sigma_{Ar}$ is scarcely evident at this coverage showing that the proportion (though possibly not the absolute number) of such abnormal intermolecular distances is less than at the lower coverage. One of the interesting conclusions from this work was that although the barriers of both P- and K-systems were sufficiently high to give essentially epitaxial adsorbed layers, nevertheless the displacements which were possible were sufficiently important for some properties to cause the two systems to show significantly different results.

It would be expected that the most interesting systems would be those in which the periodic potential is such that the tendencies to form epitaxial and non-epitaxial adsorbed layers are almost in balance. In most of the systems discussed in this chapter so far it appears that either the epitaxial (as with the P- and K-adsorbents of Nicholson *et al.*) or the non-epitaxial (as with argon on graphite, or any adsorbate on a non-periodic substrate) has a clear stability advantage. The only well studied system for which balance is approximately achieved appears to be krypton on graphite. Experiments show that usually the adsorption is epitaxial, but that at high values of the chemical potential it becomes non-epitaxial, thereby enabling ~10% more krypton atoms to be accommodated. The simulation by Spurling and Lane (1978) using a Steele periodic potential ($\Delta\psi = 40$ K) showed no evidence for epitaxial adsorption, and this has been confirmed by Whitehouse *et al.* (unpublished results) using a larger simulation box (able to accommodate ~722 molecules in a close-packed monolayer or 648 in an epitaxial one) and longer runs. Thus, some of the suspicions that the disagreement with experiment arose in a trivial way from the need to economize on computer usage have been allayed. Another possibility which has been eliminated by this later work is that the structures occurring in Spurling and Lane's simulations had in fact some epitaxial character, but that this was insufficient to enable it to be recognized by those authors from the inspection of snapshots. Whitehouse *et al.* have put the recognition of such order on a quantitative basis by employing a correlation function which is a measure of the epitaxy. Use of this confirmed that the stable phase has no sensible epitaxial character. These results force the conclusion that the model, rather than its evaluation, must be incorrect. A possible reason for the failure of the model lies in its neglect of three-body forces, and in particular those which involve two adsorbed molecules and an element of the surface. Some preliminary work shows that inclusion of a three-body potential in the Sinanoğlu and Pitzer form (Eqn (3.92)) leads to an increase in the stability of the epitaxial with respect to the non-epitaxial phase, although whether this is sufficient to explain the experimental observations remains to be seen.

The adsorption of methane on graphite has been simulated using an MD method by Severin and Tildesley (1980). The first part of this study was largely concerned with the behaviour of an isolated admolecule on the graphite surface: it corresponds therefore to the zero coverage limit. The adsorption potential was derived from an interaction site model, the sites being the atoms of the system, and the Lennard-Jones 12–6 law being assumed to hold for each site–site interaction. Two methods for the truncation of this potential were tried. In the first of these, interactions with surface carbon atoms were treated exactly up to a distance in the surface of R_c from the site closest to the admolecule. Long-range corrections were then applied for interactions with more remote carbon atoms in the surface and also carbon atoms in layers below the surface. The second, and more consistent, of these schemes took exact account of interactions with all

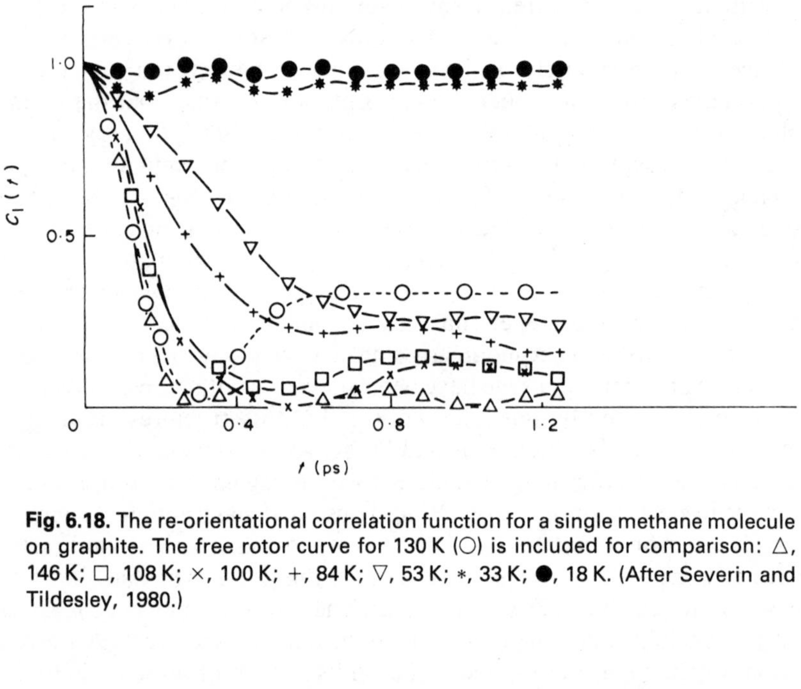

Fig. 6.18. The re-orientational correlation function for a single methane molecule on graphite. The free rotor curve for 130 K (○) is included for comparison: △, 146 K; □, 108 K; ×, 100 K; +, 84 K; ▽, 53 K; ∗, 33 K; ●, 18 K. (After Severin and Tildesley, 1980.)

carbon atoms, whether in the surface layer or within R_c of the carbon atom nearest to the admolecule; the remaining interactions were dealt with by a long-range correction. The disposable parameters in the potential energy function were chosen so as to fit the experimental values of q_{st} at $\theta = 0$ and the barriers to rotational motion of the methane molecule as determined by neutron scattering. The orientation of the methane molecule which gave the lowest energy was found to be the stable-tripod position, the barrier to rotation about the axis perpendicular to the surface being 212 cm^{-1} (experiment 205 ± 5 cm^{-1}). The equilibrium distance above the graphite surface of the methane molecule, when in the stable-tripod position, was found to be 0·33 nm, which agrees with experiment. Having established the potential, the motion of the methane molecule was followed for $\sim 10^5$ time-steps, each of $1{\cdot}25 \times 10^{-15}$ s. At 18·6 K it was found that the methane molecule could translate in the tripod position, with some rotational freedom around the axis perpendicular to the surface. Because of the very small amount of translational kinetic energy the motion was quite strongly affected by the periodicity of the lattice, with the consequence that the molecule did not travel in straight lines. On raising the temperature to 53·4 K, rotation about the axis perpendicular to the surface became fairly free and the translational motion was almost rectilinear. Just once during the run the molecule "tumbled", translational energy being converted to rotational energy about an axis parallel to the surface; for the rest of the run the molecule remained in the stable-tripod position. At a still higher temperature (108 K), vibrations perpendicular to the surface became important (+0·0102 to −0·0148 nm). As a result of these "collisions" with the surface the "tumbling", which was now quite frequent, did not correspond to free rotation. These effects can be well described in terms of the orientational autocorrelation function:

$$\mathscr{C}_1(t) = \langle \boldsymbol{a}(t) \cdot \boldsymbol{a}(0) \rangle \tag{6.9}$$

where $\boldsymbol{a}$ is the unit vector along the CH bond remote from the surface (Fig. 6.18). The frequency of the vibrations perpendicular to the surface fell gradually as the temperature was raised (98 ± 2 cm^{-1} at 0 K (extrapolated); 85·1 ± 2 cm^{-1} at 97 K; 74·5 ± 2 cm^{-1} at 170 K), these computed values being in fair agreement with the single neutron scattering result of ~100 cm^{-1} at low temperatures. Throughout this work the equations of motion used were those appropriate to a system which was classical ($h \rightarrow 0$). For a molecule such as methane with a small mass and very low moments of inertia this is difficult to justify for the lowest temperature studied (18·6 K). One may note the great difference in properties between CH_4 and CD_4 at temperatures in this range (Parsonage and Staveley, 1978). However, as Severin and Tildesley point out, it would be difficult to take account of quantum effects in the evaluation of the dynamical properties. What has been achieved, therefore, is to choose effective potentials which allow for quantum mechanical effects in that results in agreement with experiment are obtained if the potentials are used with the classical equations of mechanics.

3. Liquid-solid interfaces

(a) Introduction

More activity has been directed towards the study of this type of interface than has been the case for the previously discussed gas–solid interface. One reason for this is that with the liquid–solid interface one is not faced with simulation problems arising from large differences in density between the various regions of the system. With the gas–solid system it was this type of problem which called for the use of the GCEMC method (Ch. 4, § 2(d)). Liquid–solid systems are therefore commonly studied either by the Monte Carlo method in the canonical ensemble or by molecular dynamics.

Almost all the simulations have been carried out by the double-wall method described in the work of Lane and Spurling (1976) in the previous section. The application of this method will be discussed in more detail below. For these systems the alternative would be to have the liquid under study between a single wall, on one side, and the bulk liquid on the other. The molecules of the bulk liquid would not be moved during the simulation but would remain "frozen" in positions representing a typical "snapshot" of the liquid state. Use of a "frozen" bulk liquid region of this type to terminate a simulated liquid region has been made elsewhere by Chapela *et al.* (1975) in their study of the gas–liquid interface. Such a single wall method would appear to have the advantage of employing fewer moving particles, but this would only be so if the presence of the "frozen" bulk liquid did not alter the structure of the simulated liquid from what it would be if there were instead a steady, natural gradation to the true bulk liquid. If such a structure modification did occur, the "frozen" bulk liquid–simulated liquid interface would need to be placed further from the true wall so as to avoid interference of the surface effects.

Returning to consideration of the two-wall method, it should be realized that the walls need not be identical. If, however, they are the same then they provide a check on the results in that the two halves of the system should yield the same results. Against this, the use of different walls enables the effects of different types of wall to be surveyed more rapidly, each run giving information on two kinds of interface. Moreover, if one of the walls is hard, i.e. the molecule interacts with it according to the law $u^{(1)}=0, z \geqslant 0$ but $u^{(1)} = \infty, z < 0$, then the pressure component p_{zz} may be evaluated from the density at the wall by means of the relation $p_{zz} = kT \rho^{(1)}(z=0)$ (Ch. 3, § 4). Otherwise p_{zz} can be calculated by application of the virial theorem to the resolved parts of the forces perpendicular to the wall. As with the corresponding use of the virial theorem for bulk liquids, this latter method suffers from slow convergence. If p_{zz} is not evaluated the data obtained may, unknown to the observer, refer to a pressure which is significantly different from the normal atmospheric pressure, which is the main region of interest. Use of the GCEMC method would remove the difficulty with respect to the determination of p_{zz}, since it could be calculated from the known chemical potential, the density being automatically adjusted by addition or subtraction of molecules so as to yield the desired chemical potential. However, such simulations have rarely been used as they converge only slowly because of difficulties arising from the attempted creation of particles in an already dense fluid. Van Megen and Snook (1981) have, however, used this method in their later work on the hydrophobic effect (Ch. 8, § 1).

The majority of the work of this kind has been concerned with idealized liquids having intermolecular interactions which are represented by hard-sphere (HS), Lennard-Jones 12–6 (LJ) or Weeks–Chandler–Andersen (WCA) potentials. For the interaction between the wall and the molecules of the liquid the wall has in most cases been considered as some form of continuum and the resultant potential has been either of the hard-wall, Lennard–Jones 9–3 (Ch. 3, § 3(e)) or the Steele potential ($u^{(1)\mathrm{St}}$) (Ch. 3, § 3(e)), or a Weeks–Chandler–Andersen ($u^{(1)\mathrm{WCA}}$) version of one of the last two. Clearly one future development will be the employment of more realistic molecule–wall potentials. Indeed, Prasad and Toxvaerd (1980) for the two-dimensional system consisting of a sub-monolayer film have used a potential derived from the (111) face of an FCC crystal. The three-dimensional representations of the gas–solid interface having periodic potentials have been studied (Ch. 6, § 2(b)) and there does not seem to be any reason why the same development should not be made for the three-dimensional solid–liquid system.

In view of the large variety of information which has been obtained from the simulation of gas–solid interfaces (Ch. 6, § 2) it is surprising that for the liquid–solid systems very little other than $\rho^{(1)}$ and a smaller amount of $\rho^{(2)}$ data has been computed. Almost no attention has been given to thermodynamic or other "non-structural" quantities.

(b) Structureless solid adsorbents

A typical singlet density function $\rho^{(1)}(z)$ for these systems has marked oscillations ("ringing") near the wall, these dying out over several molecular diameters to give a smooth curve typical of the uniform liquid (Fig. 6.19). The most striking feature of these curves is the very considerable variation in amplitude of the oscillations according to the fluid density and the details of the forces in the system. Whereas the uniform fluids composed of

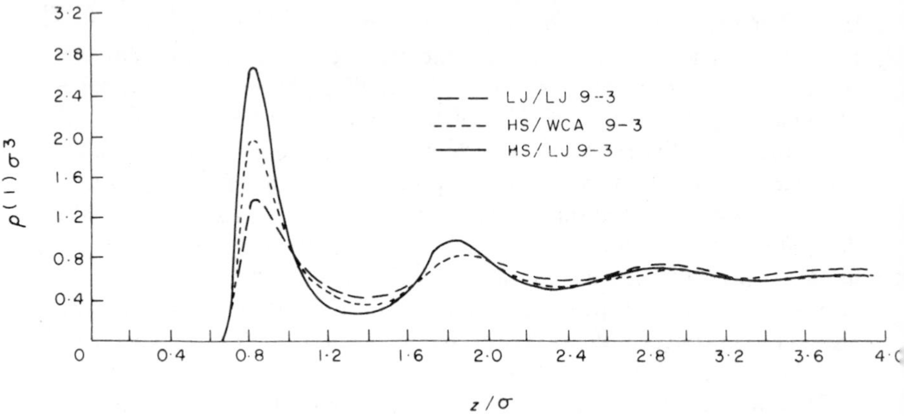

Fig. 6.19. Density profiles for three interfacial systems at $T^* = 1$ and $\rho\sigma^3 = 0{\cdot}65$. WCA 9–3 refers to the wall potential obtained by apply the WCA procedure to the Lennard-Jones 9–3 potential. (After Abraham and Singh, 1978.)

Lennard-Jones (LJ) and hard-sphere (HS) molecules have very similar structures indeed (Ch. 3, § 5(g)), the same is certainly not true when the liquids are put into contact with a wall. Figure 6.19 shows the marked difference between the LJ and HS fluids at reduced bulk density $\rho\sigma^3 = 0{\cdot}65$ and $T^* = 1$ when they are in contact with a wall imposing a Lennard-Jones 9–3 potential on the fluid. It is perhaps not surprising that the LJ fluid–wall system, for which the AA and AS interactions are more similar, both being fairly "soft", gives less "ringing". The other curve shown in the figure, that for the HS fluid in contact with a wall giving only the repulsive part of the 9–3 potential (in the Weeks–Chandler–Andersen sense)†, is also in accord with this argument if one takes the WCA 9–3 to be "harder" than the LJ 9–3 wall potential.

A more extensive examination of this problem has been made by Snook and van Megen (1979). Their simulations are all for bulk densities in the range $\rho\sigma^3 = 0{\cdot}75$–$0{\cdot}88$, which are considerably larger than those referred to in the work of Abraham and Singh, described in the previous paragraph; T^* was again 1 for most of the runs, there being two simulations at $T^* = 1{\cdot}17$. Some of their results are shown in Fig. 6.20. A striking result is the large difference in shape between the LJ/WCA and LJ/St systems ($u^{(1)\mathrm{WCA}}$ is the result of applying the WCA procedure to the Steele potential (Eqn (8.2)), i.e. $u^{(1)\mathrm{WCA}} = u^{(1)\mathrm{St}}(z) - u^{(1)\mathrm{St}}(z_{\min})$ for $z \leqslant z_{\min}$, but equals

† We use the convention that the types of AA and AS interactions are indicated by the descriptions to the left and right of the solidus, respectively.

zero for $z > z_{min}$, z_{min} being the position of the minimum in the Steele potential). This is further support for the conclusion reached earlier, that $u^{(1)WCA}$ is not a good approximation to the more realistic $u^{(1)St}$ as far as the structure of the fluid is concerned (Fig. 6.20(*a*)). Again, the softer of the two wall potentials ($u^{(1)WCA}$) leads to a smaller "ringing" effect. A similar conclusion is reached from Fig. 6.20(*b*): $u^{(1)WCA}$ is softer than HW and gives much smaller oscillations. However, a comparison of the curves for $u^{(1)St}$ and HW in Fig. 6.20(*a, b*) does not fit with these conjectures: $u^{(1)St}$ gives the larger "ringing" effect even though it is obviously softer than HW. It is difficult to imagine that this result would have been altered if the two runs had been at exactly the same density, since firstly the density difference is so small (0·84–0·80) and secondly it would be expected that the lower density of the LJ/St run would lead it to show rather less "ringing". The difference in temperature of the two runs may well be more important, but this too can be dismissed since Fig. 6.20(*c*) shows LJ/St and LJ/HW at the same temperature ($T^* = 1{\cdot}17$) and density ($\rho\sigma^3 = 0{\cdot}84$) and the greater oscillations near the wall projecting $u^{(1)St}$ are still evident. It appears, therefore, that the simple notion of similarity of the AA and AS potentials, in the way discussed here, cannot be used as a reliable predictor of the type of $\rho^{(1)}$ curve which will be obtained. Theories based on this idea of a "disturbance" projected from the interface are described in detail in Chapter 7. With regard to the oscillations, Snook and van Megen observe that a Lennard-Jones fluid seems to be much more sensitive to the type of wall potential than is the hard-sphere fluid, as seen by the similarity of the $\rho^{(1)}$ curves for an HS fluid against three types of AS potential shown in Fig. 6.20(*d*), as against the greater variation found for the LJ fluid (Fig. 6.20(*a, b, e, f*)). In particular, the LJ fluid responds quite differently to HW and WCA potentials (Fig. 6.20(*e*)), whereas the HS fluid shows stratification which is not very different for these two kinds of wall (Fig. 6.20(*e*)). The other aspect, which has already been touched on, is the effect of the density of the fluid on the oscillations. Figure 6.21 shows the variation of $\rho^{(1)}$ out to the second minimum over a wide range of bulk density ($\rho\sigma^3 = 0{\cdot}57$–$0{\cdot}91$) for a HS/HW system (Snook and Henderson, 1978). Apart from the necessary increase in $\rho^{(1)}$ at the wall as the density is increased (since $p = \rho^{(1)}kT$), the most obvious feature is the increase in sharpness of the peak centred near $1{\cdot}0\sigma$, this being accentuated by the wells on either side dipping almost to zero. The increase in sharpness of the oscillation applies also to more remote oscillations so that at the highest densities studied ($\rho\sigma^3 = 0{\cdot}81, 0{\cdot}91$), which were close to that of the melting solid (~0·94), the oscillations did not have sufficient space within the box to "die away" even though the box was long enough to accommodate ten layers. At $\rho\sigma^3 = 0{\cdot}81$ one can be fairly certain that the results correspond

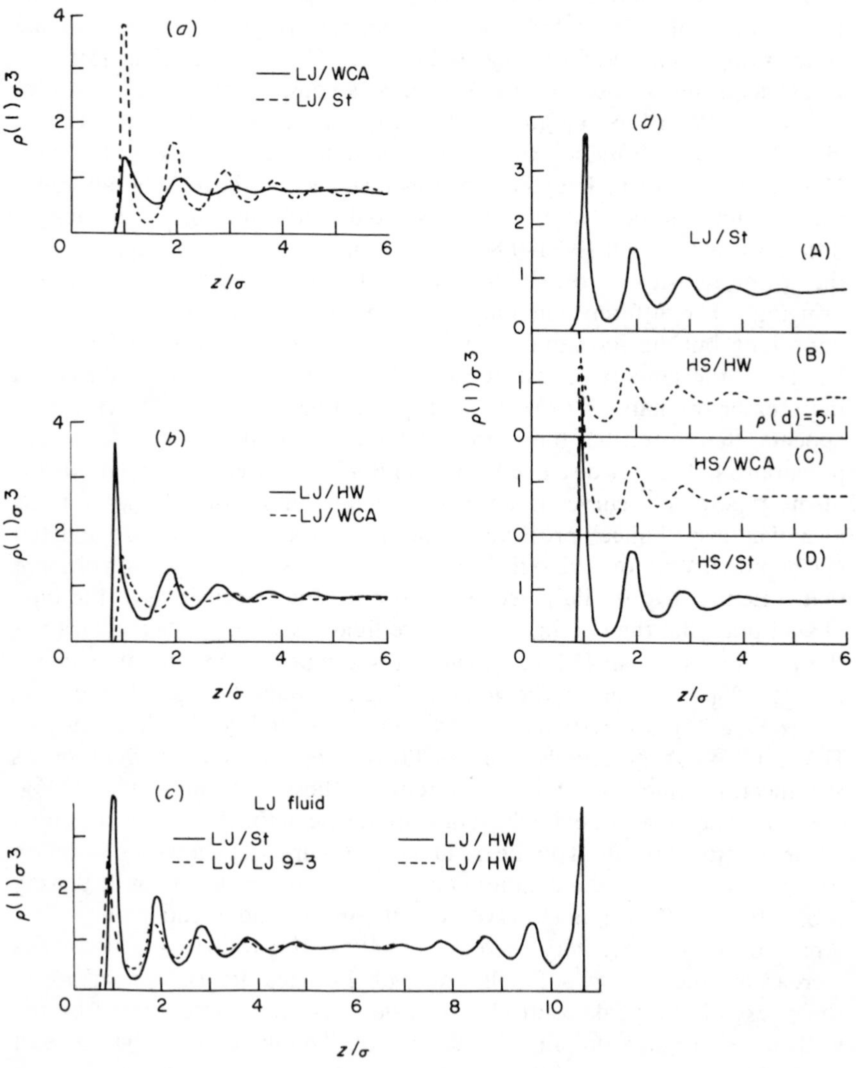

Fig. 6.20. Density profiles. (*a*) $T^* = 1$, $\rho\sigma^3 = 0{\cdot}80$. (*b*) $T^* = 1{\cdot}17$, $\rho\sigma^3 = 0{\cdot}84$. (*c*) $T^* = 1{\cdot}17$, $\rho\sigma^3 = 0{\cdot}84$. (*d*) $T^* = 1$, $\rho\sigma^3 = 0{\cdot}76$ (A), $0{\cdot}75$ (B, C, D). (*e*) $T^* = 1$, $\rho\sigma^3 = 0{\cdot}80$ (A:LJ), $0{\cdot}76$ (A:WCA), $0{\cdot}75$ (B). (*f*) $T^* = 1{\cdot}17$, $\rho\sigma^3 = 0{\cdot}84$

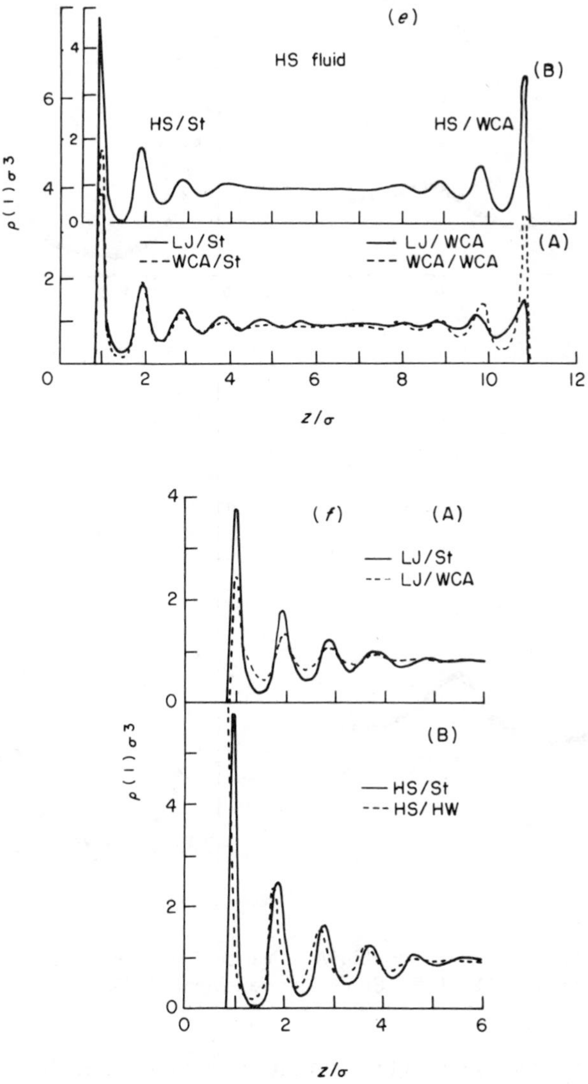

(A: LJ/St); $T^* = 1$, $\rho\sigma^3 = 0{\cdot}88$ (A: LJ/WCA), 0·90 (B). (After Snook and van Megen, 1979.)

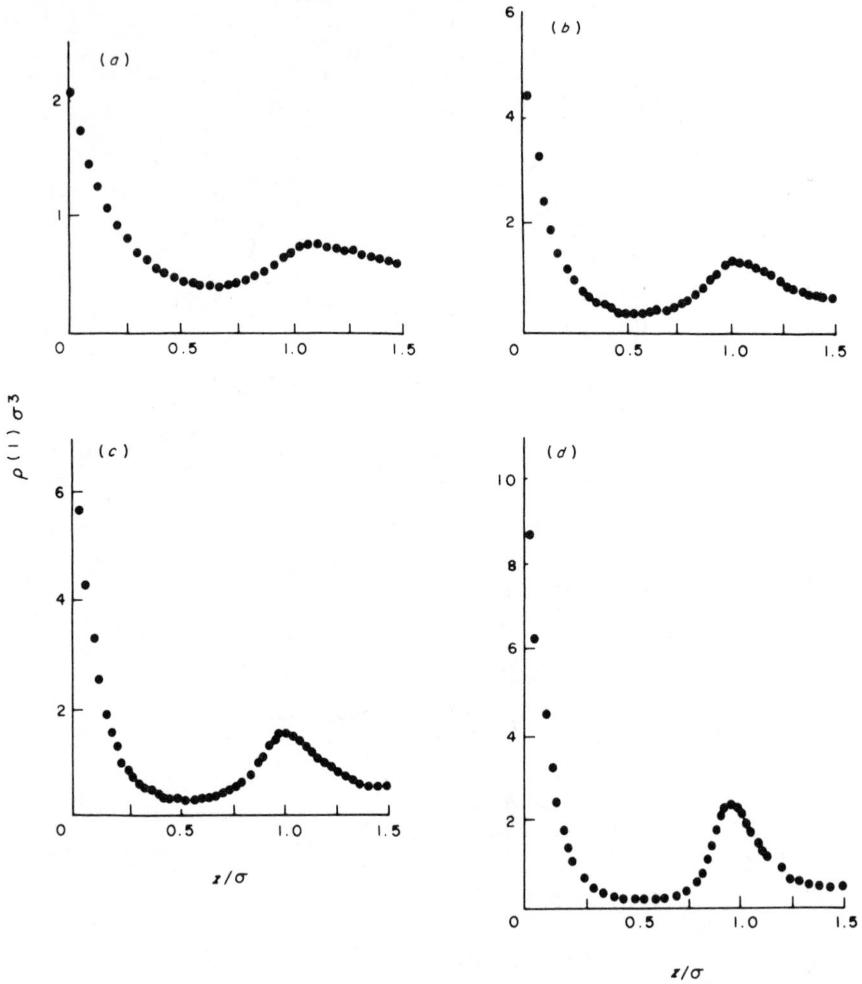

Fig. 6.21. Density profiles for HS/HW at various densities. (*a*) $\rho\sigma^3 = 0{\cdot}57$; (*b*) 0·755; (*c*) 0·81; (*d*) 0·91. (After Snook and Henderson, 1978.)

closely to those derived from two non-interfering interfaces, but it is clearly necessary at that bulk density to have a simulation box of at least 10σ in length. A more detailed discussion of the situation in which interference of the two profiles becomes large is given in Chapter 8. In spite of the very high peak values of $\rho^{(1)}$ which can occur close to the wall, when averaged over the first layer the density is appreciably less (~20%) than that in the bulk (Snook and Henderson, 1978).

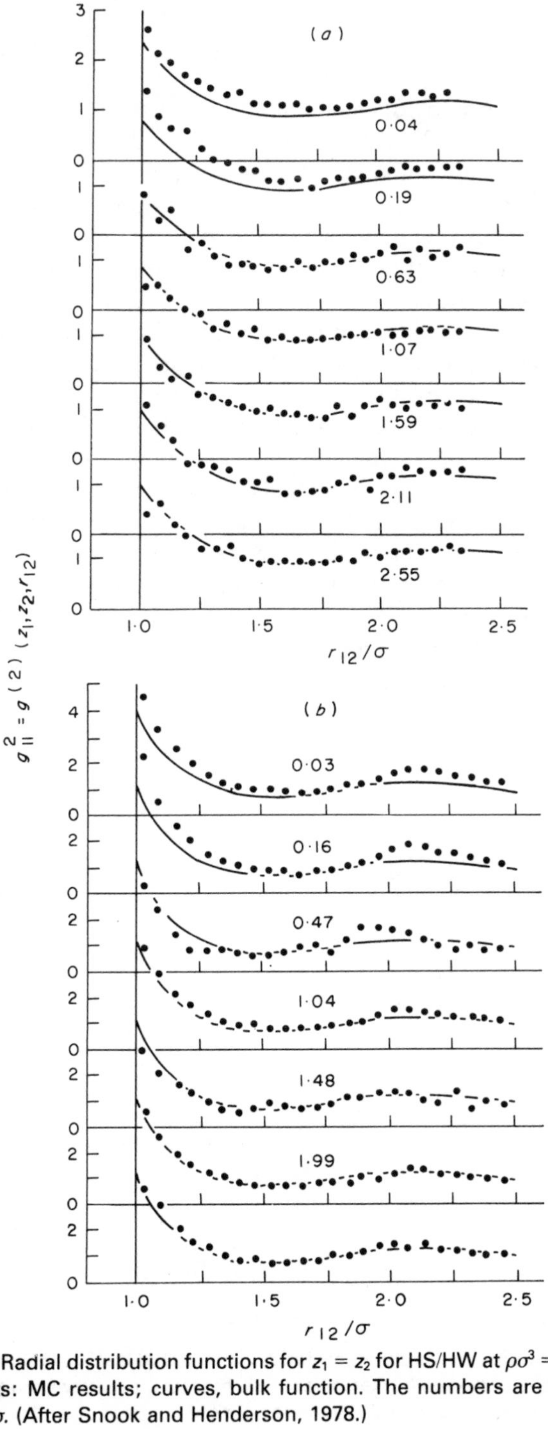

Fig. 6.22. Radial distribution functions for $z_1 = z_2$ for HS/HW at $\rho\sigma^3 = 0{\cdot}57$ (*a*), 0·81 (*b*). Points: MC results; curves, bulk function. The numbers are the values of $z_1/\sigma = z_2/\sigma$. (After Snook and Henderson, 1978.)

No such systematic study of the effect of bulk density on the profile has been presented for LJ fluids. Comparison can be made for that fluid against a wall projecting a Steele potential (Eqn (8.2)) at $\rho\sigma^3 = 0{\cdot}76$ and $0{\cdot}84$ from Fig. 6·20(d, f). The enhancement of the oscillations on increasing the density is evident, but it is difficult to say from this whether the LJ fluid is more or less sensitive in this respect to density change than the HS fluid.

The other obvious measure of liquid structure is the pair distribution function $\rho^{(2)}$, or alternatively (and better) the radial distribution function $g^{(2)}$. As mentioned earlier (Ch. 4, § 2(a)) there are severe statistical problems with regard to the evaluation of these quantities arising from their dependence on three variables, z_1, z_2 and r_{12}, even for a monatomic adsorbate and a structureless adsorbent. As a consequence of this, Snook and Henderson in their simulations of HS/HW systems restricted their evaluations to the configurations with both z_1 and $z_2 \leqslant 2\sigma$: this enabled both the first and second maxima to be included in the study. Figure 6.22 shows the radial distribution functions at two widely spaced bulk densities for $z_1 = z_2(g_{\parallel}^{(2)})$. It will be seen that only the curves for $z_1 = z_2 < 0{\cdot}5\sigma$ show significant deviations from the curves for the uniform fluid. For $z_1 = z_2 > 0{\cdot}5\sigma$ the radial distribution functions, within the limit of accuracy

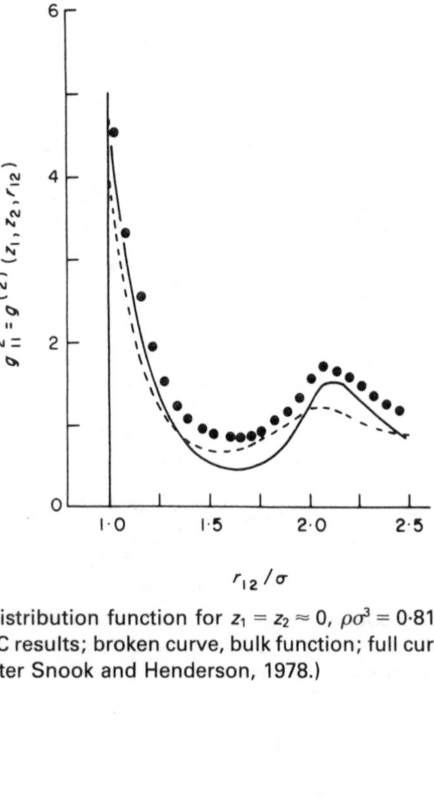

Fig. 6.23. Radial distribution function for $z_1 = z_2 \approx 0$, $\rho\sigma^3 = 0{\cdot}81$, for the HS/HW system. Points: MC results; broken curve, bulk function; full curve, bulk function for hard-discs. (After Snook and Henderson, 1978.)

of the computations, show no modification attributable to the presence of the surface. This is so in spite of the fact that $z = 0{\cdot}5\sigma$ lies well within the "layered" region of the density profile. Indeed, it corresponds to the first minimum. Thus, taking these $g_{\parallel}^{(2)}$ values rather than the density profiles, as the measure of structure it could be said that the influence of the surface is not very great. The strong layering of the fluid near the surface combined with the liquid-like radial distribution functions might suggest that the first layer could well be approximated by a hard-disc model. However, Fig. 6.23, in which the $g_{\parallel}^{(2)}$ curve for $z_1 = z_2 \approx 0$ is compared with the bulk hard-sphere and hard-disc results, shows that representation as a hard-disc system is quite good near the first and, to a lesser extent, the second maximum but is worse than the hard-sphere model near the first minimum. $g^{(2)}(z_1, z_2, r_{12})$ is very sensitive to the orientation of the inter-particle vector. In particular, it turns out that the conclusion with regard to the small effect on $g^{(2)}$ arising from the presence of the wall no longer holds if the inter-particle vector is perpendicular to the wall ($g_{\perp}^{(2)}$) (Fig. 6.24). For the higher density ($\rho\sigma^3 = 0{\cdot}81$) there is a very large reduction in $g_{\perp}^{(2)}$ at the contact distance together with a shift of the first minimum to shorter distances; at the lower density both these effects are much less marked. The reduction

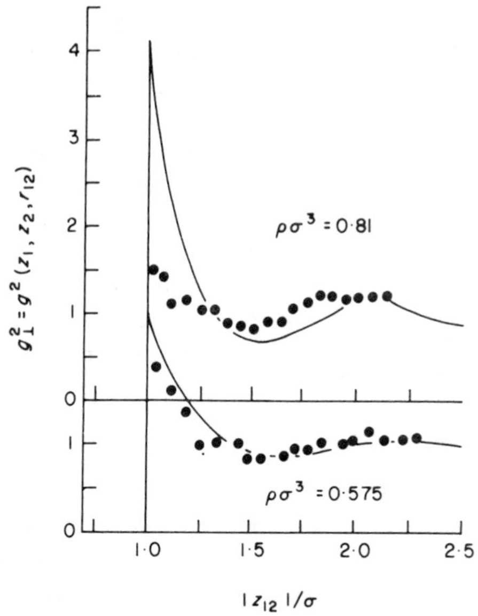

Fig. 6.24. Radial distribution function for $z_1 \approx 0$, $r_{12} = |z_{12}|$ for the HS/HW system. Points, MC results; curves, bulk functions. Upper curve, $\rho\sigma^3 = 0{\cdot}81$; lower curve, $\rho\sigma^3 = 0{\cdot}575$. (After Snook and Henderson, 1978.)

in $g_{\perp}^{(2)}$ at the contact distance may be understood in terms of the formation of an approximately close-packed layer of atoms at $z_1 \approx 0$, with the second layer now tending to pack over the interstices in the first layer and therefore being "lost" to the distribution function for inter-particle vectors perpendicular to the wall. The increase in the distribution function near $\sqrt{3}\sigma$ may also be understood in these terms, since for a quasi-CPH arrangement the third layer atoms would lie immediately over those of the first and at a distance of $\sqrt{3}\sigma$. In employing this argument it is not assumed that the structure is that of a solid. Rather, it is being assumed that the pattern of short-range distances will have some of the character of a solid, more especially at the higher densities.

When the radial distribution function for a Lennard-Jones fluid against a wall projecting the Steele potential of Eqn (8.2) was studied, it was found that for inter-particle vectors parallel to the wall there was appreciable difference from the bulk distribution function even at $2{\cdot}81\sigma$. Nevertheless, the conclusion reached for the HS/HW system, that the wall induces very much less change in $g^{(2)}$ than it does in $\rho^{(1)}$, remains true (Snook and van Megen, 1979).

In one of Abraham's simulations (of an LJ/HW system), in which the overall density is close to a coexistence value, $\rho^{(1)}$ fell smoothly to a value characteristic of a vapour phase and no oscillations were found. A similar result has been found by Sullivan *et al.* (1980) at a lower temperature and higher density (see Fig. 7.7). In this case, however, $\rho^{(1)}$ was still falling as $z \to 0$, indicating that a distinct vapour phase had not been formed. On changing the density and temperature slightly the sign of the slope rapidly changed, since the vapour phase was now being "squeezed out". Sullivan (1979) has also shown how a van der Waals model of the interface can be used to reproduce these various forms of $\rho^{(1)}$ behaviour near the wall. According to the comparative strengths of the AS and AA interactions, as represented by $\varepsilon^{(\mathrm{w,a})}/\varepsilon^{(\mathrm{a,a})}$, and the reduced temperature of the adsorptive ($T^* < T_c$), the system may have adsorption which is infinite, positive finite or negative. This work is described in §2(c) of Chapter 7.

(c) Structured solid adsorbents

Abraham (1978) has examined the effect on $\rho^{(1)}$ of introducing a more realistic potential for the AS interaction which has periodicity in the x- and y-directions. He considered a substrate having the FCC structure with the interface being a (100) face. Comparisons were made with related potentials in which the x, y dependence had been removed by averaging. The first of these was a sum of two 10–4 term pairs arising from integration over the first two planes of adsorbent atoms, whilst neglecting the remainder;

the second was a "Boltzmann-averaged" potential $(u^{(1)}(z))$ defined by

$$\exp(-\beta u^{(1)\mathrm{B}}(z)) = \mathscr{A}^{-1}\int_{\mathscr{A}} \exp[-\beta u^{(1)}(x, y, z)]\,\mathrm{d}x\,\mathrm{d}y. \qquad (6.10)$$

It is clear from Fig. 6.25 that, at least as regards the evaluation of $\rho^{(1)}$, the Boltzmann-averaged potential is superior to the double 10–4 potential. The averaging described by Abraham bears a superficial resemblance to that which was carried out in some early treatments of bulk fluids having angle-dependent intermolecular interactions, but was later discredited (Balescu, 1955; Rowlinson, 1958). However, Balescu's simplification was to perform an average over orientations in the exponent, a fault which Abraham does not make. In Abraham's paper the preliminary Boltzmann averaging applies to the xy variation of the AS interactions; the corresponding z-variation together with all the components of the AA interaction

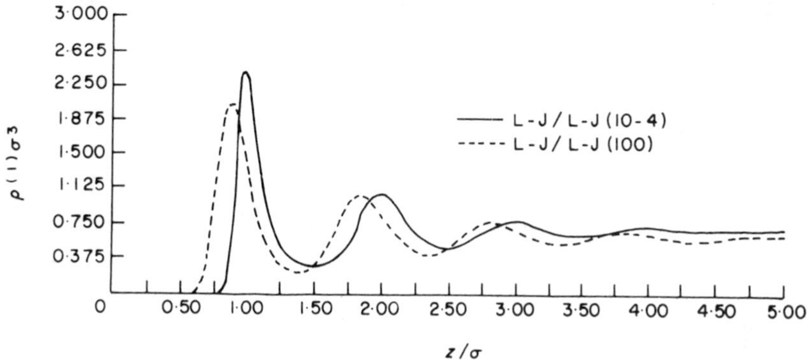

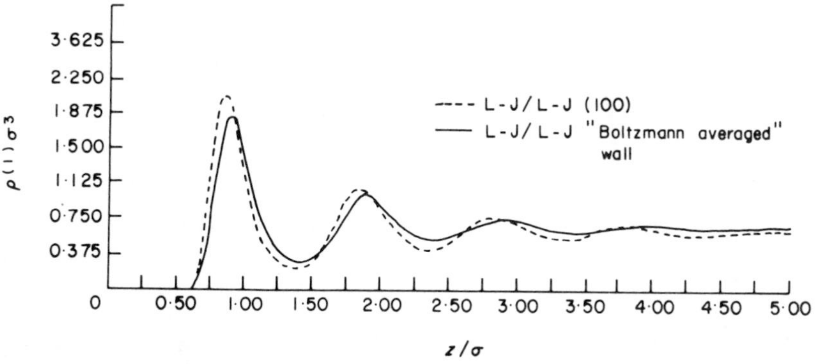

Fig. 6.25. Density profiles for a LJ fluid at $T^* = 1$, $\rho\sigma^3 = 0{\cdot}62$. (After Abraham, 1978.)

are then treated fully in a MC simulation. The comparison of this simulation with that in which the xy variation of the AS potential is retained in the simulation, which is the method used by Abraham, seems to be the only way of assessing the reasonableness of the Boltzmann-averaging procedure. Abraham's test of the approximation was for a fairly high temperature ($T^* = 1$) and a moderate density ($\rho\sigma^3 = 0{\cdot}62$). Since it would be expected

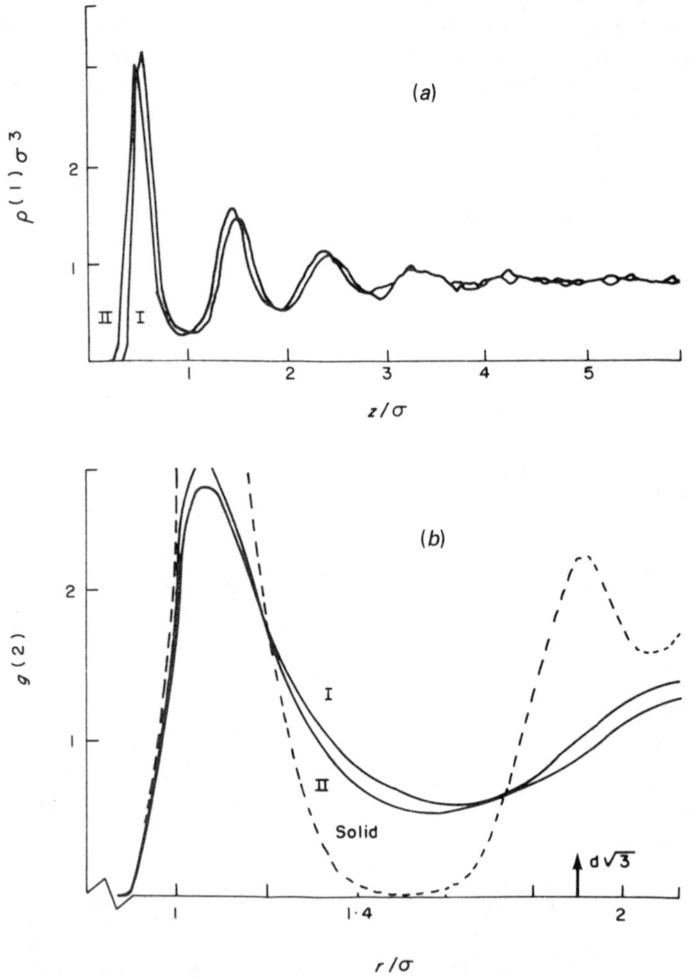

Fig. 6.26. A Lennard-Jones fluid ($\rho\sigma^3 \approx 0{\cdot}83$) in contact with a surface without (I) and with (II) a periodic potential. (*a*) The density profile. (*b*) The radial distribution function. Dotted curve, function for a (111) plane of an FCC crystal. A triangular lattice would show a peak in $g^{(2)}$ at $\sqrt{3}d$. (After Toxvaerd, 1981.)

that the validity of the approximation would vary with the relative strengths of the AS and AA interactions, it would be interesting to see some tests for some other ratios of the strengths of these interactions.

Toxvaerd (1981), using MD, has made a more extensive study of the effect of a periodic substrate field on the properties of LJ fluids with densities in the range $\rho\sigma^3 = 0{\cdot}58$–$0{\cdot}83$, thus covering some fairly high densities. All the runs were at $T^* = 1{\cdot}5$, which is the temperature found by Hansen and Verlet (1969) to be the melting temperature for the bulk solid. Figure 6.26 shows $\rho^{(1)}$ and $g^{(2)}$ (for the first layer) for systems without (curve I) and with (curve II) the periodic potential. The differences between the curves are small. In the second layer the radial distribution functions for systems I and II are indistinguishable. Calculations of the coefficients for diffusion in the surface layer parallel and perpendicular to the surface showed that at high pressure (~2 kbar) the former is less for the structured than for the unstructured surface, which is in turn very similar to that for the bulk. For diffusion perpendicular to the surface the structured and unstructured surfaces produce very similar results. The energy of adsorption is only different for the two systems when the pressure is high, the structured surface then giving the lower energy, i.e. the "stronger" adsorption. More marked changes in surface tension were found, the structured surface leading to the higher values of that quantity. The general conclusion from this work is, however, that the effects of the periodic part of the potential are essentially limited to the first layer, at least for the surface considered here, which was the (111) plane of a crystal of Lennard-Jones particles.

Computer simulations of the kind described are exceedingly demanding in computer time and so a large part of the incentive for the work on $\rho^{(1)}$ and $\rho^{(2)}$ described here has been the hope of obtaining a perturbation theory for these interfacial systems. Unfortunately, it has been found that the approximation which has been used for bulk liquids, namely that the structure of the fluid is essentially determined by the repulsive part of the interactions, is not valid for the interfacial systems studied here. Thus Fig. 6.19 shows $\rho^{(1)}$ results of Abraham and Singh (1978) for the same HS fluid against a wall giving a 9–3 potential and one in which only the repulsive part of that potential has been retained (the WCA form). $\rho^{(1)}$ is seen to be quite sensitive to the inclusion of the attractive part of the potential, at least as far as the amplitudes of the undulations are concerned. It is clear, therefore, that the repulsive part of the potential does not provide a suitable reference system for a perturbation treatment, and, indeed, no suitable simple reference system has so far been found. Perhaps the most promising attempt in this direction is that of Fischer and Methfessel, which combines features of the BBGKY and the perturbation theories (Ch. 7, §3(c)).

4. Gas–solid interfaces: small pores

All the studies mentioned so far, and most of those which will be discussed subsequently, in this chapter are concerned with plane surfaces. This is largely because these would appear to provide the simplest cases for examination, both in carrying out the simulation and in analysing the results which are produced. However, in many real systems the adsorption space is much less than infinite in either one or two directions. For most adsorbents of high specific area the major part of the accessible adsorbing surface is comprised of the walls of cylindrical pores (two non-infinite dimensions), which in many cases may be only a few molecular diameters across (micropores) or a few tens of molecular diameters across (mesopores). In other cases, the surfaces may be those on the sides of narrow slits in the adsorbent (one non-infinite dimension).

An MD study of a moderately dense ($\rho\sigma^3 = 0{\cdot}254,\ 0{\cdot}512$) HS fluid in narrow pores of square cross-section having hard walls has been made by Subramanian and Davis (1979). In the z-direction periodic boundary conditions were assumed, thereby approximating to a system infinite in that direction. When specular reflection from the other walls was assumed the results shown in Table 6.1 and Figs. 6.27 and 6.28 were obtained. $Z' = \bar{p}_{yy}V/NkT$, where $\bar{p}_{yy}$ is the pressure exerted on one of the walls of the pore (this pressure may vary across the face of the pore because the pores are non-cylindrical, so the result given is a value averaged over the wall). $Z'' = \bar{p}V/NkT$ where $\bar{p}$ is the spatial average of the trace of the fluid pressure tensor. It is seen that for the more dense system these pressures are considerably (>12%) greater than would be found for an infinite system at the same density. This discrepancy is associated with the build-up of

Table 6.1. Equilibrium results for a hard-sphere fluid in square pores of cross-section $l_x \times l_y$ and length l_z, with periodic boundary conditions in the z-direction. For comparison, data for the simulated bulk system are given (based on a box with sides $3{\cdot}75\sigma \times 3{\cdot}75\sigma \times 7{\cdot}5\sigma$ and with full periodic boundary conditions)

$\rho\sigma^3$	l_x/σ	l_y/σ	l_z/σ	Z'	Z''	Z/N^a
Pore system						
0·254	5	5	8·5	1·94	1·94	3.10
0·512	3·75	3·75	7·5	2·78	2·90	9·64
Infinite system						
0·512	–	–	–	–	2.48	7.80

[a] In units of $\sigma^{-1}(kT/m)^{1/2}$ Adapted from Subramanian and Davis, 1979.

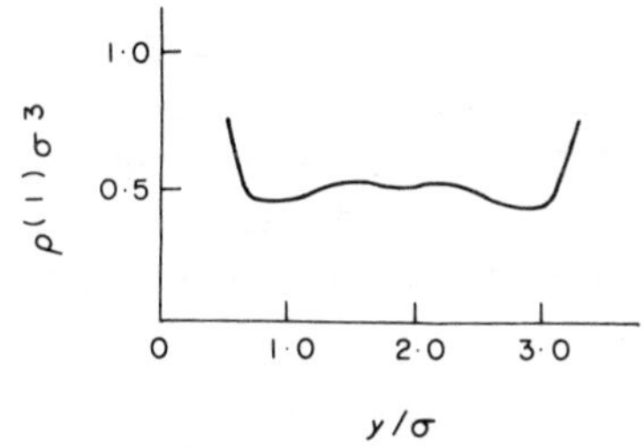

Fig. 6.27. Qualitative plot of fluid density against distance from one of the walls. (After Subramanian and Davis, 1979.)

density close to the wall at the expense of the density nearer to the centre of the pore (Fig. 6.27). Z'', which is computed as the virial for a bulk gas, is difficult to interpret, because it must be made up of unequal contributions from the three directions. Nevertheless, its deviation from the value for the infinite system is an indication of the effects which can be produced in systems of small dimension. Again, the ~20% increase in the collision

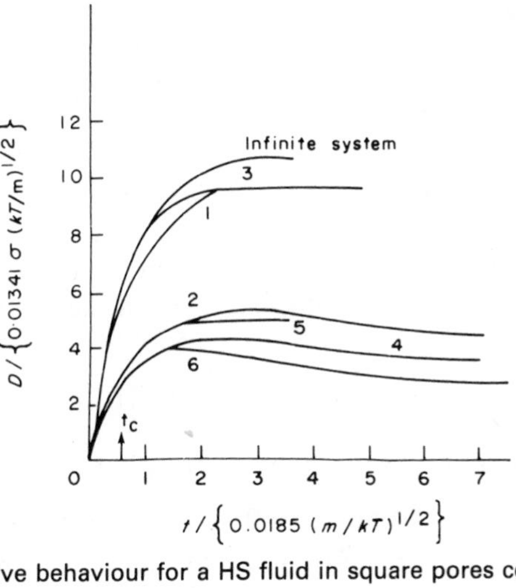

Fig. 6.28. Diffusive behaviour for a HS fluid in square pores compared with an infinite system.

$D_{\parallel}$	D_{τ}	
1	2	spec. wall I
3	4	diff. wall II
5	6	diff. wall III

Note: $D \to 0$ as $t \to \infty$.

(After Subramanian and Davis, 1979.)

rate per particle (Z/N) as compared with the infinite system is a large effect and may also be thought of as arising from unequal distribution of density: the increase arising from the regions of high density near the wall more than counterbalancing the diminished contribution from other parts.

Diffusion was also considered in this work, calculations being carried out for three types of wall:

Type I: a wall giving specular reflection.
Type II: a wall giving diffuse reflection with respect to the x- and y-directions coupled with "slip" with respect to the z-component. In other words, v_z is unchanged by the reflection, $(v_x^2 + v_y^2)^{1/2}$ is also unchanged, but v_y is chosen uniformly in the range 0 to $(v_x^2 + v_y^2)^{1/2}$ with the sign such as to indicate that the particle is reflected back into the box.
Type III: a wall with diffuse reflection with respect to all three components of velocity. The magnitude of v is unchanged; $v_z/|\boldsymbol{v}|$ is chosen uniformly on the range $(-1, +1)$, i.e. back-scattering is as likely as forward scattering; $\boldsymbol{n} \cdot \boldsymbol{v}/|v|$ (where $\boldsymbol{n}$ is the normal to the surface) is chosen uniformly on the range (0, 1), and the final component is found by difference.

The diffusion coefficient for the z-direction was computed from the velocity autocorrelation function:

$$D = \int_0^\infty \langle v_z(0) \cdot v_z(t) \rangle \, \mathrm{d}t \tag{6.11}$$

A transverse diffusion coefficient for motion in the x- and y-directions was obtained from the mean-square displacement in the x, y plane after time τ

$$D_\tau = \langle (\Delta x)^2 + (\Delta y)^2 \rangle / 4\tau \tag{6.12}$$

For an infinite system one would obtain D as $\lim_{\tau \to \infty} D_\tau$. However, because of the finite size of the system $D_\tau \to 0$ as $\tau \to \infty$. (D_τ increases with τ up to a maximum for each type of wall after which it slowly falls towards zero.) However, this quantity (D_τ) corresponds to the diffusion coefficients found by some spectroscopic methods, e.g. pulsed NMR or transient spectroscopic studies on samples in which the molecules are indeed confined in this way. Reflections from walls of types I and II do not affect v_z and so, at first sight, it might appear that the coefficient for diffusion along the axis of the pore ($D_\parallel$) would be the same for these systems as for the infinite system. This turns out to be so if times less than that between intermolecular collisions (t_c) are used in Eqn (6.11). However, for times longer than t_c (for wall I) or $2t_c$ (for wall II), as a result of back-scattering at molecule–molecule collisions, $D_{\parallel}$ falls short of the value computed for the infinite system. Wall III, which produces some back-scattering at wall–molecule collisions, not surprisingly leads to a large reduction in $D_{\parallel}$.

5. Microclusters

A considerable amount of interest has been shown in the stability and properties of small three-dimensional clusters of atoms. Many of the unique properties of such systems arise as a result of the predominance of atoms in a non-uniform or surface environment, rather than a bulk environment. This, and their importance as adsorbents as well as centres for crystal growth or catalysis, makes them a natural choice for discussion in the present context.

One item of concern is the possibility of phase transitions in such small systems. Statistical mechanics shows that, strictly speaking, singularities can only occur in the thermodynamic properties if the system is infinite; otherwise, the transition must be gradual. Nevertheless, it is still worthwhile to ask if the "transition" can be sufficiently sharp to be, for all practical purposes, a true one. Another aspect of this same problem is the question of the degree of ordering in the cluster. The absence of a true transition, as indicated above, is associated with a lack of strict long-range order at all non-zero temperatures. However, it is again worthwhile to enquire about the degree of order (short-range order) which can be found in these clusters. For example, is it possible, in some circumstances, to consider the cluster as being solid? This is important if we are considering the clusters as possible nucleation centres.

When examining the "melting" of a cluster using computer simulation, the results are not sensitive to the actual constraints on the system, for a wide range of conditions. Thus, Briant and Burton (1975) applied the MD technique to a free cluster, the atoms being able to drift away from each other to infinity. Their simulations commenced with the particles (Lennard-Jones particles with parameters appropriate to argon) in a cluster of approximately the minimum energy. The energy of the system was then increased in a step-wise fashion with virtual equilibrium being allowed to occur at each stage. This gradual procedure seems to be necessary to avoid obtaining results which refer to the system in a metastable form. Indeed, the length of run required was surprisingly great: 10^4 time-steps ($=10^4 \times 10^{-14}$ s) for clusters of 13 or fewer atoms and 3×10^3 time-steps for larger clusters. Figure 6.29 shows the temperature versus energy curves for various sizes of cluster. For seven or more atoms there seems to be a distinct first-order transition, as evidenced by the occurrence of what appear

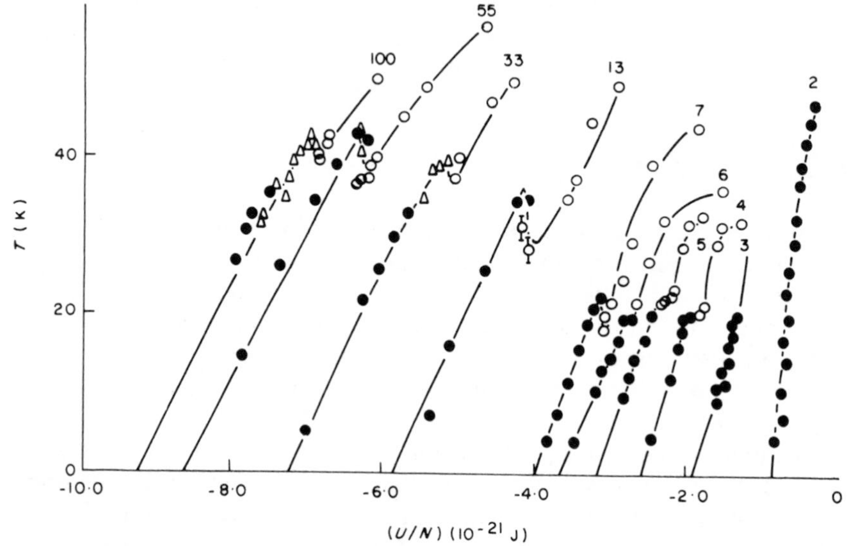

Fig. 6.29. Temperature versus energy per atom for argon microclusters. Numbers above each curve indicate the number of atoms in the cluster. ●, Solid; △, partially liquid; ○, liquid. (After Briant and Burton, 1975.)

to be van der Waals' loops. Subsequent studies on some of these clusters using MC methods show gradual, rather than first-order, transitions. It is suggested that the loops arise from non-ergodicity of the MD simulations. Much more important, however, is the fact that below the transition temperature the clusters show many features which are characteristic of solids. Firstly, the radial distribution functions show sharp peaks separated by troughs which reach down almost to zero; above the transition temperature the peaks are broader and the troughs more shallow. Secondly, below the transition temperature no noticeable diffusion of the particles occurs during the whole period of the MD run ($\sim 3 \times 10^{-11}$ s); above the transition temperature diffusion does occur and the diffusion coefficient can be evaluated. In the immediate vicinity of the transition, clusters often existed which were "crystalline" at the centre but "liquid" in the outer layers. For the partly and the wholly "liquid" clusters radial diffusion was much slower (by a factor of ~3) than diffusion perpendicular to the radius of the cluster. No detailed discussion has been given of this anisotropy. However, a resemblance can be seen with diffusion in a fluid near a solid (§3), for which again diffusion parallel to the wall is faster than diffusion perpendicular to it. Thirdly, inspection of snapshots of the clusters confirms the conclusions reached above with regard to the static structure, including

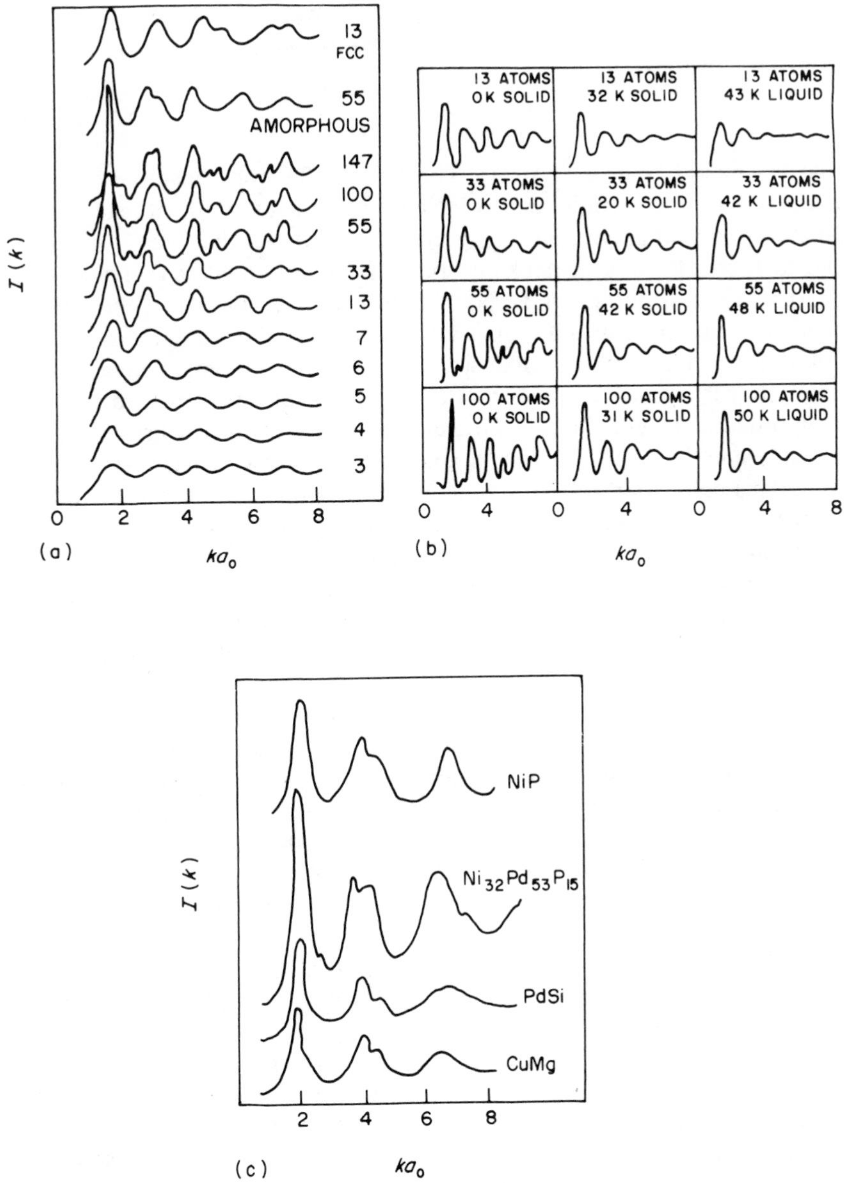

Fig. 6.30. Interference functions. (*a*) Microclusters at 0 K. Top two curves are for 13-atom FCC and quenched 55-atom microclusters, respectively. (*b*) Microclusters at various temperatures. (*c*) Experimental curves for amorphous materials. (After Briant and Burton, 1975.)

the observations about the "partly liquid" clusters. For direct comparison with experimental results the X-ray interference function

$$I(k) = 1 + 2N^{-1} \sum_{i>j} \frac{\sin(kr_{ij})}{kr_{ij}} \tag{6.13}$$

was calculated (Fig. 6.30(a, b)). The 13- and 33-atom clusters both gave shoulders on the second peak, which are often associated with substances in amorphous or glassy states (Fig. 6.30(c)). The fact that these very small clusters show the double peak may be useful in assigning the corresponding feature for amorphous solids, possibly to icosahedral units. Although this feature is absent from the curve for an "annealed" 55-atom cluster, it is found for a cluster of the same size which has been rapidly cooled. These simulations have also served to resolve a difficulty with regard to some experimental results for the lattice parameters of microclusters at 0 K, which seemed to show that the parameter decreased with decreasing cluster size. This behaviour, which was deduced from $I(k)$ values, was confirmed qualitatively if the simulation-generated $I(k)$ values were used. If, however, the radial distribution functions were employed the result was the opposite: the nearest-neighbour distance increased with decreasing cluster size. The latter conclusion is the significant one, the former being invalid as it involves the erroneous assumption that the cluster had the full symmetry of the bulk crystal.

Another interesting point for study is the way in which the melting temperature (T_m) (determined by applying the Maxwell equal area rule to the energy–temperature curves) varies with cluster size. Even the largest clusters studied here (100 atoms) had melting-points which were only about one-half of that for the bulk substance ($T_m \approx 40$ K against ≈ 85 K for bulk argon; Lee *et al.* (1973) found 40 K $< T_m <$ 60 K) (Fig. 6.31). The classical liquid-drop model leads to an expression for the decrease in the melting temperature:

$$T_m - T = \frac{(\kappa^s \gamma^s - \kappa^l \gamma^l) N^{2/3}}{\Delta H_f \cdot N / T_m} \tag{6.14}$$

where N is the number of atoms in the cluster and κ is a constant such that $\kappa N^{2/3}$ is the surface area of the sphere, ΔH_f is the enthalpy of fusion, and γ^s and γ^l are the surface tensions for the solid and the liquid, respectively, against vapour. The numerator is the change in surface free energy of the solid and liquid phases arising from the surface terms and the denominator is the entropy of the phase change. Also shown in the figure are the predictions of this model using two extreme values (0·022 and 0·030 J m^{-2}) for the surface tension of the liquid; for both the lines shown γ^s as a function of temperature is taken to be 0·0405–0·0001T J m^{-2}, which is

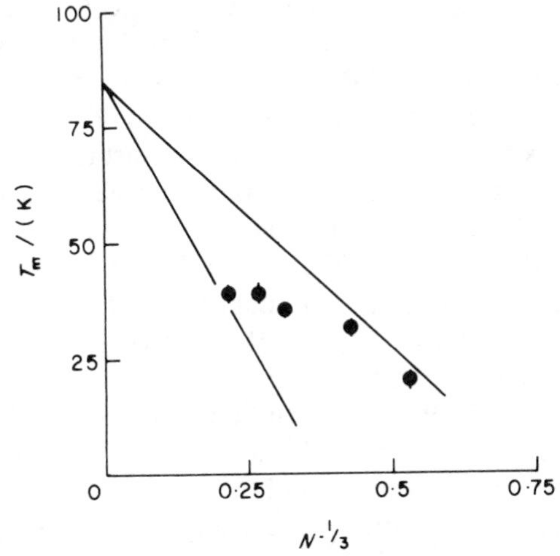

Fig. 6.31. Melting temperature (T_m) of argon microclusters versus $N^{-1/3}$, where N is the number of atoms in the cluster. Lines are predictions from the liquid drop model (Eqn (6.14)); points are MD data. (After Briant and Burton, 1975.)

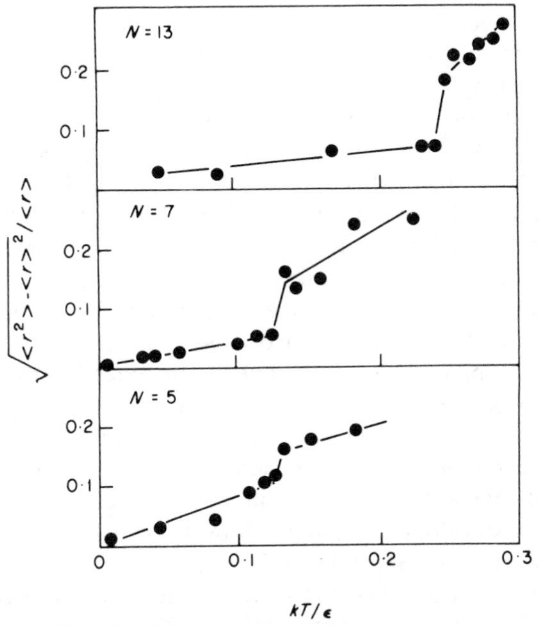

Fig. 6.32. Mean relative fluctuations of the position versus temperature for N-clusters. (After Kaelberer and Etters, 1977.)

derived from calculated values for the bulk solid (Allen and de Wette, 1969). The results from the simulation and the liquid drop model are in reasonable accord. Tolman (1949) has suggested that the "liquid drop" model formula for the surface free energy should be modified to take account of the possibility of the Gibbs dividing surface between the cluster and the vapour not being at the surface of the cluster (see also §2, Chapter 2). His proposal leads to a formula for the surface free energy of $\kappa N^{2/3}\gamma[1 + (d/N^{1/3})]$ where $d/N^{1/3}$ is equal to the ratio of the distance of the Gibbs surface from the physical boundary to the radius of that boundary. From the simulation results it appears that this correction becomes important for clusters smaller than about 13 atoms.

Kaelberer and Etters (1977) have studied very small clusters (3–13 atoms) of Lennard-Jones particles using the MC technique with runs of 10^5-10^6 trial moves. As Briant and Burton had found in their few MC calculations, fairly sudden changes in slope of the energy versus temperature curves occurred even for quite small clusters ($\geqslant$9 atoms). Changes in the shape of an equivalent of the radial distribution function also occurred at the melting point. However, a more striking change was that found in the fluctuation of the "bond distances" (inter-atomic spacings). Figure 6.32 shows the sharp change in this quantity which is found even for a cluster of as few as 5 atoms! The value of this fluctuation (expressed as a fraction of the mean "bond distance") when the solid melts is approximately the same (~0·1) for all the sizes of clusters studied, in agreement with the Lindemann theory of melting.

Because of doubt about the validity of the free cluster method used in the previous work, Nauchitel and Pertsin (1980) have made a MC study of clusters which are confined to a spherical box. They found a much greater variety of behaviour than had been reported hitherto. At low temperatures and high pressures ($T^* = 0{\cdot}25$, $p^* > 1{\cdot}2$) three kinds of solid phase could occur (FCC, icosahedral, and a structure having two shells of atoms), the actual structure obtained being dependent upon the starting configuration. In this region the icosahedral form had the greatest free energy (as determined by an integration from zero density, Eqn (4.29′)), in spite of having the lowest internal energy. Below $p^* = 1{\cdot}2$ only a two-phase state having an icosahedral core and a liquid-like outer shell was found. As the temperature was raised the relative amounts of core and "liquid" shell changed, with an accompanying large heat capacity.

Simulation results for the gas–liquid transition are less reliable, essentially because they are much more dependent upon the details of the technique used. Thus, where free clusters are studied at these temperatures (Briant and Burton, 1975; Kaelberer and Etters, 1977) the clusters will gradually disperse as the simulation proceeds and no useful results can be obtained.

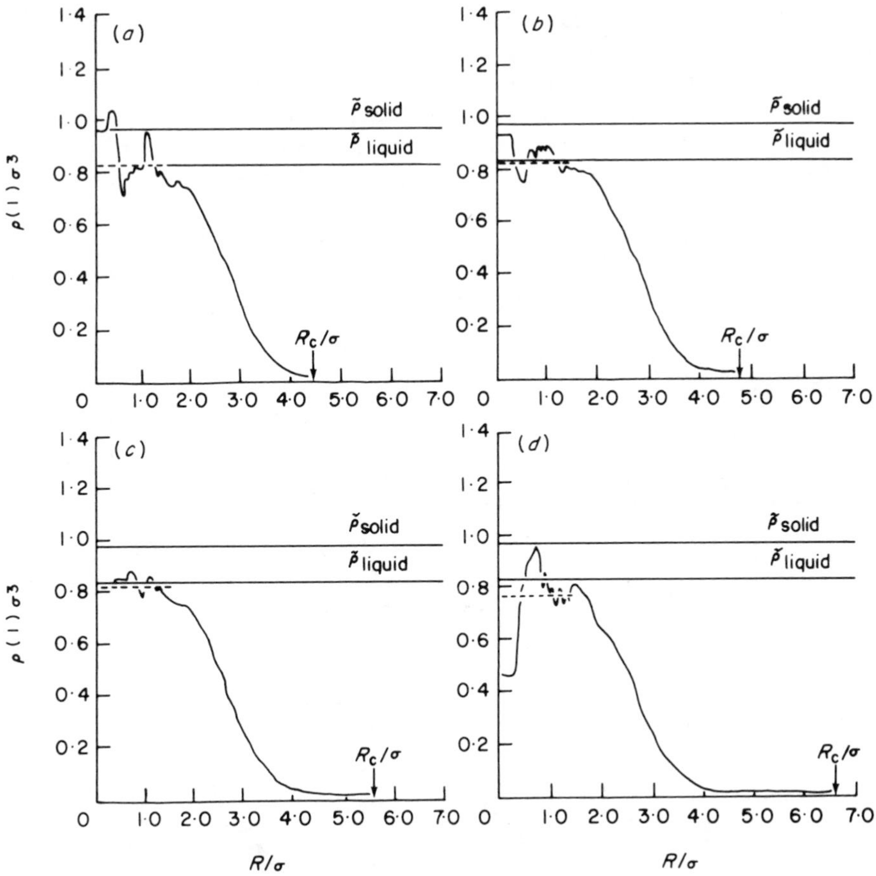

Fig. 6.33. Density ($\rho^{(1)}$) versus distance from the centre of mass (R) for an 87-atom cluster at 90 K. ρ_{solid} and ρ_{liquid} refer to bulk argon at 80 K (solid) and 84 K (liquid). Dotted line is the average density up to $R = 1{\cdot}5\sigma$. $\tilde{V}_c = \frac{4}{3}\pi R_c^3\sigma^3$: (*a*) $V_c = 374{\cdot}66$; (*b*) $V_c = 451{\cdot}82$; (*c*) $V_c = 734{\cdot}66$; (*d*) $V_c = 1200{\cdot}00$. (After Lee *et al.*, 1973.)

Lee *et al.* (1973) made a very thorough study of this transition for clusters of 13–100 Lennard-Jones argon atoms. Their definition of a cluster was all those atoms which lay within a distance R_c of the centre of mass of the cluster. It might be thought that the results would be very sensitive to the choice of R_c (for a given number of atoms). However, this turned out not to be so over wide ranges of values of R_c at moderate temperatures. In their words, all reasonable choices of cluster definition lead to almost the same result. Figure 6.33 shows the radial distribution functions for four values of R_c for an 87-atom cluster at 80 K. At this temperature the fall from liquid-like to gas-like densities occurs over about 2σ.

Since the main interest in clusters is as centres for crystal or domain growth or for catalysis, it is natural to enquire about how the properties are changed by the proximity of a surface other than that between the cluster and the vapour. Two kinds of model can be adopted to study this problem. The simplest, and most economical in computer time, is a purely two-dimensional treatment. In this the effect of the adsorbent is to restrict

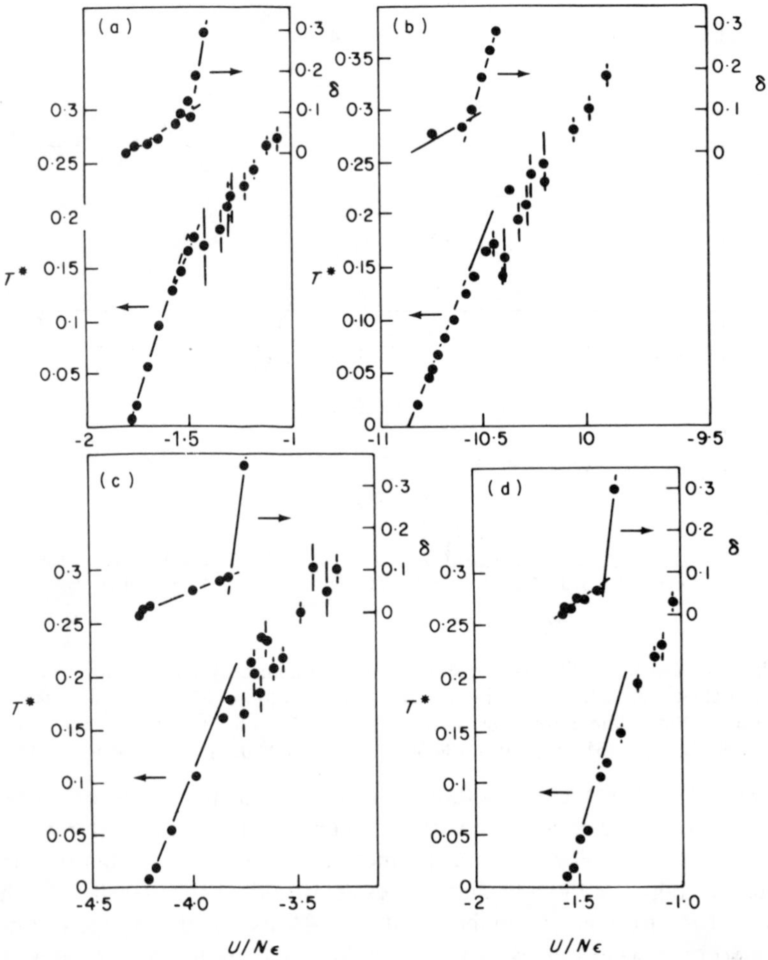

Fig. 6.34. The root-mean-square displacement of the atoms from their positions at 0 K (δ) and the temperature versus energy per atom (U/N) (in reduced units). (*a*) $N = 7$, two-dimensional model; (*b*) $N = 7$, argon-on-graphite; (*c*) $N = 7$, xenon-on-neon; (*d*) $N = 6$, two-dimensional model. (After Weissmann and Cohan, 1980.)

the molecules to a plane: no actual knowledge of the adsorbent–adatom potential is required. The other kind of model is one in which the particles can move in a semi-infinite three-dimensional space, with the wall potential being dependent upon the distance of the particle from the wall (z) and, possibly, the lateral position as given by the x, y coordinates. Weissmann and Cohan (1980) used both types of model with microclusters of 6, 7, 8 and 19 spherical particles. The inter-particle potential was taken to be of the usual Lennard-Jones type with Eqn (8.2) being used, where required, for the adsorbent–adatom potential. The actual values of the parameters were chosen to be appropriate to the systems argon-on-graphite and xenon-on-neon (100). The latter would be difficult to realize experimentally, but represents weak AS compared with AA interactions. The main result of the MD simulations was that in every case as the temperature was raised the cluster changed from being solid-like to being liquid-like. This "melting" was manifested most convincingly by the plot of δ, the root-mean square displacement of the atoms from their positions at 0 K, against energy (Fig. 6.34). Snapshots show that above the melting temperature the shapes of the clusters vary widely from that adopted at 0 K. As would be expected the melting occurs in a "transition region" rather than at a point, and throughout the region the convergence of the simulation is very sluggish (even 10^4 time-steps each of 0·005 adimensional units being inadequate). Large differences in behaviour between the true two-dimensional and the three-dimensional model-with-wall-potential are only evident if the wall potential is weak (as for xenon-on-neon). For such a system a model which is effectively two-dimensional at 0 K will become three-dimensional as the temperature is raised, with one or two particles leaving the solid surface and becoming attached to the opposite side of the cluster. For the strongly adsorbed argon-on-graphite there is little difference between the results of the two models. Some special differences were found between clusters of 7 or 19 atoms, which can take up simple non-re-entrant configurations, and clusters of 6 or 8, which cannot. For the latter it is proposed that the fluctuations in shape mainly involve configurations which are almost equivalent; for the 7- or 19-atom clusters there is a unique close-packed configuration and fluctuations necessarily involve less stable configurations.

References

Abraham, F. F. (1978). *J. Chem. Phys.* **68**, 3713.
Abraham, F. F. and Singh, Y. (1977). *J. Chem. Phys.* **68**, 4767.
Allen, R. E. and de Wette, F. W. (1969). *J. Chem. Phys.* **51**, 4820.

Balescu, R. (1955). *Bull. Cl. Sci., Acad. R. Belg.* **41**, 1242.
Baxter, R. J. (1973). *J. Phys. C: Solid State Physics* **7**, L445.
Briant, C. L. and Burton, J. J. (1975). *J. Chem. Phys.* **63**, 2045.
Chapela, G. A., Saville, G. and Rowlinson, J. S. (1975). *Faraday Disc. No. 59*, 22.
Chapela, G. A., Saville, G., Thompson, S. M. and Rowlinson, J. S. (1977). *J. Chem. Soc. Faraday Trans. II* **73**, 1133.
Finney, J. L. (1970). *Proc. R. Soc. London, Ser. A* **319**, 479.
Goellner, G. J., Daunt, J. G. and Lerner, E. (1975). *J. Low Temp. Phys.* **21**, 347.
Grillet, Y, Rouquerol, F. and Rouquerol, J. (1979). *J. Coll. Interf. Sci.* **70**, 242.
Hansen, J-P and Verlet, L. (1969). *Phys. Rev.* **184**, 151.
Hanson, F. E. and McTague, J. P. (1980). *J. Chem. Phys.* **72**, 6363.
Hanson, F. E., Mandell, M. J. and McTague, J. P. (1977). *J. de Phys.* **38**, C4–76.
Kaelberer, J. B. and Etters, R. D. (1977). *J. Chem. Phys.* **66**, 3233.
Lane, J. E. and Spurling, T. H. (1976). *Aust. J. Chem.* **29**, 2103.
Lane, J. E. and Spurling, T. H. (1978). *Aust. J. Chem.* **31**, 933.
Lane, J. E., Spurling, T. H., Freasier, B. C., Perram, J. W. and Smith, E. R. (1979). *Phys. Rev. A* **20**, 2147.
Larher (1974). *Trans. Faraday Soc.* **70**, 320.
Lee, J. K., Barker, J. A. and Abraham, F.F. (1973). *J. Chem. Phys.* **58**, 3166.
Lieb, E. H. and Wu, F. Y. (1972). In "Phase Transitions and Critical Phenomena" (C. Domb and M. S. Green, eds.) Vol. 1, p. 331. Academic Press, London and New York.
Nauchitel, V. V. and Pertsin, A. J. (1980). *Mol. Phys.* **40**, 1341.
Nicholson, D., Parsonage, N. G. and Rowley, L. A. (1981). *Mol. Phys.* **44**, 629.
Nicholson, D., Rowley, L. A. and Parsonage, N. G. (1977). *J. de Phys.* **38**, C4–69.
Osborn, T. R. and Croxton, C. A. (1977). *Mol. Phys.* **34**, 841.
Osborn, T. R. and Croxton, C. A. (1980). *Mol. Phys.* **40**, 1489.
Parsonage, N. G. (1970). *J. Chem. Soc.*, Ser.A, 2859.
Parsonage, N. G. and Nicholson, D. (1979). *CRC Crit. Revs. Solid St. Materials Sci.* **8**, 175.
Parsonage, N. G. and Staveley, L. A. K. (1978). "Disorder in Crystals", p. 572. Clarendon Press, Oxford.
Pierotti, R. A. and Thomas, H. E. (1971). *Surf. Colloid Sci.* **4**, 93.
Pierotti, R. A. and Thomas, H. E. (1974). *J. Chem. Soc. Faraday Trans. II* **70**, 1726.
Prasad, S. D. and Toxvaerd, S. (1980). *J. Chem. Phys.* **72**, 1689.
Rao, M., Berne, B. J., Percus, J. K. and Kalos, M. H. (1979). *J. Chem. Phys.* **71**, 3802.
Rowley, L. A., Nicholson, D. and Parsonage, N. G. (1976). *Mol. Phys.* **31**, 365, 389.
Rowley, L. A., Nicholson, D. and Parsonage, N. G. (1978). *J. Comput. Phys.* **26**, 66.
Rowlinson, J. S. (1958). *Mol. Phys.* **1**, 414.
Rowlinson, J. S. and Widom, B. (1982). "Molecular Theory of Capillarity". Clarendon Press, Oxford.
Saam, W. F. and Ebner, C. (1978). *Phys. Rev.* A**17**, 1768.
Sams, J. R. (1974). In "Progress in Membrane and Surface Science", Vol. 8, p. 25. Academic Press, New York.

Severin, E. S. and Tildesley, D. J. (1980). *Mol. Phys.* **41**, 1401.
Snook, I. K. and Henderson, D. (1978). *J. Chem. Phys.* **68**, 2134.
Snook, I. K. and van Megen, W. J. (1979). *J. Chem. Phys.* **70**, 3099.
Sokolowski, S. (1980). *J. Coll. Interf. Sci.* **74**, 26.
Spurling, T. H. and Lane, J. E. (1978). *Aust. J. Chem.* **31**, 465.
Spurling, T. H. and Lane, J. E. (1980). *Aust. J. Chem.* **33**, 1967.
Steele, W. A. (1973). *Surface Sci.* **36**, 317.
Steele, W. A. (1974). "The Interaction of Gases with Solid Surfaces", Chs 2–4. Pergamon Press, Oxford.
Steele, W. A. (1977). *J. Chem. Phys.* **65**, 5256.
Subramanian, G. and Davis, H. T. (1979). *Mol. Phys.* **38**, 1061.
Sullivan, D. E. (1979). *Phys. Rev. B* **20**, 3991.
Sullivan, D. E., Levesque, D. and Weis, J. J. (1980). *J. Chem. Phys.* **72**, 1170.
Thomy, A. and Duval, X. (1970). *J. Chim. Phys.* **67**, 1101.
Tolman, R. C. (1949). *J. Chem. Phys.* **17**, 333.
Toxvaerd, S. (1981). *J. Chem. Phys* **74**, 1998.
Van Megen, W. J. and Snook, I. K. (1981). *J. Chem. Phys.* **74**, 1409.
Weissmann, M. and Cohan, N. V. (1980). *J. Chem. Phys.* **72**, 4562.
Wood, W. W. and Parker, F. R. (1957). *J. Chem. Phys.* **27**, 720.

Chapter 7

Integral Equation and Model Theories

1. Introduction

The problem of developing tractable theories for uniform liquids presents statistical mechanics with one of its most formidable challenges. In recent years part of the considerable progress in this field has come from the development and careful study of model systems such as collections of hard spheres. The realization that similar models can be dealt with in the case of non-uniform fluids, using in particular the OZ equation, is of more recent origin; as might be anticipated the results are more complicated and probably of more limited application than in the corresponding uniform fluid case.

The main advantage offered by model potentials is that they present an opportunity to solve an equation exactly and rapidly and to explore a range of relevant parameters, whilst retaining some recognizable similarity to the "real" interaction between the particles. Apart from hard spheres, sticky spheres and Yukawa molecules offer this opportunity and some of the results achieved with them are discussed here.

The exploitation of models of this type is necessarily restricted. On the other hand numerical solutions can, at least in principle, be found for any potential and any type of equation. In practice numerical methods often make demands on computer time comparable to those of simulations, as well as raising other difficulties. We begin § 3 with a discussion of some of these limitations before going on to survey the results obtained with the different types of integral equation.

Although remarkable progress has been achieved in a very short time in the application of integral equations to adsorption problems, it seems unlikely that the older theories will be replaced by this type of approach for some time. If for no other reason, it is therefore worthwhile to examine some of the more popular theoretical models. These usually yield isotherms rather than distribution functions, which is perhaps one indication of their restricted nature. Nevertheless, within the context of a more general lattice model, which again can be solved numerically, this type of approach can produce some useful insight into the nature of phase transitions in multilayer adsorbates, both for planar surfaces and for the case when the adsorbate is confined within a model capillary space.

2. Exact solutions of integral equations

(a) Solution of the OZ equation using Laplace transforms

The OZ equation for uniform fluids of hard spheres was solved about 20 years ago, using the PY approximation for $c^{(2)}$, by Wertheim (1963) and Thiele (1963). In the PY approximation for this model system, $c^{(2)}$ vanishes outside the hard-sphere diameter, and $g^{(2)}$ vanishes inside this domain. It is to be noted that the MS and PY approximations are formally equivalent for this case. Solution of the OZ equation involves transformation to bipolar co-ordinates and Laplace transformation of the resulting equations. The hard-sphere pair correlation function obtained can be written as

$$\tilde{c}^{(a,a)} = -a_1^{(a,a)} \qquad r<0 \qquad (7.1a)$$

$$\tilde{c}^{(a,a)} = -a_1^{(a,a)} - 6\eta_a a_2^{(a,a)} r - \tfrac{1}{2}\eta_a a_1^{(a,a)} r^3 \qquad 0<r\leqslant 1 \qquad (7.1b)$$

$$\tilde{c}^{(a,a)} = 0 \qquad r>1 \qquad (7.1c)$$

where r is measured in σ_a units, and

$$a_1^{(a,a)} = (1+2\eta_a)^2/(1-\eta_a)^4 \qquad (7.1d)$$

$$a_2^{(a,a)} = -(1+\tfrac{1}{2}\eta_a)^2/(1-\eta_a)^4 \qquad (7.1e)$$

$$\eta_a = \pi\rho_a\sigma_a^3/6. \qquad (7.1f)$$

It can be shown (Thiele, 1963; Wertheim, 1963; Barker and Henderson, 1976) that $a_1^{(a,a)}$ is simply related to the compressibility κ of the uniform hard-sphere fluid:

$$a_1^{(a,a)} = \beta\left(\frac{\partial p}{\partial\rho_a}\right)_T = \frac{\beta}{\kappa\rho_a}. \qquad (7.1g)$$

The pair correlation can be written in the form of a Laplace transform as

$$\mathscr{L}_t(r\tilde{g}^{(a,a)}(r)) = \int r\tilde{g}^{(a,a)}(r)e^{-rt}\,dr = tL(t)/2\eta_a[S(t)e^t + L(t)] \quad (7.2a)$$

with

$$L(t) = 12\eta_a[(1 + \tfrac{1}{2}\eta_a)t + 1 + 2\eta_a] \quad (7.2b)$$

and

$$S(t) = (1 - \eta_a)^2t^3 + 6\eta_a(1 - \eta_a)t^2 + 18\eta_a^2t - 12\eta_a(1 + 2\eta_a). \quad (7.2c)$$

Inversion of Eqn (7.2a) involves integration in the complex plane and the result is therefore not straightforward. Nevertheless a hard-sphere, uniform fluid $g^{(2)}$ can be calculated fairly readily (Throop and Bearman, 1965) and with the addition of small corrections can be made to agree closely with results from computer simulations of a hard-sphere fluid (Verlet and Weiss, 1972).

Shortly after the achievement of Thiele and Wertheim, Lebowitz (1964) applied similar methods to hard-sphere mixtures, again using the PY approximation, thereby providing exact results for such systems. These can be written for the particular case of an adsorbate (a)–adsorbent (w) pair and modified by taking the limits $\rho^{(w)} \to 0$ and $\sigma_w \to \infty$, on $c^{(w,a)}$ and $g^{(w,a)}$ so that these distribution functions become singlet distributions in accord with the derivation of (3.144). The Laplace transform of the singlet density (Percus, 1976) is

$$\mathscr{L}_t(\rho^{(a)}) = \rho_a(1 + 2\eta_a)t^2e^t/[L(t) + S(t)e^t] \quad (7.3)$$

which can also be obtained as the corresponding Fourier transform (Henderson and Blum, 1978).

The limiting form of $\tilde{c}^{(w,a)}(z)$ may be written

$$\tilde{c}^{(w,a)}(z) = -a_1^{(w,a)} = -\left(\frac{\partial\beta p}{\partial\rho_a}\right) \qquad z < -1 \quad (7.4a)$$

$$= -a_1^{(w,a)} - 6\eta_a a_2^{(w,a)}(z+1)^2 - 2\eta_a a_1^{(w,a)}(z+1)^3 \qquad -1 < z < 0 \quad (7.4b)$$

where

$$a_1^{(w,a)} = (1 + 2\eta_a)^2/(1 - \eta_a)^4 \quad (7.4c)$$

$$a_2^{(w,a)} = (1 + \eta_a/2)(1 + 2\eta_a)/(1 - \eta_a)^4. \quad (7.4d)$$

Since $\tilde{g}^{(a,a)}$ for hard spheres is already well known, it is probably more straightforward to obtain $h^{(w,a)}$ by numerical integration of (3.144) than by inversion of Eqn (7.3) (Henderson *et al.*, 1976a, b). The required integrals are obtained by substitution of (7.4b) into (3.144) and may be expressed as

$$h^{(\mathrm{w,a})}(z) = -12\eta_a b_1 \int_z^\infty t\bar{h}^{(\mathrm{a,a})}(t)\{b_1[t-z]+b_2\}\,\mathrm{d}t$$

$$+\,12\eta_a b_1 \int_z^{z+1} (z+1-t)^3 t\bar{h}^{(\mathrm{w,a})}(t)\{\tfrac{1}{2}\eta_a b_1[z-t]+b_2\}\,\mathrm{d}t \quad (7.5a)$$

$$b_1 = (1+2\eta_a)/(1-\eta_a)^2 \quad (7.5b)$$

$$b_2 = -3\eta_a/2(1-\eta_a)^2. \quad (7.5c)$$

Numerical evaluation of Eqn (7.5) requires a further step because accurate values of $\bar{h}^{(\mathrm{a,a})}(r)$ as $r \to \infty$ are difficult to obtain. However, the moments of $\bar{h}^{(\mathrm{a,a})}(r)$ can be found by expanding both sides of Eqn (7.2a) and equating coefficients of t. The first two moments of $\bar{h}^{(\mathrm{a,a})}$ obtained in this way are

$$\int_0^\infty r\bar{h}^{(\mathrm{a,a})}(r)\,\mathrm{d}r = (10-2\eta_a+\eta_a^2)/20(1+2\eta_a) \quad (7.6a)$$

$$\int_0^\infty r^2\bar{h}^{(\mathrm{a,a})}(r)\,\mathrm{d}r = (\eta_a-4)(\eta_a^2+2)/24(1+2\eta_a)^2. \quad (7.6b)$$

These equations can be substituted into (7.5) and this can then be evaluated without the problem of the infinite limit.

The Laplace transform method for hard spheres makes use of the fact that $g^{(2)}$ is zero inside the core, Waisman *et al.* (1976) extended this technique to the case of hard-sphere mixtures in which the unlike molecules interact through a Yukawa potential, which has the general form

$$u^{(\mathrm{a,b})}(r) = \varepsilon^{(\mathrm{a,b})} \exp[-B^{(\mathrm{a,b})}r]/r. \quad (7.7a)$$

In the limit of infinitely large adsorbent particle (i.e. plane wall), the adsorbate–adsorbent interaction may be written

$$u^{(\mathrm{w,a})} = v^{(\mathrm{w,a})} + w^{(\mathrm{w,a})} \quad (7.7b)$$

where the hard-core part, $v^{(\mathrm{w,a})}$ is

$$v^{(\mathrm{w,a})} = \infty \qquad r < \sigma^{(\mathrm{w,a})} \quad (7.7c)$$

$$v^{(\mathrm{w,a})} = 0 \qquad r \geqslant \sigma^{(\mathrm{w,a})} \quad (7.7d)$$

and the soft attractive part of the potential becomes

$$w^{(\mathrm{w,a})}(z) = -\varepsilon \exp(-Bz). \quad (7.7e)$$

The MSA now implies that $c^{(\mathrm{w,a})} = \beta\varepsilon \exp(-Bz)$ and enables exact solutions to be found to the OZ equation.

The direct correlation function $\tilde{c}^{(w,a)}(z)$ may be expressed as the sum of a hard-sphere part, given by Eqn (7.4) and an increment $\Delta c^{(w,a)}$ given by

$$\Delta c^{(w,a)} = 0 \qquad z < -1$$

$$= 12\eta_a C_1 \left\{ -\frac{C_2(z-1)^2}{2B} + \frac{(C_2 + BC_3)}{B^2} \times \left[(z-1) + \frac{(\exp[-B(z-1)] - 1)}{B} \right] \right\}$$

$$1 < z < 0 \qquad (7.8a)$$

$$= \beta\varepsilon \exp(-Bz) \qquad z > 0$$

$$C_1 = \beta\varepsilon B^3 \exp(B)/[L(B) + S(B)\exp(B)] \qquad (7.8b)$$

$$C_2 = 1 + 2\eta_a$$

$$C_3 = 1 + \eta_a/2 \qquad (7.8c)$$

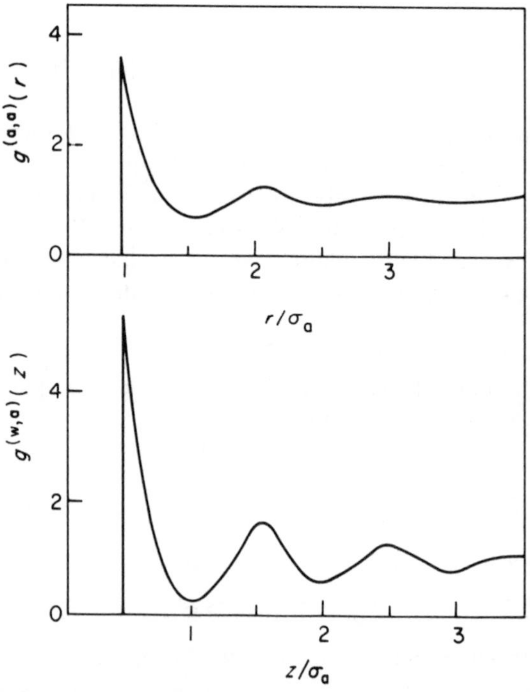

Fig. 7.1. $\tilde{g}^{(a,a)}(r)$ and $g^{(w,a)}(z)$ for hard spheres in contact with a hard wall, calculated from the Percus–Yevick equation. $\rho_a\sigma_a^3 = 0{\cdot}8$ in both cases; z is the perpendicular distance of the hard-sphere centre from the surface. (From Henderson, Abraham and Barker, 1976.)

where the functions $L(t)$ and $S(t)$ are given in Eqns (7.2b) and (7.2c). Again numerical integration of Eqn (3.144) yields the singlet distribution function.

The hard-sphere–hard-wall pair correlation, $g^{(w,a)}$, may be compared with $g^{(a,a)}$ (Fig. 7.1) and is seen to exhibit much more pronounced oscillations. A comparison with Monte Carlo data at an asymptotic bulk density of 0·609 (Liu *et al.*, 1974), illustrated in Fig. 7.2, shows rather good agreement in the range $2{\cdot}0 \geqslant z \geqslant 0{\cdot}15$ (Waisman *et al.*, 1976), but the theory fails to reproduce the density in the vicinity of the wall.

The density at the wall ($z = 0$) can be obtained by substitution of (7.6) into (7.5) making use also of the fact that $\bar{h}^{(a,a)}(r) = -1$, for $r < 1$, the result is

$$\rho^{(a)}(0) = \rho_a \tilde{g}^{(w,a)}(0) = \rho_a(1 + 2\eta_a)/(1 - \eta_a)^2. \tag{7.9}$$

On the other hand, Eqn (3.130) applies at an infinitely repulsive wall and, with the equation of state for hard spheres due to Carnahan and Starling (1969), gives the result

$$\rho_a \tilde{g}^{(w,a)}(0) = \rho_a(1 + \eta_a + \eta_a^2 - \eta_a^3)/(1 - \eta_a^3) \tag{7.10}$$

which is clearly at variance with Eqn (7.9). It is of course well known that the PY theory for a uniform hard-sphere fluid is inconsistent in giving rise

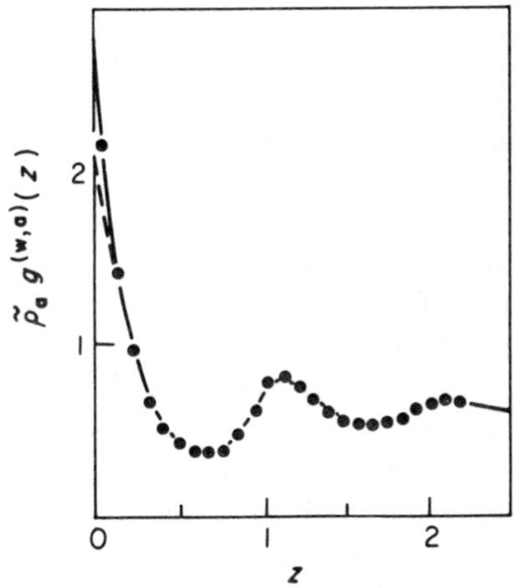

Fig. 7.2. Density profile for hard spheres of diameter σ_a near a wall. The points are the MC results of Liu *et al.*, (1974). Solid and broken curves are the GMSA and PY curves respectively. The asymptotic bulk density, $\rho_a\sigma_a^3 = 0{\cdot}609$, corresponds to the average density of 0·7 in the MC calculations. (From Waisman *et al.*, 1976.)

to two equations of state, one derived from the compressibility equation and the second derived from the pressure equation. Equation (7.10) is a combination of these results which has proved to be very accurate. Since (7.9) accords with none of these uniform fluid results, there is a further inconsistency arising from the application of the PY approximation to non-uniform systems.

An improvement can be brought about by employing the generalized mean spherical approximation (GMSA) (Waisman, 1973; Hoye *et al.*, 1974). In this a functional form is chosen for $c^{(2)}$ which renders the OZ equation soluble, but into which adjustable parameters can be introduced. Waisman *et al.* (1976) applied the GMSA to the non-uniform hard-sphere/hard-wall system with $w^{(a,w)}$ given by (7.7e) and set C_1 in (7.8a) (now an adjustable parameter) equal to $\eta_a^2(3-\eta_a)/(1-\eta_a)^5$, thus ensuring that the condition (7.10) is met and, in effect, setting the parameter ε in (7.7e). The other parameter, B, was set by using the general relationship between the integral $\int \tilde{h}^{(w,a)}\,d\boldsymbol{r}$ and the compressibilities $\partial p/\partial \rho_w$, $\partial p/\partial \rho_a$, with $p(\rho_w, \rho_a)$ chosen to meet the requirement that p be finite in the limit $\sigma_w \to \infty$ as well as satisfying Eqn (7.10) at the hard-wall surface. The resulting expression for B was

$$\frac{1}{B}=\frac{(1+2\eta_a)(1-\eta_a)}{\eta_a^2(3-\eta_a)}\left[\frac{9\eta_a(1+2\eta_a)+\eta_a^2(4-\eta_a)(1-4\eta_a)}{6[(1+2\eta_a)^2-\eta_a^3(4-\eta_a)]}-\frac{3\eta_a}{2(1+2\eta_a)}\right] \tag{7.11}$$

and excellent agreement with computed results was found for this approximation (Fig. 7.2).

Some further insight into the behaviour of the density profile in the hard-wall/hard-sphere system has been gained by developing a series expansion in powers of η_a whose coefficients can be evaluated exactly from the relevant cluster (Mayer) integrals (Fischer, 1977, 1978). Corresponding expansions can be made for the CHNC, PY and BBGKY equations (Sokolowski and Stecki, 1978).

In the latter, $\rho^{(2)}$ is first substituted by $\rho_1^{(1)}\rho_2^{(1)}\tilde{g}^{(2)}(\boldsymbol{r}_{12};\rho_a)$, often referred to as the superposition approximation by analogy with the equivalent substitution in connection with the second-order equation of the hierarchy. However, in the context of the singlet distribution function, this particular substitution is also clearly equivalent to the assumption that the non-uniform $g^{(2)}$ in Eqn (3.135) can be replaced by the uniform function $\tilde{g}^{(2)}$. With this approximation the BBGKY equation can then be put in the form (Kozak *et al.*, 1971; Berry, 1974)

$$\ln[\rho^{(a)}(\boldsymbol{r})/\rho_a] + \beta u^{[a]}(\boldsymbol{r}) = \int [\rho^{(a)}(\boldsymbol{r}+\boldsymbol{R}) - \rho_a]S(R)\,d\boldsymbol{r} \tag{7.12}$$

where

$$S(R) = -\int_R^\infty \bar{g}^{(2)}(s) \frac{\partial \beta u^{[a,a]}}{\partial s} \, ds \tag{7.13a}$$

which for hard spheres can be integrated to give (Navascués and Tarazona, 1979)

$$S(R) = y^{(2)}(\sigma_a)\delta(\sigma_a - R) \tag{7.13b}$$

and Eqn (3.135) becomes

$$\rho^{(a)}(z) = \rho_a \exp\left\{\pi y^{(a,a)}(\sigma_a) \int_{-\sigma}^{\sigma} [\rho_a - \rho^{(a)}(z + z_{12})](\sigma_a^2 - z_{12}^2) \, dz_{12}\right\} \tag{7.14}$$

which is easier to handle than an equivalent equation derived earlier (Fischer, 1977).

The coefficients of η_a obtained from the expansion and integration of the last expression are identical with the exact cluster integral expression and with that from the PY approximation. Furthermore, both approximations yield identical results for the coefficients of η_a^2 when $z \geqslant \sigma_a$. However, in the range $0 \leqslant z \leqslant \sigma_a$, the coefficient of η_a^2 from these two approximations diverge; the PY proves to be somewhat the better for $0{\cdot}5\sigma_a < z < \sigma_a$, whilst the BBGKY is superior nearer to the wall and exact at $z = 0$ where the PY approximation gives a value 30% below the exact result. This discrepancy appears to account for a major part of the failure of the PY approximation in this limit.

The CHNC approximation, as judged from terms to the third power in density, severely overestimates $\rho^{(a)}(0)$ and is decidely inferior to the PY equation in the vicinity of the wall.

Numerical solutions of Eqn (7.14) for the hard-sphere/hard-wall system confirm the superiority of the BBGKY equation at $z = 0$ where it yields results very close to those predicted by Eqn (7.11). On the other hand, the performance of this equation at $z > \sigma_a$ appears to be poor in that both the phase and the amplitude of $\rho^{(a)}$ differ from those of the PY equation and computer simulation.

(b) Solution of the OZ equation using the Weiner–Hopf method

A rather different approach to the solution of Eqn (3.142) was developed by Baxter (1967, 1968a, b, 1970) who showed that the Weiner–Hopf technique could be applied to these equations, enabling them to be solved, not only for hard spheres, but also for certain classes of attractive potential, albeit somewhat special ones. It should be mentioned, however, that the limitations which the method imposes are more to do with the short-range character of the potential than with its specific functional form. It has been suggested that even more lenient restrictions may exist which would help to extend the domain of exact solution to the OZ equation yet nearer to that of realistic potentials (Wertheim, 1980).

Baxter's method starts from the Fourier transform of the OZ equation, which can be written

$$\hat{h} - \hat{c} = \hat{c}\hat{h} \tag{7.15a}$$

or in factorized form as

$$(1 + \hat{h})(1 - \hat{c}) = 1 \tag{7.15b}$$

where, for $f^{(2)} \equiv \tilde{c}^{(2)}$ or $\tilde{h}^{(2)}$, $\hat{f} = \rho \int f^{(2)} \exp(i\boldsymbol{k} \,.\, \boldsymbol{r})\, \mathrm{d}\boldsymbol{r}$ and the product arises because $\boldsymbol{r}_{12} = \boldsymbol{r}_{13} + \boldsymbol{r}_{32}$, making the integral in the OZ equation a convolution (fältung).

Transformation to spherical co-ordinates, so that $\mathrm{d}\boldsymbol{r} = r^2\, \mathrm{d}r \sin\theta\, \mathrm{d}\theta\, \mathrm{d}\varphi$, and $i\boldsymbol{k} \,.\, \boldsymbol{r} = ikr\cos\theta$, and integration by parts gives, for $f^{(2)} = \tilde{c}^{(2)}$ or $\tilde{h}^{(2)}$,

$$\hat{f}^{(2)} = 4\pi\rho \int_0^\infty \cos(kr) F^{(2)}(r)\, \mathrm{d}r \tag{7.16a}$$

where

$$F^{(2)}(r) = \int_r^\infty t f^{(2)}(t)\, \mathrm{d}t. \tag{7.16b}$$

Now $c(r)$ is a short-ranged function and can be assumed to vanish outside some distance a of the order of a few molecular diameters. In fact, for approximations such as the PY and MSA, this condition becomes exact for those classes of potential, such as the hard-sphere potential, which are zero outside a given range. The Fourier transform $\hat{c}$ is then also confined to a finite range and is thus an entire function of k.

In a one-component system Baxter showed that an explicit factorization of $1 - \hat{c}$ could be made, and later extended this principle to mixtures. In this case Eqn (7.16) may be also extended slightly, and for a pair of components (a, b)

$$\hat{f}^{(\mathrm{a,b})} = 4\pi(\rho_\mathrm{a}\rho_\mathrm{b})^{1/2} \int_0^\infty \cos(kr) F^{(\mathrm{a,b})}(r)\, \mathrm{d}r \tag{7.17a}$$

where

$$F^{(\mathrm{a,b})} = \int_r^\infty t f^{(\mathrm{a,b})}(t)\, \mathrm{d}t. \tag{7.17b}$$

If $\hat{\boldsymbol{c}}$ is taken to represent the matrix of Fourier transformed direct correlation functions, it is possible to write the factorization

$$\boldsymbol{I} - \hat{\boldsymbol{c}} = [\boldsymbol{I} - \hat{\boldsymbol{Q}}(-k)][\boldsymbol{I} - \hat{\boldsymbol{Q}}(k)] \tag{7.18a}$$

where $\boldsymbol{I}$ is the unit matrix, and the inverse, $Q^{(a,b)}$, of an element of the matrix $\hat{\boldsymbol{Q}}$ is defined by the one dimensional Fourier transform

$$\hat{Q}^{(a,b)} = \int \exp(ikr) Q^{(a,b)}(r)\, dr. \tag{7.18b}$$

Substitution of this equation in the generalized form of Eqn (7.15) then allows $\hat{\boldsymbol{h}}$ also to be expressed solely in terms of $\hat{\boldsymbol{Q}}$,

$$[I + \hat{\boldsymbol{h}}][\boldsymbol{I} - \hat{\boldsymbol{Q}}(k)] = [\boldsymbol{I} - \hat{\boldsymbol{Q}}(-k)]^{-1}. \tag{7.18c}$$

It can be shown that on inversion of Eqn (7.18a) the terms containing $\hat{\boldsymbol{Q}}(-k)$ vanish under certain weak restrictions related to the disordered structure of the fluid at long range (Baxter, 1970). It is convenient now to define a new function $q^{(a,b)}$ by $Q^{(a,b)} = 2\pi(\rho_a\rho_b)^{1/2} q^{(a,b)}$. After using (7.17) and differentiating there results

$$r\tilde{c}^{(a,b)}(r) = -\frac{\partial q^{(a,b)}(r)}{\partial r} + 2\pi \sum_\alpha \rho_\alpha \int_{S^{(a,\alpha)}} q^{(a,\alpha)}(t) \frac{\partial q^{(\alpha,b)}(r+t)}{\partial r}\, dt \tag{7.19a}$$

$$r\tilde{h}^{(a,b)}(r) = -\frac{\partial q^{(a,b)}(r)}{\partial r} + 2\pi \sum_\alpha \rho_\alpha \int_{S^{(a,\alpha)}} [r-t] q^{(a,\alpha)}(t) \tilde{h}^{(\alpha,b)}(|r-t|)\, dt \tag{7.19b}$$

where $S^{(a,b)} = \frac{1}{2}|\sigma_a - \sigma_b|$ and the upper limit depends on the potential functions. If the repulsive region of the potential is approximated by a hard core, then $\tilde{h}^{(a,b)} = -1$ for $r < \sigma^{(a,b)}$, and (7.19b) gives a simple linear dependence of $\partial q/\partial r$ on r in the range $S^{(a,b)} < r < \sigma^{(a,b)}$, after integrating and introducing the boundary condition $q^{(a,b)}(\sigma^{(a,b)}) = 0$, there results,

$$q_0^{(a,b)} = \tfrac{1}{2} A_a [r^2 - (\sigma^{(a,b)})^2] + B_a [r - \sigma^{(a,b)}] \tag{7.20a}$$

where

$$A_a^0 = 1 + 2\pi \sum \rho_\alpha \int_{S^{(a,\alpha)}}^{\sigma^{(a,\alpha)}} q_0^{(a,\alpha)}(t)\, dt \tag{7.20b}$$

$$B_a^0 = -2\pi \sum \rho_\alpha \int_{S^{(a,\alpha)}}^{\sigma^{(a,\alpha)}} t q_0^{(a,\alpha)}(t)\, dt. \tag{7.20c}$$

The zero subscript and superscripts are to indicate that these expressions are the hard-sphere parts of the functions and the appropriate upper limit has been placed on the integrals. Equations (7.20b, c) constitute a pair of simultaneous equations for A^0_a and B^0_a which can be obtained as (Baxter, 1968, Perram and Smith, 1975)

$$A^0_a = (1 - \xi_3 + 3\sigma_a\xi_2)/(1 - \xi_3)^2 \tag{7.21a}$$

$$B^0_a = -3\sigma^2_a\xi_2/2(1 - \xi_3)^2 \tag{7.21b}$$

$$\xi_j = \sum_\alpha \eta_\alpha \sigma^{j-3}_a. \tag{7.21c}$$

Baxter extended the Weiner–Hopf factorization method to molecules interacting through a "sticky sphere" potential $u^{(2)SS}$ which is an infinitesimally narrow square well, described by the equations

$$\begin{aligned} \exp[-\beta u^{(a,b)SS}] &= \delta(r - \sigma^{(a,b)})\phi^{(a,b)} \qquad && r \leqslant \sigma^{(a,b)} \\ &= 1 && r > \sigma^{(a,b)} \end{aligned} \tag{7.22}$$

The potential well is characterized by the parameter $\phi^{(a,b)}$, which may in turn be related to a realistic pair potential $u^{(a,b)}$ through the second virial coefficient, that is,

$$[\sigma^{(a,b)}]^3\phi^{(a,b)} = \int_{\sigma^{(a,b)}}^{\infty} r^2[\exp(-\beta u^{(a,b)}) - 1]\,\mathrm{d}r. \tag{7.23}$$

Equation (7.19a), with the PY approximation (Eqn (3.181c)) for $c^{(2)}$, can now be used to derive additive terms to q_0 in Eqn (7.20a) and to A^0 and B^0 in Eqns (7.21a), (7.21b). These may be written

$$\Delta q^{(a,b)}_{\mathrm{SSPY}} = \lambda^{(a,b)}[\sigma^{(a,b)}]^2/12 \tag{7.24a}$$

$$\Delta A^{\mathrm{SSPY}}_a = -X_a/(1 - \xi_3) \tag{7.24b}$$

$$\Delta B^{\mathrm{SSPY}}_a = \sigma_a X_a/2(1 - \xi_3) \tag{7.24c}$$

where

$$X_a = \sum_\alpha \eta_\alpha \lambda^{(a,\alpha)}[\sigma^{(a,\alpha)}/\sigma_\alpha]^2 \tag{7.24d}$$

and $\lambda^{(a,b)}$ can be found by solving a set of coupled quadratic equations,

$$\frac{\lambda^{(a,b)}\sigma^{(a,b)}}{12\phi^{(a,b)}} = A_a + \frac{B_a}{\sigma^{(a,b)}} + \sum_\alpha \frac{\eta_\alpha \lambda^{(b,\alpha)}[\sigma^{(b,\alpha)}]^2 q^{(a,\alpha)}(S^{(a,\alpha)})}{\sigma^2_\alpha \sigma^{(a,b)}} \tag{7.25}$$

in which A_a, B_a and $q^{(a,\alpha)}$ are the complete functions obtained by combining (7.20), (7.21) and (7.24).

Analogous methods have been applied to the Yukawa potential with a mean spherical approximation, Eqn (3.181a), for $c^{(a,b)}$ to give a $q^{(a,b)}$ having similar structure to that for the SSPY model, but in which A_a, B_a and $q^{(a,b)}$ now include additional exponential terms (Thompson *et al.*, 1980).

The mixture equations established in this way can now be specialized, either to the pure uniform fluid, by setting $a = b$, or to a solid adsorbent–fluid adsorbate system. In the latter case the usual procedure is applied of allowing the density of the adsorbent component, ρ_w, to go to zero, and then taking the limit of infinite particle size, $\sigma_w \rightarrow \infty$; after defining z, the distance from the surface, through the relationship $r = \sigma_w + z$. The most direct route to $\rho^{(a)}$ from the Weiner–Hopf factorization would appear to be via $\tilde{c}^{(2)}$, obtained from suitable manipulation of Eqn (7.19a) and substitution into (3.144a) to obtain expressions analogous to those in Eqn (7.5). Hiroike (1969) has shown that this type of approach is feasible using a generalization of Baxter's method (Hiroike and Fukui, 1970).

An alternative method is to set up an "initial value" integral equation from Eqn (7.19b) (Perram, 1975; Perram and Smith, 1977) with the usual limits of zero density and infinite radius for the adsorbent species. In this limit, Eqn (7.23) relates the sticky-sphere parameter to any other adsorbate adsorbent potential by

$$\psi(T) = \frac{1}{\sigma_a} \int_0^\infty \{\exp[-\beta u^{(a)}(z)] - 1\}\, dz \qquad (7.26)$$

and the limiting procedure is applied to Eqn (7.25) and to (7.20a) and (7.24a) to give an expression for $q^{(w,a)}(t)$ which can be substituted in Eqn (7.19b). A similar route can be followed to arrive at $\rho^{(a)}$ for the MS–Yukawa potential model. It is to be noted that strictly speaking these are no longer exact solutions, although the resulting integral equation is now much simpler to solve than the original OZ equation, even in its limiting form. On the other hand, the mixture method and the introduction of attractive properties into the potential function permits a wide range of model systems, with single and multicomponent adsorbates, to be examined thus providing useful insight into the effect of different parameters on the shape of the density profile. For example, the behaviour of the Yukawa fluid in the MS approximation has been studied in some detail (Thompson *et al.*, 1980) and has resulted in a variety of possibilities as determined by the temperature, the energy parameters $\varepsilon^{(w,a)}$, $\varepsilon^{(a,a)}$ of the wall and fluid interactions defined in Eqns (7.7a, e), the reduced density of the bulk fluid $\rho_a\sigma_a^3$ and the range parameter B of the potential. It is possible to distinguish four types of behaviour according to thermodynamic regions determined by the co-existence curve for the bulk fluid, plotted as $\beta\varepsilon^{(a,a)}$ against density. These regions are: (a) high temperature (supercritical); (b) condensable gas (low temperature and density); (c) low temperature liquid; and (d) the co-existence region. Only the first three of these are of real physical significance, although it is sometimes possible to obtain solutions for apparently spurious states inside the co-existence region.

Figures 7.3 and 7.4 illustrate the fluid hard-wall profiles over a range of temperature for low and high densities respectively. In both figures the two lower curves correspond to solutions inside the co-existence region and are not therefore physically realistic. The gas phase ($\rho_a\sigma_a^3=0{\cdot}2$) and liquid phase ($\rho_a\sigma_a^3=0{\cdot}7$) regions are distinguished by the much stronger structuring which is observed in the denser phase. Nevertheless, all the low-density distributions show some structuring in the monolayer region. At low temperatures and low densities, $\rho^{(a)}$ is always $\leqslant\rho_a$ and it is notable that, even where this phenomenon of "depletion" does not occur, there are a number of (ρ, T) states for which the surface excess is negative. Examples of all of these types of behaviour can also be found when the SS potential is used (Perram and Smith, 1977).

These results appear to be in general qualitative agreement with simulation data for fluid–hard-wall systems which show both structuring and depletion under similar conditions (Ch. 6, § 3(b)), but the simulations do not cover the wide range of states which can be examined using integral

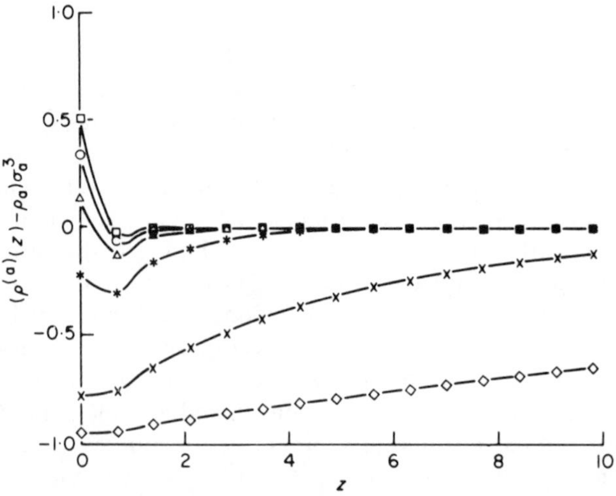

Fig. 7.3. Excess density profile for a Yukawa fluid against a hard wall on the isochore $\rho_a\sigma_a^3 = 0{\cdot}2$ for different values of $T^* = kT/\varepsilon^{(a,a)}$. □, $T^* = 10^8$; ○, $T^* = 5$; △, $T^* = 2{\cdot}5$; ★, $T^* = 1{\cdot}43$; ×, $T^* = 1$; ◇, $T^* = 0{\cdot}95$. The last two temperatures are inside the critical co-existence region at this density. (From Thompson *et al.*, 1980.)

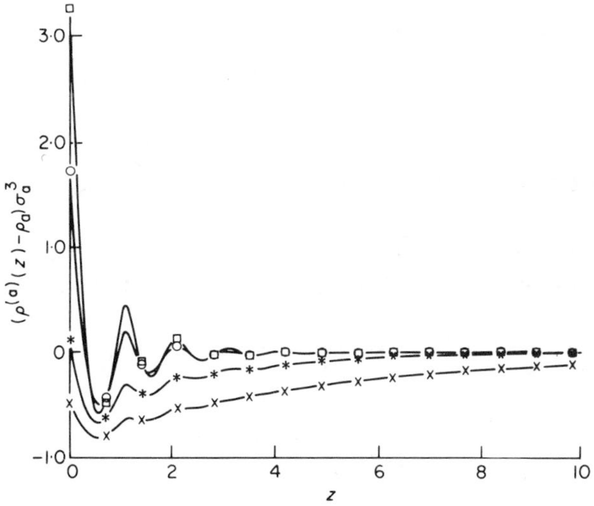

Fig. 7.4. Excess density profile for a Yukawa fluid against a hard wall on the isochore $\rho_a\sigma_a^3 = 0{\cdot}7$ for different values of $T^* = kT/\varepsilon^{(a,a)}$. □, $T^* = 10^8$; ○, $T^* = 1$; ★, $T^* = 0{\cdot}63$; ×, $T^* = 0{\cdot}59$. The last two temperatures are inside the co-existence region at this density. (From Thompson *et al.*, 1980.)

equations and no quantitative comparisons are available for these model potentials.

The effect of introducing an attractive wall potential has been systematically studied in the Yukawa model (Thompson *et al.*, 1980) and may once again be considered separately for subcritical and supercritical densities. On the liquid side the profile is only modified near to the wall, but its wavelength and amplitude remain virtually unchanged further out. At low densities, on the other hand, an increase in the wall potential can alter the profile from one of overall depletion to a region of positive surface excess with a strong adsorption into the first layer.

In discussing these results it must be borne in mind that they are subject to the inherent limitations of the PY and MS approximations, not only in respect to the density at the wall, as mentioned above, but also because of the well known failure of these approximations below the bulk density transition.

A variant of the Weiner–Hopf factorization method (Blum and Stell, 1976) has been applied to several interface problems by Sullivan and Stell (1977, 1978). In this method the direct correlation function is split into components,

$$c^{(2)} = c_0^{(2)} - c_1^{(2)} \tag{7.27}$$

such that

$$c_0^{(2)} = 0 \qquad z \geqslant \sigma/2$$

$$c_1^{(2)} = 0 \qquad z < \sigma/2$$

and the generalized matrix of form of (7.15a) expressed as $\hat{\boldsymbol{h}}[\boldsymbol{I} - \hat{\boldsymbol{c}}] = \hat{\boldsymbol{c}} = \hat{\boldsymbol{c}}_0 - \hat{\boldsymbol{c}}_1$.

After introducing the factorization given in Eqn (7.18a) and Fourier inverting, the c_0 term can be shown to vanish leaving the result

$$\tilde{h}^{(\mathrm{w,a})} - \sum_{\alpha} \rho_{\alpha} \int_{\lambda^{(\mathrm{a},\alpha)}}^{\infty} \tilde{h}^{(\mathrm{w},\alpha)}(z-r)[1 - Q^{(\mathrm{w},\alpha)}(r)]\,\mathrm{d}r$$

$$= -\sum_{\alpha} \int_{\sigma^{(\mathrm{w},\alpha)}/2}^{\infty} c_1^{(\mathrm{a},\alpha)} P^{(\mathrm{a},\alpha)}(t-z)\,\mathrm{d}t \tag{7.28}$$

where

$$P^{(\mathrm{a},\alpha)}(y) = \frac{1}{2\pi} \int_{-\infty}^{\infty} [1 - \hat{Q}^{(\mathrm{a},\alpha)}(-k)]^{-1} e^{iky}\,\mathrm{d}k.$$

Solution of this equation requires a suitable approximation for $c_1(t)$, such as the MSA, and the corresponding factorization functions which are related to c through the equations given earlier but are of course only available for certain potential models such as the sticky sphere and Yukawa potential. The method has been applied to the calculation of density profiles in the sticky-sphere/hard-wall system (Sullivan and Stell, 1977), and as might be expected, produces results similar in general pattern to those already discussed.

(c) Adsorption isotherms from solutions of the OZ equation

The OZ equation provides a convenient route to adsorption isotherms for a wide variety of circumstances including adsorption of mixtures or single components onto plane surfaces and spherical particles. From Eqn (2.2) the surface excess number of type a molecules is

$$N_{\mathrm{a}}^{\Sigma} = \int (\rho^{(\mathrm{a})} - \rho_{\mathrm{a}})\,\mathrm{d}\boldsymbol{r} = \rho_{\mathrm{a}} \int \tilde{h}^{(\mathrm{w,a})}\,\mathrm{d}\boldsymbol{r} \tag{7.29}$$

where (3.143) has been used in taking the second step. The excess number of molecules per unit area of surface Γ_{a} is

$$\Gamma_{\mathrm{a}} = \frac{N_{\mathrm{a}}^{\Sigma}}{\mathscr{A}} = \frac{\rho_{\mathrm{a}}}{\sigma_{\mathrm{w}}^2} \int_{\sigma_{\mathrm{s}}}^{\infty} \tilde{h}^{(\mathrm{w,a})}\, r^2\,\mathrm{d}r \tag{7.30}$$

which, after integration by parts and introduction of the condition $\tilde{h}^{(\mathrm{w,a})}(r) = -1, 0 < r < \sigma_{\mathrm{w}}$, becomes

$$\Gamma_{\mathrm{a}} = \frac{\rho_{\mathrm{a}}}{\sigma_{\mathrm{w}}^2}\int_0^\infty \mathrm{d}r \int_r^\infty t\,\tilde{h}^{(\mathrm{w,a})}(t)\,\mathrm{d}t + \tfrac{1}{3}\rho_{\mathrm{a}}\sigma_{\mathrm{w}}. \tag{7.31}$$

It can be seen from Eqn (7.17) that the integral in (7.31) is also that which appears in the zero-wavelength Fourier transform of $\tilde{h}^{(\mathrm{w,a})}$ and therefore

$$\Gamma_{\mathrm{a}} = \frac{1}{4\pi\sigma_{\mathrm{w}}^2}\left[\frac{\rho_{\mathrm{a}}}{\rho_{\mathrm{w}}}\right]^{1/2}\hat{h}^{(\mathrm{w,a})}(0) + \tfrac{1}{3}\rho_{\mathrm{a}}\sigma_{\mathrm{w}}. \tag{7.32}$$

This in turn can be expressed in terms of the Fourier transform of the Weiner–Hopf factorization function $q^{(\mathrm{w,a})}$ by substitution from (7.18b). In the HS or SS models, for which $q(r) = 0$ when $r > \sigma_{\mathrm{w}}$, the isotherms can now be derived by carrying out the specified integrations, although lengthy algebra is usually involved.

The surface excess for adsorption at a plane surface follows by imposing the limits $\eta_{\mathrm{w}} \to 0$, $\sigma_{\mathrm{w}} \to \infty$; for sticky spheres in the presence of a sticky wall the result may be written

$$\Gamma_{\mathrm{a}} = \rho_{\mathrm{a}}\sigma_{\mathrm{a}}\left\{\frac{(1-\eta_{\mathrm{a}})\psi}{(1-\eta_{\mathrm{a}}) + \psi[6\eta_{\mathrm{a}} - \lambda\eta_{\mathrm{a}}(1-\eta_{\mathrm{a}})]} + \frac{3\eta_{\mathrm{a}}}{2(1+2\eta_{\mathrm{a}}) - 2\lambda\eta_{\mathrm{a}}(1-\eta_{\mathrm{a}})}\right\} \tag{7.33}$$

where ψ, which increases with the strength of the adsorption potential, is given by Eqn (7.26) and λ ($\equiv\lambda^{(\mathrm{a,a})}$) is obtained from Eqn (7.25) when the usual limits have been taken there for η_{w} and σ_{w}. Under appropriate constraints Eqn (7.33) reduces to simpler expressions for the limiting cases of hard spheres ($\lambda = 0$) or hard wall ($\psi = 0$) which have both been discussed in some detail (Richmond, 1976, Mitchell and Richmond, 1976).

In models such as these, where the molecules have hard cores, adsorption in the monolayer alone can be clearly defined from the condition for contact between adsorbent and adsorbate. In the single-component plane-wall system the fractional occupation of the monolayer is found (Perram and Smith, 1976; Smith and Perram, 1977) to be

$$\theta = \frac{3}{2}\frac{\eta_{\mathrm{a}}\psi[(1+2\eta_{\mathrm{a}}) - \lambda\eta_{\mathrm{a}}(1-\eta_{\mathrm{a}})]}{(1-\eta_{\mathrm{a}})\{(1-\eta_{\mathrm{a}}) + \psi[6\eta_{\mathrm{a}} - \lambda\eta_{\mathrm{a}}(1-\eta_{\mathrm{a}})]\}} \tag{7.34}$$

and a similar but more complex expression has been derived for multi-component systems (Perram and Smith, 1977) and for binary systems in which both species are spherical particles of finite radius.

Equation (7.30) shows that Γ_{a} is directly related to the second moment of $\tilde{h}^{(\mathrm{w,a})}(r)$ and this therefore provides an alternative route to the surface excess from the Laplace transform approach. Henderson *et al.* (1976b) used this approach to obtain Γ_a for hard spheres adsorbed by a Yukawa wall in the MS approximation and found

$$\Gamma_a = \rho_a\sigma_a\left[\frac{C_1(1-\eta_a)^4}{B(1+2\eta_a)} + \frac{3\eta_a}{2(1+2\eta_a)}\right] \tag{7.35}$$

where C_1 is given by Eqn (7.8b). In the limit of vanishing adsorption potential or short-range adsorption forces, Eqns (7.33) and (7.35) both lead to equations which are identical in form (Mitchell and Richmond, 1976) and which may be written

$$\Gamma_a = \rho_a\sigma_a\left[\frac{K(1-\eta_a)^2}{(1+2\eta_a)} + \frac{3\eta_a}{2(1+2\eta_a)}\right]. \tag{7.36}$$

In the opposite limit, which is the van der Waals or long-range case, Eqn (7.33) and (7.35) lead to

$$\Gamma_a = \rho_a\sigma_a\left[\frac{K(1-\eta_a)^4}{(1+2\eta_a)^2} + \frac{3\eta_a}{2(1+2\eta_a)}\right] \tag{7.37}$$

where $K \equiv \psi$ or $\beta\varepsilon/B$ for the SS and Yukawa potentials respectively. It can be seen from Eqns (7.1g), (7.4a) and (7.4d) that the first term in Eqn (7.36) is $(\rho_a\kappa)^{1/2}$ and in (7.37) is $\rho_a\kappa$, where κ is the compressibility of the adsorptive fluid and, bearing in mind that these terms dominate the equations, a parallel may be drawn between Eqn (7.37) and (3.194), (3.195) and (3.196).

In the latter equations, $u^{[1]P}(z)$ is a perturbation potential, representing the difference between the external field at z due to the adsorbent and the same field if the adsorbent were to be replaced by adsorbate fluid. $u^{[a]P}$ is therefore small, and it is reasonable to expand (3.194) or (3.196) to give

$$\rho^{(a)}(z) - \rho_a = -\rho_a^2 u^{[a]P}(z)\kappa \tag{7.38}$$

an equation derived by Kuni and Rusanov (1968) (cf. Findenegg and Fischer, 1975) from a different route. Integration of Eqn (7.38) over z to give Γ_a reveals the close resemblance with (7.37). It must again be emphasized that it is the PY compressibility which appears in Eqns (7.4), (7.35), and (7.37) in contrast to Eqn (3.194) where no approximation has been made in relation to the compressibility for the uniform bulk fluid. It should also perhaps be remarked that, as can readily be seen from Eqn (3.196), for example, $\rho^{(a)}$ can only exhibit stationary points where $u^{[a]P}$ does so, and that this type of theory is therefore incapable of predicting oscillatory profiles.

The sticky-sphere fluid has been thoroughly studied and shown to have a critical point at $\eta_a = 0{\cdot}1213$ and $\phi = 1{\cdot}17$. In addition Baxter (1968a, b)

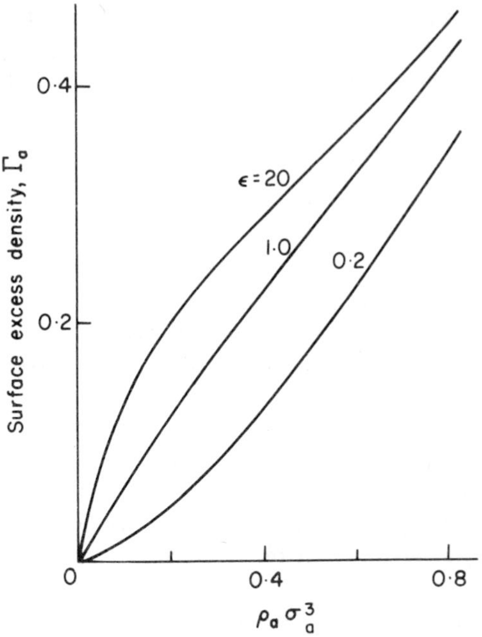

Fig. 7.5. Examples of surface excess Γ_a versus $\rho_a\sigma_a^3$ for a sticky-sphere fluid. The parameter ε is defined from $\psi(z) = \varepsilon(z - \sigma_a)/\sigma_a$, and is measured in units of kT. (From Perram and Smith, 1977.)

has shown that the requirement that the integral $\int \tilde{h}(r)\,dr$ should be absolutely convergent implies the condition $\lambda\eta_a(1 - \eta_a) < (1 + 2\eta_a)$ which in turn implies that the second term in (7.33) is always positive. On the other hand, the denominator in the first term of this equation can become zero if both ϕ and ψ are both sufficiently large (Richmond, 1976) as might occur, for example, at low temperature. This singularity corresponds to a condition for multilayer wetting; some examples are shown in Fig. 7.5. The variety of isotherms possible can be conveniently discussed by considering the properties of Eqn (7.33) at low density. In this limit the surface excess is given by

$$\Gamma_a \underset{\eta_a \to 0}{=} \rho_a\sigma_a\{\psi + 6\eta_a[\tfrac{1}{4} - (1 - 2\phi)\psi^2] + \ldots \tag{7.39}$$

where 2ϕ is the zero-density limit of λ (Richmond, 1977).

In the first place it is to be noted that this has the correct Henry law limit at $\eta_a = 0$ when Γ_a is a linear function of the bulk phase density. An initial, positive deviation from this line, which corresponds to type III behaviour in the BET classification of isotherms will be observed when

$$\psi^2(1 - 2\phi) \leqslant \tfrac{1}{4}. \tag{7.40}$$

This condition will always be met when $\phi > \frac{1}{2}$, regardless of the value of ψ, and accords with the classical association of Type III behaviour with strong adsorbate interactions. On the other hand, the model predicts that type III isotherms can occur even when there are no attractive adsorbate interactions and this observation has been attributed to hard-sphere exclusion effects (Perram and White, 1975).

The converse of the condition expressed in Eqn (7.40) corresponds to negative deviation from linearity and is therefore to be associated with type I or type II isotherms. However, in order to be identified as the latter, it is necessary at the same time that multilayer wetting should occur at saturation; and it has not been satisfactorily established that the mathematical and physical conditions necessary to meet this criterion are in fact compatible.

Finally the "high-temperature" behaviour observed experimentally, and referred to in Chapter 1, in which the isotherms show a maximum, is also predicted by Eqn (7.33). Impressive though this range of predictions appear to be, it must be stressed that superficial correspondence with experimental observations does not in itself provide strong support for the validity of the theory and its underlying assumptions as a description of physical adsorption. Considerable further work is needed before the similarities mentioned above can be regarded as being more than merely suggestive. Some effort in this direction has been made by Smith and Perram (1977), who examined the transition from a rarefied to a dense adsorbate phase for the Kr–graphite system and showed that the temperature dependence of the transition density was well represented qualitatively, although the theory does not of course account for the existence of the various dense monolayer phases observed experimentally.

Strong reservations concerning the OZ model have been voiced by Henderson and co-workers (1980), who point out that this type of theory treats the fluid at the wall as only one phase and that the singularity associated with wetting may be an artefact specifically of the SS potential.

3. Numerical solution of integral equations

The number of cases in which exact solutions can be found to the integral equations discussed in Chapter 3 is rather limited, and numerical methods open up further possibilities in both the type of interaction potential and the type of equation which can be investigated. Although such methods can often consume quite large amounts of computer time, they are almost certainly faster for a given system than the corresponding simulation would be and therefore offer the advantage that, as with the exact solutions already discussed, some insight might be gained into a wide range of physical situations. In addition, it may be possible in some cases to achieve a better understanding of the nature of the mathematical structures that underlie the system of interest, thereby providing clues to undiscovered methods for exact solutions. Moreover it seems fairly probable in some cases, for example where very long-range forces are involved, that a computer simulation on a sufficiently large scale may not be possible in the foreseeable future. The use of methods already validated in small systems may offer the best route to improved physical descriptions here.

On the other hand, some caution must be exercised in using numerical methods since it cannot be guaranteed that a particular set of equations will converge to stable solutions with given approximations and integration techniques, nor can it be guaranteed that apparently stable solutions do actually correspond to a physically real situation. In connection with this last point, it would appear that iterative methods, generally speaking, offer the most reliable route to a satisfactory solution, since no *a priori* assumption about the shape of the density profile needs to be introduced as it must be when the parameters of a trial function are adjusted. In using an iterative approach it is still necessary to choose a method of numerical integration, a suitable subdivision of the space and an initial guess for $\rho^{(1)}$, none of which should of course influence the final result, but may affect the rate of convergence. A number of authors have referred to difficulties encountered in finding stable solutions at a temperature below T_c, and it would appear that even for densities outside the coexistence region the integral equations have a tendency to diverge at $T < T_c$.

The equations which have been investigated fall roughly into three groups: the OZ equation; the density functional (DF) equations; and the BBGKY hierachy.

(a) The OZ equation

The OZ equations have already been considered in the context of exact solutions of the HAB type of equation. The exact OZ equation for non-uniform fluids, Eqn (3.141), is much more formidable since it contains unknown non-uniform correlation functions as well as the unknown singlet function. So far no attempts have been made to use this equation in adsorption problems. The HAB type of approximate equation (Eqn (3.144)) can be solved, in principle, by the introduction of a relationship between $c^{[1]}$ $(\equiv \tilde{c}^{(\mathrm{w,a})})$† and $\rho^{(1)}$ $(\equiv \rho g^{(1)} \equiv \rho_a \tilde{g}^{(\mathrm{w,a})})$. In addition $\tilde{c}^{(2)}$ $(\equiv \tilde{c}^{(\mathrm{a,a})})$

† Since the transformation from the mixture equations to adsorption equations is assumed to have been made here, we use the simpler notation $c^{[1]}$, etc.

or $\tilde{h}^{(2)}$ also needs to be known. The PY and CHNC approximations given in Eqn (3.185) provide a straightforward substitution for $c^{[1]}$. The former approximation has been used to obtain profiles for the hard-sphere/hard-wall (Fischer, 1977) and sticky-sphere/hard-wall systems (Sullivan and Stell, 1977, 1978) as already discussed, and for 12–6 fluid in the presence of an attractive wall (Ebner *et al.*, 1980). Both approximations have been applied to a 9–3 wall in contact with a 12–6 fluid (Saam and Ebner, 1978; Smith and Lee, 1979) with various choices for the parameters, corresponding approximately to fluid argon and solid graphite, argon or CO_2.

Smith and Lee (1979) investigated both low-density ($\rho^* = 0{\cdot}067$; $2{\cdot}0 < T^* < 5{\cdot}0$) and high-density ($\rho^* = 0{\cdot}65$; $T^* = 1{\cdot}0$) systems. For the latter $\tilde{c}^{(a,a)}$ was taken from the simulation data of Verlet (1968), and the singlet profiles could be compared with those from MC simulation (Abraham and Singh, 1978) at the same (ρ^*, T^*). Both approximations were found to exaggerate considerably the height of the first and second maxima as well as being out of phase with the MC profile. At the low density very little difference could be discerned between the PY and CHNC approximations, but the PY had a somewhat greater amplitude at higher density. No suitable simulation data were available for comparison at the lower densities, but considerable differences between the PY and the density functional theory, to be discussed below, have been noted (Saam and Ebner, 1978). Later work, in which the PY theory was compared with simulation data for 12–6 molecules between two walls, confirmed the inadequacy of this theory when attractive forces between molecules are present (Ebner *et al.*, 1980).

In an attempt to improve the performance of the HAB equation in these cases, Sullivan and Stell (1978) examined a number of more elaborate closures, all related to the theme of a potential subdivided into hard-sphere (HS) and attractive (A) parts. The CHNC closure can be extended in an arbitrary way by adding a term R which represents the missing terms lost in the truncation of the functional Taylor expansion. Lado (1964, 1973) introduced the conjecture that in any system, R could be approximated by its hard-sphere representation. If this conjecture is applied to $c^{[1]}$, as given in (3.183), and the term R is eliminated between the HS system and any other, there results the RHNC (renormalized hypernetted chain) approximation which can be written

$$c^{[1]} - c^{[1]\mathrm{HS}} = h^{(1)} - h^{(1)\mathrm{HS}} - \beta u^{[1]\mathrm{A}} - \ln(g^{(1)}/g^{(1)\mathrm{HS}}). \qquad (7.41)$$

A slight modification of this equation is to replace $h^{(1)}$ and $g^{(1)}$ by $h^{(1)\mathrm{N}}$ and $c^{[1]\mathrm{N}}$ defined such that

$$\left.\begin{aligned} c^{[1]\mathrm{N}} &= c^{[1]\mathrm{HS}} - \beta u^{[1]\mathrm{A}} \qquad && z > 0 \\ h^{(1)\mathrm{N}} &= -1 && z < 0 \end{aligned}\right\} \qquad (7.42)$$

which leads to the analogue of the EXP approximation of Andersen and Chandler (1972)

$$g^{(1)} = g^{(1)\mathrm{HS}} \exp[h^{(1)\mathrm{N}} - h^{(1)\mathrm{HS}}]. \tag{7.43}$$

Another approximation scheme suggested by Sullivan and Stell (1978) is to introduce a hard-sphere reference system, for which the wall–particle potential is adjusted to an effective function $u^{[1]\mathrm{E}}$, so as to make the density profile of the reference system the same as that of the actual system ($h^{(1)\mathrm{ref}} = \mathrm{h}^{(1)}$). In the CHNC approximation the effective direct correlation function $c^{[1]\mathrm{E}}$ is then given by

$$c^{[1]} - c^{[1]\mathrm{E}} = -\beta[u^{[1]} - u^{[1]\mathrm{E}}]. \tag{7.44}$$

The left-hand side of this equation can be replaced by substitution from the HAB form of the OZ equation to give

$$\beta[u_1^{[1]} - u_1^{[1]\mathrm{E}}] = \rho \int [\tilde{c}_{13}^{(2)} - \tilde{c}_{13}^{(2)\mathrm{HS}}]h^{(1)}\, \mathrm{d}\boldsymbol{r}_3. \tag{7.45}$$

In the limit of an infinitely weak long-range $u^{[1]\mathrm{A}}$, this becomes exact, and since this limit relates to a Bragg–Williams mean-field approximation for a lattice gas, Sullivan and Stell refer to this as the MF approximation. A new closure for $c^{[1]}$ can be obtained, either by using the PY relation between $h^{(1)}$ ($\equiv h^{(1)\mathrm{E}}$) and $c^{[1]\mathrm{E}}$, or the EXP relation of Eqns (7.41) and (7.42).

The above approximations and the PY closure were examined for a LJ-plus-hard-core fluid in contact with a hard wall. Only at high temperature ($T^* = 4{\cdot}0$), could convergence be achieved with the MF–PY approximation which then exhibited quite different behaviour from the other closures (Fig. 7.6). The four remaining closures were used at low temperatures ($T^* = 1{\cdot}6$, $0{\cdot}85$) and high densities ($\rho^* = 0{\cdot}6$, $0{\cdot}85$) and, in contrast to uniform fluids, showed considerable differences in behaviour, especially within the first molecular diameter or so from the wall. This difference between uniform and non-uniform fluids can be taken as a reflection of the more important role which internal attractive forces play in the non-uniform case, in contrast to the dominating effect of hard repulsive forces in determining the structure of uniform fluids.

The effect of attractive intermolecular adsorbate forces is still more emphatically revealed when the calculated results are compared with molecular simulations. The latter were carried out at $T^* = 0{\cdot}85$ ($0{\cdot}82$) and $\rho^* = 0{\cdot}85$ ($0{\cdot}83$) with 1000 particles in a box of length 8σ and a very large number of configurations were sampled ($3{\cdot}5 \times 10^7$) using CEMC. The integral equation solutions for EXP and MF-EXP closures (Fig. 7.7, Sullivan *et al.*, 1980) are strikingly different from the simulations in showing

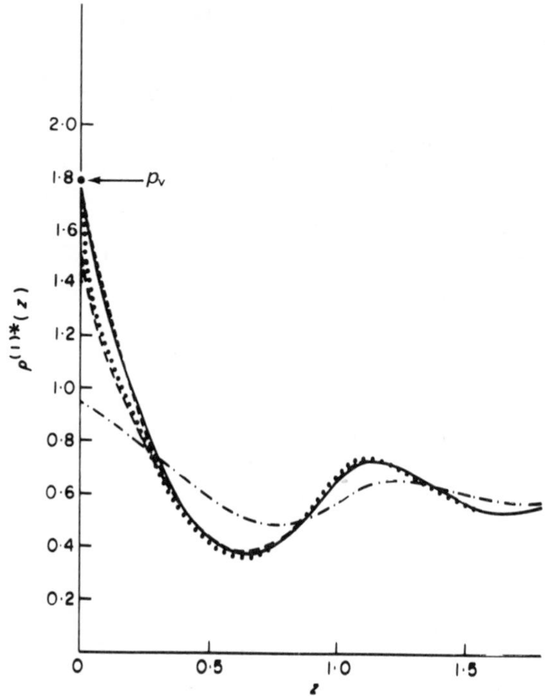

Fig. 7.6. comparison of various integral equation approximations for the density profile $\rho^{(1)*}(z) = \rho\sigma^3 g^{(1)}(z)$ of a Lennard-Jones fluid near a hard wall at $\rho^* = 0{\cdot}6$ and $T^* = 4{\cdot}0$. The distance z is in units of the hard-core diameter σ. p_v denotes the dimensionless pressure $\beta p \sigma^3$, according to the virial equation. ——, RHNC; — —, PY; . . . . ., EXP; - - -, MF–EXP; –·–·, MF–PY. (Sullivan and Stell, 1978.).

oscillations which are both too large in amplitude and too short in wavelength. It is possible that the disparities may be less severe for $T > T_c$, but the ability of the HAB equation to accurately describe the structure of non-uniform fluids of attractive particles is clearly put in doubt by these results and these doubts must of course apply with equal force to those model systems where exact results have been obtained.

As explained in Chapter 3, the expansions given in Eqns (3.177) may be considered to be equivalent to the HAB theory when the reference state is chosen to be a uniform fluid of density ρ_a. It might therefore be expected that a better description of non-uniform systems could be achieved, both by choosing a non-uniform reference state, and by improving the approximation to $c^{(2)}$. The formal expression embodying the second of these requirements is stated in (3.225), where $c^{(2)}$ is expanded in terms of higher-order correlation functions for uniform fluids. A PY variant of this

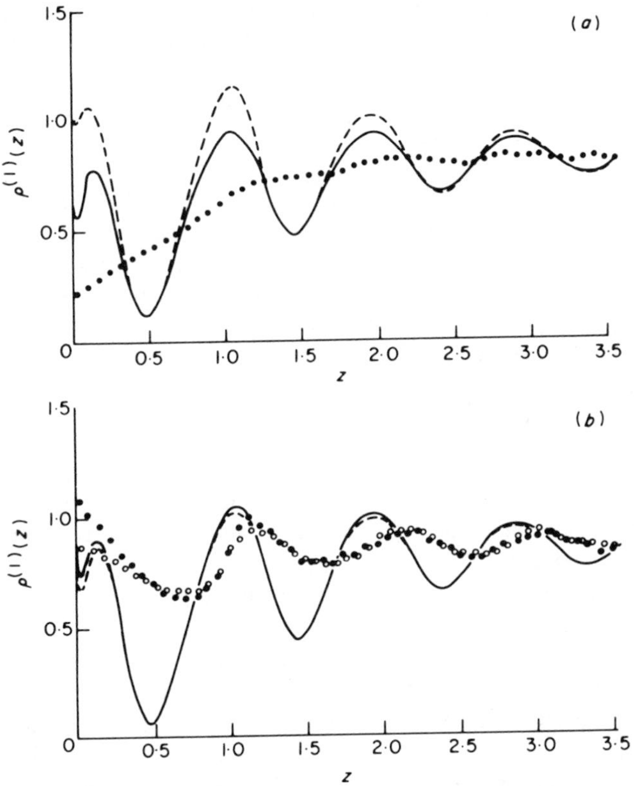

Fig. 7.7. (*a*) Comparison of Monte Carlo and integral equation (EXP, MF–EXP) density profiles at $\rho = 0{\cdot}83$, $T^* = 0{\cdot}82$. MF–EXP, ——; EXP, - - -; MC, ... (*b*) Comparison of Monte Carlo and integral equation (MF–EXP) density profiles at $\rho = 0{\cdot}85$, $T^* = 0{\cdot}85$ for two values of the truncation distance $r_c = 2{\cdot}5$ and $3{\cdot}2$. MF–EXP, $r_c = 2{\cdot}5$, ——; $r_c = 3{\cdot}2$, - - -; MC, $r_c = 2{\cdot}5$, ●●●; $r_c = 3{\cdot}2$, ○○○. (Sullivan *et al.*, 1980.)

type of theory is obtained by replacing (3.177c) (given after (3.181)) by $\tilde{c}^{(2)}(\rho)$ to give, for spherically symmetric molecules,

$$\frac{\rho^{(1)}\zeta_1^{[1]0}}{\rho^{(1)0}\zeta_1^{[1]}} = 1 + \int \tilde{c}_{12}^{(2)}[\rho_2^{(1)} - \rho_2^{(1)0}]\,\mathrm{d}\boldsymbol{r}_2. \tag{7.46}$$

The zeroth-order approximation of this form, obtained when the integral is neglected, has been found by Abraham and Singh (1977) to provide an excellent description for the hard-sphere, non-uniform fluid (the blip function approximation introduced in Eqn (3.218)). A second variant of the

PY type can be obtained by introducing an approximate resummation of the functional Taylor expansion, Eqn (3.225), to give (Lee, 1975)

$$\frac{\rho_1^{(1)}\zeta_1^{[1]0}}{\rho_1^{(1)0}\zeta_1^{[1]}} = \frac{\rho_1^{(1)} - \rho c_1^{[1]}}{\rho_1^{(1)0} - \rho c_2^{[1]}}. \tag{7.47}$$

Equations (7.46) and (7.47) were named the PYP1 and PYP approximations respectively by Lee (1980) who has evaluated them for hard-sphere and 12–6 fluids against a 9–3 wall. Two perturbation schemes were adopted:

(i) The reference system was a "WCA" wall with 9–3 potential subdivided after the method which Weeks, Chandler and Andersen applied to uniform fluids, viz

$$\begin{aligned} u^{[1]\mathrm{WCA}} &= u^{[1]\,9\text{–}3} - u_{\min} \qquad & z < z_{\min} \\ &= 0 & z \geqslant z_{\min} \end{aligned} \tag{7.48}$$

where the subscript "min" indicates the minimum in the potential. The fluid was a hard-sphere fluid, and the perturbation was from $u^{[1]\mathrm{WCA}} \to u^{[1]9\text{–}3}$.

(ii) The reference system was a hard-sphere fluid in contact with a 9–3 wall, and the perturbation was hard-sphere to 12–6 fluid.

Exact results for the two unperturbed states are available from the simulation data of Abraham and Singh (1978) at $\rho^* = 0{\cdot}65$, $T^* = 1{\cdot}00$. In scheme (i) both the blip function and the PYP theories proved to be quite inadequate. The former, as also pointed out by Abraham and Singh (1978), severely overestimates the height of the first peak in $\rho^{(1)}$, whilst in the latter

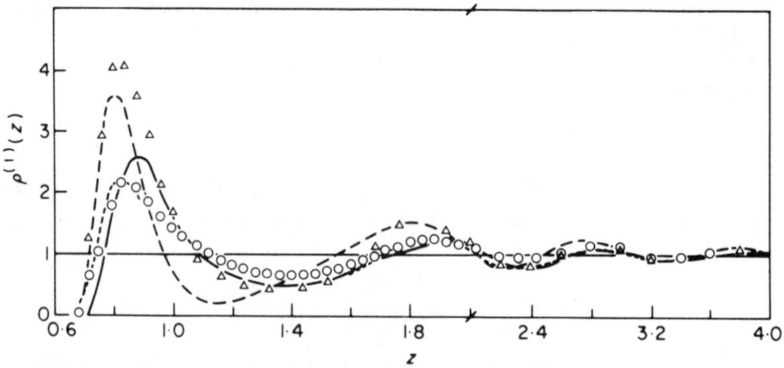

Fig. 7.8. Density profile for a 12–6 fluid against a 9–3 wall at $T^* = 1{\cdot}0$ and $\rho^* = 0{\cdot}65$. Theory: PYP1, ——; PY - - -; PYP .... MC: ○, 12–6/9–3 wall; △ hard-sphere/9–3 wall. MC data are from Abraham and Singh (1978). (Lee, 1980.)

$\rho^{(1)}$ becomes negative in the range $1{\cdot}0 < z < 1{\cdot}5$. In scheme (ii) the PY and PYP1 theories were tested but neither was found to be as accurate as the PYP theory (Fig. 7.7), which was also satisfactory in scheme (i). Unfortunately, the success of the PYP theory is marred by irregular oscillations (not shown in the figure) in the range $0{\cdot}9 < z < 1{\cdot}6$, which could be associated with uncertainties in the reference system and $\tilde{c}^{(2)}$ data.

It should be noted that the reference system simulations on which these perturbation schemes are based are themselves rather expensive to carry out, so that the route used in this work to reach a fully attractive system can only be regarded as useful in the sense of pointing to the importance of higher-order terms in the functional Taylor expansion for the description of real systems.

(b) Density functional theory

The limitations imposed by neglecting these higher-order terms can be removed in a formal way by introducing a coupling parameter, ξ, as has already been demonstrated in Chapter 3, and may be seen by comparing Eqns (3.177) and (3.179). As a purely formal manoeuvre this has no particular advantage over the methods so far discussed, but in fact a greater degree of flexibility is introduced since there is no longer any restriction on the density at which $c^{(2)}$ is to be calculated, which appears as if it were the case when the HAB equations are used. Ebner, Saam and co-workers (Ebner and Saam, 1977; Ebner *et al.*, 1976; Ebner and Punyanita, 1979; Ebner *et al.*, 1980; Saam and Ebner, 1977, 1978) have used this type of approach with some success, incorporating it with the grand free energy Ω which is a functional of $\rho^{(1)}$. Equation (3.179b) can be written in the form

$$\ln[\rho_1^{(1)}(r)/\rho_0] + \beta u_1^{[1]'} = \int_0^1 \int c_{12}^{(2)}(\xi)[\rho_2^{(1)} - \rho_0]\, d\xi\, d\boldsymbol{r}_2 \qquad (7.49)$$

where the reference state, indicated by 0, is chosen to be one of uniform density so that $u^{[1]0} = 0$, and $u^{[1]'}$ is the external potential which would produce the profile $\rho^{(1)}$. Equations (3.160) and (3.162) can be combined and written

$$\frac{\delta\beta\Omega}{\delta\rho_1^{(1)}} = \beta u_1^{[1]} + \frac{\delta\beta\Omega_{\text{int}}}{\delta\rho_1^{(1)}} = \beta u_1^{[1]} - \beta u_1^{[1]'} = 0 \qquad (7.50)$$

where the last equality expresses the minimization condition for the free energy. Equation (7.49) incorporates a linear dependence of $\rho^{(1)}$ on ξ given by Eqn (3.178). If this choice is used a second time, and (7.50) is combined with (7.49), there results

$$\beta(\Omega - \Omega_0) = \int [\beta u_1^{[1]} - 1][\rho_1^{(1)} - \rho_0]\, d\mathbf{r}_1 + \int \rho_1^{(1)} \ln(\rho_1^{(1)}/\rho_0)\, d\mathbf{r}_1$$
$$-\tfrac{1}{2}\iint c_{12}^{(2)*}[\rho_1^{(1)} - \rho_0][\rho_2^{(1)} - \rho_0]\, d\mathbf{r}_1\, d\mathbf{r}_2 \qquad (7.51)$$

where $c_{12}^{(2)*} = 2\int_0^1\int_0^1 c_{12}^{(2)}\, d\xi'\, \xi\, d\xi$, as defined earlier. By exploiting the symmetry property $c_{12}^{(2)} = c_{21}^{(2)}$, Eqn (7.51) can be put in the form

$$\beta(\Omega - \Omega_0) = \int \omega_1\, d\mathbf{r}_1 + \tfrac{1}{4}\iint c_{12}^{(12)*}[\rho_2^{(1)} - \rho_1^{(1)}]^2\, d\mathbf{r}_1\, d\mathbf{r}_2 \qquad (7.52a)$$

where

$$\omega_1 = (\rho_1^{(1)} - \rho_0)[\beta u_1^{[1]} + \ln(\rho^{(1)}/\rho_0) - 1] + \rho_0 \ln(\rho_1^{(1)}/\rho_0)$$
$$-\tfrac{1}{2}(\rho_1^{(1)} - \rho_0)^2 \int c_{12}^{(2)*}\, d\mathbf{r}_2 \qquad (7.52b)$$

is a local free energy density.

Equation (7.52) is quite similar to Eqn (3.223) which was originally reached by a more intuitive route (Ebner *et al.*, 1976). By contrast, Eqn (7.52) is exact (Saam and Ebner, 1978). However in order for it, or the earlier equation, to be useful it is necessary to remove the ξ dependence of $c^{(2)*}$ which introduces approximations. Three schemes for this purpose have been considered. In the first scheme (S1), the non-uniform OZ equation (3.141) is written with the correlation functions in a ξ-dependent form,

$$h_{12}^{(12)}(\xi) = c_{12}^{(2)}(\xi) + \int \rho_3^{(1)}(\xi) h_{12}^{(2)}(\xi) c_{32}^{(2)}(\xi)\, d\mathbf{r}_3 \qquad (7.53)$$

and the corresponding PY relationship ($c^{(2)}(\xi) = [1 + h^{(2)}(\xi)]y^{(2)}(\xi)$) is used as a closure. The second scheme (S2) can be computed more rapidly and consists of replacing $\rho_3^{(1)}$ in the OZ equation by $\bar{\rho} = \tfrac{1}{2}[\rho_1^{(1)} + \rho_2^{(1)}]$ and calculating the uniform direct correlation function $\tilde{c}^{(2)}(\xi, \bar{\rho})$. In the third scheme (S3) which has been more extensively used (Ebner *et al.*, 1976; Ebner and Saam, 1977; Ebner *et al.*, 1980), $c^{(2)*}$ is replaced by $\tilde{c}^{(2)}(\bar{\rho})$ and $\int c_{12}^{(12)*}\, d\mathbf{r}_2$ by $\int \tilde{c}_{12}^{(2)}(\rho_2^{(1)})\, d\mathbf{r}_2$.

In these calculations the form of $\Omega(\rho^{(1)})$ which satisfies the minimization condition expressed in Eqn (7.50) was sought (it may be noted that in this context Ω and Ω^{Σ} need not be distinguished). Rather than solve directly for $\rho^{(1)}$, the parameters of a trial function were determined with respect to the minimization condition.

In a study of non-uniform one-dimensional fluids, the DF theory, in conjunction with the approximations S1–S3, was compared with the HAB equation and with GEMC results (Ebner *et al.*, 1980). The external potential was in effect a model pore, symmetrical about $z = 0$, with width w and was written

$$u^{(1)} = E_0\{\exp[(z + w)/t] + 1\}^{-1} - E_0\{\exp[(z - w)/t] + 1\}^{-1} \quad (7.54)$$

where t controls the sharpness of the well and was set at 0·1, with w equal to either 0·8 or 2·5. The adsorbate fluid interacted through a 12–6 potential.

An interesting feature of these calculations was that Ω itself was found and since, according to the minimization condition, this quantity should be independent of the reference density ρ_0, its value could be used to monitor the validity of the approximation scheme. This aspect of the theory was examined in detail for scheme S2 by using high and low reference densities with four different choices for $u^{(1)}$; considerable dependence upon the choice of $u^{(1)}$ was observed. The approximations in S3 were used in conjunction with Eqn (3.223) and therefore do not contain any reference density. The remaining scheme, S1 is the most expensive to compute and was not therefore examined in great detail, but Ω was shown to exhibit dependence on the initial choice of ρ_0.

In the calculation of density profiles a number of trial functions were tested, the one finally chosen had the form

$$\rho^{(1)} = \rho^{G} + \{\exp[(|z| - \alpha)/\beta] + 1\}^{-1} \sum_j C_j \cos[(j - 1)\pi z/z_0] \quad (7.55)$$

in which α, β, C_j and z_0 are adjustable parameters. As judged by the criterion of agreement with MC calculations, S3 consistently produced the best profiles (S1 was not tested in this way) and this approximation was also compared with calculations based on the HAB equation. It may be seen from Fig. 7.9 that the latter is notably less successful than the DFS3 theory in this context.

Although the dimensionality of the systems may not be insignificant in this comparison, the results do not lend support to the earlier calculations (Ebner and Saam, 1977; Saam and Ebner, 1978) in which the DFS3 theory showed considerable discrepancy with respect to the HAB theory. Here three-dimensional systems of 12–6 molecules in contact with a single planar 9–3 adsorbent were studied. The minimization was effected by using a trial function,

$$\rho^{(1)} = \exp(-\alpha/z^9 T)\left(\beta'\{\exp[\gamma(z - \delta)] + 1\}^{-1} + \sum_j C_j z^{j-1} e^{-\varepsilon z^2} + \rho\right) \quad (7.56)$$

in which α, β', γ, δ, ε and C_j were all adjustable parameters, although at temperatures above T_c it was found that minimization was not impaired if β' was set to zero.

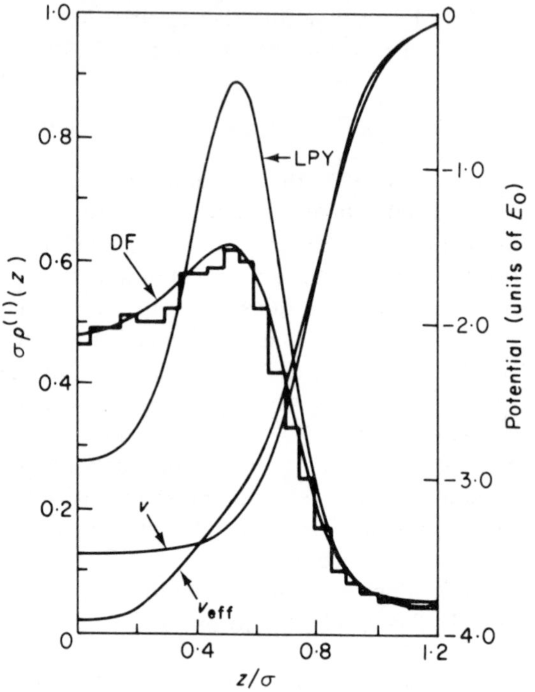

Fig. 7.9. Singlet distribution for a one-dimensional fluid at $T^* = 1{\cdot}0$ in the field v given by Eqn (7.54) with $u^{(1)} = v$. The GEMC result is shown as a histogram and v_{eff} is defined by the PY approximation $c^{(1)} = g^{(1)}[1 - \exp(-\beta v_{eff})]$. (Ebner, Lee and Saam, 1980.)

Calculations carried out for densities of 0·2, 0·3, and 0·5 at $T^* = 1{\cdot}4$ ($>T_c^*$) showed the DFS2 profiles to be somewhat greater in amplitude and shorter in wavelength than those from the HAB theory; differences which were attributed to the inability of the latter to sample correlations over a wide enough range of densities. At $T < T_c$ ($T^* = 0{\cdot}9, 1{\cdot}1$), disparities between the two theories were far more dramatic. As the vapour density approached that of a saturated vapour ($\rho_{sv}^* = 0{\cdot}04736$ at $T^* = 1{\cdot}1$, according to the PY approximation), the DFS2 theory predicts film formation as indicated by the inflation of the profiles in Fig. 7.10, which did not appear in the profiles calculated from the HAB theory. However, in this system it is the HAB theory which comes closer to simulation results, both in predicting the correct positions for the $\rho^{(1)}$ maxima and in showing no evidence of film formation, and indeed little evidence of higher layer formation under these conditions (Lane *et al.*, 1979). In respect of the last

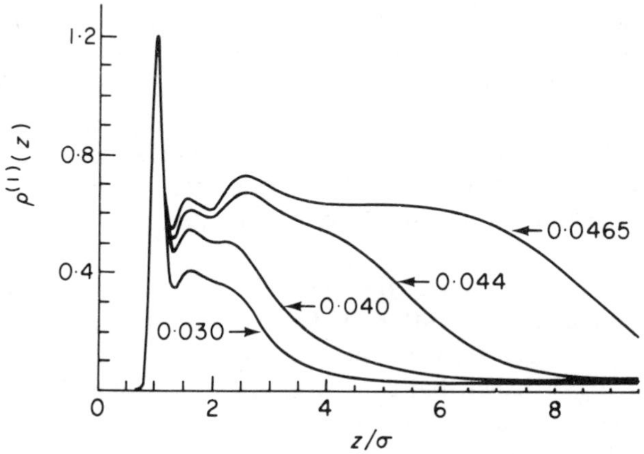

Fig. 7.10. Singlet distribution for 12–6 argon at $T^* = 1{\cdot}1$ ($T_c^* = 1{\cdot}36$) for the reduced bulk densities given on the curves. (Saam and Ebner, 1978.)

point it is to be noted that ρ_{sv}^* from simulation is well above the PY value, and that even at $\rho^* = 0{\cdot}55$ there is only a small fraction of third layer present in the simulated profiles. Furthermore it has been shown earlier that the HAB theory is able to predict multilayer wetting and is consistent with the FHH "slab" theory from which the concept of film formation arises. However, it has already been stressed that the latter involves a number of quite severe approximations and does not in fact provide any information about the shape of the density profile. With regard to the latter, GEMC simulations at high adsorbate coverage (Rowley *et al.*, 1976a, b) suggest that a multiple layer profile may persist to fairly high coverages. On the other hand if, as seems likely, there are no oscillations of the profile at liquid–vapour interfaces then profiles similar to those found in the DFS3 calculations are to be expected at densities close to ρ_{sv}^*. A weak point in the trial function of Eqn (7.56) is that it does not appear to contain the correct Henry law limit and must therefore be suspect at low adsorbate coverages (Lane *et al.*, 1979).

(c) BBGKY and van der Waals theory

The third class of equation, based on the BBGKY theory, has had the least attention paid to it so far. There are a number of reasons for this. In the first place the corresponding Born–Green theory for pair distributions, which begins from the second equation of the hierarchy, has already received considerable study in the context of uniform fluids. Here closure of the integral equation requires an ansatz for $g_{123}^{(3)}$ but the superposition approximation suggested by Kirkwood has been shown, by direct comparison with $g^{(3)}$ from computer simulation (Alder, 1964; Rahman, 1964), to be inaccurate at low densities and for configurations of triplets where the triangle joining the three molecules is not equilateral. The inadequacy of the Born–Green theory as a description of the behaviour of uniform fluids can be reasonably attributed to these limitations on the superposition approximation.

When the first equation of the hierarchy is used for the calculation of $\rho^{(1)}$ the problem is rather different. Here it is $g_{12}^{(2)}$ (or more correctly $\rho_{12}^{(2)}$) for which an ansatz must be found. The step which replaces $\rho_{12}^{(2)}$ by $\rho_1^{(1)}\rho_2^{(1)}g_{12}^{(2)}$ is exact, but $g_{12}^{(2)}$ is a non-uniform function and must be able to account for correlations between molecules 1 and 2 when they lie in regions of quite different density. It is fairly easy to propose approximation schemes which might meet this requirement; for example, the three local density approximations of Eqn (3.224) would be possible candidates. However, most of the work with this equation has been carried out using the simpler assumption that $g^{(2)}(\boldsymbol{r}_{12})$ could be replaced by $\bar{g}^{(2)}(r_{12}; \rho)$ and $\rho^{(1)}$ calculated from Eqn (7.12) (Berry, 1974; Navascués, 1976).

Unfortunately a second difficulty arises because it appears that, even with this simplification, the BBGKY equation may require particular care if stable solutions are to be achieved. Borštnik and Ažman (1975), who solved Eqn (3.139) by iteration for 12–6 molecules against a hard wall, report a divergence associated with the build up of spurious oscillations, and were only able to obtain convergent solutions for $T > T_c$. Similar difficulties were also encountered at higher densities (Navascués, 1976) and the technique of mixing the $(I-1)$th iteration with the current solution was adopted. The Ith iteration in this scheme is written

$$\rho^{(1)(I)} = (1-\alpha)\rho^{(1)(I-1)} + \alpha\mathscr{I}(\rho^{(1)(I-1)}) \tag{7.57}$$

where $\mathscr{I}$ is the operator giving $\rho^{(1)}$ and α is a mixing parameter. No formal theory exists for the determination of α, which must be adjusted by trial and error.

Profiles obtained from these calculations show the anticipated oscillatory character with maxima separated by approximately the distance of equilibrium separation for the 12–6 potential. Comparisons with simulation data and with other theories so far discussed have not been made for molecules with attractive forces.

An application of the BBGKY equation to 2-D systems with a lateral periodic potential was made by Prasad and Toxvaerd, (1980). They applied a Fourier transform technique of the kind described in § 3(e) of chapter 3 to reduce the equation to a set of non-linear algebraic equations, and used data from MD simulations of 2-D systems both with and without the periodic external potential, to obtain $g^{(2)}$. In the latter case an averaging over positions and angles was applied. Agreement between the singlet distributions (in this case $\rho^{(1)}(x, y)$) from BBGKY and MD calculations was very close for both approximations to $g^{(2)}$, showing that with an accurate estimate of this quantity, the equation is capable of handling the effect of an external field on fluid structure quite satisfactorily. Furthermore, since results obtained with the two approximations to $g^{(2)}$ were virtually indistinguishable, it was concluded that, as with uniform fluids, repulsive forces play a major role in determining correlations. This contrasts with earlier conclusions from studies of non-uniform 3-D fluids using other integral equations which suggested that a perturbation theory approach along these lines was unsatisfactory.

A promising development which incorporates the ideas of the uniform fluid perturbation theory into the BBGKY equation has been made by Fischer and Methfessel (1980). In this the pair potential is split according to the prescription of (3.214) and the repulsive part replaced by a hard-sphere interaction for spheres of diameter d, determined by the Barker and Henderson criterion, (see for example, Barker and Henderson, 1976). For this part of the interaction the pair correlation function is replaced by a uniform, hard-sphere function $\bar{g}^{(2)\mathrm{HS}}(\langle\rho_{12}\rangle)$ where the density $\langle\rho_{12}\rangle$ is calculated on a coarse graining principle from

$$\langle\rho_{12}\rangle = \frac{1}{v}\int \rho^{(1)}(\boldsymbol{r} + r_c)\,\mathrm{d}\boldsymbol{r} \tag{7.58}$$

in which v is the volume of a sphere of diameter d centred at the contact separation $r_c = \frac{1}{2}(r_1 + r_2)$. The attractive part of the divided pair interaction was treated as a mean field ($g^{(2)} = 1$), so that the BBGKY equation finally appears in the form,

$$\frac{\partial \ln \rho_1^{(1)}}{\partial \boldsymbol{r}_1} = -\beta\frac{\partial u_1^{[1]}}{\partial \boldsymbol{r}_1} - \beta\frac{\partial}{\partial \boldsymbol{r}_1}\int \rho_2^{(1)} u^{(2)\mathrm{A}}\,\mathrm{d}\boldsymbol{r}_2 - \int \rho_2^{(1)}\bar{g}^{(2)\mathrm{HS}}(d;\langle\rho_{12}\rangle)\left[\frac{\boldsymbol{r}_{12}}{r_{12}}\right]\delta(r_{12} - d)\,\mathrm{d}\boldsymbol{r}_2 \tag{7.59}$$

where Eqn (7.13) has been used to obtain the last integral, and it is to be noted that only the contact value of $\bar{g}^{(2)\mathrm{HS}}$ is needed.

This equation was solved by iteration at low temperature and density ($T^* = 1{\cdot}002$, $\rho^* = 0{\cdot}02$) and was reported to be in close agreement with the simulation data of Rowley *et al.*, (1976).

The underlying character of this theory is to represent the structure of the non-uniform fluid by a non-uniform hard-sphere fluid which is held together by long-range attractive forces. The essential requirement is that the correct structure is established for the non-uniform hard-sphere fluid, and this is assured by the coarse graining feature which feeds back the singlet density into this structure through the density $\langle \rho \rangle$. It is interesting to note that the BBGKY theory, without this coarse graining feature, does not successfully predict the $\rho^{(1)}$ profile for a non-uniform fluid of hard spheres.

A less accurate but very revealing analysis in a somewhat similar spirit has been made by Sullivan (1979) using a non-uniform van der Waals theory. Here again the potential is subdivided into a short-range part $u^{(2)\mathrm{HS}}$ and an attractive part $u^{(2)\mathrm{A}}$ where the latter is now characterized by an inverse range parameter γ such that, in the limit $\gamma \to 0$, the interaction of a molecule with the rest of the assembly is given by an effective mean-field term u_{eff}. If the density varies only in the z-direction these requirements are specified by

$$u^{(2)\mathrm{A}} = \gamma\phi(\gamma r) \tag{7.60a}$$

$$u^{(2)}_{\mathrm{eff}}(\gamma z) = \int \phi(\gamma|\boldsymbol{r}_1 - \boldsymbol{r}_2|)\rho^{(1)\mathrm{L}}(\gamma z_2)H(\gamma z_2)\,\mathrm{d}(\gamma z_2) \tag{7.60b}$$

where $H(x)$ is a step function which ensures cut-off at the hard-wall surface. Here $\rho^{(1)\mathrm{L}}(\gamma z)$ represents that part of $\rho^{(1)}$ which varies slowly on the scale γ^{-1} and the remainder of the profile is represented by a short-range term, $\Delta\rho^{(1)\mathrm{S}}$ which has been neglected in Eqn (7.60b), and which can be written (Sullivan and Stell, 1978),

$$\Delta\rho^{(1)\mathrm{S}} = \rho^{(1)\mathrm{S}}(z) - \rho^{(1)\mathrm{L}}(0) \tag{7.61}$$

where $\rho^{(1)\mathrm{S}}$ is the density profile for the hard sphere at a hard wall when the bulk density is $\rho^{(1)\mathrm{L}}(0)$. This last equation demonstrates the importance of $\rho^{(1)\mathrm{L}}$ in determining the form of $\rho^{(1)\mathrm{S}}$, and accounts for the effectiveness of the feedback mechanism in the theory of Fischer and Methfessel.

The more approximate van der Waals theory can now be stated by a local density approximation to Eqn (3.164) as

$$\mu - u_{\mathrm{eff}}(\gamma z) = \mu(\rho^{(1)\mathrm{L}}(\gamma z)) \tag{7.62a}$$

$$u_{\mathrm{eff}} = u^{(2)}_{\mathrm{eff}} + u^{[1]\mathrm{A}}. \tag{7.62b}$$

Using exponential forms of the Yukawa type for the external and molecular interaction potentials, Sullivan has solved this equation, which is first-order in the range parameter, and has shown that it is able to predict the three classes of film formation discussed in § 5 of Chapter 2. These are

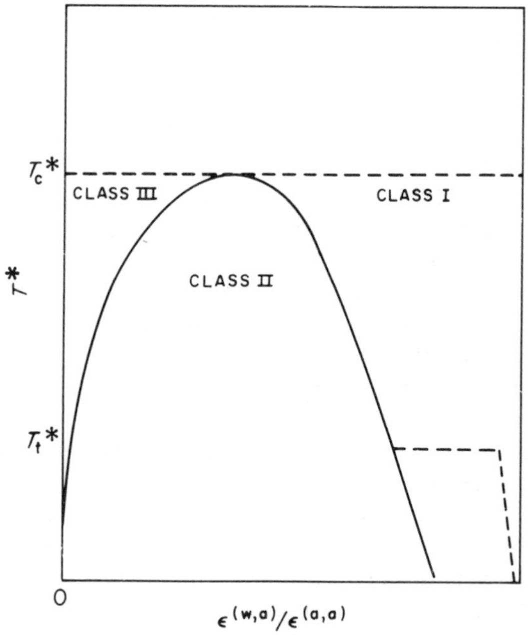

Fig. 7.11. Schematic coexistence curve given by the van der Waals model (solid curve). The dashed line shows the conjectured modification of the boundary between classes I and II below the reduced triple point temperature T_t^*. (Sullivan, 1979.)

shown to be determined by the parameters $\varepsilon^{(w,a)}$ and $\varepsilon^{(a,a)}$ characterizing the interactions. When $T^*(=kT/\varepsilon^{(a,a)})$ is below T_c^* all three classes of film can exist in the different regions of the coexistence curve illustrated in Fig. 7.11. As might be anticipated, the class I films are associated with relatively strong wall forces.

This work helps to resolve the controversy over the possible thin-to-thick film transition which has been predicted from DF theory (Ebner *et al.*, 1976), by showing that, in common with some other theories of the van der Waals type, the DF theory is equivalent to the $\gamma = 0$ limit of Eqn (7.62a) and that this type of approximation inevitably predicts class I film formation no matter how weak the external field.

On the other hand, the van der Waals theory serves to underline a limitation in computer simulation imposed by the finite potential cut-off which removes the long-range part of the potential and may thereby affect the critical temperature.

4. Theoretical models

(a) A general lattice model for multilayer adsorption

The primary object of attention for the integral equation theories is the singlet distribution function $\rho^{(1)}$, from which other properties of the system can be determined. Distribution functions have attracted increasing attention in recent years, not only because they are the natural quantity to determine from integral equations, but also because they can readily be found from simulations and because direct measurements from scattering experiments are possible. Many of the earlier theories were directed towards more easily determined experimental properties such as adsorption isotherms, heats and heat capacities. For many applications these theories are widely used and it is therefore important to reassess them in the light of more recent investigations.

Prominent among the models proposed to account for adsorption phenomena are the lattice models. These have a long history of discovery and rediscovery going back to Langmuir (1918) who, in addition to proposing the well known monolayer model which bears his name, also proposed the main features of, and discussed the problems associated with, the multilayer model of Brunauer, Emmett and Teller (1938) (BET). Originally this theory was derived by a kinetic route which tends to obscure much of the significance of the assumptions introduced. In a long series of papers Hill (see for example, Hill, 1960) discussed many of the statistical mechanical aspects of the theory and attempted to introduce various refinements. In the main, this work showed up the difficulty of improving the theory in a rational way without, at the same time, making it computationally inaccessible. The simple closed form of the BET equation has ensured its continued popularity, and its extensive use in surface area measurement seems, on the whole, to be justified (Gregg and Sing, 1982).

Later developments of the model (Champion and Halsey, 1953, 1954) attempted to improve the treatment of intermolecular interactions, but only rather limited calculations were possible before modern computing facilities became readily available. More recent work (Nicholson, 1975; De Oliveira and Griffiths, 1978; Ebner, 1980) has shown that lattice models can be quite useful in at least a qualitative way for the discussion of multilayer adsorption. This is especially the case where low-temperature adsorption is of interest and much of the adsorbate may be solid-like in structure. Refinements to the model, such as the inclusion of vibration frequencies dependent on local structure, or the extension to a Lennard-Jones cell model, or significant structure theory, have not been very fully explored outside the context of monolayer adsorption. In strictly two-dimensional systems it is of course possible to obtain exact solutions in certain cases, and renormalization group theory has contributed much recently to the understanding of phase diagrams in lattice systems (Ch. 2, § 7).

To establish a general multilayer model we begin by considering a set of two dimensional lattices stacked normal to the adsorbent surface and numbered $1, 2, \ldots, K$ from the surface outwards. Since the model has been applied both to adsorption on a plane surface and to adsorption inside cylinders, it is useful to bear in mind that for the latter case the lattices comprise annuli of successively decreasing radius. The adsorbate atoms are placed on the 2-D lattice sites in a square-packed or close-packed arrangement so that, in the case of a planar adsorbent the 3-D array is either simple cubic or cubic close packed.

We suppose that the Ith layer contains M_I sites, of which N_I are occupied by adsorbate atoms, and introduce the assumption that within a layer all atoms are equivalent. This means that they will have the same site partition function q_I and the same adsorption energy u_I which is a function only of the distance of layer I from the adsorbent. The interaction between adsorbate atom i at a site within the Ith layer, and an atom $j(\neq i)$ in the Jth layer $(J = 1, K)$ is $u_{ij}^{[2]}(\mathbf{r}_{ij})$ where $\mathbf{r}_{ij}$ is now determined by the geometry of the lattice. The interaction u_{IJ} between all the atoms in a completely filled Ith layer and those in a completely filled Jth layer is likewise a property only of the lattice geometry and can be found by summation as $u_{IJ} = \frac{1}{2}\sum u_{ij}^{(2)}$ where the sum is over pairs of sites in the two layers and the factor $\frac{1}{2}$ corrects for double counting. In many lattice model theories it is customary to take into account nearest-neighbour interactions only and this limitation may offer some advantage in opening up possibilities for the exact treatment of the lattice statistics. However, as will be seen below, distant-neighbour interactions can have an important influence on co-operative phenomena in non-uniform systems, especially when capillary condensation is being considered (Nicholson, 1975, 1976).

If each atom is now assumed to experience the average interaction due to those on surrounding sites, the pairwise term in U_N can be replaced by a mean-field approximation,

$$\sum u_{ij}^{(2)} \cong \sum_{I\geqslant 1} \sum_{J\geqslant 1} N_I \theta_I u_{IJ} \tag{7.63}$$

in which $\theta_J = N_J/M_J$, is the fractional occupancy of layer J.

With these approximations the canonical partition function can be written

$$Q_N = \sum_{\{N\}} \prod_I g_I q_I^{N_I} \exp\left[-\beta N_I\left(u_I + \sum_{J\geqslant 1} \theta_J u_{IJ}\right)\right] \tag{7.64}$$

where the summation is over all occupation sets. The degeneracy factor g_I in this equation can be treated in a number of ways. Here we confine attention to the zeroth-order approximation, consistent with the mean-field assumption for the energy, in which the sites are occupied in a random way irrespective of the energy at a particular site; g_I is then given by

$$g_I = M_I!/(M_I - N_I)!N_I!, \tag{7.65}$$

and if the lattice is planar, $M_I = M_J = M$, the same for all layers. M_I will of course vary in a cylindrical or spherical geometry.

In the BET model, it is assumed that adsorption in the Ith layer ($I > 1$) can only occur over those sites in the ($I - 1$)th layer which are already occupied, and g_I then depends on the supposed lattice structure. In the original BET model the 3-D array of sites is simple cubic and atoms adsorb normal to the surface, giving

$$g_I = N_{I-1}!/N_I!(N_{I-1} - N_I)! \tag{7.66}$$

with $N_{(0)} = M$, the number of sites in a layer. If, on the other hand, the lattice is close packed then the equivalent assumption is that adsorption will occur over triangular sites in the lower layer, and N_{I-1} in Eqn (7.66) would be replaced by $N_{I-1}\theta_{I-1}^2$ (Taylor *et al.*, 1965).

The chemical potential is given by $\beta\mu = -\partial \ln Q_N/\partial N_I$. If we choose the most probable occupation set in Eqn (7.66) and carry out the operation, there results a set of equations of which the Ith member is

$$\beta\mu + \frac{\partial \sum \ln g_I}{\partial N_I} + N_I \frac{\partial \sum \ln q_I}{\partial N_I} + \ln q_I - \beta u_I - \beta \sum \theta_J u_{IJ} - \beta \sum (M_J \theta_J u_{JI}/M_I) = 0. \tag{7.67}$$

Exactly the same result can be obtained by starting from Ξ and by minimizing the grand free energy (Ebner, 1980).

Given the mean-field approximation, Eqn (7.67) applies to a three-dimensional lattice of vibrators in an external field and is valid for any adsorbent geometry. Considerable simplification results when certain special assumptions are introduced. For example, if q_I is taken to be independent of N_I, the third term vanishes, as also does the fourth term if vibrations are ignored altogether. In the case of a planar adsorbate, $M_J = M_I$, and the last two terms in Eqn (7.67) can be amalgamated. Thus if a plane surface is considered, Eqn (7.65) is used for g_I and Q_I is independent of N_I, Eqn (7.67) simplifies to

$$\beta\mu + \ln[(1 - \theta_I)/\theta_I] + \ln q_I - \beta u_I - 2\beta \sum_J \theta_J u_{IJ} = 0. \tag{7.68}$$

A reference state can now be introduced by writing Eqn (7.68) for a uniform phase in the absence of an external field

$$\beta\mu_r + \frac{\partial \ln g_r}{\partial N_r} + \ln q_r - \alpha\theta_r = 0 \qquad (7.69)$$

in which the parameter α relates to the strength of the intermolecular interaction and is given by

$$\alpha = 2\beta \sum_J u_{1J}. \qquad (7.70)$$

If $T < T_c$ we can write, without serious approximations,

$$\beta(\mu - \mu_r) = \ln(p/p_0) \qquad (7.71)$$

where p_0 is the saturated vapour pressure and is related to θ_r through an equation of state (Nicholson, 1975). By coupling this equation of state with Eqn (7.69), the fractional occupations θ_r^D and θ_r^R for the co-existing dense and rarefied phases of the uniform system can be determined and it can be shown that these are subject to the condition $\theta_r^D + \theta_r^R = 1$ when q_r is the same in both phases. For a planar adsorbate the limit

$$\theta_J \underset{J \to K}{\to} \theta_r^R \qquad (7.72)$$

can be imposed in order to determine an upper bound for the summation in (7.68).

It is of interest at this point to consider the BET theory in the light of the more general lattice model developed here. One of the consequences of the special assumption for g_I which appears in this theory is that $\theta_r^D = 1$ and $\theta_r^R = 0$ in the reference state, implying from Eqn (7.69), that this is solid-like with

$$\beta\mu_r = \alpha - \ln q_r. \qquad (7.73)$$

Equation (7.67), with (7.66) and (7.73) for a planar adsorbate, simplifies to

$$\ln(p/p_0) + \ln[(\theta_{I-1} - \theta_I)/(\theta_I - \theta_{I+1})] + \ln(q_I/q_r)$$
$$-\beta u_I - 2\beta \sum_J \theta_J u_{IJ} = 0 \qquad (7.74)$$

where $\theta_{(0)} = 1$.

The set of equations represented by (7.74) still contains a difficult non-linear term, because of the summation over J, as well as the complication of a layer-dependent site partition function q_I and external field term u_I. Three further assumptions are therefore introduced:

(i) $u_I = 0; I \geqslant 2$. This is equivalent to stating that the layers above the first are adsorbed by attraction to the lower layer, and is quite reasonable for a plane surface. It has already been shown in the previous section that if the adsorbate interactions are very much less than the adsorbate–adsorbent interactions, class II adsorption occurs; chemisorption is an extreme example where usually only a monolayer is adsorbed. It may be noted, however, that in small enough pores, because of overlap effects, the adsorbent field can make a more significant contribution beyond the first layer.

(ii) $q_I = q_a$, independent of θ_J. This implies that the vibrational partition function is not modified by the inhomogeneity of the surrounding field. To remove this assumption would introduce an additional and rather complicated θ_I dependence in Eqns (7.67) and (7.74). In view of the artificial nature of the lattice model, it is questionable to what extent this can be justified. However, it is interesting to note that the cell model of Lennard-Jones and Devonshire, when appropriately modified for non-uniformity, should be able to account for this term. An elementary theory of this type which was developed some time ago has been shown to compare quite favourably with MC data (Pace, 1957; Rowley *et al.*, 1976b).

(iii) Only nearest-neighbour adsorbate interactions in a direction normal to the surface of the adsorbent are included. This effectively removes the θ_J dependence in the summation over J and implies that the term remaining here is equal to α.

With these assumptions, Eqn (7.74) gives,

$$\frac{\theta_{(\mathrm{i})} - \theta_{(\mathrm{ii})}}{1 - \theta_{(\mathrm{i})}} = c\left(\frac{p}{p_0}\right) \qquad \text{for } I = 1 \tag{7.75a}$$

$$\frac{\theta_I - \theta_{I+1}}{\theta_{I-1} - \theta_I} = \frac{p}{p_0} \qquad \text{for } I \geqslant 2 \tag{7.75b}$$

where $c = (q_a/q_r)\exp[-\beta u_{(\mathrm{i})} + \alpha/2]$ is the BET constant, and it is customary to assume that $\alpha/2$ is equivalent to the experimental value of the latent heat for the adsorptive.

From these equations it is easily shown that

$$\theta_I = (\theta_{I-1} - \theta_K)p/p_0 \qquad I \geqslant 2 \tag{7.76a}$$

and, for the first layer,

$$\theta_{(\mathrm{i})}[1 + (c-1)p/p_0] = (c + \theta_K)p/p_0. \tag{7.76b}$$

When the adsorption is onto a single plane surface and the filling in the outermost layer $\theta_K = \theta_r^R = 0$, these equations lead to the BET isotherm

$$\theta = \frac{cp/p_0}{(1 - p/p_0)[1 + (c - 1)p/p_0]}. \tag{7.77}$$

It is interesting to note that the form of Eqn (7.76a) is a direct consequence of the assumption leading to Eqn (7.66) for g_l, so that the envelope of the density profile is determined by this assumption in the BET theory.

This is not the case when Eqn (7.68) is used with (7.69), but no closed form of the solution to the equations is then available and the set of non-linear equations in θ_l must be solved numerically. In this, fully interacting, lattice model (LM theory) the reference state is less simple than for the BET theory and comparison with simulation can be made by introducing a reduced temperature $\tau = T/T_c$, where T_c is an appropriate critical temperature.

Since van der Waals loops may occur, more than one solution to the equations is possible in general. A simple method for selecting the correct solution which does not require the calculation of loop areas has been given (Hill, 1956; Nicholson, 1975). The transition which corresponds to these loops introduces a new feature, not encountered in the theories considered up to now, nor in the BET lattice model. This phenomenon of stepwise multilayer isotherms is associated with condensation in successive adsorbate layers, and the steepness of the steps is therefore related to the strength of the adsorbate interactions. Isotherm steps have been observed experimentally on well characterized graphite surfaces (see § 2(a), Ch. 1) and were found in MC calculations for Ar on graphite at low temperature ($T^* = 0{\cdot}667$) (Rowley *et al.*, 1976a).

The full curves in Fig. 7.12 illustrate a set of isotherms calculated from the LM theory (Nicholson and Silvester, 1977) at $\tau = 0{\cdot}42, 0{\cdot}51, 0{\cdot}64, 0{\cdot}77$ compared with points from MC calculations at $\tau = 0{\cdot}51$ and $0{\cdot}77$ and the BET theory at these two temperatures. In the LM theory the lattice was CCP, which is the structure closest to that found in the MC calculations. The LM curves illustrate the way in which steps become smoother as the strength of the adsorbate interactions is reduced. It also appears that the LM tends to underestimate somewhat the tendency to step formation and finds the transitions at higher values of p/p_0 (a displacement of about 0·1) compared with MC data. As already mentioned above, the BET theory makes no allowance for "lateral" interactions and therefore does not show isotherm steps, but as can be seen from the figure, agreement with MC calculations can be very close, especially when it is borne in mind that no validity is claimed for the theory for $p/p_0 > 0{\cdot}35$. The tendency for the BET

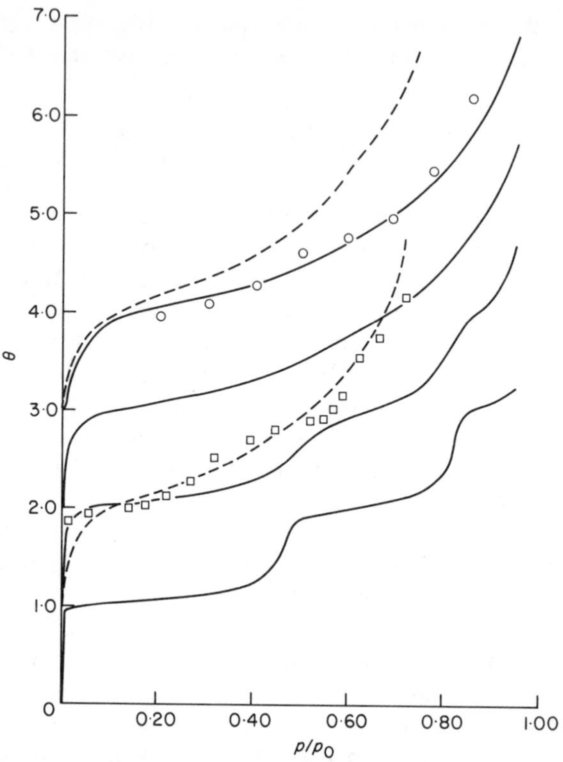

Fig. 7.12. Multilayer adsorption isotherms at reduced temperatures of 0·41, 0·51, 0·64, 0·77. ○, □, MC data at $\tau = 0{\cdot}51$ and 0·77. ——, Lattice model with full interaction (LM theory). ----, BET theory at $\tau = 0{\cdot}51$ and 0·77. The parameters were chosen to correspond to 12–6 argon adsorption on 9–3 graphite. (Rowley *et al.*, 1976a). (From Nicholson and Sylvester, 1977.)

theory to overestimate the adsorption at higher relative pressures is well known.

It is interesting to note that because of the weaker intermolecular attractions between nitrogen molecules compared with Ar ($\varepsilon^{(N_2,N_2)*} \cong 95$ K, $\varepsilon^{(Ar,Ar)*} \cong 120$ K), isotherms for the former adsorbate are likely to show steps which are much less well defined than argon at comparable temperatures (Nicholson and Silvester, 1977). The position of the steps is also affected by the ratio $\varepsilon^{(w,a)}/\varepsilon^{(a,a)}$ of the potential minimum of the adsorbate–adsorbent interaction to that of the adsorbate–adsorbate interaction as well as by the surface heterogeneity. Theories such as the lattice model theory, in which "lateral" interactions are included, are also able to account for the waves in isosteric heat curves (Rowley *et al.*, 1976b).

As judged by a comparison of isotherms and heat curves a lattice model of elementary type discussed here describes quite well the main features of experimental systems and MC data at $T < T_c$. It is therefore of interest to consider some of the other consequences of this type of model. For example, phase diagrams provide a useful guide to the properties of the system. These have been studied, using the LM theory, in relation to the ratio $\varepsilon^{(w,a)}/\varepsilon^{(a,a)}$ mentioned above (Ebner, 1980). It was found that qualitatively different types of behaviour could be identified as this ratio was varied. When it is large, the phase diagram corresponds to adsorption increasing either smoothly or in steady increments of monolayer steps as p/p_0 increases to saturation. On the other hand, if the ratio $\varepsilon^{(w,a)}/\varepsilon^{(a,a)}$ is sufficiently small, coverage does not extend much beyond a monolayer until condensation occurs at $p/p_0 = 1$. Finally in the intermediate range, it is possible for large discontinuities to occur in θ from below monolayer to several multilayers coverage at certain temperatures. These three categories appear to be consistent with class I, III and II types of behaviour respectively, and the calculations lend support to the view that a thin-to-thick phase transition, such as that found from the DF theory, can occur for suitably chosen temperatures and energies (Ebner, 1980).

(b) The slab theory

The BET theory is one way in which simplifying assumptions can be introduced in order to produce an isotherm equation which is tractable and, although not accurate, is serviceable when certain limitations are imposed. The BET theory fails badly when $p/p_0 > 0{\cdot}3$; a dense film model of the adsorbate might be expected to do better in this region. The essential ingredient of this type of theory, already discussed in Chapter 3, is the proposal that the adsorbate film differs from an identical film in a slab of liquid because it now interacts with adsorbent instead of adsorbate molecules. The perturbation energy experienced by a molecule at $\boldsymbol{r}$ in the film due to this replacement is $\rho^{(a)}(\boldsymbol{r})u^{[1]P}(\boldsymbol{r})\,d\boldsymbol{r}$, where, as before

$$u^{[1]P} = u^{[1]SOL} - u^{[1]LIQ}. \tag{7.78}$$

Two assumptions are now introduced:

(i) The partial molar entropy in the adsorbate film is unaltered by the replacement process.

(ii) The change in the partial molar enthalpy due to the replacement process can be substituted by the change in partial molar energy, ΔU.

With these assumptions we can write

$$\mu_a - \mu_r = \frac{\partial \Delta U}{\partial N_a^{\Sigma}} \tag{7.79}$$

and, for a field varying in the z-direction only,

$$\beta(\mu_a - \mu_r) = \ln(p/p_0) = \int \beta \frac{\partial[\rho^{(a)}(z)u^{[1]P}(z)]}{\partial \Gamma_a} dz. \tag{7.80}$$

From Eqn (2.5), the thickness t of the film is related to Γ_a by

$$\Gamma_a = t(\rho_a^F - \rho_a^G). \tag{7.81}$$

If $\rho^{(a)}$ changes sharply from high density to low density, it can be written in terms of the Heaviside step function $H(x)$, defined to be unity for $x < 0$ and zero for $x > 0$, as

$$\rho^{(a)} = \rho^F H(z - t) + \rho^G H(t - z) \tag{7.82}$$

from which, $\partial\rho^{(a)}/\partial t = \rho^F \delta(z - t) + \rho^G \delta(t - z)$, and Eqn (7.80) becomes

$$\ln(p/p_0) = \beta u^{[1]P}(t). \tag{7.83}$$

Equation (7.83) is the fundamental equation of the FHH or slab theory and it can be seen from the foregoing derivation to be a consequence of the dense film model of the adsorbate under the two assumptions stated above. If the film thickness is more than about two layers, and only dispersion forces are involved, it is reasonable to use a continuum approximation for $u^{[1]P}(t)$ and to ignore the effects of short-range repulsion, giving

$$u^{[1]P} = -a_1[\varepsilon^{(w,a)} - \varepsilon^{(a,a)}]/t^3 \tag{7.84}$$

which could also be written in terms of Γ_a using Eqn (7.81). Here a_1 is a constant and the term in square brackets measures the relative importance of adsorbate–adsorbent and adsorbate interactions. When this is positive the isotherm increases monotonically and approaches p_0 asymptotically. When $\varepsilon^{(w,a)} < \varepsilon^{(a,a)}$, no adsorption can occur and class III isotherms result.

Equation (7.83) may also be regarded as a special form of the Polanyi theory which proposes that the thermodynamic potential, $kT \ln(p/p_0)$ is equal to a universal function $\phi(\Gamma_a)$ independent of the temperature. From this point of view, Eqn (7.84) is among a number of possible forms which $\phi(\Gamma_a)$ might take. Other possibilities (Dubinin, 1967), although they do not relate to any clearly definable molecular model, have found considerable popularity in the description of experimental adsorption data.

According to the FHH theory, embodied in Eqns (7.83) and (7.84), graphs of $\ln[-\ln(p/p_0)]$ against $\ln\theta$ should be linear with a slope s of -3 in the region $\theta > 2$. Although linearity is often very closely obeyed, it has been found that s is usually in the range $-2{\cdot}8 < s < -2{\cdot}0$ for the majority of experimental isotherms. For a long time these values were attributed to the artefacts associated with real adsorbents, especially surface heterogeneity. However, MC data, for which the adsorbent is indisputably homogeneous and the potential function inverse cubic for distances greater than $\sim 2\sigma$ from the surface, also produce values of $|s|$ which are very much below 3 and which vary with T (Rowley *et al.*, 1976).

Steele (1980) has re-examined the FHH theory in the light of these findings, and with the aid of detailed data available from the simulation work. Here it will be convenient to set the modified FHH theory in the context of the lattice model.

If we consider the adsorbate interaction in Eqn (7.64) it can be seen that it is consistent with the model to write this as

$$u_I = \sum_{J<1} \theta_J^{\mathrm{SOL}} u_{IJ}^{(\mathrm{a,w})} \tag{7.85}$$

where θ_J^{SOL} is the fractional occupation of the layers in the solid adsorbent which is in the domain $I < 1$, and the superscripts indicate the type of atoms interacting in layers I and J. If the adsorbent is replaced by liquid adsorbate the above term would have a different value

$$u_I^{\mathrm{LIQ}} = \sum_{J<1} \theta_J^{\mathrm{LIQ}} u_{IJ}^{(\mathrm{a,a})} \tag{7.86}$$

where θ_J^{LIQ} is the same in every layer. Within the adsorbate film we could write a similar term

$$u_I^{\mathrm{F}} = \sum_{J\geqslant 1} \theta_J u_{IJ}^{(\mathrm{a,a})} \tag{7.87}$$

but here the occupancy may vary from layer to layer. Eqn (7.64), written for the most probable occupation set (*), can now be put in the form

$$Q_N^* = \prod_I g_I q_I^{N_I} \exp[-\beta N_I(u_I + u_I^{\mathrm{F}})] \tag{7.88}$$

and for the liquid reference state

$$Q_{\mathrm{r},N}^* = \prod_I g_{\mathrm{r},I} q_{\mathrm{r},I}^{N_I} \exp[-\beta N_I(u_I^{\mathrm{LIQ}} + u_I^{\mathrm{F}'})]. \tag{7.89}$$

We introduce the assumption (i) above and assume also that $u^{\mathrm{F}} = u^{\mathrm{F}'}$, so that Eqns (7.88) and (7.89) lead to

$$\partial \frac{\ln (Q_N^*/Q_{\mathrm{r},N}^*)}{\partial N_{\mathrm{a}}^{\Sigma}} = -\ln(p/p_0) = -\sum_I \beta \frac{\partial[N_I(u_I - u_I^{\mathrm{LIQ}})]}{\partial N_{\mathrm{a}}^{\Sigma}} \tag{7.90}$$

which is the equation derived by Steele using a different route. The usual FHH theory is retrieved if it is assumed that $(\partial N_I/\partial N_{\mathrm{a}}^{\Sigma}) = 1$ when the Ith layer is filling and zero otherwise. At low temperatures and for $\varepsilon^{(\mathrm{w,a})} \gg \varepsilon^{(\mathrm{a,a})}$ this might be expected to be a reasonable approximation; under these conditions the probability of finding many atoms in layer I when there are few in layer $(I+1)$, which is roughly proportional to $(\exp(-\beta u_I)/\exp(-\beta u_{I+1}))$, is higher, and layer-by-layer filling would be anticipated. However the MC data show that quite a different situation prevails for the simulated argon–graphite system, especially at the higher temperature $T^* = 1{\cdot}002$. In this case the curve of θ_I, the fractional occupation of the Ith layer having a maximum capacity $\theta_{I,\mathrm{m}}$, versus θ can be closely fitted to the equation

$$\theta_I = \theta_{I,\mathrm{m}}\{1 + \tanh[B_I(\theta - \eta_I)]\} \tag{7.91}$$

in which B_I and η_I are parameters. After differentiation of this expression and substitution in Eqn (7.90), the modified FHH equation (FHHS theory) takes the form,

$$\ln(p/p_0) = (\beta/2)\sum_I B_I\theta_{I,\mathrm{m}}u_I^{[1]\mathrm{P}}\,\mathrm{sech}^2[B_I(\theta - \eta_I)]. \tag{7.92}$$

To compare the theory with the MC data it is necessary to take into account the correct layer spacing, $(I - 0{\cdot}30) \times 3{\cdot}3/3{\cdot}504$ in σ_{Ar} units, and the fact that the z^{-9} term affects the energy of the first layer. The constant of proportionality in the interaction energy was left as a fitting parameter. The theory was tested by plotting $\ln[-\ln(p/p_0)]$ versus Γ_{a}.

Although the FHHS theory fits the MC data much better than the FHH theory (with $s = -3$), the agreement is rather poorer than that of a straight line whose slope gives the index $s = -2{\cdot}1$. The fitting parameter was estimated as, $1{\cdot}80 \pm 0{\cdot}05$, compared with $1{\cdot}12$ calculated from the MC data. It is interesting to note that, for the LM theory itself (and also for the argon–graphite parameters) a similar plot is almost exactly linear in the range $1{\cdot}8 < \theta < 6{\cdot}0$ with $s = -2{\cdot}88$ (Nicholson, 1976). Steele has shown that values of $|s|$ close to this figure are to be expected even for very thick films.

In its general form the FHHS theory requires values for the parameters B_I and η_I which must be prescribed from elsewhere, and even when these are taken directly from computed data the theory falls short of an exact quantitative fit. This work nevertheless sheds considerable light on the limitations and nature of the FHH type of theory in showing that the model of a thick film with a sharp boundary is never likely to be a realistic representation of the actual situation, and that values of $s < 3$ are to be expected on homogeneous as well as heterogeneous surfaces even for very thick films (Lando and Slutsky, 1970a, b).

(c) Adsorption in model pores

The lattice model has been shown to provide a useful reference framework in which to discuss several theories relating to adsorption on a planar surface. The LM theory is also suitable for discussing adsorption into capillary spaces for which a cylinder model provides a useful representation. Some justification for employing here a LM theory again comes from MC calculations (Richards, 1973) which indicate that an arrangement of concentric annuli inside a cylinder is a reasonable choice, although agreement between MC and LM theory isotherms for cylinders of about 5σ across is qualitative rather than quantitative. The general features of isotherms measured on porous materials can thus be accounted for. It should be mentioned that adsorbents well characterized by independent means are not available in the intermediate (meso) pore size range (roughly $1{\cdot}6\,\text{nm} < r_p < 50{\cdot}0\,\text{nm}$) and that, although the dimensions of zeolite pore spaces are well established to a high degree of accuracy, the potential fields in zeolite cavities are complicated by the presence of ions. Indeed, it is common to use adsorption methods in order to characterize the pore properties of mesoporous or microporous materials. These measurements must therefore rely heavily on the validity of the theories which they adopt in processing experimental data.

Figure 7.13 shows a set of isotherms calculated for cylinders of different radii at $\tau = 0{\cdot}77$; Θ here is defined as the number of molecules adsorbed, divided by the number which would be required to fill the same volume in the bulk reference phase, and is thus a measure of the fractional filling of the space. The energy parameters are again chosen to be typical for the argon–graphite system. Both type I and type IV isotherms in the BET classification are evident, with the large transition in the latter occuring when capillary condensation takes place at a relative pressure $(p/p_0)^*$ characteristic of the radius of the cylinder. The type IV isotherms also show humps associated with the filling of the second layer and similar to the steps in the plane surface isotherms. The capillary condensation transition in this type of isotherm links an unfilled state, in which the adsorbate tends to be concentrated close to the adsorbent walls, with a filled state. As the cylinder radius is increased the stability of the unfilled state is extended to higher p/p_0, and the Gibbs thickness of the adsorbate, as defined in Eqn (2.7), which can be tolerated without condensation also increases (very approximately as $r_p^{1/2}$). Conversely the coexistence region for this transition is diminished as r_p decreases, as can be seen in Fig. 7.13, and eventually disappears at some critical radius where the isotherms change from type IV to type I in character. For a N_2 adsorbate on an adsorbent with a rather stronger field than graphite, this critical radius is

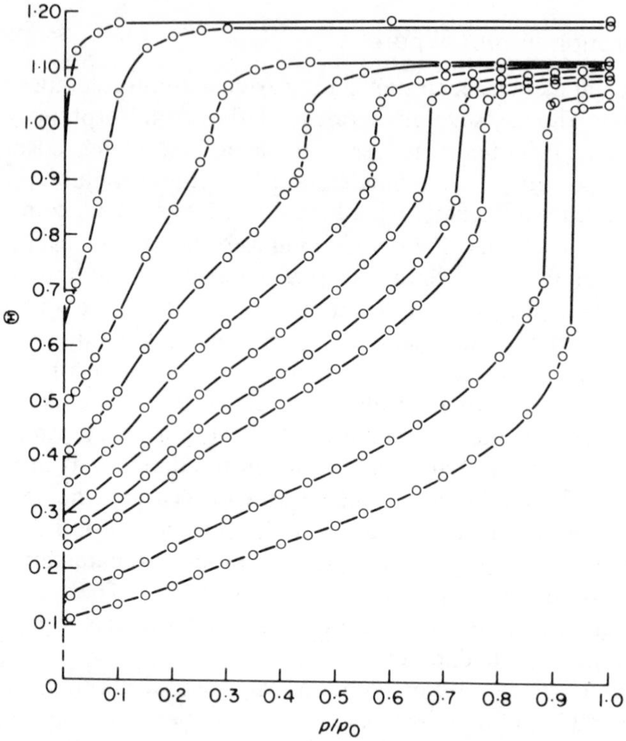

Fig. 7.13. Ar adsorption in cylinders at $\tau = 0{\cdot}77$ ($T^* = 1{\cdot}00$) calculated using the LM theory, in the same reduced units as τ, $\varepsilon^{(a,a)} = 0{\cdot}77$ and $\varepsilon^{(w,a)} = 4{\cdot}5$. Cylinder radii in units of σ_{Ar} are shown on the isotherms, the fractional filling Θ is defined in the text. (Nicholson, 1975.)

~2·0 nm which is in satifactory agreement with other estimates of the pore size below which "micropore filling" occurs. The LM theory also permits an evaluation to be made of currently used thermodynamic theories discussed in § 3(b) of Chapter 2. This prediction is important because the measurement of pore size distribution in real porous materials must rely on a calculation of this type. A number of theories have been proposed and they can be summarized by rewriting Eqn (2.39) as

$$\ln(p/p_0) + \beta \bar{V}^F p_0(1 - (p/p_0)) = \frac{\beta \phi^F \bar{V}^F}{(r_p - t)} \tag{7.93}$$

where the earlier equation has been modified by putting $K = 1$ for cylinders and identifying $\phi_C \equiv \phi_{CYL}$ with $\phi^F(p)$, the spreading pressure of the film discussed in § 3(b) of Chapter 2. As before, r_p is the cylinder radius and

$\bar{V}^F$ the partial molar volume of the dense phase; the pressure p^G in Eqn (2.39) has now been identified with the external gas phase pressure p, although, as pointed out earlier, this is an approximation. If the Gibbs isotherm equation (2.60) is integrated between limits p and p_0, where it is assumed that $\phi^F(p_0) = -\gamma^{FG}$, the surface tension of the bulk liquid phase, $\phi^F(p)$, can be related to a function $f(t)$ which describes the development of the film of thickness t, by

$$\phi^F(p) = -\gamma^{FG} - (r_p - t)f(t)/\bar{V}^F. \tag{7.94}$$

It should be mentioned that the condition for class I, II and III film formation will of course still hold in cylinders, and that the present analysis assumes that class I films are formed. Substitution of (7.94) into (7.93) gives

$$\ln(p/p_0) + \beta\bar{V}^F p_0(1 - (p/p_0)) = \frac{-\beta\bar{V}^F\gamma^{FG}}{(r_p - t)} - \beta f(t). \tag{7.95}$$

Typically $f(t)$ might be chosen as the FHH equation or a related expression (Broekhoff and de Boer, 1967; Nicholson, 1968). Alternatively an empirical "t-curve" can be used, determined from adsorption on a planar adsorbent which is assumed to have the same surface properties as its porous counterpart. However, it must be stressed that substitution of an expression relating to adsorption on a plane surface introduces a further approximation into the thermodynamic theory. Equation (7.94) can be derived from the rigorous hydrostatic result (Eqn (2.33)) discussed earlier. The replacement of the first term in this equation by one containing γ^{FG} was discussed in Chapter 2, § 3(b), and the second term can be transformed by introducing just those assumptions which were shown above to lead to the FHH adsorption equation. It is of course possible to write (7.95) formally without the final term, but, as the discussion in chapters 2 and 3 has shown, it is then necessary to replace γ^{FG} by a different γ which takes into account the adsorbent–adsorbate interactions. This is certainly not the surface tension of an unperturbed liquid–vapour interface.

Broekhoff and de Boer (1967) suggested that $(p/p_0)^*$, and hence the critical thickness t_E, could be found from the condition $\partial(p/p_0)/\partial t = 0$ at the transition. This may be compared with a Gibbsian thickness t_G, estimated from the LM theory isotherms, which is always lower than t_E. The difference between them increases with r_p. The LM theory also permits γ^{FG} and the constants in $f(t)$ (assumed to be of the form, $-\text{const}/t^s$) to be accurately calculated in a way which is consistent with the molecular model. The prediction of the thermodynamic theory can thus be tested directly against a molecular model. The comparison shows considerable discrepancy over a wide range of r_p; the poorest agreement being obtained when the

$f(t)$ term was omitted from Eqn (7.95) (the standard equation used in capillary condensation calculations). The inclusion of a "correction" term involving $f(t)$ brought about a marked improvement in the ability of the theory to predict $(p/p_0)^*$, but even so the left-hand side of Eqn (7.95) was overestimated by a factor of 1·5 to 2·0. Although the LM theory itself involves severe approximations, it has been seen that it can provide, at least qualitatively, an acceptable account of the MC calculations for plane surfaces. Suitable simulation data for enclosed spaces is lacking at present, but the foregoing critique suggests that the present practice in the application of thermodynamic theory to capillary condensation must be regarded with some caution.

References

Abraham, F. F. and Singh, Y. (1977). *J. Chem. Phys.* **67**, 2384.
Abraham, F. F. and Singh Y. (1978). *J. Chem. Phys.* **68**, 4767.
Alder, B. J. (1964). *Phys. Rev. Lett.* **12**, 317.
Andersen, H. C. and Chandler, D. (1972). *J. Chem. Phys.* **57**, 1918.
Barker, J. A. and Henderson, D. (1976). *Rev. Mod. Phys.* **48**, 625.
Baxter, R. J. (1967). *Phys. Rev.* **154**, 170.
Baxter, R. J. (1968a). *J. Chem. Phys.* **49**, 2770.
Baxter, R. J. (1968b). *Aust. J. Phys.* **21**, 563.
Baxter, R. J. (1970). *J. Chem. Phys.* **52**, 4559.
Berry, M. V. (1974). *J. Phys. A: Mathematical and General* **7**, 231.
Blum, L. and Stell, G. (1976). *J. Stat. Phys.* **15**, 439.
Borštnik, B. and Ažman, A. (1975). *Mol. Phys.* **30**, 1565.
Broekhoff, J. C. P. and de Boer, J. H. (1967). *J. Catalysis.* **9**, 8.
Brunauer, S., Emmett, P. H. and Teller, E. (1938). *J. Amer. Chem. Soc.* **60**, 309.
Carnahan, N. F. and Starling, K. E. (1969). *J. Chem. Phys.* **51**, 635.
Champion, W. M. and Halsey Jr., G. D. (1953). *J. Phys. Chem.* **57**, 646.
Champion, W. M. and Halsey, Jr., G. D. (1954). *J. Amer. Chem. Soc.* **76**, 974.
De Oliveira, M. J. and Griffiths, R. B. (1978). *Surf. Sci.* **71**, 687.
Dubinin, M. M. (1967). *J. Colloid Interface Sci.* **23**, 487.
Ebner, C. (1980). *Phys. Rev. A.* **22**, 2776.
Ebner, C. and Punyanita, C. (1979). *Phys. Rev. A* **19**, 856.
Ebner, C. and Saam, W. F. (1977). *Phys. Rev. Lett.* **38**, 1486.
Ebner, C., Saam, W. F. and Stroud, D. (1976). *Phys. Rev. A* **14**, 2264.
Ebner, C., Lee, M. A. and Saam, W. F. (1980). *Phys. Rev. A* **21**, 959.
Findenegg, G. H. and Fischer, J. (1975). *Disc. Faraday Soc.* **59**, 38.
Fischer, J. (1977). *Molec. Phys.* **33**, 75.
Fischer, J. (1978). *Molec. Phys.* **35**, 897.
Fischer, J. and Methfessel, M. (1980). *Phys. Rev. A* **22**, 2836.
Gregg, S. J. and Sing, K. S. W. (1982). "Adsorption Surface Area, and Porosity," 2nd edn. Academic Press, London and New York.
Henderson, D. and Blum, L. (1978). *J. Chem. Phys.* **69**, 5441.
Henderson, D., Waisman, E. and Lebowitz, J. L. (1976). *In* "Colloid and Interface

Science, "Vol VIII, Proceedings of the International Conference on Colloids and Surfaces, Puerto Rico, 1976, (M. Kerker ed). Academic Press, New York and London.
Henderson D., Abraham, F. F. and Barker, J. A. (1976). *Molec. Phys.* **31**, 1291.
Henderson, D., Lebowitz, J. L., Blum, L. and Waisman, E. (1980). *Molec. Phys.* **39**, 47.
Hill, T. L. (1956). "Statistical Mechanics." McGraw-Hill, New York.
Hill, T. L. (1960). "Introduction to Statistical Thermodynamics." Addison-Wesley, Cambridge, Mass.
Hiroike, K. (1969). *J. Phys. Soc. Japan.* **27**, 1415.
Hiroike, K and Fukui, Y. (1970). *Progr. Theor. Phys.* **43**. 660.
Hoye, J., Lebowitz, J. L. and Stell, G. (1974). *J. Chem. Phys.* **61**, 5233.
Kozak, J. J., Rice, S. A. and Weeks, J. D. (1971). *Physica* **54**, 573.
Kuni, F. M. and Rusanov, A. I. (1968). *Russ. J. Phys. Chem.* **42**, 443.
Lado, F. (1964). *Phys. Rev.* **135**, 1013.
Lado, F. (1973). *Phys. Rev. A* **8**, 2548.
Lando, D. and Slutsky, L. (1970a). *Phys. Rev. B* **2**, 2863.
Lando, D. and Slutsky, L. (1970b). *J. Chem. Phys.* **52**, 1510.
Lane, J. E., Spurling, T. H., Freasier, B. C., Perram, J. W. and Smith, E. R. (1979). *Phys. Rev. A* **20**, 2147.
Langmuir, I. (1918). *J. Amer. Chem. Soc.* **40**, 1361.
Lebowitz, J. L. (1964). *Phys. Rev.* **133**, 895.
Lee, L. L. (1975). *J. Chem. Phys.* **62**, 4436.
Lee, L. L. (1980). *J. Chem. Phys.* **73**, 4050.
Liu, K. S., Kalos, M. H. and Chester, G. V. (1974). *Phys. Rev. A* **10**, 303.
Mitchell, D. J., and Richmond, P. (1976). *J. Chem. Soc. Faraday Trans. II* 72, 1613.
Navascués, G. (1976). *J. Chem. Soc. Faraday Trans. II* **72**, 2035.
Navascués, G. and Tarazona, P. (1979). *Molec. Phys.* **37**, 1077.
Nicholson, D. (1968). *Trans. Faraday Soc.* **64**, 3416.
Nicholson, D. (1975). *J. Chem. Soc. Faraday Trans. I* **71**, 238.
Nicholson, D. (1976). *J. Chem. Soc. Faraday Trans. I* **72**, 29.
Nicholson, D. and Silvester, R. G. (1977). *J. Colloid Interface Sci.* **62**, 447.
Pace, E. L. (1957). *J. Chem. Phys.* **27**, 1341.
Percus, J. K. (1976). *J. State. Phys.* **15**, 423.
Perram, J. W. (1975). *Molec. Phys.* **30**, 1505.
Perram, J. W. and Smith, E. R. (1975). *Chem. Phys. Lett.* **35**, 138.
Perram, J. W. and Smith, E. R. (1976). *J. Colloid Interface Sci.* **59**, 198.
Perram, J. W. and Smith, E. R. (1977). *Proc. Roy. Soc. A* **353**, 193.
Perram, J. W. and White, L. R. (1975). *Disc. Faraday Soc.* **54**, 29.
Prasad, S. D. and Toxvaerd, S. (1980). *J. Chem. Phys.* **72**, 1689.
Rahman, A. (1964). *Phys. Rev. Letts.* **12**, 575.
Richards, E. L. (1973). Ph.D. Thesis, London University.
Richmond, P. (1976). *Phys. Chem. Liq.* **5**, 251.
Richmond, P. (1977). *J. Chem. Soc. Faraday Trans. II* **73**, 316.
Rowley, L. A., Nicholson, D. and Parsonage, N. G. (1976a). *Molec. Phys.* **31**, 365.
Rowley, L. A., Nicholson, D. and Parsonage, N. G. (1976b). *Molec. Phys.* **31**, 389.
Saam, W. F. and Ebner, C. (1977). *Phys. Rev. A* **15**, 2566.

Saam, W. F. and Ebner, C. (1978). *Phys. Rev. A* **17**, 1768.
Smith, E. R. and Perram, J. W. (1977). *J. Stat. Phys.* **17**, 47.
Smith, L. S. and Lee, L. L. (1979). *J. Chem. Phys.* **71**, 4085.
Sokolowski, S. and Stecki, J. (1978). *Molec. Phys.* **35**, 1483.
Steele, W. A. (1980). *J. Colloid Interface Sci.* **75**, 13.
Sullivan, D. E. (1979). *Phys. Rev. B.* **20**, 3991.
Sullivan, D. E. and Stell, G. (1977). *J. Chem. Phys.* **67**, 2567.
Sullivan, D. E. and Stell, G. (1978). *J. Chem. Phys.* **69**, 5450.
Sullivan, D. E., Levesque, D. and Weis, J. J. (1980). *J. Chem. Phys.* **72**, 1170.
Taylor, L. H., Langley, W. W. and Bryant, P. J. (1965). *J. Chem. Phys.* **43**, 1184.
Thiele, E. (1963). *J. Chem. Phys.* **38**, 1959.
Thompson, N. E., Isbister, D. J., Bearman, R. J. and Freasier, B. C. (1980). *Molec. Phys.* **39**, 27.
Throop, G. J. and Bearman, R. J. (1965). *J. Chem. Phys.* **42**, 240.
Verlet, L. (1968). *Phys. Rev.* **165**, 201.
Verlet, L. and Weis, J. J. (1972). *Phys. Rev. A* **5**, 939.
Waisman, E., Henderson, D. and Lebowitz, J. L. (1976). *Molec. Phys.* **32**, 1373.
Waisman, E. (1973). *J. Chem. Phys.* **59**, 495.
Wertheim, M. S. (1963). *Phys. Rev. Lett.* **8**, 321.
Wertheim, M. S. (1980). *J. Chem. Phys.* **73**, 1393.

Chapter 8

Effects of Liquid Structuring: Hydrophobic and Related Effects

In the preceding chapters the simulation studies all used two surfaces, with the length of the box being sufficiently large to prevent any interference between the liquid structural changes induced by the two walls. However, in many real systems in colloid science the force between the walls arising from such interference is of very considerable interest. For this reason we now examine some of the results which have been obtained for these systems together with the methods which have been used to obtain them.

1. Systems involving simple fluids

Liquids composed of spherical molecules which interact via Lennard-Jones 12–6 potentials are the natural choice for a starting point in making a study of liquid structuring effects. However, there are no experimental data which can be compared directly with the theoretical predictions for these systems, and so the amount of attention which has been given to them by theoreticians is very limited indeed.

The first quantitative theoretical study of this kind of problem was made by Ash, Everett and Radke (1973) using normal thermodynamic arguments. Quite generally, for parallel plates the extra force per unit area is

$$f_s = f - f^0 = -2\{\partial(\gamma - \gamma^0)/\partial h\}_{T,\mu} = 2\int_{-\infty}^{\mu} (\partial\Gamma/\partial h)_{T,\mu}\,\mathrm{d}\mu \qquad (8.1)$$

where f and f^0 are the force/unit area in the presence and absence of adsorption, γ and γ^0 are the corresponding surface tensions, and Γ is the surface excess. They were able to apply this equation to a gaseous medium for which adsorption on the parallel plates obeyed Henry's Law. When they attempted to extend their treatment to condensed fluid phases by considering multilayer adsorption they were only able to reach qualitative conclusions of a very limited kind.

Subsequently the corresponding system with a liquid of spherical molecules interacting according to a Lennard-Jones 12–6 potential (truncated at $\sim 2{\cdot}5\,\sigma$) has been studied using the canonical ensemble Monte Carlo method (Van Megen and Snook, 1979). In this work each of the two walls was taken to be a (100) plane of an FCC lattice, and it was assumed that there was a Lennard-Jones 12–6 interaction for each molecule of the fluid with each molecule of the wall. By integrating over each layer of wall atoms the interaction potential for a fluid molecule a distance z from the plane containing the outermost layer of wall atoms took a form due to Steele (1973) (cf. Eqn. 3.81a),

$$u^{(1)}(z) = \frac{2\pi[\sigma^{(\mathrm{w,a})}]^2\,\varepsilon^{(\mathrm{w,a})}}{a_c}\left[\frac{2}{5}\left(\frac{\sigma^{(\mathrm{w,a})}}{z}\right)^{10} - \left(\frac{\sigma^{(\mathrm{w,a})}}{z}\right)^{4} - \frac{\sigma^{(\mathrm{w,a})4}}{3\Delta(z+0{\cdot}61\Delta)^3}\right] \tag{8.2}$$

where a_c is the area per atom of the surface and Δ is the interlayer spacing for the solid. The force between the walls other than that arising from the usual pressure term was then given by

$$f_s(h) = \mathscr{A}\left\{\int_0^h F_s(z)\rho^{(1)}(z)\,\mathrm{d}z - \int_0^\infty F_s(z)\rho^{(1)}(z)\,\mathrm{d}z\right\} \tag{8.3}$$

where $\mathscr{A}$ is the area of each wall, h is the distance between the walls, and $F_s(z) = \partial u^{(1)}(z)/\partial z$. The singlet density profile $\rho^{(1)}(z)$ was obtained from the simulation. Unfortunately, because this simulation was carried out in the canonical ensemble the chemical potential to which the results correspond could not easily be obtained. A knowledge of the bulk density would have sufficed, since previous machine calculations on Lennard-Jones 12–6 systems would have provided the link with the chemical potential. The need to know μ arises because $f_s(h)$ is the force between the plates less the hydrostatic pressure which would be apparent in the absence of these structural effects and with a fluid density equal to that which a bulk system would have at the same chemical potential. Another difficulty was that in order to determine the force as a function of box-length (h) at fixed μ it would be necessary to know the $\rho^{(1)}$ profile for various values of h *at fixed* μ. This creates a severe problem for most simulations for which $N =$ constant is used. Faced with these problems, van Megen and Snook used data for the density profile pertaining to an isolated solid surface, that is, obtained from a long box, and assumed that when the two walls were brought to a distance h apart the profile would be given by the simple overlap function:

$$\rho^{(1)}(z; h) = \{\rho^{(1)}(z; \infty) + \rho^{(1)}(h - z; \infty)\}/2 \qquad (8.4)$$

Using this assumption, they obtained values for $f_s(h)$ which showed strong oscillations, dying away beyond about $h = 6\sigma$. As would be expected, the oscillations were of greater amplitude the larger the fluid density. These findings have been confirmed by simulations in the grand canonical ensemble (Snook and van Megen, 1980; van Megen and Snook, 1981). The AA potential in this study was cut off at $3{\cdot}5\sigma$, but the model was otherwise the same. Figure 8.1(a) shows the results for $f_s(h)$ at $T^* = 1{\cdot}2$ and three different chemical potentials, corresponding to bulk densities $\rho\sigma^3 = 0{\cdot}593$, $0{\cdot}705$ and $0{\cdot}736$. For the last two densities the difference in the amplitude of the oscillations is not very large. Figure 8.1(b) shows how the number

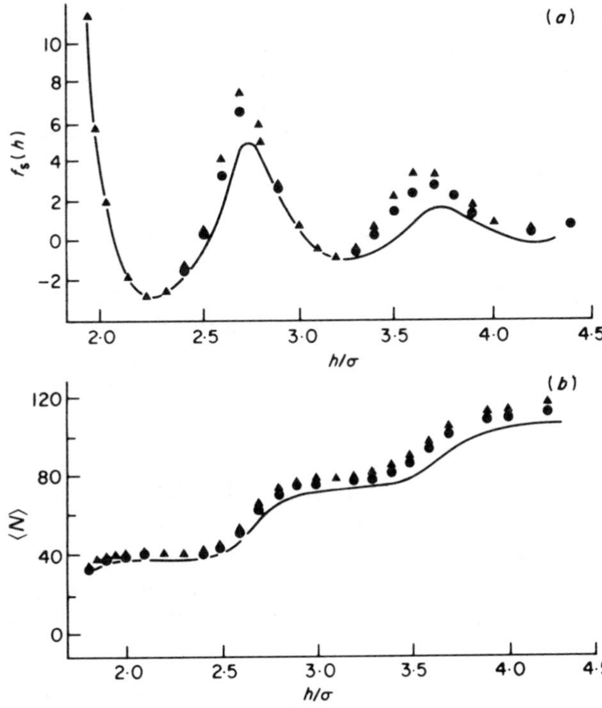

Fig. 8.1. (a) Solvation force $f_s(h)$ for $\mu^* = -2{\cdot}477(\rho\sigma^3 = 0{\cdot}593)$ (solid line), $\mu^* = -1{\cdot}786(\rho\sigma^3 = 0{\cdot}705)$ (●), and $\mu^* = -1{\cdot}237(\rho\sigma^3 = 0{\cdot}736)$ (▲) at $T^* = 1{\cdot}2$. (b) Average number of molecules $\langle N \rangle$ in the volume (=$49h$) between the plates. Symbols as for (a). (From van Megen and Snook, 1981.)

of molecules between the plates increases in an approximately step-wise manner as h is increased, or, expressing the same data in a different way, how the surface excess (Γ) oscillates as h is increased. Whether this extra force component is attractive or repulsive is determined by the way in which the oscillations in $\rho^{(1)}$ fit into the box. For $h \gtrsim 2$ nm ($\gtrsim 6\sigma$) the force becomes fairly insignificant. With water as the fluid the corresponding distance seems to be larger, and may be >10 nm: this is consistent with the much greater tendency of water to form an extensive ordered three-dimensional network (Christou, Nicholson, Parsonage and Whitehouse, unpublished results).

2. Systems involving water: hydrophobic effects

(a) Historical and experimental review

In this section we shall frequently use the term "structure" when referring to water. This is normal practice, although very few workers have made a serious attempt to define the term. Ben-naim (1980) has defined a single scalar quantity to represent the "structure". This quantity is the mean number of hydrogen bonds in which each molecule is involved (ν): it can take values in the range 0–4. More often the term "structure" has been associated with the pair distribution functions, as is common for simple fluids, with the recognition of change depending upon an overall assessment of the changes in the distribution function taking account of all inter-particle distances. It would be expected that modification of the pair distribution functions from normal shapes for uniform systems, would be associated with non-uniform characteristics in the singlet distribution functions such as those discussed for simple fluids (Ch. 6, § 3). Our use of the term "structure" will correspond to the latter, rather than to Ben-naim's, practice.

Discussion of the forces between colloidal particles has been dominated for many years by the Derjaguin–Landau–Verwey–Overbeek (DLVO) theory. According to that theory there is a strong and moderately long-ranged repulsion between charged double-layers surrounding the particles and a shorter-ranged van der Waals attractive force, the potential being given by

$$E(d) = \frac{64nkT}{\kappa}\exp(-\kappa d)\frac{\exp(ze\psi_\delta/2kT) - 1}{\exp(ze\psi_\delta/2kT) + 1} - \frac{A}{12\pi d^2} \tag{8.5}$$

for two parallel plates a distance d apart, where ψ_δ is the potential at a distance from the plates equal to the Stern layer thickness δ, κ is the

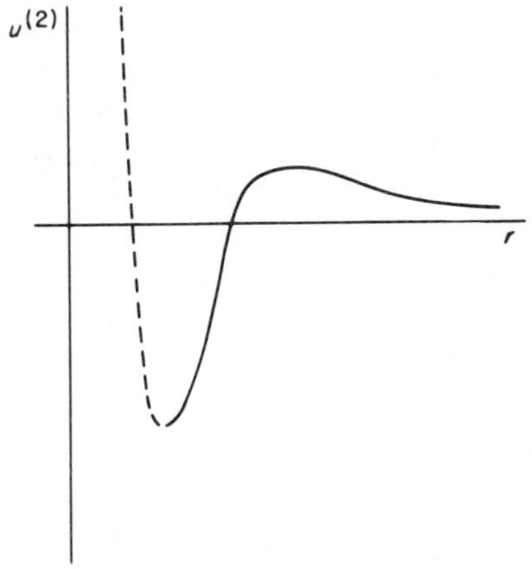

Fig. 8.2. Interaction potential for a pair of colloidal particles (DLVO theory).

reciprocal of the double-layer thickness (equal to $(e^2(2nz^2)/\varepsilon_0\varepsilon_r kT)^{1/2}$), z is the charge and n the concentration of the ions (all assumed to be the same), e is the electronic charge, A is a constant characteristic of the plates (known as the Hamaker constant), and ε_r is the relative permittivity. This involves a Debye–Hückel (Gouy-Chapman) approximation for electrostatic repulsion, and a continuum solid approximation (dispersion forces only) for the van der Waals attraction. As the water is treated as a continuous medium no account is taken of any effects which might arise from changes in its structure. When the inter-particle distance is $\gtrsim$70 nm the neglect of structural effects is almost certainly correct, although Drost-Hansen (1977) has claimed that these are important even at 100 nm. In direct attempts to measure these forces the results are expressed in terms of the disjoining pressure (Π), a term originally introduced by Derjaguin as the pressure required to prevent a liquid from disjoining (breaking contact) with a surface. Alternatively, it can be defined as the pressure which must be applied to a thin film in order to keep it in equilibrium with the bulk liquid. Thus

$$RT \ln (p/p_0) = \Pi V_m \tag{8.6}$$

where p and p_0 are the vapour-pressures of the film and the bulk liquid, respectively and V_m is the molar volume of the liquid. An especially careful study of disjoining pressures was made by Read and Kitchener (1969), who observed the thickness of a liquid film between two bubbles or a bubble and an optically flat plate as a function of the pressure in the bubble. In spite of the great care exercised no meaningful results could be obtained for film thicknesses of less than 20 nm, the region of interest to us.

Derjaguin and Churaev (1974) have analysed data on the disjoining pressure for water against quartz and mica at distances 20–2400 nm in order to obtain the structural component. Their analysis confirms that the electrostatic term is dominant for thicknesses $\gtrsim 70$ nm. Below this the van der Waals interaction also becomes important. Finally, the structural component, which is negligible above 10 nm, increases rapidly and becomes dominant on going to smaller distances.

Some very careful direct measurements of the force between mica sheets immersed in aqueous electrolyte solutions have disclosed that in addition to the double-layer repulsions and van der Waals attractions, both of which are well described by normal theories (DLVO), there is a repulsive force, which, for parallel plates, would correspond to an extra pressure of $10^7 \exp(-d/(1{\cdot}0\ \mathrm{nm}))\ \mathrm{N\ m^{-2}}$. That this additional effect arises from hydration forces was suggested by the observations that it is largely independent of electrolyte concentration and that the decay length of the force is always the same (Israelachvili and Adams, 1978). The extra contributions became evident at plate separations of ~6 nm, below which it increased rapidly. Israelachvili and co-workers have subsequently improved their technique to such an extent that they were able to observe six oscillations in $f_s(h)$ with c-C_6H_{12} as the medium between the plates and ten oscillations with the large quasi-spherical molecule $[(CH_3)_2SiO]_4$, the wavelength of the oscillations being approximately equal to the diameters of the molecules (1·0 and 0·6 nm, respectively) (Horn and Israelachvili, 1980). The observation of oscillations with water as the medium still presents formidable problems, as the wavelength expected would be only ~0·3 nm (Israelachvili, 1981).

Because of the difficulties in performing direct measurements on the structural effects, most of the experimental information has come from analysis of the thermodynamic data for the aqueous solutions of substances which are neither non-polar or at least contain large non-polar sections (Némethy and Scheraga, 1962a, b, 1963). These authors found it convenient to divide the field of study into two parts: hydrophobic hydration and hydrophobic interaction. The former is concerned with the effect of introducing an isolated solute molecule into the solvent; the latter, which is directly related to the force between solute molecules, involves the bringing of two (or more) solute particles to the chosen relative position. Much of the interest in the hydrophobic interaction has arisen because of its applicability in the study of the three-dimensional structure of biopolymers in aqueous solution (Kauzmann, 1959; Némethy, 1967). We shall follow the procedure of treating the hydrophobic hydration and interaction separately, although recent theoretical studies have usually led to information on both topics. However, before examining these we must consider the models which have been proposed to represent water and the interactions between it and other molecules.

(b) Interaction between water molecules

In recent years great progress has been made in obtaining a usable potential and much of this can be attributed directly to the increased availability of large-scale computing capacity. The present discussion centres on uniform bulk water since it is only following success in this field that studies of interfacial systems can be contemplated. It should be stressed, however, that most potentials are "effective" ones and the caveat that this effectiveness may not transfer from uniform to non-uniform systems must be kept in mind.

An important class of potential models is that in which separated point charges surround the oxygen atom in a rigid arrangement (Bernal and Fowler, 1933; Rowlinson, 1951; Ben-naim and Stillinger, 1969; Stillinger and Rahman, 1974). An early potential of this type was due to Rowlinson, who took into account virial coefficients and dipole moment data and considered separate contributions from E_R, E_D and E_c (defined in § 3(a) of Chapter 3). Initially, a partial positive charge was placed on each of the hydrogen atoms and chosen so that, with the experimental bond-angle (104·5°), the measured dipole moment (1·83 D) was obtained; the balancing negative charge was concentrated on the oxygen atoms. It was found that the multipole energy of the crystal lattice could be better accounted for if this charge was split and placed either side of the oxygen in a plane normal to that containing the hydrogen atoms (Fig. 8.3(*a*)). The dispersion energy term was incorporated into an empirical inverse sixth power term whose coefficient was also intended to include allowance for mean polarization effects ($-2\mu^2\alpha/4\pi\varepsilon_0 r^6$) and contributions from higher-order dispersion terms. E_R was written as an inverse 12th power term so that the final potential took the form of a 12–6 function augmented by a summation over the 16 separate contributions to the coulombic interaction. This basic form

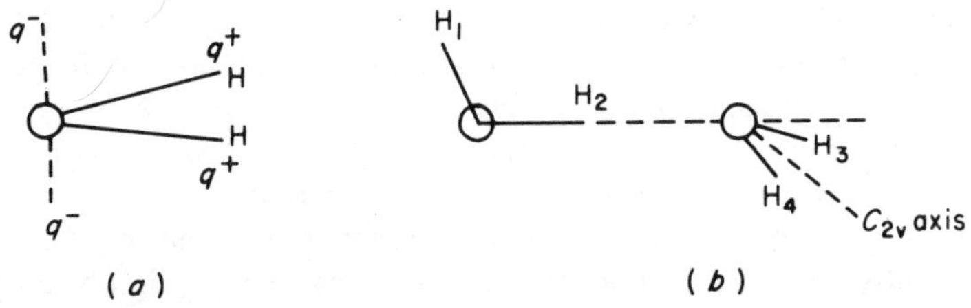

Fig. 8.3. (*a*) The Rowlinson model for water. (*b*) A favoured configuration for two water molecules according to the BNS model.

of potential function is also to be found in the BNS (Ben-naim and Stillinger, 1969) and ST2 (Stillinger and Rahman, 1974) potentials. However, the starting point in these models is the electronic structure in common between neon and water and consequently the 12–6 part of the potential is based on accepted values of ε and σ for the neon–neon interaction. In the BNS potential the charges were distributed symmetrically at the corners of a tetrahedron, which has the disadvantage that it gives rise to an intense directionality along the H-bond for configurations such as those in Fig. 8.3(*b*), which is liable to induce librations between two strong minima (Weres and Rice, 1972). The ST2 potential was designed with a built-in asymmetry in order to overcome this problem, which does not of course occur in the highly unsymmetrical Rowlinson potential. The partial charges in the BNS and ST2 potentials were somewhat larger than Rowlinson's choice, giving rise to a dipole moment greater than that found experimentally for isolated water molecules (Table 8.1); However, this is probably more realistic in the mutually polarizing environment of liquid water (Barnes *et al.*, 1979).

An attractive alternative to the semi-empirical approach typified by this class of potential functions, which is again available through the increase in large-scale computing capacity, is to carry out quantum mechanical calculations on the H_2O dimer system (Clementi and Popkie, 1972; Popkie *et al.*, 1973; Lie and Clementi, 1975; Matsuoka *et al.*, 1976). Jorgensen (1979a, b) has identified four levels of sophistication for this type of calculation: (i) minimal basis set, employed in his calculations; (ii) double ζ basis, (iii) double ζ plus polarization or Hartree–Fock limit (HF); (iv) double ζ plus polarization plus correlation correction via configuration interaction (CI). Work of this type was initiated by Clementi and Popkie (1972) using HF calculations and was subsequently refined and developed by Clementi and co-workers. In their early work they found that the Bernal and Fowler expression provided the most suitable fitting function, and have retained this basic form for fitting the later calculations. In this expression the water molecule is represented by four centres, three of them lying on the HOH triangle, with a partial charge of $+q$ on each hydrogen atom, and the fourth, which carries a charge of $-2q$, lying on the C_{2v} axis and within the HOH triangle. The mathematical expression for the potential comprises the appropriate coulombic terms and a sum over terms having the form $a_k \exp(-b_k r_{ij})$, where r_{ij} represents the distance between all possible atomic centres and a_k and b_k are fitting parameters.

In the original potential, dispersion forces were ignored, but these were subsequently incorporated separately as inverse sixth, eighth and tenth power terms with coefficients estimated from a new perturbation technique developed by Kolos (Table 8.1) (Lie and Clementi, 1975). In the most

Table 8.1. Comparison of properties of various water models (All values of thermodynamic properties refer to 25°C unless otherwise stated)

Model	Row[1]	BNS[1]	ST2[2]	MCY[3]	CF[4]	STO-3G[b]5	PE[6]	Expt/Theor
$-U$/(kJ mol^{-1})	33–35	28–33	33–35	28–29	33	29	21	34[e][1]
C_V/(J K^{-1} mol^{-1})	63–92	71–84	71	75–84	—	53	—	75[e][1]
C_6/atomic units	60	8·0	8·0	—	—	—	—	67[t]
μ/Debye units	1·84	2·17	2·35	2·26	1·86	1·22	2·50	1·86[a][1]
PV/NkT for $\rho = 1$ g cm^{-3}	< -4[1]	$\sim -3{\cdot}1$ (0°)[7] < -4[1]	0·09 (10°)[7] −0·01 (41°)[7]	$>2{\cdot}20$[8]	0·1 (29·5°)[4]	—	—	~0·003 (0°)[9] ~0·05 (25°)[9] ~0·07 (30°)[9]

(a) Experimental, for an isolated molecule; (b) fitted to 4-point inverse 12–6–3–1 potential; (e) experimental; (t) theoretical (see text); (1) Watts (1974); (2) Ladd (1977); (3) Mezei *et al.* (1979); (4) Stillinger and Rahman (1978); (5) Jorgensen (1979a); (6) Barnes *et al.* (1979); (7) Stillinger and Rahman (1974); (8) Lie *et al.* (1976); (9) CRC Handbook of Chemistry and Physics, 58th edition.

recent version of the potential (the MCY potential) a very detailed evaluation of the dimer interaction was made, using the CI method with a large basis set of Gaussian orbitals (Lie *et al.*, 1976; Matsuoka *et al.*, 1976). Although allowance for dispersion forces was included in this work, the related terms were not explicitly exhibited in the fitting function, which had the Bernal–Fowler form. Like its predecessors, the MCY potential has been tested by comparison between MC simulation of the uniform fluid and experimental data. Jorgensen used the minimal basis set of Pople (STO-3G) to calculate 266 points on the potential energy surface (compared with about 60 in Clementi's work). He found that an inverse 12–6–3–1 function gave a more accurate fit than the Bernal–Fowler form. Both 3-centre and 4-centre separated charge models were tried, and the latter found to give better properties for the linear dimer, although giving a very low value for the dipole moment. Higher-order multipoles were not considered.

A weakness of the rigid molecule models is their inability to account for possibly highly significant many-body effects arising from polarization. These could be of even greater importance in non-uniform environments than in a uniform one, where a large amount of field cancellation would be expected (Barnes, 1978; Barnes *et al.*, 1979). Barnes has proposed a polarizable electropole (PE) model to overcome this problem in which the multipole tensors and polarizability are initially set to experimentally determined values, but are subsequently readjusted during a simulation by an iterative calculation. Expansions as far as the quadrupole term were mainly employed and it has been pointed out that none of the rigid molecule models is capable even of predicting the correct sign for the latter. Alternative approaches have been proposed by Berendsen (1972), whose starting point was a four-charge model in which the magnitude, rather than the position, of the charges is varied, and by Stillinger and David (1978) who developed a polarization model in which the basic structural elements are bare protons and O^{2-} units which possess a non-local polarizability. Extensive testing of these models by simulation is exceptionally demanding, and only in the first case has some progress been made (Barnes *et al.*, 1979).

An earlier model which has some of the qualities required to account for distortions of one molecule by its neighbours is the central force model (CF) also developed by Stillinger and co-workers (Lemberg and Stillinger, 1975; Rahman, Stillinger and Lemberg, 1975; Stillinger and Rahman, 1978). Here the interactions are obtained from a set of three central potentials related to OO, OH and HH interactions, each of which is written as a sum of coulombic terms with additional, empirically determined, functions involving exponentials. In this model each molecule is able to change both the OH bond-lengths and the HOH angle.

Testing of these models by computer simulation has employed a number of systems, temperatures, methods (MC and MD) and different corrections for long-range effects. It would be inappropriate to discuss these exhaustively here, but suffice it to say that recent work suggests that, provided attention is confined to atom–atom pair distributions and thermodynamic properties, variations among these different techniques should not be too great to render comparisons meaningless (Impey *et al.*, 1980; Mezei *et al.*, 1979). Some details relating to the simulations are important in choosing among the potentials: thus the ST2 potential needs to incorporate a switching function to avoid complications from coincidence of charges at small intermolecular separations. Watts (1974) found that this complication could be overcome in the case of the Rowlinson potential (ROW) simply by introducing a cut-off radius. In a comparison between these two potentials and the MCY potential (Matsuoka *et al.*, 1976) quite significant differences in computing speed were noted (Christou, unpublished; Jorgensen, 1979a), the order of fastest to slowest being ROW > ST2 > MCY, which is as might be expected since the evaluation of exponentials is expensive. The above three potentials, and also the BNS, the Barnes potential and the revised CF model are all able to give a satisfactory account of the $\tilde{g}^{(O,O)}$ distribution although some of the detail, especially at large separations, is missing. $\tilde{g}^{(O,H)}$ and especially $\tilde{g}^{(H,H)}$ are less satisfactory although it must be borne in mind that the experimental data for these functions can only be obtained by indirect means and may themselves be subject to considerable uncertainties. A convenient criterion for comparison is provided by the internal energy and heat capacity; some values for these are collected in Table 8.1.

Also collected in Table 8.1 are values for the dipole moment μ and the coefficient C_6 of r^{-6} in the dispersion part of the interaction (in atomic units). Although the total interaction is of course dominated by coulombic terms the dispersion contribution is the most important term where water interacts with a non-polar substrate (Pratt and Chandler, 1980). This point is, of course, highly relevant in the study of hydrophobic effects. The theoretical value quoted is from the HF calculations of Kolos mentioned earlier and expected to be quite accurate; other theoretical estimates range upwards from ~40 atomic units (Amos and Crispin, 1976).

A notable weakness of these models is in their predictions for the pressure. Indeed, the Rowlinson and BNS models both give large negative pressures for water at 25°C and 1 g cm^{-3}. This does not create any special difficulties when the topic of interest is the "structure" of the uniform fluid, since the use of periodic boundary conditions in all three directions together with the smallness of the simulation systems enables a tension to be sustained without separation into rarefied and dense phases. This is also true for the studies on hydrophobic systems in this chapter, but serious problems can arise if simulations of the type described in §3 of Chapter 6 are attempted for water using either the Rowlinson or BNS models. In the latter, phase separation will occur if the interactions with the walls are insufficiently attractive to enable the required tension to be applied.

It is interesting to note that, among these models, only those derived directly from quantum mechanical calculations can be considered to be true pair potentials. The others, with the exception of the PE model, are closer to being effective pair potentials. the small values of $-U$ from these potentials can therefore be confidently associated with the missing higher-order interactions, and indeed preliminary calculations have been claimed to support this opinion (Lie *et al.*, 1976). The PE model allows for many-body interactions and the rather small value of $-U$ here may be misleading since in other respects its properties appear to be superior to those of the rigid molecule models (Barnes *et al.*, 1979).

(c) Hydrophobic hydration

In this section we look at the structural changes and associated thermodynamic effects arising from the introduction of isolated solute molecules. The models used are essentially of the type suggested by Frank and Evans (1945), in which the solvent surrounding each solute molecule is "ice-like", that is it has to some extent the ordered structure of ice. Frank (1958) proposed that this ordering was quite strongly cooperative and that, as a consequence, there would be fairly large ordered regions separated by disordered regions. He gave the name "flickering clusters" to these ordered regions to indicate that he considered that these structures were continually changing. As will be seen, Frank's idea of "icebergs", as his large ordered regions were often called, has not been supported by more recent work. However, the general idea of regions of enhanced order surrounding the solute molecules is now accepted. A related concept, that of "clathration", has also been carried over into the field of aqueous solutions. In the true clathrate hydrates the "host" water structures used are not those of any of the known forms of ice. The "host" structures involved are such that the water molecules are able to form strong hydrogen-bonds and yet at the same time to provide approximately spherical cages of sizes suitable to accommodate small molecules. Two "host" structures occur, each having two kinds of cavity. The two kinds of cavity in Structure I hydrates can take molecules of radius up to ~0·52 and ~0·59 nm, respectively, whilst the corresponding values for Structure II hydrates are ~0·48 and ~0·69 nm. Which of these is formed depends upon the size of the molecules to be encaged (the "guest" molecules). When the term "clathrate" is used for solution phenomena the existence of large regions of such ordered structure is not implied. Rather, it is envisaged that the *immediate* environment of a solute molecule may be rather like that of a clathrated molecule in a hydrate.

The first computer study on this type of problem was by Dashevsky and Sarkisov (1974), who employed a Monte Carlo procedure for the simulation of $CH_4 + H_2O$. They used a central force model, with the force centres on the O and H atoms of each H_2O molecule, together with an added H-bond term and with fractional point charges on the O and H atoms. Using a canonical ensemble they were not able to obtain easily the free energy or entropy values. They therefore resorted to the procedure of evaluating $\langle \exp(U_N/kT) \rangle$, which is correct in principle but which suffers from serious statistical weaknesses (Ch. 4, §2(b)). In spite of this, the agreement between their results for the solubility of methane and those from experiment was quite satisfactory.

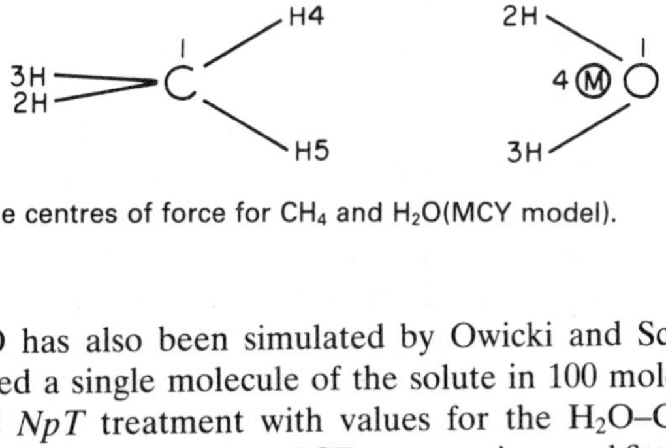

Fig. 8.4. The centres of force for CH_4 and H_2O(MCY model).

$CH_4 + H_2O$ has also been simulated by Owicki and Scheraga (1977), who considered a single molecule of the solute in 100 molecules of water using an MC *NpT* treatment with values for the H_2O–CH_4 interaction parameters derived from *ab initio* SCF computations and fitted to an inverse 12–6–1 potential function with 9 centres corresponding to the C, O and 6H atoms plus one additional centre (M), this being the site of the negative charge in the MCY model of water (Fig. 8.4). The "coulombic" terms (in r^{-1}) are merely fitting functions and it is not claimed that they have any special physical significance. The H_2O–H_2O potential was the MCY function of Clementi and co-workers (1976).

The results for $\Delta U^0_{\mathrm{soln}}$ and $\Delta V^0_{\mathrm{soln}}$ were consistent with experiment, though the uncertainties in the theoretical values were very large indeed: $\Delta U^0_{\mathrm{soln}} = -10{\cdot}9$ kJ mol^{-1} (experiment), -46 ± 63 kJ mol^{-1} (theory); $\Delta V^0_{\mathrm{soln}} = 37$ cm^3 mol^{-1} (experiment), 25 ± 34 cm^3 mol^{-1} (theory). The main reason for the wide error bands is that the whole of the uncertainty in the energy or volume (at standard pressure) is "assigned" to the single CH_4 molecule. Furthermore, by more recent standards the simulation runs were rather short, $\sim 0{\cdot}7 \times 10^6$ configurations being used for data compilation, and it would now be felt that, especially for a hydrogen-bonded solvent such as water, at least twice that number would be needed. Owicki and Scheraga attempted to make better use of their computing time by adopting a biasing scheme for the MC moves. The latter were chosen so as to favour the sampling of the space close to the solute molecule ("the shell"), this bias being very great in the initial stages of the equilibration, being then gradually reduced, and being about a factor of two during the data-collection part of the run. The interactions between "shell" water molecules were found to be more stable and sharply distributed with respect to energy than was the case for molecules in the bulk. The energy distribution found was, indeed, very similar to that subsequently obtained by Pangali, Rao and Berne (1979a, b) (Fig. 8.15) and discussed below. The work of Owicki and Scheraga was important in indicating a direction for future profitable research and in obtaining correctly the main features of the hydration shells. They did not, as subsequent workers have done (Geiger *et al.*, 1979), obtain data on the orientations of the "shell" water molecules.

A very similar study of argon in water has been carried out by Alagona and Tani (1980). Again, their H_2O-solute potential was derived by making a fit to SCF results of Clementi (private communication). The potential included r^{-1} and r^{-6} attractions and r^{-12} repulsions for O–Ar, r^{-6} attractions and r^{-12} repulsions for H–Ar, and r^{-4} and r^{-12} repulsions for M–Ar, where M lies on the C_{2v} axis of the H_2O molecule (Fig. 8.4). The H_2O–H_2O potential was the same as that used both by Owicki and Scheraga and Swaminathan *et al.* (1978) and originally due to Clementi (MCY). The computed value for the energy of solution (ΔU_{soln}) was $-37{\cdot}8\ kJ\ mol^{-1}$, which is close to the values previously obtained by simulation for $CH_4 + H_2O$, but is about three times the experimental value ($-12{\cdot}6\ kJ\ mol^{-1}$). The partial molar internal energy of the argon can be broken down into a direct term ($9{\cdot}2\ kJ\ mol^{-1}$), arising from summing the argon–water interaction terms, and a relaxation term ($-46{\cdot}8\ kJ\ mol^{-1}$), from the induced change in the H_2O–H_2O interaction energy. As would be expected, the relaxation term is dominant. No heat capacity results were reported, evidently because of the considerable uncertainty in them. The length of the run ($0{\cdot}25 \times 10^6$ for equilibration, $0{\cdot}5 \times 10^6$ for data collection) is, again, probably not adequate to give reliable values for a "fluctuation quantity" such as the heat capacity. The fact that in the Ar–O and Ar–H distribution functions the first maxima occurred within $\sim 0{\cdot}02$ nm of each other supports the conclusion reached by other workers (Geiger *et al.*, 1979) that some of the OH bonds are directed tangentially to the argon sphere. Indeed, the distribution of angles between successive H-bonds shows the similarity to the clathrate hydrate structures (Fig. 8.5).

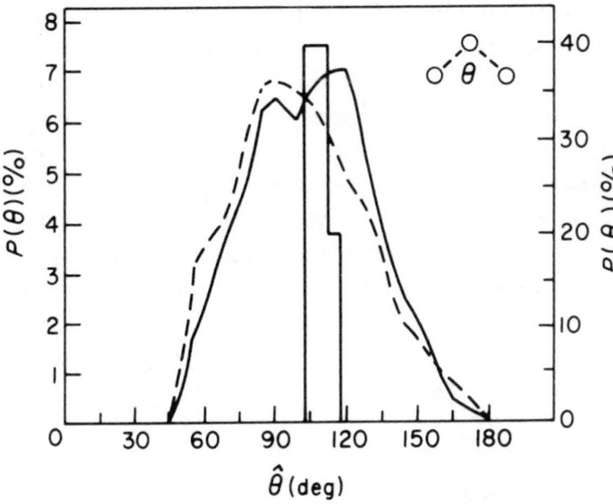

Fig. 8.5. The probability density function of the angle θ formed by sets of three water molecules. The histogram is for a crystalline clathrate hydrate (right-hand scale). The curves are for the solution (left-hand scale): "shell" molecules (full curve), bulk molecules (dashed curve). (From Alagona and Tani, 1980.)

Nevertheless, Alagona and Tani conclude that the energy distribution curves are more decisive than the various distribution functions in distinguishing "shell" from bulk water (cf. Fig. 8.15).

Geiger *et al.* (1979) have made a far-reaching study of a system consisting of two spherical Lennard-Jones particles in 214 water molecules. Since the non-polar particles were, in fact, derived from the H_2O molecules by eliminating the electrostatic charges from the ST2 water model whilst retaining the Lennard-Jones parameters at their values for water, they correspond to a noble gas intermediate between neon and argon. By choosing this procedure much of their original program for bulk ST2 water could remain intact. Because of the long times over which it was necessary to follow the MD simulation an economy on the time spent in computing interactions and forces was made by using a small cut-off (0·707 nm, or 2·28 times the σ for the Lennard-Jones interactions). Bearing in mind the long-range nature of the electrostatic interactions, there must be considerable reservations about some of the results obtained. It is recognized that different properties vary greatly in their sensitivity to such simplifications of the electrostatic energy terms (Adams and Adams, 1981; Pangali *et al.*, 1980), and it is reasonable to claim that the topic of main interest, the structure and properties very near to the solute molecules, would not be too seriously affected by these approximations. The other point made by

Geiger *et al.* in justifying their procedure is that many of the properties of interest are obtained as differences between the properties of the solvent and those of bulk water and that this leads to a cancellation of errors. However, it can just as well be argued that where one is interested in small differences between large quantities it is more (not less) important to be able to determine the latter with accuracy. The authors, and also Pangali *et al.* (1979a, b), whose work on a very similar system is discussed later, are especially careful to emphasize the "healthy scepticism" which should be adopted towards these results. Their MD study was concerned with

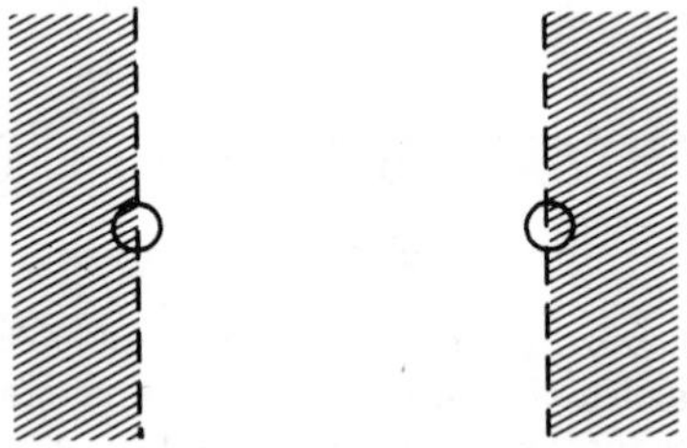

Fig. 8.6.

both hydrophobic hydration and interaction. For this reason they had two apolar solute atoms in the simulation system, these being initially in contact ($r = 0{\cdot}28$ nm) although free to drift apart. The solute atoms remained near to contact ($r \approx 0{\cdot}28$–$0{\cdot}48$ nm) for the first $2{\cdot}5 \times 10^{-12}$ s, after which they quickly moved apart to a distance at which there could be a single water molecule between them (0·54–0·65 nm). The results on hydrophobic hydration were obtained by considering only those water molecules in the shaded regions of Fig. 8.6. The assumption implied here is that the water in these regions was not affected by the more remote solute atoms. This assumption has been tested by Pangali *et al.* (1979b), who found that it was not entirely valid, particularly when the distance between the solute atoms is near to the value at which there is a maximum in the hydrophobic interaction, i.e. the distance approximately mid-way between those corresponding to direct and solvent-separated contact. At both larger and smaller distances the assumption seemed to be a good one. Since the solute atoms spent only a very small amount of their time at this distance, the assumption of Geiger *et al.* with regard to the insensitivity of the structure in the outer regions to the presence of the second solute molecule is seen to be reasonable.

Figure 8.7 shows the "neon"–O pair distribution function found by Geiger *et al.*, together with the accumulated hydration number. Choosing the boundary between first and second shell neighbours as 0·48 nm, the

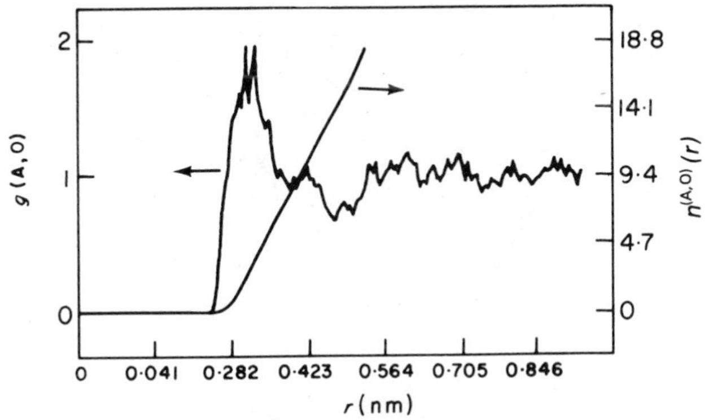

Fig. 8.7. "Neon"–oxygen pair distribution function $g^{(A,O)}$ for the "undisturbed" part of the water environment (shaded in Fig. 8.6). $n^{(A,O)}(r)$ (right-hand scale) gives the running coordination number of a complete hydration shell composed of two "undisturbed" halves. (From Geiger *et al.*, 1979.)

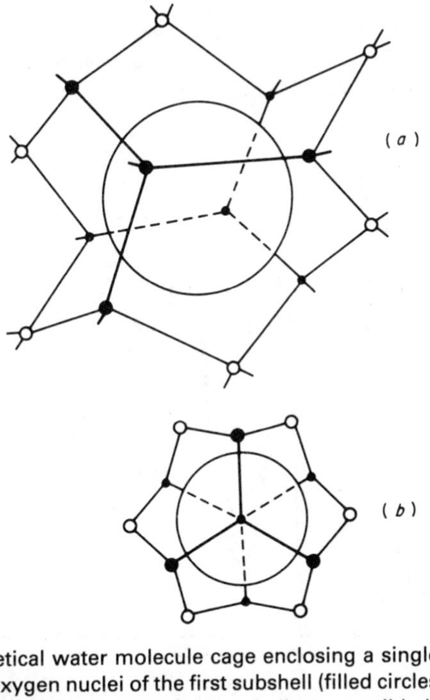

Fig. 8.8. A hypothetical water molecule cage enclosing a single Lennard-Jones solute molecule. Oxygen nuclei of the first subshell (filled circles), oxygen nuclei of the second subshell (open circles). Lines indicate possible hydrogen bonds. (*a*) perspective drawing, (*b*) projection along the axis of highest symmetry. (From Geiger *et al.*, 1979.)

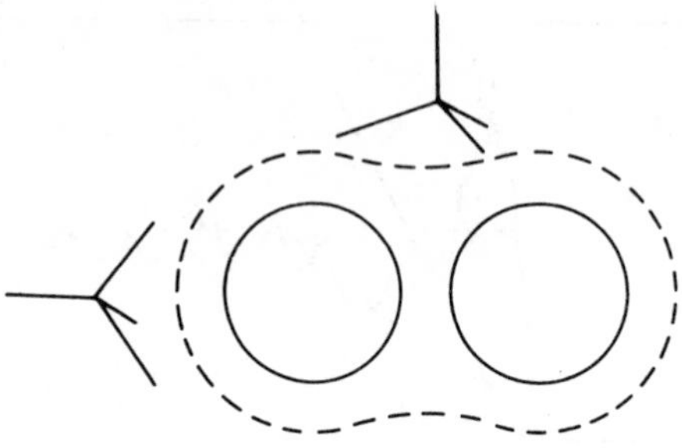

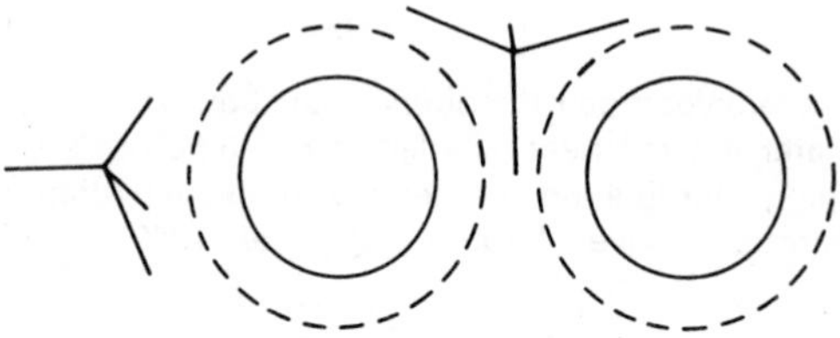

Fig. 8.9. Schematic representation of the orientations of the water molecules for (*a*) contact pair of solute molecules, (*b*) solvent-separated pair of solute molecules. (From Geiger *et al.*, 1979.)

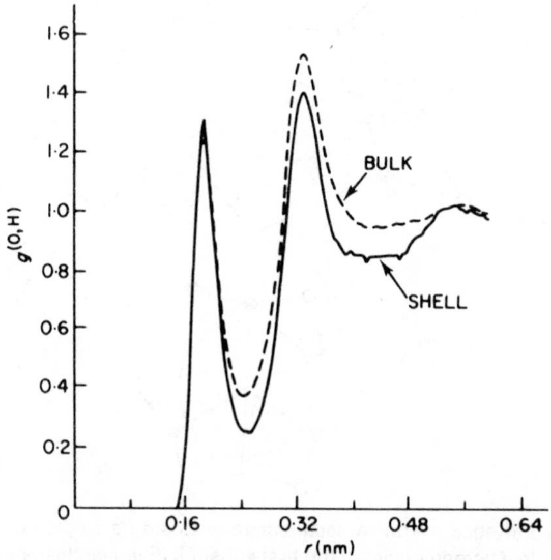

Fig. 8.10. The oxygen–hydrogen radial distribution function $g^{(O,H)}$. All water pairs containing at least one "shell" member contribute to the "shell" average; all other pairs contribute to the bulk average. (From Geiger *et al.*, 1979.)

distance at which a distinct minimum occurs in the correlation function, the first shell was found to contain 14 water molecules, 8 of these being in an inner sub-shell. From these data and the corresponding distribution functions for the orientation of the water molecules, Geiger *et al.* concluded that the basic (idealized) structure of the water surrounding a solute atom was as in Fig. 8.8, but that only fragments of this structure would exist at any one time. They found, therefore, that the clathrate structure was a "goal" which was far from being achieved. (In the later part of the run, when the solute molecules were separated by a water molecule, the suggested ideal cage structure was one with the two cages, of the kind just described, sharing a single water molecule at the overlap point. Near the median plane between the apolar atoms, the water molecules have one bond tangential to both spheres (Fig. 8.9). It appears that the water molecules try to avoid having any of their tetrahedrally disposed bond directions pointing directly into a cavity.) Other quantitative evidence for the structure-enhancing effect of the apolar molecules comes from a comparison of the maximum/minimum ratio of the three atom–atom distribution functions for the "shell" and bulk water molecules. Figure 8.10 shows the $g^{(O,H)}$ function and Table 8.2 summarizes the enhancement data for all three functions. The increase in structure is very marked. An interesting feature is that, in spite of the increase in "ice-likeness" (and in H-bonding) there is no noticeable change in the position of the first maximum in $g^{(O,H)}$. If it is supposed that this distance should be dependent upon the amount of H-bond character in each bond then it would have been expected that in the more strongly bonded "shell" region the distance would have been smaller than in bulk water. It may be relevant that NMR studies of this type of system, although with larger solute molecules (1,4-dioxan, tetrahydrofuran, t-butyl alcohol) have also led to results which are difficult to

Table 2. Ratios of the heights of the first maximum and the following minimum (g_{max}/g_{min}) for various water–water pair correlation functions in bulk and shell, and the enhancement factor in proceeding from bulk to shell.

		g_{max}	g_{min}	g_{max}/g_{min}	Shell/bulk enhancement
$g^{(OO)}$	Shell	2·96	0·52	5·69	1·35
	Bulk	3·03	0·72	4·21	
$g^{(O,H)}$	Shell	1·33	0·25	5·32	1·57
	Bulk	1·25	0·37	3·38	
$g^{(H,H)}$	Shell	1·40	0·62	2·26	1·30
	Bulk	1·42	0·82	1·73	

(After Geiger *et al*, 1979).

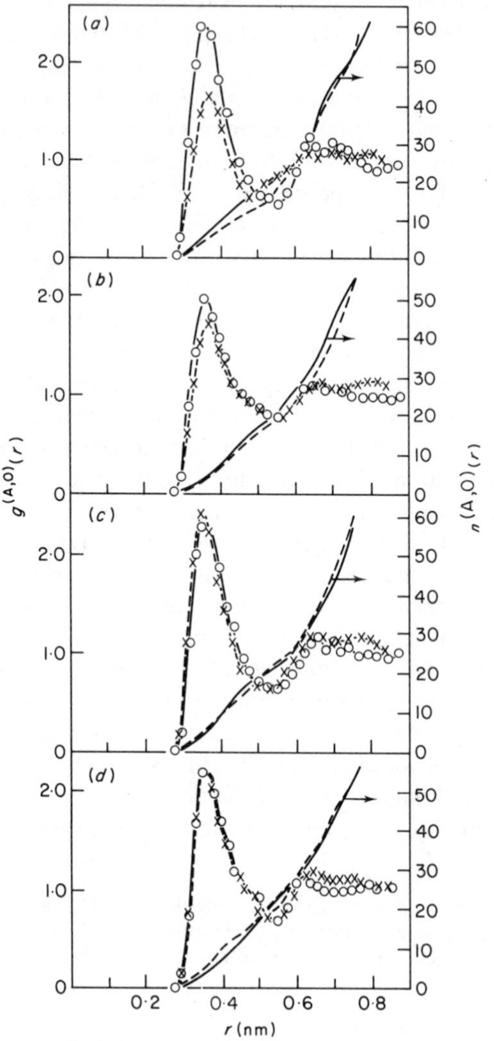

Fig. 8.11. "Xenon"–oxygen radial distribution function $g^{(A,O)}$ for four values of the inter-"xenon" distance (*a, b, c, d*; see Fig. 8.12). The running coordination number $n^{(A,O)}(r)$ is also shown (right-hand scale). Exterior functions (full curves), interior functions (dashed curves). (From Pangali *et al.*, 1979.)

understand. At low temperatures (0°C) and concentrations ($<\sim 2$ mole %) the proton resonance for "shell" water molecules lies on the down-field side of that for bulk water molecules, as would be expected, since the more strongly H-bonded "shell" protons would be less well shielded. However, at higher concentrations and room temperature the reverse is found to be the case, the shift being to higher magnetic fields (Glew *et al.*, 1968). No adequate explanation of this observation has been found. It has been suggested that at low solute concentrations each solute molecule can induce a cage of water molecules to form around it, with an increase in H-bond character, but that at higher concentrations this structure is no longer possible and, for some incompletely explained reason, the outcome is a diminution in H-bonding. It is believed that the speed of H-bond breaking and re-making is sufficiently great that the resonance shift observed is, indeed, the time-averaged value, as is assumed in the foregoing discussion.

Pangali *et al.* (1979b) have studied a similar system to that of Geiger *et al.*, using their force-bias MC technique (Ch.4, §2(a)). This work is part of the same study of which their hydrophobic interaction results are another

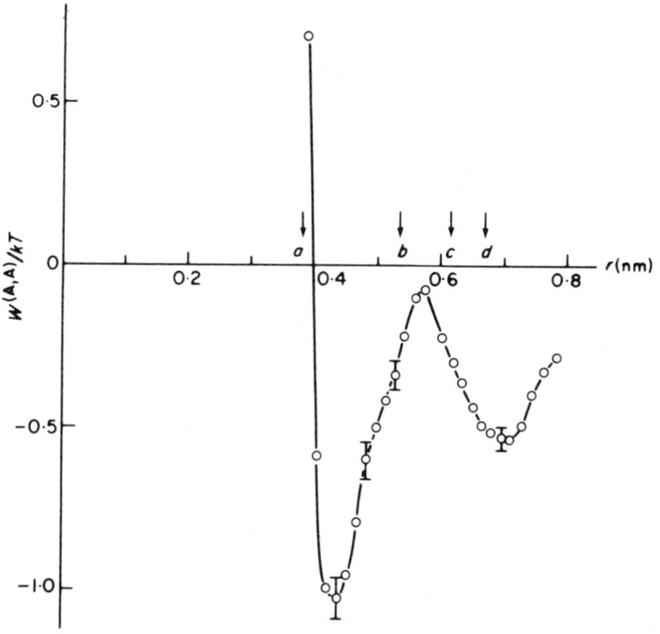

Fig. 8.12. The potential of mean force $W^{(A,A)}(r)$ for two "xenon" molecules dissolved in water. The separations *a, b, c* and *d* correspond to 0·387 nm, 0·527 nm, 0·608 nm and 0·660 nm, respectively. (From Pangali *et al.*, 1979.)

part. The solution again involves Lennard-Jones molecules of type (A) and of size 0·412 nm and intermolecular energy parameter $\varepsilon^{(A,A)}/k = 170{\cdot}1$ K, these values corresponding roughly to the accepted values for xenon. In analysing their data Pangali *et al.* divided the space available as in Fig. 8.6. When the inter-solute distance (r_{AA}) is large, the populations of the interior and exterior regions become equal. However, when r_{AA} is small (Fig. 8.11(*a*)) the A–O distribution function near the first peak is much smaller for the interior region than for the exterior one. The differences are much less pronounced for larger r_{AA}, as would be expected. Figure 8.11(*b*) refers to a distance r_{AA} which is close to the maximum in the potential of mean

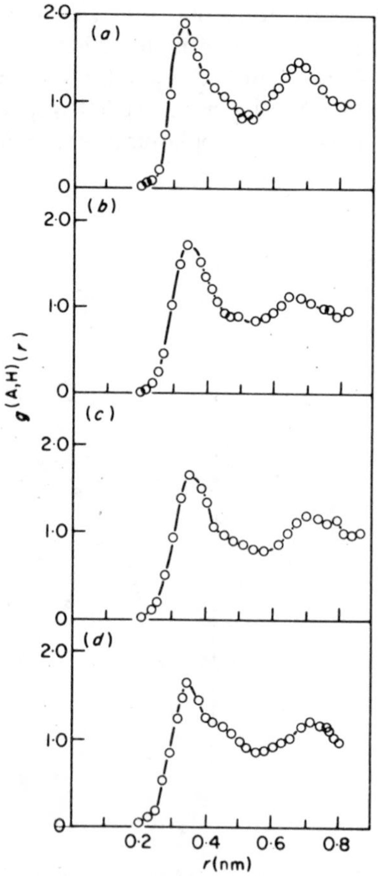

Fig. 8.13. "Xenon"–hydrogen radial distribution function $g^{(A,H)}$ for the four inter-"xenon" spacings defined in Fig. 8.12. (From Pangali *et al.*, 1979.)

force (Fig. 8.12), and the instability of this position for the two solute molecules is thought to be associated with the width of the first peak in Fig. 8.11(*b*). Figure 8.13 shows the corresponding pair correlation functions for the A–H distances of the water molecules in the exterior region for the same four A–A distances as in Fig. 8.11. The curves are very similar, as was found to be the case for the A–O distances in the exterior region. The idea of a cage-like structure for the nearest water molecules is supported by these curves, since the first maximum for A–H occurs at ~0·330 nm,

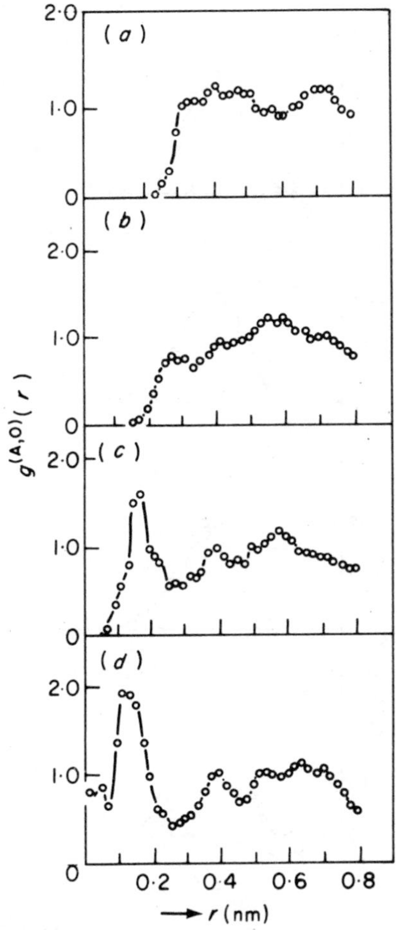

Fig. 8.14. Centre of mass of "xenon"–oxygen radial distribution function $g^{(A,O)}$ for the four inter-"xenon" spacings defined in Fig. 8.12. (From Pangali *et al.*, 1979.)

which is very close to that found for the A–O distances (~0·360 nm), suggesting that the OH bonds lie parallel to the surface of a sphere centred on the solute particle. This is in line with the suggested idealized structure of Geiger *et al.* Figure 8.14 shows correlation functions for the distance of the oxygen atoms from the centre of mass of the A–A pair for the same four A–A distances. Only (*d*) shows a non-zero value at $r_{AA} = 0$ (the centre of mass of the A–A pair) and that has an inter-solute distance equal to that of the second minimum in the $W(r_{AA})$ curve. Analysis of the energy

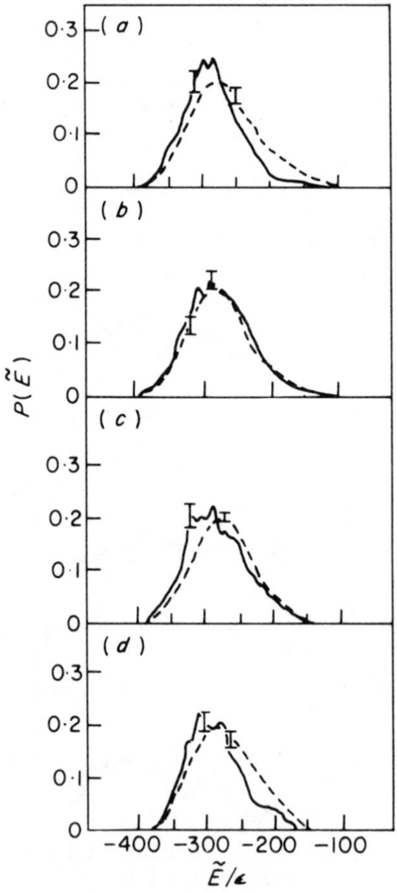

Fig. 8.15. Probability distribution function for water–water binding energies ($\tilde{E}$) for the "shell" (full curve) and bulk (dashed curve) regions with the inter-"xenon" spacings at the four values defined in Fig. 8.12. ε is the Lennard-Jones energy parameter for water. (From Pangali *et al.*, 1979.)

contributions shows that the binding energy ($\tilde{E}$) (the sum of the interaction energies of a molecule with all other molecules) of the water molecules which abutt onto the solute molecules is in fact 2–4 kJ mol^{-1} higher than for the bulk water (Fig. 8.15). This is so in spite of the weakness of the A–H_2O interaction, and is attributed to the relatively stronger H-bonds which are found in that region. A similar statement has been made by Fetterly (1964) with regard to the H-bonds in the adducts of urea with hydrocarbons and related long-chain compounds, for which the term "supported hydrogen bonds" was coined.

In this kind of work doubts always arise as to the effect which boundary conditions and the cut-off in the inter-molecular potential may have had on the final results. Pangali *et al.* (1979) used a spherical cut-off at $2.47\sigma^{(A,\mathcal{W})}$, which is similar to those customarily used in studies of non-polar systems. Because of the long-range nature of the electrostatic forces involved in aqueous systems a cut-off as small as this may introduce serious errors. Pangali *et al.* recognize that this may be so for the orientational order. However, they argue that the estimates of the interaction, the potential of mean force, should not be so seriously affected because it depends primarily on $\rho^{(A,O)}$ and $\rho^{(A,A)}$, and is less sensitive to the orientational order of the water molecules. This is in accord with the observations of Adams *et al.* (1979) for bulk systems. In all the simulation studies of gases in water there is one obvious way in which the experimental conditions are not being reproduced: the concentration of the solution is always very much larger than can be achieved experimentally, the latter being limited by the insolubility of the gases. Thus, for argon in water, Alagona and Tani (1980) had 124 H_2O to 1 Ar as against ~40 000 H_2O per argon atom in the experimental saturated solution.

As an indication of the amount of computing required to obtain data of this kind, Pangali *et al.* used more than 100 hours of CPU time on an IBM 360/91.

(d) Hydrophobic interaction

The origin of this interaction is qualitatively fairly straightforward. When the apolar solute is dissolved in water it causes a re-structuring of the solvent which results in the latter becoming more ordered than it was originally, as has been described in the previous section. Thus there is an entropy decrease which in the case of CH_4 is ~ -72 J K^{-1} (mol CH_4)$^{-1}$ (Pangali *et al.* (1979) quote -30 J K^{-1} mol^{-1} for argon, which seems rather low), with an internal energy contribution of ~10·5 kJ mol^{-1}. When two solute molecules approach each other the amount of solvent ordering which occurs is less than that for two isolated solute molecules. Thus, bringing the two solute particles together from infinity results in a positive entropy change. For small solute molecules the measured ΔH_{soln} is generally negative, but very small in magnitude compared with $T\Delta S_{soln}$, so that the entropy term is the dominant factor in favouring the self-attraction of the solute. For large solute molecules, ΔH_{soln} may even be positive, so that it would then support the entropy term in favouring clustering (Franks, 1973). The sign of the enthalpy change on bringing the solute molecules together (ΔH_{HI}) determines whether the clustering tendency will increase or decrease with temperature. Ben-naim (1980) insists that $\Delta H_{HI} > 0$ so that clustering should increase with rise in temperature.

Ben-naim (1980) defines the hydrophobic interaction in terms of the change in Gibbs free energy ($\Delta G(r_{AA})$) when the two solute molecules are brought from an infinite distance apart, but both in the solvent, up to the distance r_{AA}. Part of this free energy change arises from the change in the energy of the direct interaction between the solute molecules ($u^{(A,A)}(r_{AA})$) and does not involve the solvent in any way. The free energy change associated with hydrophobic interaction is obtained by deducting from the total amount that part due to the direct interaction:

$$\delta G_{HI}(r_{AA}) = \Delta G(r_{AA}) - u^{(A,A)}(r_{AA}). \tag{8.7}$$

In many cases it is the value of this quantity at the contact distance i.e. $\delta G_{HI}(\sigma_A)$ which is of interest, but the function has meaning and use at other values of r_{AA} also. By means of a thermodynamic cycle $\Delta G(\sigma_A)$ and $\delta G_{HI}(\sigma_A)$ can be related to the difference in the standard chemical potential for solution of the "dimer" (subscript D, $r_{AA} = \sigma_A$) and two "monomers" (subscript M, $r_{AA} = \infty$):

$$\Delta G_{HI}(\sigma_A) = \Delta\mu_D^\circ - 2\Delta\mu_M^\circ + u^{(A,A)}(\sigma_A) = \delta G_{HI}(\sigma_A) + u^{(A,A)}(\sigma_A). \tag{8.8}$$

The probability of the pair of solute molecules being found at distance apart r_{AA}, and the radial distribution function for a system of such particles, may then be represented in terms of $\delta G_{HI}(r_{AA})$:

$$\begin{aligned} g^{(A,A)}(r_{AA}) = \exp(-\beta\Delta G(r_{AA})) &= \exp[-\beta(\delta G_{HI}(r_{AA}) \\ &\quad + u^{(A,A)}(r_{AA}))] \\ &= y^{(A,A)}(r_{AA})\exp(-\beta u^{(A,A)}) \end{aligned} \tag{8.9}$$

where the cavity function

$$y^{(A,A)}(r_{AA}) = \exp(\beta u^{(A,A)})g^{(A,A)}(r_{AA}).$$

Data, both experimental and theoretical, on these distribution functions over a range of r_{AA} are difficult to obtain, and so $y^{(A,A)}(r_{AA})$, which is a smoother function than $g^{(A,A)}(r_{AA})$ at $r_{AA} < \sigma$, is useful, particularly where interpolation may be necessary.

The osmotic second virial coefficient defined by the equation:

$$\pi/kT = \rho_M + B_2^*\rho_M^2 + B_3^*\rho_M^3 + \ldots \tag{8.10}$$

can be related to $g^{(A,A)}(r_{AA})$ by the equation:

$$B_2^* = -\tfrac{1}{2}\lim_{\rho\to 0}\int_0^\infty [g^{(A,A)}(r_{AA}) - 1]\, 4\pi r_{AA}^2\, dr_{AA}. \tag{8.11}$$

B_2^* is a quantity which can be used for assessing the implications of a simulation. By statistical mechanical arguments it can be shown that:

$$\delta G_{HI}(r_{AA}) = -kT\ln[\langle\exp(-\beta B_D(r_{AA}))\rangle_0/\langle\exp(-\beta B_M)\rangle_0^2] \tag{8.12}$$

where B_M and B_D are binding energies to water of the monomer and the dimer, respectively. $\langle\ \rangle_0$ indicates that the average is to be taken over all configurations of the water molecules. If the solute molecules are considered to be hard spheres, then for $r_{AA} = 0$ we are concerned with the superimposition of two identical cavities. The "double cavity" has the same interaction with the water molecules as the "single cavity", and therefore

$$\begin{aligned}\delta G_{HI}(r_{AA} = 0) &= -kT\ln[\langle\exp(-\beta B_M)\rangle_0/\langle\exp(-\beta B_M)\rangle_0^2]\\ &= -\Delta\mu_{HS}^{\circ}.\end{aligned} \tag{8.13}$$

With regard to the obtaining of experimental data, Ben-naim has made much use of the comparison of the properties of one molecule of C_2H_6 and two molecules of CH_4. The argument used is that, since we are only concerned with the structural changes in the surrounding water molecules, C_2H_6 can be taken to be two CH_4 molecules at a distance apart equal to the C–C bond length ($r_{AA} = 0{\cdot}153$ nm). However, of much greater interest is δG_{HI} for $r_{AA} \approx \sigma \approx 0{\cdot}38$ nm, and so the assumption is here being made that the trends in $\delta G_{HI}(r_{AA} = 0{\cdot}153\text{ nm})$ are similar to those for $\delta G_{HI}(r_{AA} = 0{\cdot}38\text{ nm})$.

Dashevsky and Sarkisov (1974) were the first to attempt a computer simulation of this problem. Unfortunately, to obtain the hydrophobic interaction free energy of the potential of mean force they used a very unsatisfactory computer method (Eqn (4.22)). Their results, which showed a single minimum in the potential of mean force, should not, therefore, be given too much weight. The MD simulation carried out by Geiger *et al.*, which has also been discussed in connection with hydrophobic hydration, showed clearly that the solvent-separated solute pair was a fairly stable configuration. Figure 8.16 shows some of the H_2O structures found between

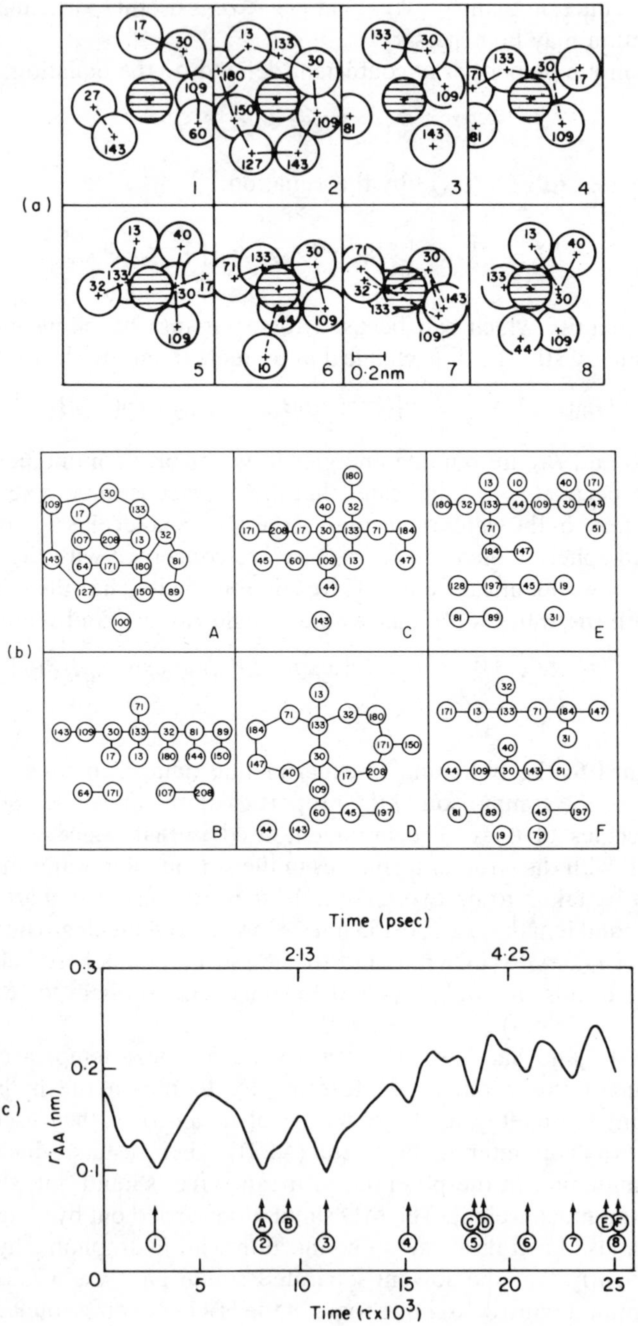
(a)
0·2nm
(b)
(c)
Time (psec)
2·13
4·25
r_{AA} (nm)
0·3
0·2
0·1
0
5
10
15
20
25
Time (τx10³)

the apolar molecules. It is clear that the simple views represented by the existence of "icebergs" or a rigid clathrate structure are not in accord with these simulations. Clark *et al.* (1977) had previously suggested from an analysis of experimental data that such a structure might be more stable than one with apolar pairs in contact.

A much more detailed and far-reaching theoretical treatment of hydrophobic interaction has been carried out by Pratt and Chandler (1977, 1980), who found that the contact and the solvent-separated "dimers" were of comparable stability before the computer simulations arrived at the same conclusion. They took full account of all the forces between pairs of water molecules by using X-ray scattering data for the distribution functions of water. For the solute–water and solute–solute interactions, on the other hand, they took the repulsive part (in the sense of Weeks, Chandler and Andersen) of each as the main determining factors with the corresponding attractive parts as perturbations:

$$
\begin{aligned}
g^{(\mathrm{A},\mathcal{W})}(r; u^{(\mathrm{A},\mathrm{A})}, u^{(\mathrm{A},\mathcal{W})}, u^{(\mathcal{W},\mathcal{W})}) &\approx g^{(\mathrm{A},\mathcal{W})}(r; u^{(\mathrm{A},\mathrm{A})^\circ}, u^{(\mathrm{A},\mathcal{W})^\circ}, u^{(\mathcal{W},\mathcal{W})}) \\
g^{(\mathrm{A},\mathrm{A})}(r; u^{(\mathrm{A},\mathrm{A})}, u^{(\mathrm{A},\mathcal{W})}, u^{(\mathcal{W},\mathcal{W})}) &\approx g^{(\mathrm{A},\mathrm{A})}(r; u^{(\mathrm{A},\mathrm{A})^\circ}, u^{(\mathrm{A},\mathcal{W})^\circ}, u^{(\mathcal{W},\mathcal{W})}) \qquad (8.14)
\end{aligned}
$$

where superscript ° denotes that only the repulsive part of the potential was included. The justification for this procedure is that the attractive forces are extremely (and unusually) important in deciding the structure of water, and cannot, therefore, be relegated to a perturbative role. The solute–solute (A, A) and solute–water (A, $\mathcal{W}$) interactions, on the other hand, do not include attractive electrostatic forces which are so important in water–water interactions, and this is, therefore, the normal situation in which the structure is largely determined by the repulsive parts of the interactions. Pratt and Chandler obtained very good agreement with experimental results for the solubility of CH_4(Me) in H_2O($\mathcal{W}$) over the range 20–50°C with the parameters $\sigma^{(\mathrm{Me},\mathcal{W})} = \frac{1}{2}\sigma_{\mathrm{Me}} + 0{\cdot}135 = 0{\cdot}32$ nm and $\varepsilon^{(\mathrm{Me},\mathcal{W})} = (\varepsilon^{(\mathrm{Me},\mathrm{Me})}\varepsilon^{(\mathcal{W},\mathcal{W})})^{1/2}$, with $\varepsilon^{(\mathrm{Me},\mathrm{Me})}/k = 148$ K and $\varepsilon^{(\mathcal{W},\mathcal{W})}/k = 192$ K giving $\varepsilon^{(\mathrm{Me},\mathcal{W})}/k = 168$ K. The change in chemical potential on transferring

Fig. 8.16. (a) Configuration of the water molecules in the layer between the two apolar atoms as seen by looking along the apolar axis at inter-"neon" separations indicated by the labels in (c). The "neon" atoms are shaded. Water pairs with interaction energies $< -40\ \varepsilon$ (full lines); pairs with interaction energies between $-20\ \varepsilon$ and $-40\ \varepsilon$ (dashed lines); $\varepsilon = 316{\cdot}9$ J mol^{-1}. (*b*) H-bond connectivities among water molecules with distance of the centre of mass less than 0·42 nm to at least one "neon"; lines shown are between pairs with interaction energies $< -40\ \varepsilon$. (*c*) Inter-"neon" distance as a function of simulated time during the MD run. Arrows indicate the points at which the snapshots shown in (*a*) and (*b*) were taken with *d* in units of 0·282 nm. (From Geiger *et al.*, 1979a.)

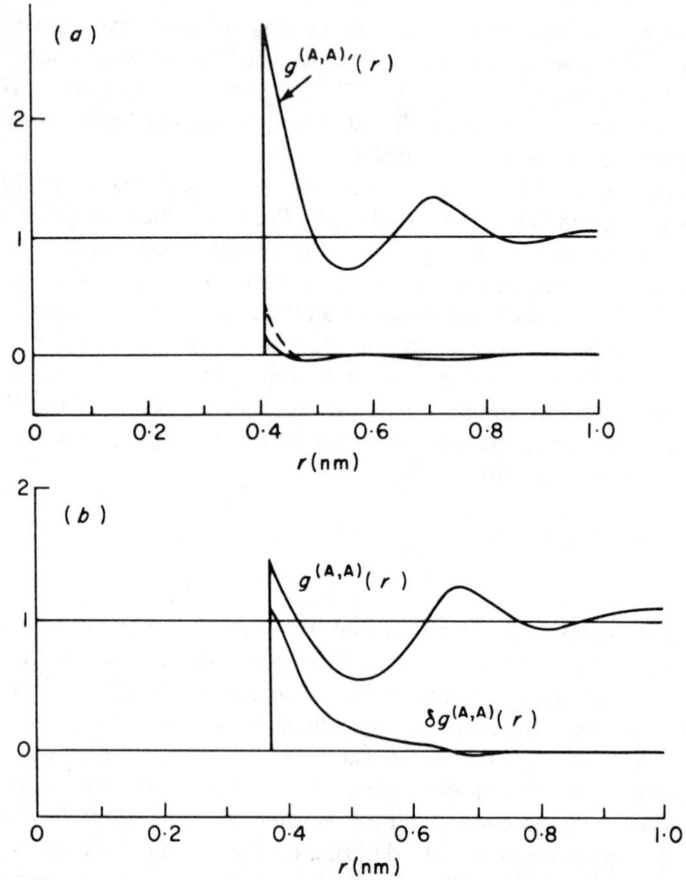

Fig. 8.17. Contributions to the AA radial distribution function. (*a*) "Xenon" in water. $g^{(A,A)\prime}$ is for the system with the "xenon"–"xenon" and "xenon"–water attractive interactions turned off. The two lower curves are two estimates of the correction to be expected when these interactions are turned on. (*b*) Methane in water. $g^{(A,A)}$ is the usual radial distribution function, and $\delta g^{(A,A)}$ is the change in that quantity when the attractive parts of the methane–methane and methane–water interactions are turned off. (From Pratt and Chandler, 1980.)

the substance from the gas phase to aqueous solution ($\Delta\mu_A$) is sensitive to the attractive perturbation, because for this quantity the contributions from the repulsive and attractive parts are nearly equal but of opposite sign. The effect on the potential of mean force between solute molecules seems to be very dependent on the size of the spherical solute molecule: for methane ($\sigma = 0{\cdot}37$ nm) the effect is large, whereas for "xenon" ($\sigma = 0{\cdot}412$ nm) it is slight (Fig. 8.17). This sensitivity to the size of the

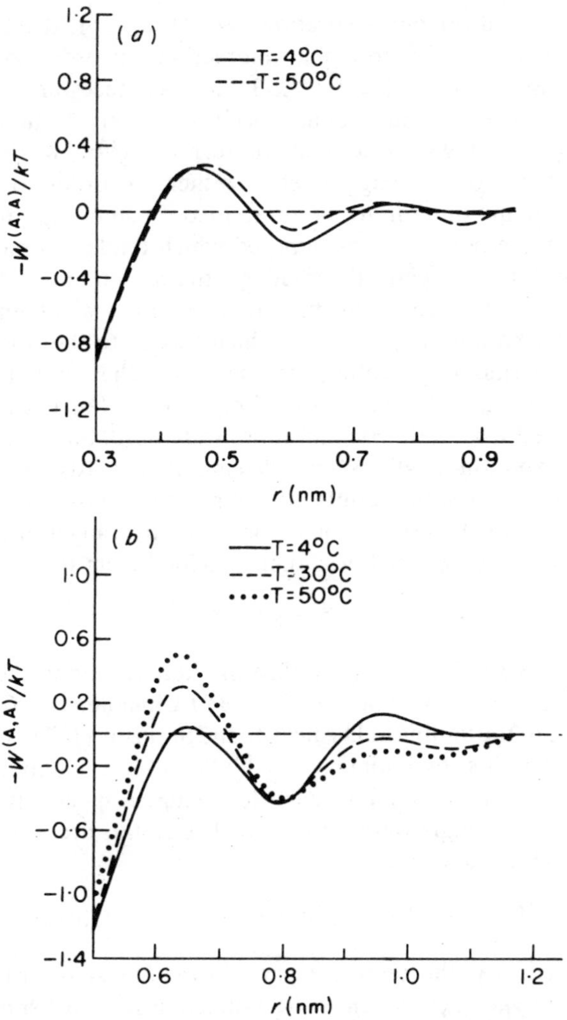

Fig. 8.18. Potential of mean force between hard spheres of diameter (*a*) 0·3 nm (*b*) 0·5 nm at infinite dilution in water at 1 atm and several temperatures. (From Pratt and Chandler, 1977.)

solute molecules may well be related to the similar sensitivity to size which is found for the clathrate hydrates (Barrer, 1964). A somewhat similar effect occurs for solutions of hard spheres in water at various temperatures. In this case, it is the solutions of the larger solute particles which show the greater sensitivity to temperature (Fig. 8.18).

The work of Pangali *et al.* (1979a), which has been discussed previously with respect to hydrophobic hydration (§2(c)), also yielded a good deal of information on the hydrophobic interaction. In order to obtain the potential of mean force ($W^{(A,A)}(r_{AA})$) it was necessary to compute $g^{(A,A)}(r_{AA})$. A somewhat unusual approach was taken: a number of short ranges of r_{AA} ("windows") were taken in turn and $g^{(A,A)}$ was computed in each range, these then being joined together. In order to restrict the sampling within any one run to the prescribed range of r_{AA}, an additional artificial harmonic potential was imposed which tended to pull the inter-particle distance back towards the middle of the chosen range. A correction to eliminate the influence of this harmonic potential was then applied so as to give information on $g^{(A,A)}(r_{AA})$ which relates to the original system (without the harmonic potential), but in which the sampling has been concentrated into the chosen region of r_{AA} values. For each of the four ranges examined 1.08×10^6 configurations were used. Since the uncertainties in the distribution functions increased greatly towards the edge of each "window", the fitting of the results for one region to those for the next was done at points which were reliable for both the ranges involved. The potential of mean force which was then obtained from the relation

$$W^{(A,A)}(r_{AA}) = -kT \ln g^{(A,A)}(r_{AA}) \tag{8.15}$$

is shown in Fig. 8.19. $W^{(A,A)}(r_{AA})$ displays marked oscillations, in agreement with the earlier results obtained by Pratt and Chandler and Geiger *et al.*, but in contrast to the finding of Dashevsky and Sarkisov (1974) and Marcelja *et al.* (1977). The first two minima in $W^{(A,A)}(r_{AA})$ are readily identified as corresponding to the two solute particles being, respectively, in contact and separated by a single water molecule. The results are a good fit to the predictions of Pratt and Chandler:

$$W^{(A,A)}(r_{AA}) = -kT \ln y_{HS}^{(\circ)}(r_{AA}) + u^{(A,A)\circ}(r_{AA}) \tag{8.16}$$

where $u^{(A,A)\circ}(r_{AA})$ is the repulsive part (in the WCA sense) of the AA interaction and $y_{HS}^{(\circ)}(r_{AA})$, which is the *y*-function for hard spheres having a radius related to the WCA repulsion by a construction similar to that of Barker and Henderson (1967), was evaluated by an approximate method. In computing $W^{(A,A)}(r_{AA})$ for Fig. 8.19, a disposable constant was chosen so as to ensure agreement with the equation of Pratt and Chandler at the first minimum.

The stability of the arrangement with the two apolar molecules separated by a molecule of water is contrary to the "accepted wisdom" of colloid chemistry, and is, therefore, an important contribution to the shaping of thought on these matters. Thus, the Derjaguin–Landau–Verwey–Overbeek theory of colloids, which treats the water as a continuum, would not predict

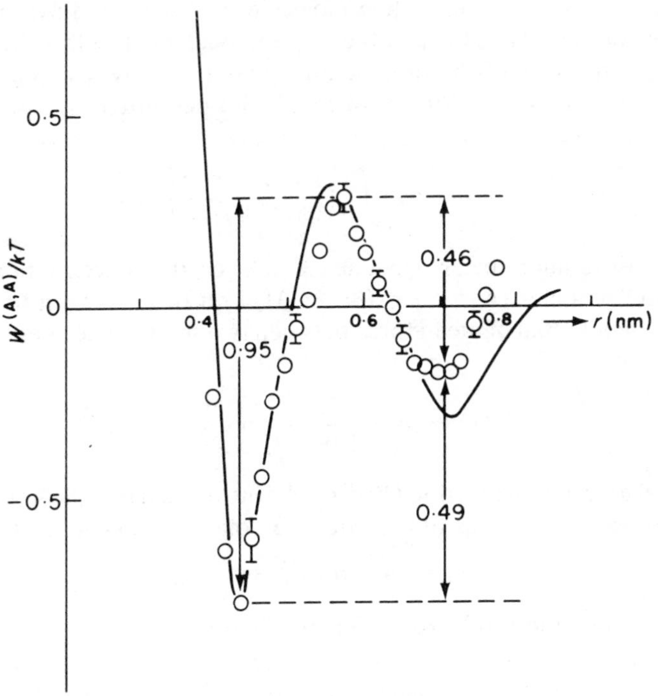

Fig. 8.19. Potential of mean force between two "xenon" atoms in water. Predictions of Pratt and Chandler (solid curve), simulation results of Pangali *et al.* (circles). (From Pangali, Rao and Berne, J. Chem. Phys. (1979) **71**, 2975.)

the relative stability of the solvent-separated state. It is true that Kitchener (1963), Derjaguin and Churaev (1974) and Lyklema (1977) had already cast some doubt on the DLVO theory with respect to this particular process. A one-dimensional model proposed by Ninham and co-workers (Chan *et al.*, 1978) and chosen to mimic the hydrophobic effect also has a potential of mean force with a strong minimum at the point corresponding to the solvent-separated state. This model has $u^{(A,A)}(x) = \infty, -\varepsilon_A$, or 0, according as $x < R_1, R_1 < x < 2R_1$, or $x > 2R_1$, and $u^{(A,\mathscr{W})}(x) = \infty$ or 0 according as $x < \frac{1}{2}(R_1 + R_2)$ or $x > \frac{1}{2}(R_1 + R_2)$, with $R_2 < R_1$ so that next-nearest-neighbour solvent molecules can interact. Almost the only similarity of this model with real aqueous solutions is that "clathrates" can be formed.

Mitchell *et al.* (1978) have also employed the Ornstein–Zernike equation to demonstrate the relationship between the Lifshitz approach traditionally favoured by colloid scientists and the molecular theories which are finding increasing acceptance in this field. They first consider the interaction per unit area (E) between two solute molecules (A) separated by a region of solvent molecules (B). Lifshitz theory can be adapted to this situation by applying it to two half-spaces of solution with permittivity ε' separated by a region of pure solvent of length l and with permittivity ε, for which E can be expressed as

$$E = -\frac{\hbar}{16\pi^2 l^2}\int_0^\infty \left(\frac{\varepsilon'(\omega) - \varepsilon(\omega)}{\varepsilon'(\omega) + \varepsilon(\omega)}\right)^2 \mathrm{d}\omega. \tag{8.17}$$

After substituting a linear dependence of ε' on the concentration (c_A) of the solute particles ($\varepsilon' = \varepsilon + c_A(\partial\varepsilon'/\partial c_A)$), Eqn (8.17) is seen to be analogous to a potential of mean force between two solute molecules separated by a distance r:

$$W^{(A,A)}(r) = -\frac{3\hbar}{16\pi^3 r^6}\int_0^\infty \left(\frac{\partial\varepsilon'}{\varepsilon\partial c_A}\right)^2 \mathrm{d}\omega. \tag{8.18}$$

The total correlation function $h^{(A,B)}$ can be introduced into the last equation by expressing ε' in terms of the effective polarizabilities α_A and α_B,

$$\varepsilon'/\varepsilon_0 = 1 + 4\pi c_A \alpha^{(A)} + 4\pi c_B \alpha^{(B)} \tag{8.19}$$

and employing the result from standard thermodynamics

$$\partial c_B/\partial c_A = c_B \int h^{(A,B)}\, \mathrm{d}\mathbf{r} \tag{8.20}$$

giving

$$W^{(A,A)}(r) = -\frac{3\hbar}{\pi r^6}\int_0^\infty \frac{1}{\varepsilon^2}\left(\alpha^{(A)} + c_B\alpha^{(B)}\int h^{(A,B)}\, \mathrm{d}\mathbf{r}\right)^2 \mathrm{d}\omega. \tag{8.21}$$

This illustrates that Lifshitz theory does indeed include structural effects through $h^{(A,B)}$ but does not of course give any information about how this correlation is to be determined.

A second approach to the same problem can be made through the OZ equation written for the special case in which the molecules are hard spheres of diameters σ_A, σ_B with attractive pairwise interactions,

$$u^{(A,B)}(r) = -C_6^{(A,B)}/r^6 \qquad r > \tfrac{1}{2}(\sigma_A + \sigma_B) \tag{8.22}$$

where $C_6^{(A,B)}$, from Eqn (3.40) can be written in terms of polarizabilities as

$$C_6^{(A,B)} = \frac{3}{\pi}\int_0^\infty \alpha^{(A)}(i\omega)\alpha^{(B)}(i\omega)\, \mathrm{d}\omega. \tag{8.23*}$$

Using the approximation $-\beta W^{(A,A)} \sim h^{(A,A)}$ and the mean spherical (MS) approximation $-\beta u^{(A,B)} = c^{(A,B)}$ in the binary form of the OZ equation Eqn (3.142), this leads to the result:

$$W^{(A,A)} \cong -\frac{3}{\pi r^6}\int_0^\infty \left[\alpha^{(A)} - \frac{4\pi}{3} c_B \alpha^{(B)} (\sigma^{(A,B)})^3\right]^2 d\omega. \qquad (8.24^*)$$

In this equation the ε^{-2} term in Eqn (8.21) does not appear since, unlike the Lifshitz theory, only the two-body interactions have been taken into account.

Thirdly, a wall–solvent interaction was considered, again in terms of the OZ equation, but here the limiting "infinite particle" approach already discussed in Chapter 7 was employed to determine the interaction of two half-spaces of A-molecules interacting across a medium of B-molecules. The potential of mean force $W^{(A,A)}$ is related to the free energy of interaction/unit area between the half-spaces when their separation is x by

$$A(x) = -\frac{2}{\pi \sigma_A}\frac{\partial}{\partial x} W^{(A,A)}(x) \qquad (8.25)$$

where σ is finally to be taken at the infinite limit. We omit the details of the analysis, which is lengthy, but the form of the OZ equation is clearly evident in the result:

$$\beta A(x) = \beta E^{(A,A)} - \rho_B \int h^{(A,B)}(\sigma_A + x) c^{(A,B)}(\sigma_A + x - y - \sigma_B)\, dy \qquad (8.26)$$

where $E^{(A,A)}$ is a derivative of the potential $u^{(A,A)}$ analogous to the expression for $A(x)$ (Eqn (8.25)). This last equation may be further transformed into an equation containing as its leading term the classical Hamaker expression which depends upon the inverse square of the separation between the walls like the Lifshitz expression (cf. Eqn (8.17) above). The second term in the expansion, however, contains an inverse cube dependence on this separation and is

$$\frac{4(A^{(A,B)} - A^{(B,B)})}{x^3}\int_0^\infty h^{(A,B)}(y)\, dy \qquad (8.27)$$

where $A^{(\alpha,\beta)}$ are Hamaker constants (proportional to the interaction coefficient $C_6^{(\alpha,\beta)}$). This term represents a significant modification of the Lifshitz–Hamaker theories, which are based on a continuum model, and can be interpreted as being due to the effect of solvent structuring or adsorption of the solvent induced by the solid walls. Calculations based on this result indicate that this structuring is expected to be very important for separations up to $\sim 20\sigma_B$ and that continuum theories are unreliable at shorter distances. The conclusion is indeed not too surprising, following the detailed discussion of density profiles which has already been given

(Chapter 6). Nevertheless, it is important to stress the severe limitation of traditional theory, and to have a quantitative estimate of where this can be expected to fail. Further development of these ideas has naturally centred around the determination of the density profile and can therefore be considered a part of the discussion devoted to that topic in Chapter 7.

(e) Hydrophobic effects for solutes with non-spherical molecules

All the theoretical work described so far on the hydrophobic effect has assumed the solute molecules to be spherical. However, many of the systems of greatest interest and importance to which the theories would ultimately be applied have elongated molecules. Pratt and Chandler (1980) have extended their earlier approach for spherical molecules by means of the RISM method introduced by Chandler and co-workers (Lowden and Chandler, 1973, 1974; Ladanyi and Chandler, 1975; Chandler, 1978). In this approach an ethane molecule, for example, is treated as being two interaction sites (α, β) each with the same interaction parameters as methane and with a distance apart of 0·154 nm. More generally, if the sites are denoted by the Greek letters (α, β, γ, . . .), then a pair of equations similar to the OZ equation (Eqn 3.86) is obtained:

$$h^{(\alpha,\mathcal{W})}(r) = w^{(\alpha,\gamma)} * c^{(\gamma,\mathcal{W})}(r) + \rho_{\mathcal{W}} w^{(\alpha,\gamma)} * c^{(\gamma,\mathcal{W})} * h^{(\mathcal{W},\mathcal{W})}(r) \qquad (8.28)$$

and

$$h^{(\alpha,\gamma)}(r) = w^{(\alpha,\gamma)} * c^{(\gamma,\nu)} * w^{(\nu,\gamma)}(r) + \rho_{\mathcal{W}} w^{(\alpha,\nu)} * c^{(\nu,\mathcal{W})} * h^{(\mathcal{W},\gamma)}(r) \qquad (8.29)$$

where $*$ indicates a convolution of the adjacent functions, repeated Greek superscripts indicate summation,

$$w^{(\alpha,\gamma)}(r) = \delta_{\alpha\gamma}\delta(r) + (1 - \delta_{\alpha\gamma})\langle\delta(r - r_1^{\alpha} + r_1^{\gamma})\rangle \qquad (8.30)$$

r_1^{α} is the position of site α in the solute molecule 1 and $\langle\ \rangle$ denotes an ensemble average. If the usual closure relations

$$c^{(\alpha,\mathrm{M})}(r) = 0 \qquad r > \sigma \qquad (8.31)$$

and

$$g^{(\alpha,\mathrm{M})}(r) = 0 \qquad r < \sigma \qquad (8.32)$$

where $\mathrm{M} = \gamma$ or $\mathcal{W}$ are invoked it is possible to solve these equations by a Fourier transform technique (Weiner-Hopf, Chapter 7, §2(b)). The radial distribution function determined is shown in Fig. 8.20(a) for the ethane + water system and also for the cases of two methane molecules in contact ($l = 0{\cdot}37$ nm) and at a distance which would just allow a water

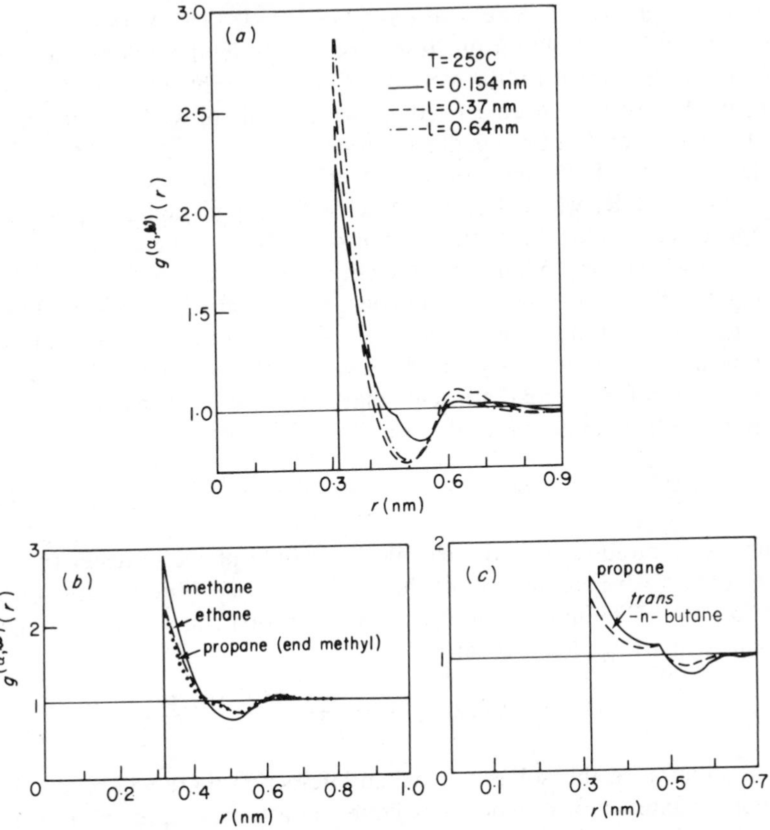

Fig. 8.20. Radial distribution functions for water around alkane residues for dilute solutions of the alkanes in water at 25°C and 1 atm. (*a*) for methane residues in a methane dimer with bond length *l*; (*b*) for primary methyl residues in CH_4, C_2H_6 and C_3H_8; (*c*) for methylene groups in C_3H_8 and n-C_4H_{10}. (From Pratt and Chandler, 1980.)

molecule to be present between the solute molecules ($l = 0{\cdot}64$ nm). It will be seen that the two longer distances lead to rather similar distribution functions, but that there is a big difference on going to the ethane interatomic distance (0·154 nm): the first peak is reduced in size and, even more strikingly, the first well becomes more shallow. This suggests that the first "shell" of water molecules is breaking up. These observations cast some doubt on Ben-naim's use of experimental data for ethane for the discussion of the hydrophobic interaction between methane molecules at much larger distances. On going to longer hydrocarbon chains the distribution function must be broken into two parts, one relating to end(methyl) groups and the other to inner(methylene) groups. Figures 8.20(b, c) show that the methyl–water distribution function is only slightly affected by the length of the chain, whereas for the methylene groups the change is much greater. No simulations have been carried out which could provide a check on this extension of the Pratt–Chandler theory.

De Gennes (1976) has considered effects which may arise in mixed solvents (A + B) when they are close to the liquid–liquid critical point. Large concentration fluctuations then occur with a characteristic length $\xi(c_A, T)$ which goes to infinity at the critical point as $|T - T_c|^{-\nu}$ ($\nu \approx \frac{2}{3}$) and $|\Delta c_A|^{-\nu/\beta}$ ($\beta \approx \frac{1}{3}$). In this case the concentration changes induced by one solute (polymer) molecule would affect another, since the mean distance between solute molecules would be less than the correlation length ξ. In the absence of the second molecule the concentration change at a distance r from the first solute molecule would be of the form

$$\delta c_A(r) = -\varepsilon_1 \frac{\bar{\chi}}{4\pi\xi^2} \frac{\exp(-r/\xi)}{r} \tag{8.33}$$

where $\varepsilon_1 = d\mu_p/dc_A$, μ_p is the chemical potential of the polymer (p), and $\bar{\chi} = \int(\delta c_A/\varepsilon_1)\, d\mathbf{r}$ is a measure of the total response of the solvent concentration to ε_1. The interaction free energy for a pair of solute molecules a distance apart r_{pp} then becomes:

$$W(r_{pp}) = \varepsilon_1 \delta c_A(r_{pp}) = -\frac{\varepsilon_1^2 \pi^2}{4\pi\xi^2} \frac{\exp(-r_{pp}/\xi)}{r_{pp}}. \tag{8.34}$$

In addition to this solvent-mediated interaction there will be the usual repulsive interactions. When ξ is large the attractive interaction will be dominant at most distances and a polymer chain in the medium would tend to coil or precipitate out. Nether simulation nor experimental results are available for comparison with these predictions. In any case, the simulation of systems near to critical points, and having large correlation lengths, is notoriously difficult. If one adds to this the difficulties associated with simulating polymer chains, then it can be seen that the problem of providing a check on de Gennes' predictions by this type of approach is a formidable one.

(f) Hydrophobic interactions in biological systems

The importance of these interactions with regard to the unfolding of proteins has been emphasized by Kauzmann (1959) and Némethy (1967). Of particular interest to biochemists is the large effect which dissolved urea can have on the tendency of the protein to unfold (degrade). This effect seems to be well correlated with the structure-breaking effect of the urea.

The "weakened" water structure causes a reduction in the advantage to be gained by protein chains in lying side-by-side.

The repulsion of bilayers in lecithin–water dispersions, which was studied experimentally by Le Neveu *et al.* (1976), has been satisfactorily treated by Marcelja and Radic (1976) using a Landau model for the ordering of the water:

$$g = g_0 + a(\eta(x))^2 + b(\eta(x))^3 + \ldots + c(\partial\eta/\partial x)^2 \tag{8.35}$$

where g is the free energy density and η is the order parameter, which is a function of the distance (x) from the mid-plane. The boundary conditions (at the walls) are set as $\eta(d/2) = -\eta(-d/2) = \eta_0$, since it is supposed that at the boundaries ($x = d/2$ and $-d/2$) the molecules are ordered in opposite directions. From the solution of these equations the excess free energy per unit area and the pressure on the walls can be found:

$$p = -\varepsilon_0 \operatorname{cosech}^2(\xi_0 d/2) \tag{8.36}$$

where $\xi_0 = (c/a)^{1/2}$ and $\varepsilon_0 = a\eta_0^2$. For $d > \xi_0$, we have

$$p \sim -4\varepsilon_0 \exp(-d/\xi). \tag{8.37}$$

The experimental data correspond to $\xi_0 = 0{\cdot}193$ nm and $4\varepsilon_0 = 1{\cdot}0 \times 10^8$ $\mathrm{J\,m^{-2}}$ or $\varepsilon_0 = 40{\cdot}4\ \mathrm{kJ\,mol^{-1}}$. This does not seem unreasonable when one considers for the energy scale of the system that the energy of the H-bonds in ice is $28{\cdot}1\ \mathrm{kJ\,mol^{-1}}$ and ΔH_f of ice is $5{\cdot}9\ \mathrm{kJ\,mol^{-1}}$.

Marcelja (1976) has used a somewhat similar mean-field approach in a theory of lipid-mediated protein–protein interaction in bilayer membranes. The order parameter was here taken to be:

$$\eta = \left\langle \frac{1}{n} \sum_m \left(\tfrac{3}{2}\cos^2\nu_m - \tfrac{1}{2}\right) \right\rangle \tag{8.38}$$

where ν_m gives the orientation of the mth segment of the chain and the sum is taken over all segments of that chain. The results showed that a minimum in the free energy occurred at a certain protein separation.

There has been increasing interest in computer studies of the arrangement of water molecules around protein molecules. Because of the innate complexity of the protein molecules, until recently it was still only possible to study a single protein molecule together with a "sheath" of ~50 water molecules. Thus, the water environment of nucleic acid bases and base pairs has been simulated by Clementi and Corongiu (1980). They examined the single bases adenine (A), cytosine (C), guanine (G), thymine (T) and uracil (U); there was one molecule of the base fixed at the centre of the simulation box and 40 water molecules which were free to translate and

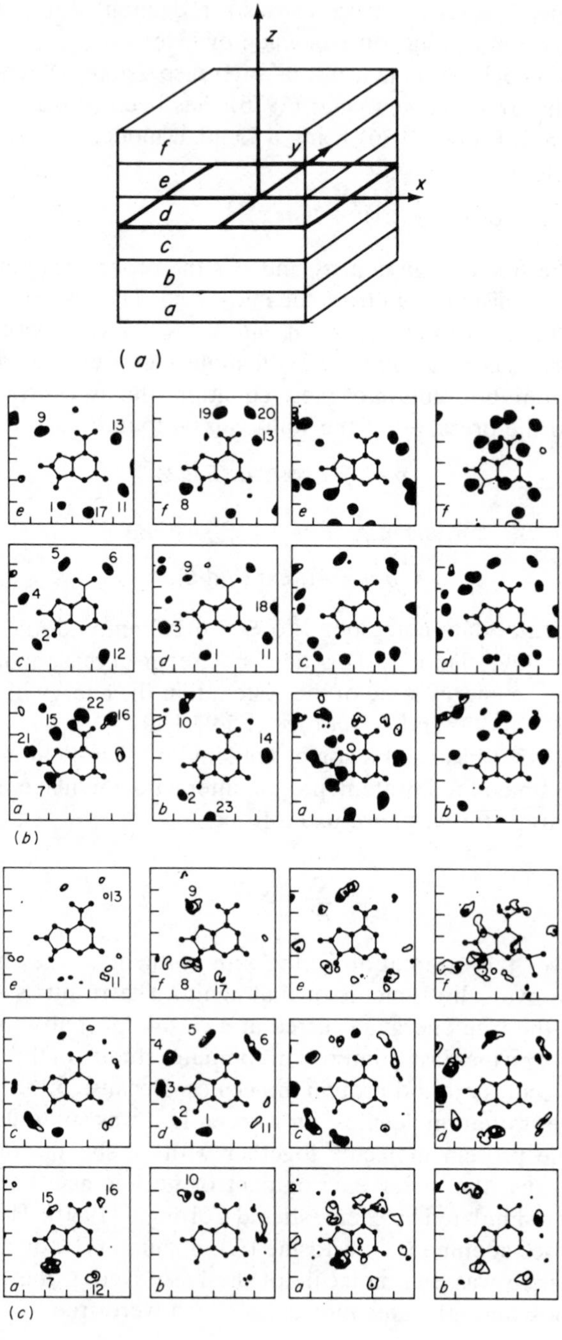
z
y
x
f
e
d
c
b
a
(a)
(b)
(c)

rotate. Typical structures obtained at 300 K and 100 K are shown in Fig. 8.21. As would be expected, the water structure is much sharper at the lower temperature. Examination of the base-pairs showed up interesting energy differences. For example, the mean energy of water–water interactions is $-23{\cdot}0 \pm 0{\cdot}25\ \mathrm{kJ\,mol^{-1}}$ for the G–C base pair, which is of smaller magnitude than the values obtained for either of the single constituents (−23·8 for G, −26·8 for C). Similar effects are found for A–T and A–U pairs. Also, the mean interaction energy between protein and water is found to be smaller in magnitude for the pair than for the two separate constituents, this difference being 100, 150 and 180 kJ $(\text{mol base-pair})^{-1}$ for G–C, A–T and A–U, respectively. This reduction is not unexpected, since formation of the pair reduces the amount of protein–water interface.

Clementi (1981) has extended this work to the study of DNA in its B-conformation; single crystals have not yet been obtained for this and as a consequence the structure is very imperfectly known. The simulations involved a length of DNA consisting of 12 base pairs, which is somewhat more than a full turn of the B–DNA helix, together with up to 400 water molecules and, in some cases, ~20 Na^+ ions. From this study it was possible to suggest the structure of the B–DNA in the aqueous environment and also the probable positions of the counter ions. Work of this nature is, of course, very expensive, but it is capable of yielding very valuable information which, if obtained by other means, would involve a similarly large (or larger) expense.

References

Adams, D. J. and Adams, E. M. (1981). *Mol. Phys.* **42**, 907.
Adams, D. J., Adams, E. M. and Hill, G. J. (1979). *Mol. Phys.* **38**, 387.
Alagona, G. and Tani, A. (1980). *J. Chem. Phys.* **72**, 580.
Amos, A. T. and Crispin, R. J. (1976). *In* "Theoretical Chemistry: Advances and Perspectives", Vol. 2, Ch. 1. Academic Press, New York and London.
Ash, S. G., Everett, D. H. and Radke, C. (1973). *J. Chem. Soc. Faraday Trans. II* **69**, 1256.
Barker, J. A. and Henderson, D. (1967). *J. Chem. Phys.* **47**, 4714.
Barnes, P. (1978). *In* "Progress in Liquid Physics" (C. A. Croxton, ed.), p. 405.
Barnes, P., Finney, J. L., Nicholas, J. D. and Quinn, J. E. (1979). *Nature* **282**, 459.

Fig. 21. Probability distribution maps for water around a molecule of adenine (*b*) at 100 K, (*c*) at 300 K. The adenine molecule lies in the boldly marked *xy*-plane of (*a*). The maps refer to the regions a–f shown in (*a*). For each of (*b*) and (*c*) the left-hand (right-hand) set of six maps is for oxygen (hydrogen). (From Clementi and Corongiu, 1980.)

Barrer, R. M. (1964). *In* "Non-stoichiometric Compounds" (L. Mandelcorn, ed.), p. 413. Academic Press, New York and London.
Ben-naim, A. (1980). "Hydrophobic Interactions", p. 225. Plenum Press, New York.
Ben-naim, A. and Stillinger, F. H. (1969). *In* "Structure and Transport Processes in Water and Aqueous Solutions" (R. A. Horne, ed.), Ch. 3. Wiley-Interscience, New York.
Berendsen, H. J. C. (1972). *CECAM Report of Molecular Dynamics and Monte Carlo Calculations in Water*, p. 63.
Bernal, J. D. and Fowler, F. D. (1933). *J. Chem. Phys.* **1**, 515.
Chan, D. Y. C., Mitchell, D. J., Ninham, B. W. and Pailthorpe, B. A. (1978). *J. Chem. Soc. Faraday Trans. II* **74**, 2050.
Chandler, D. (1978). *Ann. Rev. Phys. Chem.* **29**, 441.
Christou, N., Nicholson, D., Parsonage, N. G. and Whitehouse, J. S. (1981). *Chem. Soc. Faraday Symp. No. 16*, (in press).
Clark, A. H., Franks, F., Pedley, M. D. and Reid, D. S. (1977). *J. Chem. Soc. Faraday Trans. I* **73**, 290.
Clementi, E. (1981). *IBM J. Res. Develop.* **25**, 315.
Clementi, E. and Corongiu, G. (1980). *J. Chem. Phys.* **72**, 3979.
Clementi, E. and Popkie, H. (1972). *J. Chem. Phys.* **57**, 1077.
Dashevsky, V. G. and Sarkisov, G. N. (1974). *Mol. Phys.* **27**, 1271
De Gennes (1976). *J. de Physique* **37**, L59.
Derjaguin, B. V. and Churaev, N. V. (1974). *J. Colloid Interface Sci.* **49**, 249.
Drost-Hansen, W. (1977). *J. Colloid Interface Sci.* **58**, 251.
Fetterly, L. C. (1964). *In* "Non-stoichiometric Compounds" (L. Mandelcorn, ed.), p. 503. Academic Press, New York and London.
Frank, H. S. (1958). *Proc. Roy. Soc. London, Ser. A* **247**, 481.
Frank, H. S. and Evans, M. W. (1945). *J. Chem. Phys.* **13**, 507.
Franks, F. (1973). *In* "Water: A Comprehensive Treatise" (F. Franks, ed.), Vol. 2, Ch. 1. Plenum Press, New York and London.
Geiger, A., Rahman, A. and Stillinger, F. H. (1979). *J. Chem. Phys.* **70**, 263.
Glew, D. N., Mak, H. D. and Roth, N. S. (1968). *Chem. Comm.*, 264.
Horn, R. G. and Israelachvili, J. N. (1980). *Chem. Phys. Lett.* **71**, 192.
Impey, R. W., McDonald, I. R. and Klein, M. L. (1980). *J. Chem. Phys.* **74**, 647.
Israelachvili, J. N. (1981). *Phil. Mag. A* **43**, 753.
Israelachvili, J. N. and Adams, G. E. (1978). *J. Chem. Soc. Faraday Trans. I* **74**, 975.
Jorgensen, W. L. (1979). *J. Am. Chem. Soc.* **101** (a) 2011, (b) 2016.
Kauzmann, W. (1959). *Adv. Protein Chem.* **14**, 1.
Kitchener, J. A. (1963). *Endeavour* **22**, 118.
Ladanyi, B. M. and Chandler, D. (1975). *J. Chem. Phys.* **62**, 4308.
Ladd, A. J. C. (1977). *Mol. Phys.* **33**, 1039.
Lemberg, H. L. and Stillinger, F. H. (1975). *J. Chem. Phys.* **62**, 1677.
Lie, G. C. and Clementi, E. (1975). *J. Chem. Phys.* **62**, 2195.
Lie, G. C., Clementi, E. and Yoshimine, M. (1976). *J. Chem. Phys.* **64**, 2314.
Lowden, L. J. and Chandler, D. (1973). *J. Chem. Phys.* **59**, 6587.
Lowden, L. J. and Chandler, D. (1974). *J. Chem. Phys.* **61**, 5228.
Lyklema, J. (1977). *J. Colloid Interface Sci.* **58**, 242.
Marcelja, S. (1976). *Biochem. Biophys. Acta* **455**, 1.
Marcelja, S. and Radic, N. (1976). *Chem. Phys. Lett.* **42**, 129.

Marcelja, S., Mitchell, D. J., Ninham, B. W. and Sculley, M. J. (1977). *J. Chem. Soc. Faraday Trans. II* **73**, 630.
Matsuoka, O., Clementi, E. and Yoshimine, M. (1976). *J. Chem. Phys.* **64**, 1351.
Mezei, M., Swaminathan, S. and Beveridge, D. L. (1979). *J. Chem. Phys.* **71**, 3366.
Mitchell, D. J., Ninham, B. W. and Pailthorpe, B. W. (1978). *J. Chem. Soc. Faraday Trans. II* **74**, 1098.
Némethy, G. (1967). *Angew. Chem. Int. Ed.* **6**, 195.
Némethy, G. and Scheraga, H. A. (1962a). *J. Chem. Phys.* **36**, 3401.
Némethy, G. and Scheraga, H. A. (1962b). *J. Phys. Chem.* **66**, 1773.
Némethy, G. and Scheraga, H. A. (1963). *J. Phys. Chem.* **67**, 2888.
Le Neveu, D. M., Rand, R. P. and Parsegian, V. A. (1976). *Nature* **259**, 601.
Owicki, J. C. and Scheraga, H. A. (1977). *J. Am. Chem. Soc.* **99**, 7413.
Pangali, C., Rao, M. and Berne, B. J. (1979a). *J. Chem. Phys.* **71**, 2975.
Pangali, C., Rao, M. and Berne, B. J. (1979b). *J. Chem. Phys.* **71**, 2982.
Pangali, C., Rao, M. and Berne, B. J. (1980). *Mol. Phys.* **40**, 661.
Popkie, H., Kistenmacher, H. and Clementi, E. (1973). *J. Chem. Phys.* **59**, 1325.
Pratt, L. R. and Chandler, D. (1977). *J. Chem. Phys.* **67**, 3683.
Pratt, L. R. and Chandler, D. (1980). *J. Chem. Phys.* **73**, 3434.
Rahman, A., Stillinger, F. H. and Lemberg, H. L. (1975). *J. Chem. Phys.* **63**, 5223.
Read, A. D. and Kitchener, J. A. (1969). *J. Colloid Interface Sci.* **30**, 391.
Rowlinson, J. S. (1951). *Trans. Faraday Soc.* **47**, 120.
Snook, I. K. and van Megen, W. J. (1980). *J. Chem. Phys.* **72**, 2907.
Steele, W. A. (1973). *Surf. Sci.* **36**, 317.
Stillinger, F. H. and David, C. W. (1978). *J. Chem. Phys.* **69**, 1473.
Stillinger, F. H. and Rahman, A. (1974). *J. Chem. Phys.* **60**, 1545.
Stillinger, F. H. and Rahman, A. (1978). *J. Chem. Phys.* **68**, 666.
Swaminathan, S., Harrison, S. W. and Beveridge, D. L. (1978). *J. Am. Chem. Soc.* **100**, 5705.
Van Megen, W. J. and Snook, I. K. (1979). *J. Chem. Soc. Faraday Trans. II* **75**, 1095.
Van Megen, W. J. and Snook, I. K. (1981). *J. Chem. Phys.* **74**, 1409.
Watts, R. O. (1974). *Mol. Phys.* **28**, 1069.
Weres, O. and Rice, S. A. (1972). *J. Am. Chem. Soc.* **94**, 8983.

Appendix 1

Units and Dimensions for Electrical Quantities

In the non-rationalized system the force between two charges, $q_{1\mathrm{NR}}$, $q_{2\mathrm{NR}}$, separated by a distance r, is given by

$$F = \frac{q_{1\mathrm{NR}} q_{2\mathrm{NR}}}{r^2} \tag{A1.1}$$

and the charges would usually be expressed in ESU (electrostatic units) which have the dimensions $M^{1/2}\, L\, T^{-1}$.

In the rationalized system

$$F = \frac{q_1 q_2}{4\pi\varepsilon_0 r^2} \tag{A1.2}$$

where ε_0 is the permittivity of free space, and charge is given by the dimensions of (current $\times$ time). This implies that ε_0 has the dimensions $M^{-1}\, L^{-3}\, T^4$ (current)^2, in which the current is a fourth fundamental dimension in the SI convention. The value of ε_0 in SI units is $8{\cdot}854 \times 10^{-12}$ kg^{-1} m^{-3} s^4 A^2. An mth order multipole ($m = 1, 2, \ldots,$ for dipole, quadrupole, etc.) has the dimensions of (charge) $\times\, L^m$ (cf. Eqn (3.59)), and the relationship between rationalized and non-rationalized multipoles is therefore simply that implied by (A1.1), (A1.2) for rationalized and non-rationalized charges, i.e.

$$q_{\mathrm{NR}} = q/(4\pi\varepsilon_0)^{1/2} \tag{A1.3}$$

Since the dimensions of the energy remain unchanged in the rationalized system, those of the polarizability must be adjusted. For example, the components of the first-order polarizability $\boldsymbol{\alpha}$ can be calculated from

$$\alpha_{\alpha\beta} = 2 \sum_{i \neq 0} \frac{\langle \psi_0 | \mu_\alpha | \psi_i \rangle \langle \psi_i | \mu_\beta | \psi_0 \rangle}{W_0 - W_i} \tag{A1.4}$$

where W_i is the energy of the ith state of the molecule and μ_α a component of its dipole in the distorting electric field. Thus, α has the dimensions of $(\text{charge})^2 L^2 T^{-2}$. In the non-rationalized system this reduces to L^3, i.e. a volume. From (A1.3) it follows that

$$\alpha = 4\pi\varepsilon_0\alpha_{NR}$$

and its units in the SI convention are C kg^{-1} s^2 or $\text{C}^2\,\text{m}^2\,\text{J}^{-1}$.

The definition of the "fourth quantity", the ampere in the rationalized system, derives from Ampères law of force, which gives the force between two current-carrying conductors of unit length separated by a distance d as

$$F = \frac{\mu_0 I_1 I_2}{2\pi d} \quad \text{(A1.5)}$$

From which it follows that the permeability of free space μ_0 has the dimensions of (force)/(current)^2, i.e. $MLT^{-2}/(\text{current})^2$. The product $\mu_0\varepsilon_0$, therefore, has the dimensions $L^{-2}T^2$ and is equal to c_0^{-2}, where c_0 is the speed of light in free space.

The volume magnetic susceptibility $\chi = \mu_r - 1$, where μ_r is the relative permeability μ/μ_0, is a dimensionless quantity, but differs from the non-rationalized quantity χ_{NR} by a factor of 4π i.e,

$$\chi = 4\pi\chi_{NR} \quad \text{(A1.6)}$$

Atomic units

Where equations are starred in the text, this implies that quantities are expressed in dimensionless form in terms of atomic units. These derive from the Bohr radius a_0 which is $(4\pi\varepsilon_0\hbar^2/m_e e^2)$ in rationalized form $(\equiv \hbar/m_e e^2$ in non-rationalized form).

The unit of energy, the hartree, then follows from the coulombic energy derivable from (A1.2), as

$$E_h = \frac{e^2}{4\pi\varepsilon_0 a_0} \quad \text{(A1.7)}$$

with obvious modification for the "non-rationalized hartree". In these expressions e is the charge on the electron ($1{\cdot}6022 \times 10^{-19}$ C) and m_e is its mass ($9{\cdot}1095 \times 10^{-31}$ kg); a_0 and E_h therefore have the values $5{\cdot}2918 \times 10^{-11}$ m and $4{\cdot}3598 \times 10^{-18}$ J. Other atomic units are readily obtained from these expressions, for example polarizability in the rationalized system will have atomic units $e^2 a_0^2 E_h^{-1}$ and time has the units $\hbar E_h^{-1}$. A complete table of atomic units and their corresponding values in SI units, will be found in D. H. Whiffen, *Pure and Applied Chem.* (1978) **50**, 75.

Appendix 2

Functional Differentiation

Functions such as Ξ, g, h, c, etc., are defined for certain specified parameters, e.g. T, ρ over a range of values of n specified vectors and may perhaps be regarded as a "shape" in $3n$- or $6n$-dimensional space. A variation of these functions implies a variation of this "shape"; the relevant functional derivatives may be defined through a Taylor expansion. Thus, for a function $F_{1,\ldots,m}$ which depends on the functions $f_i (\equiv f(q_i))$, $i = 1, \ldots, m$,

$$F_{1,\ldots,m}(f + \delta f) = F(f) + \int \frac{\delta F}{\delta f_1} \delta f_1 \, \mathrm{d}\boldsymbol{q}_1 + \frac{1}{2!} \iint \frac{\delta^2 F}{\delta f_1 \delta f_2} \delta f_1 \, \delta f_2 \, \mathrm{d}\boldsymbol{q}_1 \, \mathrm{d}\boldsymbol{q}_2 \quad \text{(A2.1)}$$

where the integrals are over the appropriate range of vectors q_i. The following examples are of importance to the material in the main text.

(i) If

$$F(f) = \int y(q) f(q) \, \mathrm{d}q$$

$$F(f + \delta f) = \int y(q)[f + \delta f] \, \mathrm{d}q$$

$$= F(f) + \int y(q) \delta f \, \mathrm{d}q$$

A comparison of the last equation with the Taylor expansion shows that

$$\frac{\delta F}{\delta f} = y(q)$$

(ii) Consider

$$F = \iint f_1 f_2 \, \mathrm{d}\boldsymbol{q}_1 \, \mathrm{d}\boldsymbol{q}_2$$

where f_i means the value of f at q_i but F is of course defined over a range of values, then

$$F(f + \delta f) = \iint (f_1 + \delta f_1)(f_2 + \delta f_2)\, \mathrm{d}\boldsymbol{q}_1\, \mathrm{d}\boldsymbol{q}_2$$
$$= F(f) + \iint f_1 \delta f\, \mathrm{d}\boldsymbol{q}_1\, \mathrm{d}\boldsymbol{q}_2 + \iint f_2 \delta f\, \mathrm{d}\boldsymbol{q}_1\, \mathrm{d}\boldsymbol{q}_2 + \mathcal{O}(\delta f^2)$$
$$= F(f) + 2 \int \delta f\, \mathrm{d}q \int f_2\, \mathrm{d}q_2$$

The choice of f_2 rather than f_1 in the second integral is arbitrary. Comparison with the Taylor series shows

$$\frac{\delta F}{\delta f_1} = 2 \int f_2\, \mathrm{d}\boldsymbol{q}_2$$

(iii) These ideas may be extended to higher-order products and the process of obtaining higher-order derivatives is quite analogous to that used in ordinary function calculus. For example, if

$$F = \underbrace{\int \cdots \int}_{\leftarrow N \rightarrow} f_1 \ldots f_N\, \mathrm{d}q_1 \ldots \mathrm{d}q_N$$

then

$$\frac{\delta F}{\delta f_1} = N \underbrace{\int \cdots \int}_{\leftarrow (N-1) \rightarrow} f_2 \ldots f_N\, \mathrm{d}q_2 \ldots \mathrm{d}q_N$$

$$\frac{\delta^2 F}{\delta f_1 \delta f_2} = N(N-1) \underbrace{\int \cdots \int}_{\leftarrow (N-2) \rightarrow} f_3 \ldots f_N\, \mathrm{d}q_3 \ldots \mathrm{d}q_N \text{ etc.}$$

So that, multiplying both sides by $f_1 f_2$

$$\frac{\delta^2 F}{\delta \ln f_1 \delta \ln f_2} = N(N-1) \underbrace{\int \cdots \int}_{\leftarrow (N-2) \rightarrow} f_1 f_2 \ldots f_N\, \mathrm{d}q_3 \ldots \mathrm{d}q_N$$

(iv) The inverse derivative of $\delta X(\boldsymbol{q}_1)/\delta X(\boldsymbol{q}_3)$ is defined by

$$\int \frac{\delta X(\boldsymbol{q}_1)}{\delta X(\boldsymbol{q}_3)} \frac{\delta X(\boldsymbol{q}_3)}{\delta X(\boldsymbol{q}_2)}\, \mathrm{d}\boldsymbol{q}_3 = \delta_{12}$$

Appendix 3

The Singlet CHNC Equation from the Addition of a Wall Particle to a Uniform Fluid of Adsorptive Particles

Consider a fluid of a-type particles, containing n particles numbered $2 \rightarrow N+1$. We introduce an extra particle, which is the "adsorbent" (s) particle, at 1.

Without the new particle the singlet distribution is everywhere the same and we can write for position $\boldsymbol{q}_i$.

$$\rho_i^{(a)}(0) \equiv \rho = \langle N \rangle / V \qquad \text{(A3.1)}$$

where (0) means a zero adsorbent field.

The new particle interacts with all the others contributing φ to the total potential, where

$$\varphi = \sum_i u_{1i}^{[s,a]}$$

Before the addition of the adsorbent particle, the potential was

$$U_N = \sum_{i>j} u_{ij}^{(a,a)}$$

The grand partition functions before and after the addition of the adsorbent particle are:

$$\Xi(0) = \sum \frac{\lambda_a^N}{\Lambda_a^{3N} N!} \underbrace{\int \cdots \int}_{N} \exp(-\beta U_N)\, d\boldsymbol{q}_2 \ldots d\boldsymbol{q}_{N+1} \qquad \text{(A3.2)}$$

$$\Xi(\varphi) = \sum \frac{\lambda_s \lambda_a^N}{\Lambda_s^3 \Lambda_a^{3N} N!} \underbrace{\int \cdots \int}_{\leftarrow (N+1) \rightarrow}$$

$$\exp(-\beta U_N) \exp(-\beta\varphi)\, d\boldsymbol{q}_1 \ldots d\boldsymbol{q}_{N+1} \qquad \text{(A3.3)}$$

The pair distribution between s and fluid particles is

$$\rho_{12}^{(s,a)} = \frac{1}{\Xi(\varphi)} \sum \frac{\lambda_s \lambda_a^N}{\Lambda_s^3 \Lambda_a^{3N}(N-1)!} \int_{(N-1)} \cdots \int \exp(-\beta(U_N + \varphi))\, d\boldsymbol{q}_3 \ldots d\boldsymbol{q}_{N+1}$$

$$= \frac{\exp(-\beta\varphi)\lambda_s}{\Xi(\varphi)\Lambda_s^3} \sum \frac{\lambda_a^N}{\Lambda_a^{3N}(N-1)!} \int \cdots \int \exp(-\beta U_N)\, d\boldsymbol{q}_3 \ldots d\boldsymbol{q}_{N+1}$$

$$= \frac{\exp(-\beta\varphi)\lambda_s}{\Xi(\varphi)\Lambda_s^3} \Xi(0)\rho_2^{(a)}(0) \tag{A3.4}$$

Now consider the group of factors from the last expression.

$$\frac{\lambda_s \exp(-\beta\varphi)\Xi(0)}{\Lambda_s^3} = \sum \frac{\lambda_s \lambda_a^N}{\Lambda_s^3 \Lambda_a^{3N} N!} \int \cdots \int \exp(-\beta U_N) \exp(-\beta\varphi)\, d\boldsymbol{q}_2 \ldots$$

$$= \rho_1^{(s)}(\varphi)\Xi(\varphi)$$

Substitution of this result into (A3.4) gives

$$\rho_{12}^{(s,a)} = \rho_1^{(s)}(\varphi)\rho_2^{(a)}(0)$$

This may be interpreted as an identification for $\rho_2^{(a)}(0)$ when the field due to the adsorbent particle is turned on, i.e.

$$\rho_2^{(a)}(0) \rightarrow \rho_{12}^{(s,a)}(\varphi)/\rho_1^{(s)}(\varphi) \tag{A3.5}$$

Now (3.156) written for particles 2 and 3 is

$$\frac{\delta \ln(\rho_2^{(a)}/\zeta_2^{[1]})}{\delta\rho_3^{(a)}} = c_{23}^{(a,a)}$$

When (A3.5) is introduced, the functional Taylor expansion to the first term may be written

$$\ln\left[\frac{\rho_{12}^{(s,a)}(\varphi)}{\rho_1^{(s)}(\varphi)\varepsilon_{12}^{(s,a)}}\right] = \ln\left[\frac{\rho_2^{(a)}}{\varepsilon_{12}^{(s,a)}}\right]^{\circ} + \int \tilde{c}_{23}^{(a,a)\circ}[\rho_3^{(a)\circ} - \rho_3^{(a)}]\, d\boldsymbol{q}_3$$

in which the superscript ° indicates the reference state and

$$\varepsilon_{12}^{(s,a)} = \frac{\lambda_s}{\Lambda_s^3} \exp(-\beta u_{12}^{[s,a]})$$

With the usual identification of the (s, a) pair functions with the (a) singlet functions and $\rho^{(a)} \leftrightarrow \rho_a$, (A3.6) leads to

$$\ln g_{12}^{(s,a)} + \beta u_{12}^{(s,a)} = \ln \rho_a + \rho_a \int \tilde{c}_{23}^{(a,a)} h_3^{(a)}\, dq_3$$

from which the singlet CHNC equation, Eqn (3.185b), follows by comparison with the HAB equation.

Subject Index